Probability and Mathematical Statistics (Continued)
RAO • Linear Statistical Inference and Its Applications, *Second Edition*
ROBERTSON, WRIGHT, and DYKSTRA • Order Restricted Statistical Inference
ROGERS and WILLIAMS • Diffusions, Markov Processes, and Martingales, Volume II: Îto Calculus
ROHATGI • Statistical Inference
ROSS • Stochastic Processes
RUBINSTEIN • Simulation and the Monte Carlo Method
RUZSA and SZEKELY • Algebraic Probability Theory
SCHEFFE • The Analysis of Variance
SEBER • Linear Regression Analysis
SEBER • Multivariate Observations
SEBER and WILD • Nonlinear Regression
SEN • Sequential Nonparametrics: Invariance Principles and Statistical Inference
SERFLING • Approximation Theorems of Mathematical Statistics
SHORACK and WELLNER • Empirical Processes with Applications to Statistics
STAUDTE and SHEATHER • Robust Estimation and Testing
STOYANOV • Counterexamples in Probability
STYAN • The Collected Papers of T. W. Anderson: 1943–1985
WHITTAKER • Graphical Models in Applied Multivariate Statistics
YANG • The Construction Theory of Denumerable Markov Processes

Applied Probability and Statistics
ABRAHAM and LEDOLTER • Statistical Methods for Forecasting
AGRESTI • Analysis of Ordinal Categorical Data
AGRESTI • Categorical Data Analysis
AICKIN • Linear Statistical Analysis of Discrete Data
ANDERSON and LOYNES • The Teaching of Practical Statistics
ANDERSON, AUQUIER, HAUCK, OAKES, VANDAELE, and WEISBERG • Statistical Methods for Comparative Studies
ARTHANARI and DODGE • Mathematical Programming in Statistics
ASMUSSEN • Applied Probability and Queues
*BAILEY • The Elements of Stochastic Processes with Applications to the Natural Sciences
BARNETT • Interpreting Multivariate Data
BARNETT and LEWIS • Outliers in Statistical Data, *Second Edition*
BARTHOLOMEW • Stochastic Models for Social Processes, *Third Edition*
BARTHOLOMEW and FORBES • Statistical Techniques for Manpower Planning
BATES and WATTS • Nonlinear Regression Analysis and Its Applications
BECK and ARNOLD • Parameter Estimation in Engineering and Science
BELSLEY • Conditioning Diagnostics: Collinearity and Weak Data in Regression
BELSLEY, KUH, and WELSCH • Regression Diagnostics: Identifying Influential Data and Sources of Collinearity
BHAT • Elements of Applied Stochastic Processes, *Second Edition*
BHATTACHARYA and WAYMIRE • Stochastic Processes with Applications
BIEMER, GROVES, LYBERG, MATHIOWETZ and SUDMAN • Measurement Errors in Surveys
BLOOMFIELD • Fourier Analysis of Time Series: An Introduction
BOLLEN • Structural Equations with Latent Variables
BOX • R. A. Fisher, the Life of a Scientist
BOX and DRAPER • Empirical Model-Building and Response Surfaces
BOX and DRAPER • Evolutionary Operation: A Statistical Method for Process Improvement
BOX, HUNTER, and HUNTER • Statistics for Experimenters: An Introduction to Design, Data Analysis, and Model Building
BROWN and HOLLANDER • Statistics: A Biomedical Introduction
BUCKLEW • Large Deviation Techniques in Decision, Simulation, and Estimation
BUNKE and BUNKE • Nonlinear Regression, Functional Relations and Robust Methods: Statistical Methods of Model Building
BUNKE and BUNKE • Statistical Inference in Linear Models, Volume I
CHAMBERS • Computational Methods for Data Analysis
CHATTERJEE and HADI • Sensitivity Analysis in Linear Regression
CHATTERJEE and PRICE • Regression Analysis by Example, *Second Edition*

*Now available in a lower priced paperback edition in the Wiley Classics Library.

Applied Probability and Statistics (Continued)

CHOW · Econometric Analysis by Control Methods
CLARKE and DISNEY · Probability and Random Processes: A First Course with Applications, *Second Edition*
COCHRAN · Sampling Techniques, *Third Edition*
COCHRAN and COX · Experimental Designs, *Second Edition*
CONOVER · Practical Nonparametric Statistics, *Second Edition*
CONOVER and IMAN · Introduction to Modern Business Statistics
CORNELL · Experiments with Mixtures, Designs, Models, and the Analysis of Mixture Data, *Second Edition*
COX · A Handbook of Introductory Statistical Methods
COX · Planning of Experiments
CRESSIE · Statistics for Spatial Data
DANIEL · Applications of Statistics to Industrial Experimentation
DANIEL · Biostatistics: A Foundation for Analysis in the Health Sciences, *Fourth Edition*
DANIEL and WOOD · Fitting Equations to Data: Computer Analysis of Multifactor Data, *Second Edition*
DAVID · Order Statistics, *Second Edition*
DAVISON · Multidimensional Scaling
DEGROOT, FIENBERG, and KADANE · Statistics and the Law
*DEMING · Sample Design in Business Research
DILLON and GOLDSTEIN · Multivariate Analysis: Methods and Applications
DODGE · Analysis of Experiments with Missing Data
DODGE and ROMIG · Sampling Inspection Tables, *Second Edition*
DOWDY and WEARDEN · Statistics for Research, *Second Edition*
DRAPER and SMITH · Applied Regression Analysis, *Second Edition*
DUNN · Basic Statistics: A Primer for the Biomedical Sciences, *Second Edition*
DUNN and CLARK · Applied Statistics: Analysis of Variance and Regression, *Second Edition*
ELANDT-JOHNSON and JOHNSON · Survival Models and Data Analysis
FLEISS · The Design and Analysis of Clinical Experiments
FLEISS · Statistical Methods for Rates and Proportions, *Second Edition*
FLEMING and HARRINGTON · Counting Processes and Survival Analysis
FLURY · Common Principal Components and Related Multivariate Models
FRANKEN, KÖNIG, ARNDT, and SCHMIDT · Queues and Point Processes
GALLANT · Nonlinear Statistical Models
GIBBONS, OLKIN, and SOBEL · Selecting and Ordering Populations: A New Statistical Methodology
GNANADESIKAN · Methods for Statistical Data Analysis of Multivariate Observations
GREENBERG and WEBSTER · Advanced Econometrics: A Bridge to the Literature
GROSS and HARRIS · Fundamentals of Queueing Theory, *Second Edition*
GROVES · Survey Errors and Survey Costs
GROVES, BIEMER, LYBERG, MASSEY, NICHOLLS, and WAKSBERG · Telephone Survey Methodology
GUPTA and PANCHAPAKESAN · Multiple Decision Procedures: Theory and Methodology of Selecting and Ranking Populations
GUTTMAN, WILKS, and HUNTER · Introductory Engineering Statistics, *Third Edition*
HAHN and MEEKER · Statistical Intervals: A Guide for Practitioners
HAHN and SHAPIRO · Statistical Models in Engineering
HALD · Statistical Tables and Formulas
HALD · Statistical Theory with Engineering Applications
HAND · Discrimination and Classification
HEIBERGER · Computation for the Analysis of Designed Experiments
HELLER · MACSYMA for Statisticians
HOAGLIN, MOSTELLER, and TUKEY · Exploring Data Tables, Trends and Shapes
HOAGLIN, MOSTELLER, and TUKEY · Understanding Robust and Exploratory Data Analysis
HOAGLIN, MOSTELLER, and TUKEY · Fundamentals of Exploratory Analyses of Variance
HOCHBERG and TAMHANE · Multiple Comparison Procedures
HOEL · Elementary Statistics, *Fourth Edition*

Continued on back end papers

*Now available in a lower priced paperback edition in the Wiley Classics Library.

Design and Inference in
Finite Population Sampling

Design and Inference in Finite Population Sampling

A.S. HEDAYAT
University of Illinois
Chicago, Illinois

BIKAS K. SINHA
Indian Statistical Institute
Calcutta, India

A Wiley-Interscience Publication
JOHN WILEY & SONS, INC.
New York • Chichester • Brisbane • Toronto • Singapore

Library of Congress Cataloging in Publication Data:

Hedayat, A.
Design and inference in finite population sampling / A. S. Hedayat, Bikas K. Sinha.
p. cm. -- (Wiley series in probability and mathematical statistics. Probability and mathematical statistics, ISSN 0271-6232)
"A Wiley-Interscience publication."
Includes bibliographical references and index.
ISBN 0-471-88073-6
1. Sampling (Statistics) I. Sinha, Bikas Kumar. II. Title. III. Series.
QA276.6.H43 1991
001.4'222--dc20 90-23349
CIP

Printed in the United States of America

10 9 8 7 6 5 4 3 2 1

To my wife, Batool, and my children, Leyla and Yashar

—A.S.H.

To my wife, Pritha, and my children, Karabi and Kuver

—B.K.S.

Preface

This book is an introduction to design and inference in survey sampling for a two-semester or three-quarter course at the senior or graduate level in statistics. By a judicious selection of topics, one could also teach a one-semester course based on this book. A serious practitioner of survey sampling is expected to benefit from the theoretical results discussed here.

We have tried to elucidate in a unified way the basic concepts and results from a mature point of view, to present fundamental topics with thoroughness, and to reach the frontiers of present-day research. We hope that the serious reader will find the book stimulating. From the outset we assume an appreciation for the subtleties of mathematical and statistical reasoning.

We have divided the material into 12 chapters. Each chapter contains its separate bibliography and exercises. They are guides for further study and do not aim at completeness.

The subject of survey sampling holds rich rewards for those who learn its secrets.

ACKNOWLEDGMENTS

The book has developed out of lecture materials on sampling theory that we have given to our graduate students at various institutions, including Florida State University, Tallahassee; University of Illinois, Chicago; North Carolina State University, Raleigh; Institute of Mathematics, Federal University of Bahia, Salvador, Brazil; Calcutta University and Indian Statistical Institute, Calcutta. We thank our students whose excellent thought-provoking questions and suggestions shaped the format of the book.

We are grateful to our friends and colleagues Drs. Dibyen Majumdar, Samindranath Sengupta, and John Stufken, who have all in various degrees contributed to this book by their careful reading of portions of the book and whose critical comments and suggestions resulted in numerous significant improvements.

Special thanks go to our graduate students, in particular to Weiguang Zhang, Gerhardt Pohl and Peter Meyer, who have read through the book and offered suggestions for better presentation. Our heartfelt thanks go to Ms. Mahalia Triplett, who has typed and retyped various versions of the manuscript with extreme patience and courtesy at all stages.

Our sincere appreciation is expressed to the Air Force Office of Scientific Research for partially sponsoring this project.

A.S. Hedayat
Bikas K. Sinha

Chicago, Illinois
Calcutta, India

Contents

Design and Inference in Finite Population Sampling

CHAPTER ONE

A Unified Setup for Probability Sampling

In this chapter we present the basic terminology and notation to be used. We introduce the general notion of a sampling design and describe an algorithmic method for implementing it. Finally, we define some structural parameters of sampling designs and study their interrelations.

1.1. PRELIMINARIES

There are two generally accepted options in studying the characteristics of finite populations. The first option is a study in which every unit of the population is examined, called a *census*. Use of a census to study a population is time-consuming, expensive, often impossible, and, strangely enough, often inaccurate. The other option is to study the characteristics of a population by examining a part of it. The theory of survey sampling as developed during the past several decades provides us with various kinds of reasonable scientific tools for drawing samples and making valid inference about the population parameters of interest. To acquire a taste of the historical development of the subject, we refer the reader to Hansen et al. (1985), Johnson and Smith (1969), Krishnaiah and Rao (1988). We now proceed to present the basic theory of design and inference in sampling from finite populations.

Let $\mathscr{U} = \{u_1, u_2, \ldots, u_i, \ldots, u_N\}$ denote a finite population of N *distinct* and *identifiable* units. We refer to N as the *size* of the finite population or finite universe $\mathscr{U}$ and N is generally assumed to be known. The elements of $\mathscr{U}$ namely $u_1, u_2, \ldots, u_N$ are known as the *sampling units*. Sometimes we refer to the sampling units simply by their indices. Thus, we may talk about u_i as *unit i* and represent the population as $\mathscr{U} = \{1, 2, \ldots, i, \ldots, N\}$. Here identifiability means that there is a 1:1 correspondence between the units and the indices $1, 2, \ldots, N$. For example, in a study of a block of 150 villages, the population units (villages) are distinct and readily identifiable.

However, not every natural finite population admits the identifiability of its units. The fish population in a lake is an example of one which does not admit identifiability.

There are underlying characteristics of interest (also called *survey variables*) whose values, qualitative or quantitative, are well defined on every unit of the universe and are unknown a priori but these may be obtained on any unit, with or without error, through an evaluation effort. The basic purpose of survey sampling is to draw inferences about some selected functions (called *parameters* of the population) of the survey variables of interest, based on their values on some units which are said to constitute a *sample* from the population. For practical convenience, the units in a sample are usually drawn from the population one at a time, according to a definite set of rules, as we shall see later.

Formally defined, a sample, denoted by s, from a finite labelled population $\mathcal{U} = \{1, 2, \ldots, N\}$ is an ordered collection of distinct units from $\mathcal{U}$. The number of units in a sample is called the *sample size* and is usually denoted by $n(s)$ for a sample s. Thus, for example, $s = (i_1, i_2, \ldots, i_{n(s)})$ represents a sample, where $1 \le i_1 \ne i_2 \ne \cdots \ne i_{n(s)} \le N$. It should be noted that the units in a sample are generally shown in an *ordered* form, where the ordering corresponds to the order of the draws.

Example 1.1. Let $\mathcal{U} = \{1, 2, 3, 4, 5\}$ refer to a population of five units. Then $s_1 = (1, 2, 4)$ and $s_2 = (5, 1)$ could be regarded as samples from $\mathcal{U}$. We have $n(s_1) = 3$ and $n(s_2) = 2$.

Two samples s_1 and s_2 are said to be *identical* if they are equal in the sense of vectors $\mathbf{s}_1$ and $\mathbf{s}_2$. Otherwise, they are said to be *different*. Two samples s_1 and s_2 are said to be *equivalent* if they involve the same set of units from the population. We write $s_1 \approx s_2$ to denote equivalence of s_1 and s_2. Clearly, for a sample of size n, there are $n!$ different but equivalent forms, namely, those obtained by permuting its members in all possible ways. If two samples are not equivalent, then they are said to be *distinct*.

Example 1.2. In Example 1.1, there are $3! = 6$ different but equivalent forms of s_1: (1, 2, 4), (1, 4, 2), (2, 1, 4), (2, 4, 1), (4, 1, 2), (4, 2, 1). Also there are two equivalent forms of s_2: (5, 1) and (1, 5).

There are as many as $\Sigma_{j=1}^{N} N^{(j)}$ different samples that can be formed from a population of N units. Here $N^{(j)} = N(N-1)\ldots(N-j+1)$, $j = 1, 2, \ldots, N$. However, not all of the $N^{(j)}$ different samples of size j are distinct. They are equivalent in sets of $j!$ and, hence, there are $N^{(j)}/j! = \binom{N}{j}$ *distinct* samples, a sample of size j having $j!$ equivalent forms, each different.

Example 1.3. Let $\mathcal{U} = \{1, 2, 3\}$. Then there are seven *distinct* samples: {(1), (2), (3), (1, 2), (1, 3), (2, 3), (1, 2, 3)} even though altogether there are as many as 15 *different* samples that can be formed of $\mathcal{U}$. These are: (1), (2), (3), (1, 2), (2, 1), (1, 3), (3, 1), (2, 3), (3, 2), (1, 2, 3), (1, 3, 2), (2, 1, 3), (2, 3, 1), (3, 1, 2), and (3, 2, 1).

We denote by S the collection of all possible samples from a given finite population $\mathcal{U}$. Much of the customary inference in finite population sampling is drawn from the data provided by what is termed a *probability sample*, that is a sample selected via a certain known chance mechanism. In the terminology of survey sampling, the phrase chance mechanism is formalized through the concept of a *(probability) sampling design*.

Definition 1.1. A sampling design d, based on $\mathcal{U}$, is a pair (S_d, P_d) where S_d is a subset of S and P_d is a probability distribution on S_d such that

(i) $P_d(s) > 0$ for each s in S_d,
(ii) for every unit in the population, there exists at least one sample $s \in S_d$ containing the unit.

In the above definition, we have stipulated the condition (ii) as, otherwise, the sampling design would be based only on a subpopulation or a subset of $\mathcal{U}$.

Example 1.4. Let $\mathcal{U} = \{1, 2, 3\}$. Define $d_1 = (S_1, P_1)$ with $S_1 = \{(1), (2), (1,3)\}$, $P_1((1)) = \frac{1}{3}$, $P_1((2)) = a$, $P_1((1,3)) = \frac{2}{3} - a$, $0 < a < \frac{2}{3}$. Then d_1 is a sampling design. Again, define $d_2 = (S_2, P_2)$ with $S_2 = \{(1,2), (1,3), (2,3)\}$, $P_2((1,2)) = \frac{1}{4}$, $P_2((1,3)) = \frac{1}{4}$, $P_2((2,3)) = \frac{1}{2}$. Then d_2 is also a sampling design. The sampling designs d_3: $P_3((1,2,3)) = 1$ and d_4: $P_4((1,2,3)) = p_1$, $P_4((1,3,2)) = p_2, \ldots, P_4((3,2,1)) = p_6$ where $0 < p_i < 1$, $\Sigma_{i=1}^{6} p_i = 1$ both refer to a *census* (or *complete enumeration* of the population). The sampling design d_2 is, in some sense, a *reduced version* of the following sampling design d_5: $P_5((1,2)) = q_1$, $P_5((2,1)) = \frac{1}{4} - q_1$, $P_5((1,3)) = q_2$, $P_5((3,1)) = \frac{1}{4} - q_2$, $P_5((2,3)) = q_3$, $P_5((3,2)) = \frac{1}{2} - q_3$, $0 < q_1, q_2 < \frac{1}{4}$, $0 < q_3 < \frac{1}{2}$.
Indeed, we have in the above,

$$P_2((i, j)) = P_5(i, j)) + P_5((j, i)) , \qquad 1 \le i < j \le 3 .$$

Generally a sampling design d can be brought to its reduced form d^* by defining $P_{d^*}(s^*) = \Sigma_{s \approx s^*} P_d(s)$ over all $s^* \in S_{d^*}$ where in S_{d^*} all samples are distinct and in each sample s^*, the units involved are arranged in *increasing order* of their indices. Consequently, for any sampling design d, not in its reduced form, d^* is uniquely defined. The sample s^* is also referred to as an *unordered* sample while s is referred to as *ordered*. Further, the *reduced* sampling design d^* is termed as the *symmetrized version* of the sampling design d. When d and d^* are understood from the context, we may use the notations $P(s)$ and $P^*(s^*)$ for $P_d(s)$ and $P_{d^*}(s^*)$, respectively. (The reader will learn more about d^* in Theorem 2.9 and Remark 2.2 following the theorem).

So far as the inferential aspect of finite population sampling, based on the logic of probability sampling, is concerned, we shall see in Section 2.2 that the reduced form d^* of a sampling design d is *at least as useful* as d in making inferences about any population parameter. It is only at the stage of implementing a sampling design that we might prefer a *suitable* unreduced form. This latter point will be discussed further in Section 1.2.

Definition 1.2. S_d is called the *support* of the design d and the cardinality of S_d is called the *support size* of d.

Definition 1.3. A sampling design d is said to be *uniform* if P_d is uniform on S_d. Otherwise, we call it *nonuniform*.

Definition 1.4. A sampling design d is said to be a *fixed sample size design* of size n if $n(s) = n$ for all s in S_d. We call such a sampling design as a *fixed-size* (n) design. Otherwise, it is a variable-size design.

Example 1.5. The sampling designs in Example 1.4 are classified as in Table 1.1.

Table 1.1. Classification of Sampling Designs d_1 to d_5

Uniform and Fixed Size	Nonuniform but Fixed Size	Uniform but Not Fixed Size	Neither Uniform nor Fixed Size
d_3	d_2	d_1	d_1
d_4	d_4		
(provided p_i's are all equal)	(if p_i's are unequal)	(when $a = \frac{1}{3}$)	(when $a \neq \frac{1}{3}$)
	d_5		

Remark 1.1. In the literature, most often a sample is defined as an ordered collection of units, *with or without repetitions*, from a given population and a sampling design is defined in terms of a probability measure on a finite or a countably infinite subset of such samples. Thus, for example, $\tilde{d}$: $\tilde{P}((1,1,2)) = 0.3$, $\tilde{P}((1,2,3,2,1)) = 0.4$, $\tilde{P}((3,2,1,2,2)) = 0.2$, $\tilde{P}((1,1,3,2,3,1,2)) = 0.1$ has been traditionally regarded as an example of a sampling design defined on a population of three units. In the present approach however, samples of the type $(1,1,2)$, that is, those including repetitions of the units, have been deleted outright. Our description of a sampling design might be regarded as the one arising out of a *first step* towards reduction of a traditional sampling design. For example, $(3,2,1,2,2)$ might be reduced at the first step to $(3,2,1)$, just ignoring the repetitions. This sort of reduction of the other samples in $\tilde{d}$ would reduce $\tilde{d}$ at the first stage to d where d is: $P((1,2)) = 0.3$, $P((1,2,3)) = 0.4$, $P((3,2,1)) = 0.2$, and $P((1,3,2)) = 0.1$. The final reduction is, of course, achieved by using d^*: $P^*((1,2)) = 0.3$, $P^*((1,2,3)) = 0.7$.

In our present approach, we start directly with sampling designs of the type d instead of designs of the type $\tilde{d}$, as is usually discussed in the literature. *No further explicit mention of designs of the type $\tilde{d}$ will be made in the book.*

1.2. THE IMPLEMENTATION OF SAMPLING DESIGNS: NOTION OF SAMPLING SCHEMES

By implementation of a sampling design $d = (S_d, P_d)$, we mean the selection of a sample according to d. In practice, this may not *always* be an easy task. When the cardinality of S_d is large (that is, the sampling design d attributes positive probability to a large number of samples), the task of selecting a sample according to d is quite difficult even when d is a uniform sampling design. One general procedure which applies to any arbitrary sampling design is to list all the samples and the corresponding probabilities, and then to select one of these samples by a mechanism which guarantees the probability $P_d(s)$ of selection of s for every s in the support of the sampling design. Once the listing is complete, this latter task is usually accomplished with the help of a *table of random numbers* by the *cumulative total method* (Cochran, 1977). At this stage, if necessary, the stated probabilities are reduced to rational fractions of a sufficient degree of accuracy. Operationally, however, such a procedure becomes very inconvenient in most cases. Take, for example, $\mathcal{U} = \{1, \ldots, 20\}$ and let d be a uniform design defined over $S_d = \{(i, j, k)\} \mid 1 \le i < j < k \le 20\}$. Clearly there are $\binom{20}{3} =$ 1140 samples each with equal probability in d. A complete listing of these samples would itself be prohibitive. On the other hand, since any sample is, in effect, an ordered collection of some units of the population, it would be desirable to have some sort of mechanism which would generate the units, *one by one*, of different samples (according to certain chance rules) such that the *overall* chance of selecting a sample s would coincide with $P_d(s)$. In other words, given a sampling design, we would like to discover an underlying *scheme of sampling* that would dictate the operational steps involved in generating the units of all samples one by one in a way that would eventually result in the given sampling design. Below we discuss in detail the concept of the *unit drawing chance mechanism* due to Hanurav (1966), which results in such a sampling scheme.

In its most general form, a unit drawing chance mechanism can be formally defined as an algorithm $A = A\{q_1(\cdot), q_2(s), q_3(\cdot \mid s)\}$ where

(i) $(S_1, q_1(\cdot))$ is a sampling design, defined on $\mathcal{U}$ or a subset of $\mathcal{U}$, with $S_1 \subseteq \{(i) \mid 1 \le i \le N\}$, that is, $0 < q_1((i)) \le 1$, $(i) \in S_1$, $\Sigma_{(i) \in S_1} q_1((i)) = 1$,

(ii) $q_2(s)$ is a nonnegative fraction defined for all samples $s \in S$, that is, $0 \le q_2(s) \le 1$ for all $s \in S$,

(iii) whenever $q_2(s) > 0$, $(S_2(s), q_3(\cdot|s))$ is a sampling design defined on $\mathscr{U} - s$ or a subset of it, with $S_2(s) \subseteq \{(j) | j \in \mathscr{U} - s\}$, that is, $0 < q_3((j)|s) \leq 1$, $(j) \in S_2(s)$, $\Sigma_{(j) \in S_2(s)} q_3((j)|s) = 1$.

The sampling scheme arising out of the above algorithm is described as follows:

(a) Use $(S_1, q_1(\cdot))$ to select one unit from $\mathscr{U}$. This can be easily implemented by using a table of random numbers. A method due to Lahiri (1951) is also useful in this context (see Section 5.5). Let i_1 be the unit selected. From $s_1 = (i_1)$.
(b) Check the value of $q_2(s_1)$. If $q_2(s_1) = 0$, stop sampling and decide on s_1 as the ultimate sample. If $q_2(s_1) = 1$, pass on to the next step. However, if $0 < q_2(s_1) < 1$, perform a Bernoulli experiment with success probability $q_2(s_1)$. If it results in failure (with probability $(1 - q_2(s_1))$), stop sampling and again decide on s_1 as the ultimate sample. If the Bernoulli experiment yields success, pass on to the next step.
(c) Use $\{S_2(s_1), q_3(\cdot|s_1)\}$ to select another unit from $\mathscr{U} - s_1$. This is also carried out using a table of random numbers. Let $i_2(\neq i_1)$ be the unit selected. Now form $s_2 = (i_1, i_2)$ and go back to step (b), replacing s_1 with s_2. Continue this procedure until sampling is stopped and a sample is drawn.

Each time a unit is selected, the sample s is formed and the value of $q_2(s)$ is checked to decide whether to stop or to sample the next unit.

Example 1.6. Let $\mathscr{U} = \{1, 2, 3\}$. An example of a sampling algorithm follows: $A = A\{q_1((1)) = q_1((2)) = 0.5,\ q_2((2)) = 0.7,\ q_2((2,3)) = 0.2,$ $q_2(s) = 0$ for the remaining samples in S, $q_3((1)|(2)) = 0.2$, $q_3((3)|(2)) = 0.8$, $q_3((1)|(2,3)) = 1\}$.

Readers may visualize such a sampling algorithm using a tree diagram and verify that the scheme yields the sampling design in Table 1.2.

Certain basic questions naturally arise. Does a sampling scheme *always* result in a sampling design? Given a sampling design, can we *always* get hold of a scheme underlying it? Does there exist a 1:1 correspondence between a sampling design and a sampling scheme? The answers to all these questions are affirmative. We investigate them below.

Table 1.2

Samples	(1)	(2)	(2, 1)	(2, 3)	(2, 3, 1)	Total
Probability	0.5	0.15	0.07	0.224	0.056	1.00

Theorem 1.1. *Every scheme of sampling (based on a unit drawing chance mechanism) results in a unique sampling design.*

Proof. Let $A = A\{q_1(\cdot),\ q_2(s),\ q_3(\cdot|s)\}$ be an algorithm describing a scheme of sampling. Then, for a sample $s = (i_1, i_2, \ldots, i_{n(s)})$ drawn according to the scheme A, we find

$$P_A(s) = q_1((i_1)) \prod_{k=1}^{n(s)-1} \{q_2((i_1, i_2, \ldots, i_k))q_3((i_{k+1}) \mid (i_1, i_2, \ldots, i_k))\} \times Q \tag{1.1}$$

where $Q = \{1 - q_2((i_1, i_2, \ldots, i_{n(s)}))\}$. It may be verified that $P_A(s) > 0$ and $\Sigma_s P_A(s) = 1$ so that the scheme results in a sampling design, say, d (Exercise 1.5). The uniqueness follows from the expression (1.1). □

Theorem 1.2. *Let $d = (S_d, P_d)$ be a sampling design. Then there exists a unique scheme of sampling, underlying d, defined by an algorithm of the type $A\{q_1(\cdot),\ q_2(s),\ q_3(\cdot|s)\}$.*

Proof. First we show the existence of an algorithm underlying the design d. Let $s = (i_1, \ldots, i_{n(s)})$ be a typical sample in S_d. Define $W_i = \{s \in S_d \mid i_1 = i\}$, $W_{ij} = \{s \in S_d \mid i_1 = i,\ i_2 = j\}$, etc.; and, moreover, let

$$\alpha_i = \sum_{s \in W_i} P_d(s)\,, \qquad \alpha_{ij} = \sum_{s \in W_{ij}} P_d(s), \qquad \ldots,$$

$$\beta_i = P_d((i)) \qquad \beta_{ij} = P_d((i, j)) \qquad \ldots.$$

Note that $1 = \Sigma_i\, \alpha_i$, $\alpha_i = \beta_i + \Sigma_j\, \alpha_{ij}$, $\alpha_{ij} = \beta_{ij} + \Sigma_k\, \alpha_{ijk}$, etc. Further, $\beta_i = 0$ if $(i) \not\in S_d$, etc. We now construct an algorithm A by defining the q_i's in the following way:

$$q_1((i)) = \alpha_i\,, \qquad 1 \le i \le N\,,$$

$$q_2((i_1, \ldots, i_k)) = \begin{cases} 1 - \dfrac{\beta_{i_1, \ldots, i_k}}{\alpha_{i_1, \ldots, i_k}} & \text{if } \alpha_{i_1, \ldots, i_k} > 0\,, \\ 0 & \text{otherwise}\,, \end{cases} \tag{1.2}$$

$$q_3((j) \mid (i_1, \ldots, i_k)) = \frac{\alpha_{i_1, \ldots, i_k j}}{\alpha_{i_1, \ldots, i_k} - \beta_{i_1, \ldots, i_k}}\,, \qquad j \ne i_1, \ldots, i_k\,.$$

provided that $q_2((i_1, \ldots, i_k)) > 0$.

It can be verified that $A\{q_1, q_2, q_3\}$ defines an algorithm (Exercise 1.6). We may now use the relation (1.1) and the above description of the algorithm A to demonstrate further that $P_A((i_1, \ldots, i_{n(s)})) = \beta_{i_1, \ldots, i_{n(s)}} =$

$P_d((i_1, \ldots, i_{n(s)}))$ for every $s \in S_d$. Thus the above algorithm $A\{q_1, q_2, q_3\}$ describes one scheme of sampling underlying the design d. If possible, let $A^*\{q_1^*, q_2^*, q_3^*\}$ be a different algorithm describing another scheme of sampling underlying the same design d. Then we must have $P_A(s) = P_d(s) = P_{A^*}(s)$ for every $s \in S_d$ where $P_A(s)$ is as in (1.1) and $P_{A^*}(s)$ is to be obtained from (1.1) by replacing the q_i's by q_i^*'s. Now equating $P_A(s)$ to $P_{A^*}(s)$ for all $s \in S_d$, we will end up with a system of equations from which it can be shown that $q_i = q_i^*$, $i = 1, 2, 3$ (Exercise 1.7). This illustrates the uniqueness of a sampling algorithm of the type A underlying a sampling design d. □

Remark 1.2. The proof of Theorem 1.2 is constructive and allows us to write out a scheme of sampling corresponding to *any* sampling design. The scheme is described by an algorithm $A(d)$ derivable through (1.2) and it gives the operational steps involved in selecting the units of a sample one by one, ultimately resulting in the sampling design d we start with. Because of the 1:1 correspondence, *very often* only the sampling schemes are presented and discussed, rather than describing the underlying sampling designs. Usually such sampling schemes result in unreduced forms of sampling designs which can be subsequently brought to their reduced forms.

Example 1.7. We take up some of the sampling designs of Example 1.4 and present the algorithms underlying them. As for the sampling design d_1, we choose $a = 1/6$.

$$d_1:\quad A(d_1) \equiv \{q_1((1)) = 5/6,\ q_1((2)) = 1/6,\ q_2((1)) = 3/5,\ q_3((3) \mid (1)) = 1,$$
$$q_2(s) = 0 \text{ for every } s \text{ different from } (1)\}\,.$$

$$d_2:\quad A(d_2) \equiv \{q_1((1)) = 1/2,\ q_1((2)) = 1/2,\ q_2((1)) = q_2((2)) = 1,$$
$$q_3((2) \mid (1)) = q_3((3) \mid (1)) = 1/2,\ q_3((3) \mid (2)) = 1,$$
$$q_2(s) = 0 \text{ for every } s \text{ different from } (1) \text{ and } (2)\}\,.$$

$$d_3:\quad A(d_3) \equiv \{q_1((1)) = q_2((1)) = q_3((2) \mid (1)) = q_2((1, 2))$$
$$= q_3((3) \mid (1, 2)) = 1,\ q_2(s) = 0$$
$$\text{for every } s \text{ different from } (1) \text{ and } (1, 2)\}.$$

For the sake of illustration, we derive the algorithm for the sampling design d_1.

We have $W_1 = \{(1), (1, 3)\}$, $W_2 = \{(2)\}$, $W_3 = \emptyset$ so that $\alpha_1 = 5/6$, $\alpha_2 = 1/6$, $\alpha_3 = 0$, $\alpha_{12} = 0$, $\beta_1 = 2/6$, $\beta_2 = 1/6$, $\beta_3 = 0$. Next, $W_{13} = \{(1, 3)\}$ so that $\alpha_{13} = \beta_{13} = 3/6$. Using (1.2), we derive explicitly

$$q_1((1)) = \alpha_1 = \frac{5}{6}, \qquad q_1((2)) = \alpha_2 = \frac{1}{6}, \qquad q_1((3)) = \alpha_3 = 0;$$

$$q_2((1)) = 1 - \frac{\beta_1}{\alpha_1} = \frac{3}{5}, \qquad q_2((2)) = 1 - \frac{\beta_2}{\alpha_2} = 0;$$

$$q_3((2)|(1)) = \frac{\alpha_{12}}{\alpha_1 - \beta_1} = 0, \qquad q_3((3)|(1)) = \frac{\alpha_{13}}{\alpha_1 - \beta_1} = 1;$$

$$q_2((1,3)) = 1 - \frac{\beta_{13}}{\alpha_{13}} = 0.$$

Example 1.8. Consider the following algorithm.

$$\begin{aligned}
A \equiv \{ & q_1((i_1)) = \frac{1}{N}, \ 1 \le i_1 \le N, \\
& q_2((i_1)) = 1, \ 1 \le i_1 \le N, \\
& q_3((i_2)|(i_1)) = \frac{1}{N-1} \text{ for all } i_2(\neq i_1), \\
& q_2((i_1, i_2)) = 1 \text{ for all } 1 \le i_1 \neq i_2 \le N, \\
& q_3((i_3)|(i_1, i_2)) = \frac{1}{N-2} \text{ for all } i_3(\neq i_1, \ \neq i_2), \\
& \vdots \\
& q_2((i_1, \ldots, i_{n-1})) = 1 \text{ for all } 1 \le i_1 \neq \cdots \neq i_{n-1} \le N, \\
& q_3((i_n)|(i_1, \ldots, i_{n-1})) = \frac{1}{N-n+1} \text{ for all } i_n(\neq i_1, \ldots, \neq i_{n-1}), \\
& q_2((i_1, \ldots, i_n)) = 0 \text{ for all } 1 \le i_1 \neq \cdots \neq i_n \le N\}.
\end{aligned}$$

We can easily verify that the underlying sampling design is given by d: $P_d((i_1, \ldots, i_n)) = 1/N^{(n)}$, $1 \le i_1 \neq \cdots \neq i_n \le N$. Further, the reduced form of d is

$$d^*: \quad P_{d^*}((i_1, \ldots, i_n)) = \frac{1}{\binom{N}{n}}, \qquad 1 \le i_1 < i_2 < \cdots < i_n \le N.$$

The sampling design d (or its reduced form d^*) is well known in the literature as the *simple random sampling without replacement* design and is abbreviated as SRSWOR(N, n) design. The sampling scheme underlying the above design is very popular and the name SRSWOR technically describes the scheme as one of selecting the units *one at a time without replacement* and *with equal probability*. The sample thus generated is referred to as a *simple random sample*. In this book we will use the notation SRS(N, n) design. In practice, of course, we may draw the units *with equal probability* and *with replacement* until n *distinct* units are recorded and these then constitute the units in the random sample (repetitions are ignored) underlying the sampling design (Exercise 1.21). As regards the sampling design d^*, the simplest way to implement d^* in practice would involve a two-stage process. First, we implement d and generate a sample $s = (j_1, \ldots, j_n)$, $1 \le j_1 \ne j_2 \ne \cdots \ne j_n \le N$. Next we transform $(j_1, \ldots, j_n)$ in d to $(i_1, \ldots, i_n)$ in d^* with $1 \le i_1 < \cdots < i_n \le N$ where $(i_1, \ldots, i_n)$ refers to the ordered permutation (in ascending order) of the units in the sample $(j_1, \ldots, j_n)$. We take $s^* = (i_1, \ldots, i_n)$ as the sample under d^*. We could, of course, write out an algorithm directly for d^*, using the constructional steps described in the proof of Theorem 1.2, and implement it, but this would in general be complicated.

Example 1.9. Consider the following scheme of sampling, known as *Midzuno's scheme* (Midzuno, 1951).

(i) Draw one unit from $\mathcal{U}$, the chance of selection of the unit i being p_i, $0 < p_i < 1$, $\Sigma\, p_i = 1$.
(ii) Remove the unit selected and, from the remaining $(N-1)$ units, draw $(n-1)$ units by an SRS procedure.

In terms of the algorithm $A(\cdot)$, the above scheme is represented as follows:

$$q_1((i)) = p_i\,, \qquad 1 \le i \le N\,,$$

$$q_2((i_1, i_2, \ldots, i_m)) = \begin{cases} 1 & \text{for } 1 \le m < n\,, \\ 0 & \text{for } m \ge n\,, \end{cases}$$

$$q_3((j)|(i_1, i_2, \ldots, i_m)) = (N-m)^{-1}\,, \qquad j \ne i_1, \ne i_2, \ldots, \ne i_m$$
$$\text{for } m = 1, 2, \ldots, n-1\,.$$

The underlying sampling design is given by

$$d\colon\ P_d((i_1, \ldots, i_n)) = p_{i_1} \cdot \frac{1}{(N-1)^{(n-1)}}\,, \qquad 1 \le i_1 \ne i_2 \ne \cdots \ne i_n \le N\,.$$

The reduced form of d is

$$d^*\colon\ P_{d^*}((i_1, \ldots, i_n)) = \left(\sum_{j=1}^{n} p_{i_j}\right) \Big/ \binom{N-1}{n-1}\,, \qquad 1 \le i_1 < i_2 < \cdots < i_n \le N\,.$$

Remark 1.3. At this stage, it may be emphasized that the unit drawing mechanism as explained above serves as one method of implementation of a sampling design. Sometimes it may be operationally more convenient to follow other sampling procedures. We refer the reader to some rejective sampling schemes described in Hájek (1981). See also Exercises 1.21, 1.24, and 1.25 and some methods discussed in Chapter Five. A brief history of random sampling methods has been reported by Bellhouse (1988).

1.3. INCLUSION PROBABILITIES OF THE UNITS AND THEIR RELATIONSHIP WITH MEAN AND VARIANCE OF SAMPLE SIZE DISTRIBUTION

The customary theory of sampling relies heavily on the inclusion probabilities of the sampling units associated with sampling designs. These inclusion probabilities are defined and studied in this section.

Definition 1.5. For each unit i in $\mathcal{U}$ the *first-order inclusion probability* $\pi_i(d)$ under the sampling design $d = (S_d, P_d)$ is defined by

$$\pi_i(d) = \sum_{s \ni i} P_d(s) , \tag{1.3}$$

where the sum extends over all samples $s \in S_d$ containing i. The *second-order (joint) inclusion probability* $\pi_{ij}(d)$ of the units i and j in $\mathcal{U}$ under d is defined to be

$$\pi_{ij}(d) = \sum_{s \ni i,j} P_d(s) , \tag{1.4}$$

where the sum extends over all samples $s \in S_d$ containing i and j simultaneously. Note that $\pi_{ij}(d) = \pi_{ji}(d)$.

For simplicity of notation, we shall denote $\pi_i(d)$ and $\pi_{ij}(d)$ by π_i and π_{ij}, respectively, when the sampling design d is understood from the context.

Example 1.10. Consider the sampling design d_1 of Example 1.4. We have

$$\pi_1 = P_{d_1}((1)) + P_{d_1}((1,3)) = 1/3 + 2/3 - a = 1 - a ,$$

$$\pi_2 = P_{d_1}((2)) = a ,$$

$$\pi_3 = P_{d_1}((1,3)) = 2/3 - a .$$

Similarly, $\pi_{12} = 0$, $\pi_{13} = 2/3 - a$, and $\pi_{23} = 0$.

In fact, the first- and second-order inclusion probabilities underlying a sampling design d are the probabilities of obtaining the unit i and units i and j, respectively, while sampling from the population according to the sampling design d. Notice also that for each unit i in $\mathcal{U}$, π_i is positive because of condition (ii) in Definition 1.1. The higher order inclusion probabilities are also similarly defined. The inclusion probabilities of various orders are known as the *structural constants* of a sampling design.

If the sampling design d is a fixed-size (n) design, then the sample size $n(s)$ assumes the value n for every sample s in the support S_d of d. If again d is a variable sample size design, then the sample size $n(s)$ is a random variable whose mean and variance can be computed in a routine manner. In usual notations,

$$E(n(s)) = \sum_s n(s)P_d(s)\,, \qquad E(n^2(s)) = \sum_s n^2(s)P_d(s)\,,$$

$$V(n(s)) = E(n^2(s)) - E^2(n(s))\,.$$

Theorem 1.3. *The following relations involving π_i's and π_{ij}'s and the moments of the sample size hold under the sampling design d:*

$$\sum_{i=1}^{N} \pi_i = \sum_{s \in S_d} n(s)P_d(s)\,, \tag{1.5}$$

$$\sum\sum_{i \neq j} \pi_{ij} = \sum_{s \in S_d} n(s)(n(s)-1)P_d(s)\,. \tag{1.6}$$

Proof

(i)
$$\sum_i \pi_i = \sum_i \Big(\sum_{s \ni i} P_d(s)\Big) = \sum_s \sum_{i \in s} P_d(s)$$
$$= \sum_s \Big[P_d(s)\Big(\sum_{i \in s} 1\Big)\Big] = \sum_s n(s)P_d(s)\,.$$

(ii)
$$\sum\sum_{i \neq j} \pi_{ij} = \sum\sum_{i \neq j}\Big(\sum_{s \ni i,j} P_d(s)\Big) = \sum_s \Big(\sum_{\substack{i,j \in s \\ i \neq j}} P_d(s)\Big)$$
$$= \sum_s \Big[P_d(s)\Big(\sum_{\substack{i,j \in s \\ i \neq j}} 1\Big)\Big]$$
$$= \sum_s n(s)(n(s)-1)P_d(s)\,. \qquad \square$$

For a fixed-size (n) design since $n(s) = n$ for all $s \in S_d$ and $\Sigma_{s \in S_d} P_d(s) = 1$, (1.5) and (1.6) reduce to

$$\sum_{i=1}^{N} \pi_i = n \tag{1.7}$$

and

$$\sum\sum_{i \neq j} \pi_{ij} = n(n-1)\,. \tag{1.8}$$

Some other useful quantities in terms of π_i's and π_{ij}'s are given below. For a fixed unit i,

$$\sum_{j(j\neq i)} \pi_{ij} = \sum_{j(j\neq i)} \Big(\sum_{s\ni i,j} P_d(s)\Big) = \sum_{s\ni i} (P_d(s)) \sum_{\substack{j(\neq i)\\ i,j\in s}} 1$$

$$= \sum_{s\ni i} (n(s)-1)P_d(s)\,. \tag{1.9}$$

For a fixed-size (n) design, (1.9) reduces to

$$\sum_{j(\neq i)} \pi_{ij} = (n-1)\sum_{s\ni i} P_d(s) = (n-1)\pi_i\,. \tag{1.10}$$

Let us define the symmetric matrix

$$\Pi = ((\pi_{ij}))\,, \qquad 1 \le i,\, j \le N$$

with $\pi_{ii} = \pi_i$. Some results for an arbitrary sampling design and the corresponding results for a fixed-size (n) design are as follows.

	Arbitrary Sampling Design	**Fixed-Size (n) Design**
(i)	$\text{trace}(\Pi) = \sum_{i=1}^{N} \pi_i = \sum_{s\in S_d} n(s)P_d(s)$	$(=n)$
(ii)	$\mathbf{1}'\Pi\mathbf{1} = \sum_{i=1}^{N}\sum_{j=1}^{N} \pi_{ij} = \sum_{s\in S_d} n(s)[n(s)-1]P_d(s) + \sum_{s\in S_d} n(s)P_d(s)$	
	$= \sum_{s\in S_d} [n(s)]^2 P_d(s)$	$(=n^2)$
(iii)	$\mathbf{e}_i'\Pi\mathbf{1} = \sum_{j=1}^{N} \pi_{ij} = \sum_{s\ni i} n(s)P_d(s)$	$(=n\pi_i)$

where $\mathbf{1}$ is the $N\times 1$ column vector with all 1's and $\mathbf{e}_i$ is the $N\times 1$ column vector with 1 at the ith position and 0 elsewhere, $\mathbf{e}_i'$ being the transpose of $\mathbf{e}_i$.

The expressions (1.3) and (1.4) also suggest that the following probability-inequality must hold for the pair (i, j):

$$\max\{0, \pi_i + \pi_j - 1\} \le \pi_{ij} \le \min\{\pi_i, \pi_j\}\,, \qquad 1 \le i \ne j \le N\,.$$

It is interesting to note that corresponding to a given set of values of π_i's and π_{ij}'s, consistent with the probability-inequalities, there does *not always* exist a sampling design attaining these specified values of the first- and second-order inclusion probabilities (Exercise 1.8).

From (1.5) and (1.6) it again follows that the mean and the variance of the sample size can be expressed in terms of the first- and second-order inclusion probabilities as follows.

$$E(n(s)) = \sum_{i=1}^{N} \pi_i , \tag{1.11}$$

$$\operatorname{Var}(n(s)) = \sum_{i=1}^{N} \sum_{j=1}^{N} \pi_{ij} - \left[\sum_{i=1}^{N} \pi_i\right]^2 . \tag{1.12}$$

Equation (1.11) is obvious and (1.12) follows from (1.11) and (1.6). It is seen that

$$\begin{aligned}
\operatorname{Var}(n(s)) &= E(n^2(s)) - E^2(n(s)) \\
&= E\{n(s)(n(s)-1)\} + E(n(s)) - E^2(n(s)) \\
&= \sum_s n(s)[n(s)-1]P_d(s) - E(n(s))[E(n(s))-1] \\
&= \sum_{i \neq j}\sum \pi_{ij} - \left[\sum_i \pi_i\right]\left[\left(\sum_i \pi_i\right) - 1\right] \\
&= \sum_i \sum_j \pi_{ij} - \left[\sum_i \pi_i\right]^2 .
\end{aligned}$$

Note that, for a fixed-size (n) design, $n(s)$ is equal to n for every s and thus it should not be surprising that, when d is a sampling design of fixed size n, we have

$$\sum_{i=1}^{n} \pi_i = n$$

whether or not d is uniform on its support.

We now introduce the notion of *indicator random variables*. This provides another interesting way of deducing (1.5), (1.6), (1.11), and (1.12).

For every unit i of $\mathcal{U}$, let $f_i(s)$ stand for the indicator random variable defined as $f_i(s) = 1$ if $i \in s$, $f_i(s) = 0$, otherwise, Clearly, $n(s) = \Sigma_i f_i(s)$ for every $s \in S_d$. As regards the f_i's, we have the following result.

Theorem 1.4. *Under $d = (S_d, P_d)$ with π_i's and π_{ij}'s as the first- and second-order inclusion probabilities respectively, we have*

$$E(f_i(s)) = \pi_i , \qquad \operatorname{Var}(f_i(s)) = \pi_i(1-\pi_i)$$

$$\operatorname{Cov}(f_i(s), f_j(s)) = \pi_{ij} - \pi_i\pi_j , \qquad 1 \le i \neq j \le N . \tag{1.13}$$

Proof. By (1.3) and (1.4), $E(f_i(s)) = 1 \cdot \Sigma_{s \ni i} P_d(s) + 0 \cdot \Sigma_{s \not\ni i} P_d(s) = \pi_i = E(f_i^2(s))$ and

$$E(f_i(s) \cdot f_j(s)) = 1 \cdot \sum_{s \ni i,j} P_d(s) + 0 \cdot \sum_{\substack{s \ni i \\ \not\ni j}} P_d(s) + 0 \cdot \sum_{\substack{s \not\ni i \\ \ni j}} P_d(s) + 0 \cdot \sum_{\substack{s \not\ni i \\ \not\ni j}} P_d(s) = \pi_{ij} .$$

Equation (1.13) is now evident. □

The results (1.5), (1.6), (1.11), and (1.12) now follow from (1.13) using the relation $n(s) = \Sigma_i f_i(s)$, which holds for all $s \in S_d$.

Example 1.11. Let $\mathcal{U} = \{1, 2, 3, 4\}$ and let $d = (S_d, P_d)$ be the following design:

$$S_d = \{s_1 = (1), \quad s_2 = (1,2), \quad s_3 = (2,3,4), \quad s_4 = (2,3)\}$$
$$P_d(s_1) = \frac{1}{8}, \quad P_d(s_2) = \frac{1}{8}, \quad P_d(s_3) = \frac{4}{8}, \quad P_d(s_4) = \frac{2}{8} .$$

Then the matrix Π is given by

$$\Pi = \begin{bmatrix} \frac{1}{4} & \frac{1}{8} & 0 & 0 \\ & \frac{7}{8} & \frac{3}{4} & \frac{1}{2} \\ & & \frac{3}{4} & \frac{1}{2} \\ & & & \frac{1}{2} \end{bmatrix}$$

Further, $E(n(s)) = \frac{19}{8}$ and $V(n(s)) = \frac{31}{64}$.

EXERCISES

1.1. For each of the following sampling designs, find the corresponding sampling algorithm:

(a) $\mathcal{U} = \{1, 2\}$; d: $P((1)) = 0.2$, $P((1,2)) = 0.3$, $P((2,1)) = 0.5$;

(b) $\mathcal{U} = \{1, 2, 3, 4\}$; d: $P((1)) = 0.1$, $P((1,2)) = 0.2$, $P((1,2,3)) = 0.3$, $P((1,2,3,4)) = 0.4$;

(c) $\mathcal{U} = \{1, 2, 3, 4\}$; d: $P((1)) = 0.1$, $P((2)) = 0.2 = P((3))$, $P((4)) = 0.3$, $P((1,2,3,4)) = 0.2$;

(d) $\mathcal{U} = \{1, 2, \ldots, 8\}$; d: $P((1,2,3)) = 0.4$, $P((4,5,6)) = 0.3$, $P((7,8)) = 0.3$;

(e) $\mathcal{U} = \{1, 2, \ldots, N\}$; d: $P((i)) = p$, $1 \le i \le N$, $P((1, 2, \ldots, N)) = 1 - Np$, $0 < p < 1/N$;

(f) $\mathcal{U} = \{1, 2, \ldots, N\}$; d: $P((N)) = p_1$, $P((N, N-1)) = p_2, \ldots$, $P((N, N-1, \ldots, 1)) = p_N$, $0 < p_i < 1$, $1 \le i \le N$, $\Sigma_i p_i = 1$.

(g) $\mathcal{U} = \{1, 2, \ldots, N\}$; d: $P(1, 2)) = p_1$, $P((2, 3)) = p_2, \ldots$, $P((N-1, N)) = p_{N-1}$, $P((N, 1)) = p_N$, $P((1, 2, \ldots, N)) = q$, $p_1 + p_2 + \cdots + p_N + q = 1$, $0 < q$, $p_i < 1$; $1 \leq i \leq N$.

(h) $\mathcal{U} = \{1, 2, \ldots, N\}$; d: $P((1)) = p_1$, $P((1, 2)) = p_2$, $P((3)) = p_3$, $P((3, 4)) = p_4, \ldots$, $P((N-1)) = p_{N-1}$, $P((N-1, N)) = p_N$, $p_1 + p_2 + \cdots + p_N = 1$, N even; $0 < p_i < 1$, $1 \leq i \leq N$.

1.2. For each of the sampling designs in the previous exercise, calculate the values of the first- and second-order inclusion probabilities. In each case, verify the relations (1,5), (1.6), (1.11), (1.12).

1.3. Take $\mathcal{U} = \{1, 2, 3\}$. Give examples of sampling designs (in reduced forms) such that

(i) $\pi_1 = 1$, $\pi_{12} = \pi_{13} = 0.5$, $\pi_{23} = 0$;
(ii) $\pi_3 = 1$, $\pi_{23} = 0.5$, $\pi_{13} = \pi_{12} = 0.25$;
(iii) $\pi_i = 0.7$, $1 \leq i \leq 3$, $\pi_{ij} = 0.5$, $1 \leq i \neq j \leq 3$.

1.4. Give an example of a sampling algorithm. Show the resultant sampling design. Draw a random sample based on this sampling design.

1.5. With reference to Theorem 1.1, verify that $\Sigma_s P_A(s) = 1$.

1.6. With reference to Theorem 1.2, verify that $A\{q_1, q_2, q_3\}$ defines an algorithm.

1.7. With reference to Theorem 1.2, verify that $q_i = q_i^*$, $i = 1, 2, 3$.

1.8. Take $\mathcal{U} = \{1, 2, 3\}$. Verify whether there exist sampling designs satisfying

(i) $\pi_1 = \pi_3 = 0.5$, $\pi_2 = 0.6$, $\pi_{12} = \pi_{13} = \pi_{23} = 0.1$;
(ii) $\pi_1 = 0.6$, $\pi_2 = \pi_3 = 0.8$, $\pi_{12} = 0.2$, $\pi_{13} = 0.4$, $\pi_{23} = 0.5$;
(iii) $\pi_1 = \pi_3 = 0.47$, $\pi_2 = 0.57$, $\pi_{12} = \pi_{13} = \pi_{23} = 0.20$.

Comment on the results.

1.9. Prove or disprove that if in a sampling design all the second-order inclusion probabilities are the same then all the first-order inclusion probabilities are the same.

1.10. In Exercise 1.9, if the sampling design is a fixed-size (n) design, do you still have the same conclusion?

1.11. Prove or disprove that if in a sampling design all the π_i's are the same, then all the π_{ij}'s are also the same.

1.12. If in a fixed-size (n) design with $n < N$, $\pi_{ij} > 0$ for all i and j, then find a lower bound on h, the support size of the design.

1.13. Is it possible to construct uniform fixed-size sampling designs with (a) $\pi_{ij} = \pi_i \pi_j$, (b) $\pi_{ij} < \pi_i \pi_j$, and (c) $\pi_{ij} > \pi_i \pi_j$ for all $1 \le i \ne j \le N$? If yes, exhibit the design. If no, why not?

1.14. The same problem as in Exercise 1.13, except for the change of uniform to nonuniform.

1.15. Is it true that if in a sampling design $\pi_{ij} = \pi_i \pi_j$ for a given pair (i, j), then $\pi_{ij} = \pi_i = \pi_j = 1$?

1.16. For the SRS(N, n) design of Example 1.8, show that $\pi_i = n/N$, $\pi_{ij} = n(n-1)/N(N-1)$, $1 \le i \ne j \le N$. Take $N = n^2$. Can you suggest any other fixed-size (n) sampling design which will also yield the above values of π_i's and π_{ij}'s?

1.17. (a) Let π_i^c and π_{ij}^c denote respectively the probabilities that a sample selected according to a sampling design will not include the unit i, and the units i and j. Show that (i) $\pi_i^c = 1 - \pi_i$, (ii) $\pi_{ij}^c = 1 - \pi_i - \pi_j + \pi_{jj}$, and (iii) $\pi_i \pi_j - \pi_{ij} = \pi_i^c \pi_j^c - \pi_{ij}^c$.

(b) Prove that a fixed-size (2) or a fixed-size ($N - 2$) sampling design in its reduced form is uniquely determined by its second-order inclusion probabilities.

(c) Show that the inclusion probabilities of all orders remain the same whether the sampling design is in its reduced form or not.

1.18. Given a set of numbers π_i^0's such that $0 < \pi_i^0 \le 1$, $1 \le i \le N$, with $\Sigma\, \pi_i^0 \ge 1$, show that there always exists a design attaining these values for the first-order inclusion probabilities.

1.19. In the previous exercise, if in addition the π_i^0's satisfy $\Sigma\, \pi_i^0 = n$, a positive integer, does there exist a fixed-size (n) design attaining these values?

1.20. If $E(n(s)) = \nu$ admits the representation $\nu = [\nu] + \theta$, $0 \le \theta < 1$, show that $\theta(1-\theta) \le V(n(s)) \le (N - \nu)(\nu - 1)$ and, hence, that $\nu(\nu - 1) + \theta(1 - \theta) \le \Sigma \Sigma_{i \ne j}\, \pi_{ij} \le N(\nu - 1)$.

1.21. Show that the SRS sampling design (see Example 1.8) can also be based on the following scheme of sampling. Draw units one at a time, at random and *with replacement*, *until n distinct units* are recorded. Regard these units as forming the *sample* (*repetitions ignored*).

1.22. For the sampling scheme of the previous exercise, work out the mean and the variance of the *number of times* the drawings are made.

1.23. Take $\mathcal{U} = \{1, 2, \ldots, N\}$ and let n be a fixed positive integer. Consider the following sampling design: d: $P_d((j_1, j_2, \ldots, j_\nu)) \propto \Delta^\nu 0^n$, $1 \le j_1 < j_2 < \cdots < j_\nu \le N$, $1 \le \nu \le \min(n, N)$ where $\Delta^\nu 0^n = \Delta^\nu x^n|_{x=0}$, Δ being the "difference" operator (see Section 4.8 for the definition and properties of Δ).

(a) Find the constant of proportionality.

(b) Show that the following sampling method results in the above design.

(i) Draw an SRS sample of one unit from the population. Call it i_1. Draw a fresh SRS sample of one unit again from the whole population. Call it i_2. Repeat this process n times and let $i_1, i_2, \ldots, i_n$ be the units thus chosen from the population. Note that *not* all of $i_1, i_2, \ldots, i_n$ are necessarily distinct.

(ii) Suppose $1 \le j_1 < j_2 < \cdots < j_\nu \le N$ are *all* the distinct units among $i_1, i_2, \ldots, i_n$. Declare $s = (j_1, j_2, \ldots, j_\nu)$ to be the ultimate sample. (In the literature, the sampling method used in (i) above is commonly referred to as *simple random sampling with replacement*).

(c) Show that

$$\pi_i = 1 - \left(\frac{N-1}{N}\right)^n, \qquad \pi_{ij} = 1 - 2\left(\frac{N-1}{N}\right)^n + \left(\frac{N-2}{N}\right)^n,$$

$1 \le i \ne j \le N$.

(d) Compute the mean and the variance of $n(s)$.

1.24. Show that each of the following two schemes of sampling results in an SRS(N, n) design in its reduced form.

(a) Start with the first unit of the population and include it in the sample with probability n/N. Next consider the second unit of the population and include it in the sample with probability $(n - n_1)/(N-1)$, where $n_1 = 1$ or 0 according to whether the first unit is included in the sample. In general, if n_k is the number of units selected after considering the kth unit, stop sampling if $n_k = n$. Otherwise, pass on to the next unit and include it in the sample with probability $(n - n_k)/(N - k)$.

(b) Start with the sample s_n consisting of the first n units of the population. Consider then the $(n+1)$th unit of the population and include it with probability $n/(n+1)$. If it is included, discard one unit of the n units in s_n with equal probability; otherwise retain s_n. Call the resulting sample of n units s_{n+1}.

In general, if s_k is the sample of n units selected after considering the kth unit, $k \geq n$, consider the $(k+1)$th unit of the population and include it with probability $n/(k+1)$. If it is included, discard one unit from s_k with equal probability and, if not, retain s_k to obtain the sample s_{k+1}. Continue this way until all N units have been checked one at a time. Finally, take s_N as the selected sample.

1.25. Show that the following is an alternative method of implementing a sampling design (S_d, P_d) through the implementation of a sampling design uniform in S_d. Let $S_d = \{s_1, s_2, \ldots, s_k\}$ and $p_d(s_j)$ be a rational number for each j, $1 \leq j \leq k$. Set, specifically, $p_d(s_j) = X_j/\mathbf{X}$, $1 \leq j \leq k$, where X_j's are positive integers, $\mathbf{X} = \Sigma_{j=1}^{k} X_j$. Let also $X_0 = \max\{X_1, X_2, \ldots, X_k\}$. Then draw one sample, say s_j, from S_d with equal probability and one number, say R, at random from 1 to X_0. If $R \leq X_j$, select s_j as the sample. Otherwise, discard the pair (s_j, R) and repeat the procedure. Continue this way until a sample is selected.

1.26. For the following scheme of sampling, compute the values of π_i's and π_{ij}'s for the resulting fixed-size (2) sampling design. Verify (1.7) and (1.8). Also examine whether $0 < \pi_{ij} < \pi_i \pi_j$ for all pairs of units or not. The reference population has only 5 units. Consider 10 possible partitions of the population into two groups of sizes 2 and 3, respectively. Select one of these partitions with probability 1/10 for each one of them. Let $G_1 = \{a, b\}$, $G_2 = \{c, d, e\}$ be the selected partition in terms of the labels a, b, c, d, e of the population units. Next, select one unit from G_1 with probability $a/(a+b)$ for unit a and $b/(a+b)$ for unit b. Further, select another unit from G_2 using selection probabilities $c/(c+d+e)$, $d/(c+d+e)$, and $e/(c+d+e)$, respectively, for the units c, d, and e.

REFERENCES

Bellhouse, D. R. (1988). A brief history of random sampling methods. In: *Handbook of Statistics, Vol. 6, Sampling* (Krishnaiah, P.R., and Rao, C.R., Eds.), 1–14. North-Holland, Amsterdam.

Cochran, W. G. (1977). *Sampling Techniques*. Third Edition. Wiley, New York.

Hájek, J. (1981). *Sampling from a Finite Population*. Marcel Dekker, New York.

Hansen, M. H., Dalenius, T., and Tepping, B.J. (1985). The development of sample surveys of finite populations. In: *A Celebration of Statistics* (Atkinson, A. C., and Fienberg, S. E., Eds.), 327–354. Springer-Verlag, New York.

Hanurav, T. V. (1966). Some aspects of unified sampling theory. *Sankhyā*, **A28**, 175–203.

Johnson, N. L., and Smith, H., Jr., Eds. (1969). *New Developments in Survey Sampling*. Wiley-Interscience, New York.

Krishnaiah, P. R. and Rao, C. R., Eds. (1988). *Handbook of Statistics, Vol. 6, Sampling*. North-Holland, Amsterdam.

Lahiri, D. B. (1951). A method of sample selection providing unbiased ratio estimates. *Bull Internat. Statist. Inst.*, **33**, 133–140.

Midzuno, H. (1951). On the sampling system with probability proportionate to sum of sizes. *Ann. Inst. Statist. Math.*, **2**, 99–107.

CHAPTER TWO

Inference in Finite Population Sampling

We pose some specific problems of statistical inference in finite population sampling. Referring to the standard nomenclature in the statistical theory of inference, we formulate the usual decision-theoretic notions for comparing different sampling strategies. Then we present the basic existence and nonexistence results pertinent to the estimation of finite population totals. In the process, we introduce the celebrated Horvitz–Thompson Estimator (HTE).

2.1. PRELIMINARIES

Let Y be a study variable or characteristic of interest related to a finite population $\mathcal{U}$ of N units, labelled $\{1, 2, \ldots, N\}$. We denote by Y_i the value of Y, assumed to be quantitative, on the unit labelled i, that is, the ith unit of $\mathcal{U}$, $1 \leq i \leq N$. We write $\mathbf{Y} = (Y_1, \ldots, Y_N)$ to denote the vector of Y-values. A priori, $Y_1, \ldots, Y_N$ are all unknown. However, we assume that the survey statistician, engaged in studying the behavior of the particular characteristic in the given population, can observe the value Y_i (with or without error) by contacting the ith unit, through a survey.

In any specific field of study, it is not enough to talk about the reference population in a casual sense. The notion of population *frame* is needed. A frame is a clear and concise description of the population under study, by virtue of which the population units can be identified unambiguously and contacted, if desired, for the purpose of the survey. We emphasize that, in practice, the formation of a frame is a difficult issue and it may pose difficult problems depending on the nature and the complexity of the survey population.

Another important practical aspect of data collection is the manner of approach. Before approaching the selected units in the sample by mail, telephone, or personal interview for the purpose of collecting information, we prepare a survey form which is commonly known as a *questionnaire*. In many instances, the quality of the survey data depends on the way the questionnaire is prepared. While this task is nonstatistical in nature, we must emphasize that the validity of the statistical inference drawn on a given occasion depends largely on the quality of the data and this in turn depends on the quality of the questionnaire. This is even more important in multipurpose surveys. Several books as well as articles have been written on the art and science of preparing questionnaires. We refer the reader to Chapter 15 of Levy and Lemeshow (1980), Oppenheim (1966), Dillman (1978), and Sudman and Bradburn (1982). See also Deming (1960).

The purpose of the survey is to draw inference about some well-defined function(s) of $Y_1, \ldots, Y_N$—called *parameter*(*s*) of the finite population. We might be asked to infer about the parameter $\theta(\mathbf{Y})$, say, after observing the Y_i-values for the units included in a sample, say s, suitably chosen and surveyed. The common functions of interest are $T(\mathbf{Y}) = \Sigma_1^N Y_i$, the *finite population total*, $\mu_Y = T(\mathbf{Y})/N = \bar{\mathbf{Y}}$, the *finite population mean*, and $\sigma_Y^2 = 1/N\, \Sigma_1^N Y_i^2 - \mu_Y^2$, the *finite population variance*. The customary inference restricts itself to *point estimates* and *interval estimates*.

Throughout we shall assume Y to be a quantitative real-valued survey variable and regard $\mathbf{Y} = (Y_1, \ldots, Y_N)$ as a point in R^N, the N-dimensional Euclidean space. Further, we assume that any Y_i can be observed without error. Of course, this is not the case in practice. In actual surveys, the common sources of error include respondent error, the interviewer bias, the organizational structures of the data collection machinery, etc. These may be grouped into basically two types: errors of measurement and errors of nonresponse. Chapters Eleven and Twelve are devoted to a discussion of these problems.

With respect to inference problems in finite population sampling, the survey statistician has to both choose and survey a sample s and also form his estimate of the parametric function of interest. The solution to this twin problem leaves room for arbitrariness and personal judgement unless, of course, one is guided by certain definite inference rules. The early stage of development of this subject has witnessed only one approach to inference based on the general Statistical Theory of Inference or Decision-Making, suitably modified for finite population inference (see Remark 2.2). However, some other approaches have recently been presented and discussed.

2.2. DECISION-THEORETIC APPROACH: ROLE OF SAMPLING DESIGNS

In this approach the sample to be chosen is assumed to be a *probability* sample based on a certain sampling design. To fix ideas, let $d = (S_d, P_d)$ be the sampling design adopted.

Let s be a typical sample and $\theta(\mathbf{Y})$ be the parametric function of interest, again assumed to be real-valued, that is, a point on R^1, the real line.

Definition 2.1. An estimator $e(s, \mathbf{Y})$ is a real-valued function, defined on $S \times R^N$, which, for each $s \in S$, depends on $\mathbf{Y}$ only through those Y_i's for which the unit i occurs in the sample s.

If $s = (i_1, \ldots, i_{n(s)})$ is a sample actually chosen and surveyed, the survey statistician calculates the value of a proposed estimator $e(s, \mathbf{Y})$ based on the observed quantities $Y_{i_1}, \ldots, Y_{i_{n(s)}}$, which then serves as a point estimate for $\theta(\mathbf{Y})$. The following are some illustrative examples of estimators. Let $s = (1, 5, 3)$ be a typical sample chosen according to a sampling design. Then $(Y_1 + Y_3 + Y_5)/3$, $(Y_1^2 + Y_3^2 + Y_5^2)/3$, $(Y_1Y_3 + Y_1Y_5 + Y_3Y_5)/3$, etc., are examples of estimates of various parametric functions.

In estimating the finite population total $T(\mathbf{Y})$, for example, we could employ a homogeneous linear estimator of any one of the forms given in Table 2.1, as suggested by Koop (1963) and Prabhu-Ajgaonkar (1965).

It can be verified that, of all the forms in Table 2.1, the most general form is given by $e_5(s, \mathbf{Y})$ (Exercise 2.1). Exercise 2.2 can help the reader to become familiar with the computations of these estimators.

Definition 2.2. The estimator

$$l(s, \mathbf{Y}) = \sum_{i \in s} l_{si} Y_i \tag{2.1}$$

Table 2.1. Different Types of Homogeneous Linear Estimators

Estimators	Characteristics
$e_1(s, \mathbf{Y}) = \sum_{t=1}^{n(s)} l_t Y_{i_t}$	The coefficients are *independent* of the actual units drawn but *dependent* on the order of selection
$e_2(s, \mathbf{Y}) = \sum_{t=1}^{n(s)} l_{i_t t} Y_{i_t}$	The coefficients are *dependent* on the units as well as on the order of selection
$e_3(s, \mathbf{Y}) = l_s\left\{\sum_{t=1}^{n(s)} Y_{i_t}\right\}$	The coefficients depend only on the sample and are the same for all the units in the sample
$e_4(s, \mathbf{Y}) = \sum_{t=1}^{n(s)} l_{i_t} Y_{i_t}$	The coefficients are independent of the sample, depending only on the units (but not on the order of the selection)
$e_5(s, \mathbf{Y}) = \sum_{t=1}^{n(s)} l_{si_t} Y_{i_t}$	The coefficients depend on the sample as well as on the unit drawn

is said to define the most general form of a homogeneous *linear* estimator (hle).

Here l_{si} is the coefficient of Y_i and generally it depends on both the sample s and the unit $i \in s$. In general, a nonhomogeneous linear estimator may be obtained by adding a quantity like a_s to $l(s, \mathbf{Y})$ in (2.1).

Definition 2.3. The estimator

$$q(s, \mathbf{Y}) = \sum_{i \in s} q_{si} Y_i^2 + \sum_{i \in s, j \in s, i \neq j} q_{sij} Y_i Y_j \tag{2.2}$$

is said to define the most general form of a homogeneous *quadratic* estimator (hqe).

Here the coefficients q_{si}, q_{sij} generally depend on both the sample s and the unit i or the units i, j.

For a sampling design d and an estimator e, the pair (d, e) is said to form a *sampling strategy* and is also denoted by (p, e) where p refers to P_d. The performance of the strategy (p, e) is *not* judged by the *estimate* is provides on a given occasion based on the *realized* data set $\{(i_1, Y_{i_1}), (i_2, Y_{i_2}), \ldots, (i_{n(s)}, Y_{i_{n(s)}})\}$. Instead, reference is made to the fact that the chosen sample is only a realization of the adopted sampling design d. Therefore, the behavior of the strategy (p, e) is studied through the performance of the estimator $e(s, \mathbf{Y})$ in terms of its long-term average over repeated sampling. One performance criterion is (design-) unbiasedness of the estimator $e(s, \mathbf{Y})$.

Definition 2.4 Based on a sampling design d, an estimator $e(s, \mathbf{Y})$ is said to be *(design-) unbiased* for $\theta(\mathbf{Y})$ if the design expectation of $e(s, \mathbf{Y})$ equals $\theta(\mathbf{Y})$ uniformly in $\mathbf{Y} \in R^N$.

This means that $e(s, \mathbf{Y})$ is an unbiased estimator for $\theta(\mathbf{Y})$ with respect to the sampling design d if and only if

$$E_p(e(s, \mathbf{Y})) = \sum_{s \in S_d} e(s, \mathbf{Y}) P_d(s) = \theta(\mathbf{Y}) \text{ uniformly in } \mathbf{Y} \in R^N . \tag{2.3}$$

In the above, we used the notation $E_p(e(\cdot))$ which is consistent with existing literature, while $E_d(e(\cdot))$ might have been more meaningful.

Whenever the sampling design is understood from the context, we will conveniently call an estimator *unbiased* if it is (design-) unbiased. Unbiased estimators are not, however, unique unless the sampling design is essentially a census. One may see this easily by referring to nonhomogeneous estimators. On the other hand, even with homogeneous estimators, in most cases unbiased estimators are not unique. To see this, let d be a sampling design such that $P_d(s') > 0$, $P_d(s'') > 0$, and $s' \cap s'' \neq \emptyset$, the null subset of $\mathcal{U}$.

Let $i \in s' \cap s''$. In other words, we take i to be one unit common to both s' and s''. Form an estimator $\tilde{e}(s, \mathbf{Y})$ such that

$$
\begin{aligned}
\tilde{e}(s, \mathbf{Y}) &= 0\,, && \text{for all } s \neq s', s'' \\
&= f(Y_i)P_d(s'')\,, && \text{for } s = s' \\
&= -f(Y_i)P_d(s')\,, && \text{for } s = s''\,.
\end{aligned}
$$

Then $E_p(\tilde{e}(s, \mathbf{Y})) = 0$ uniformly in $\mathbf{Y} \in R^N$. Here $f(\cdot)$ refers to any finite real-valued function. Certainly this $\tilde{e}$, the so-called *error function* in the terminology of statistical estimation theory, can be combined with any unbiased estimator of $\theta(\mathbf{Y})$ to yield another unbiased estimator for the same parameter $\theta(\mathbf{Y})$.

The above concepts can be explained as follows. Suppose we have the sampling design d: $P((1)) = 0.2$, $P((2)) = 0.5$, $P((1, 2)) = 0.3$. Then, among the estimators listed in Table 2.2, e_1, e_2, and e_3 are unbiased for the population total while e_4 is not.

Having defined the concept of unbiasedness, we need tools to construct unbiased estimators and to verify whether a given estimator is unbiased or not for a given parametric function. We address these problems below. At the very outset we mention that the general problem of constructing an unbiased estimator for an arbitrary function $\theta(\mathbf{Y})$ is difficult. However, we provide a general tool for constructing unbiased estimators for a class of parametric functions $\theta(\mathbf{Y})$ under an arbitrary sampling design d. This tool is applicable in many cases of interest.

Theorem 2.1. *Suppose with respect to a given sampling design d, $\theta(\mathbf{Y})$ admits the representation*

$$\theta(\mathbf{Y}) = \sum_{s \in S_d} P_d(s)\delta(s, \mathbf{Y}) \tag{2.4}$$

uniformly in $\mathbf{Y} \in R^N$, where $\delta(s, \mathbf{Y})$ is a function of $\{Y_i \mid i \in s\}$ for every s in S_d. Then

$$\hat{\theta}(\mathbf{Y}) = \delta(s, \mathbf{Y}) \tag{2.5}$$

serves as an unbiased estimator of $\theta(\mathbf{Y})$.

Table 2.2. Examples of Estimators for the Population Total

	Estimators			
Samples	e_1	e_2	e_3	e_4
(1)	$2Y_1$	$2Y_1 + 3Y_1^2$	$5 + Y_1/2$	$3Y_1$
(2)	$5Y_2/4$	$5Y_2/4$	$2Y_2 - 2$	$7Y_2$
(1, 2)	$2Y_1 + 5Y_2/4$	$2Y_1 - 2Y_1^2 + 5Y_2/4$	$3Y_1$	$Y_1 - Y_2$

Proof. By definition,

$$E_p(\hat{\theta}(\mathbf{Y})) = \sum_{s \in S_d} P_d(s)\delta(s, \mathbf{Y}) = \theta(\mathbf{Y})$$

uniformly in $\mathbf{Y} \in R^N$, and hence the result. □

The reader will note that we have exploited the definition of unbiasedness above in arriving at an unbiased estimator of $\theta(\mathbf{Y})$. However unrevealing the constructional procedure may look, the idea presented above provides a very powerful and extremely useful tool for constructing unbiased estimators on many occasions. This is particularly true with reference to linear and quadratic functions of $Y_1, \ldots, Y_N$. Now we apply the above tool to present a result that applies when $\theta(\mathbf{Y})$ is linear.

Theorem 2.2. *Given known real numbers $a_1, \ldots, a_N$, not all zero, an unbiased estimator of*

$$\theta(\mathbf{Y}) = \sum_{i=1}^{N} a_i Y_i$$

based on any sampling design is given by

$$\delta(s, \mathbf{Y}) = \sum_{j \in s} a_j Y_j / \pi_j \tag{2.6}$$

where π_j's refer to the first-order inclusion probabilities of the units in the chosen sampling design.

Proof. We apply the tool of Theorem 2.1. To do so we need a representation of $\theta(\mathbf{Y})$ in the form (2.4). This is easily achieved as follows.

$$\begin{aligned}\theta(\mathbf{Y}) &= \sum_{i=1}^{N} a_i Y_i = \sum_{i=1}^{N} a_i Y_i(\pi_i/\pi_i) \\ &= \sum_{i=1}^{N} \frac{a_i Y_i}{\pi_i}\Big(\sum_{s \ni i} P(s)\Big) = \sum_{s} P(s)\Big[\sum_{j \in s} a_j Y_j/\pi_j\Big].\end{aligned}$$

This yields the estimator in (2.6) which is, therefore, unbiased for estimating $\theta(\mathbf{Y}) = \Sigma_{i=1}^{N} a_i Y_i$. □

Corollary 2.3. *Every sampling design provides one homogeneous linear unbiased estimator of $T(\mathbf{Y})$ given by*

$$l(s, \mathbf{Y}) = \sum_{i \in s} Y_i / \pi_i\,. \tag{2.7}$$

Proof. In the notation of Theorem 2.2, $T(\mathbf{Y}) = \theta(\mathbf{Y})$ with $a_1 = a_2 = \cdots = a_N = 1$. Clearly then (2.6) simplifies to (2.7). □

For a sampling design of the type discussed below (2.3), we can combine the estimator in (2.7) with a homogeneous linear error function and thus construct one and, hence, an infinite number of homogeneous linear unbiased estimator(s) of $T(\mathbf{Y})$. The estimator in (2.7) is the celebrated *Horvitz–Thompson Estimator* (HTE) proposed by Horvitz and Thompson (1952). The HTE has been so extensively studied in the literature and used in practice, that it has been the objective of almost all researchers in this area to relate their work, in some form or other, to it. Chapter Three will be devoted mainly to a study of some aspects of the HTE.

With respect to the question of unbiased estimation of quadratic and other functions of $Y_1, Y_2, \ldots, Y_N$ utilizing the tool developed in Theorem 2.1, we postpone its discussion to Chapter Three where, among other things, the questions of unbiased estimation of the population variance will be studied. The reader will find another highly nontrivial and novel application of Theorem 2.1 in Section 5.5.

We now address the question regarding verification of unbiasedness of a given estimator. Whereas in general we have to be guided by the definition, at least in one important special case we can proceed one step further and provide a satisfactory result. This relates to linear homogeneous and nonhomogeneous unbiased estimators of the population total.

Theorem 2.4. *The estimator*

$$l(s, \mathbf{Y}) = \sum_{i \in s} l_{si} Y_i \,, \qquad s \in S$$

is unbiased for $T(\mathbf{Y})$ if and only if the coefficients l_{si} satisfy

$$\sum_{s \ni i} l_{si} P(s) = 1 \,, \qquad 1 \le i \le N \,. \tag{2.8}$$

Proof. We first compute the (design-) expectation of $l(s, \mathbf{Y})$.

$$E_p[l(s, \mathbf{Y})] = \sum_{s} P(s) \left(\sum_{i \in s} l_{si} Y_i \right) = \sum_{i=1}^{N} Y_i \left[\sum_{s \ni i} l_{si} P(s) \right].$$

For $l(s, \mathbf{Y})$ to be unbiased for $T(\mathbf{Y})$ it is necessary and sufficient that $E_p(l(s, \mathbf{Y})) = T(\mathbf{Y})$ uniformly in $\mathbf{Y} \in R^N$. This yields

$$\sum_{i=1}^{N} Y_i \left(\sum_{s \ni i} l_{si} P(s) \right) = \sum_{i=1}^{N} Y_i$$

uniformly in $\mathbf{Y} \in R^N$. This results in (2.8). □

If, on the other hand, we use a nonhomogeneous linear estimator by adding a quantity a_s to $l(s, \mathbf{Y})$, then the conditions of unbiasedness will also include, apart from those in (2.8), the following:

$$\sum_s a_s P(s) = 0 . \tag{2.9}$$

Recalling that $l_{si} = 0$ whenever $s \not\ni i$, it is evident that (2.8) cannot hold if $\pi_i = 0$ for any i, $1 \le i \le N$. In other words, the conditions $\pi_i > 0$, $1 \le i \le N$ turn out to be necessary for linear unbiased estimation of the population total. We recall that Definition 1.1 of a sampling design incorporates this feature. That these conditions are sufficient as well has already been established in Corollary 2.3.

Remark 2.1. We would like to emphasize that even though the tool presented in Theorem 2.1 is generally applicable to any problem of unbiased estimation, the choice of the estimator $\delta(s, \mathbf{Y})$ is left too much open. A desirable and natural guideline for a choice of $\delta(s, \mathbf{Y})$ would be to mimic, in some sense, the functional form of $\theta(\mathbf{Y})$ in the sample, as traditionally done in statistical estimation theory. Murthy (1963) and Srivastava (1985) took this approach and introduced a class of unbiased estimators for functions of parameters of a finite population, under an arbitrary sampling design.

Specializing to the problem of estimation of a finite population total, we will present, following Srivastava (1985), a general solution to $\delta(s, \mathbf{Y})$ in (2.4) leading to a class of HT-type estimators. Towards this end, we introduce the following concept.

Let $\phi(s)$ be a real-valued function defined for *all* samples of the form $s = (i_1, i_2, \ldots, i_{n(s)})$ where $1 \le i_1 < i_2 < \cdots < i_{n(s)} \le N$, $1 \le n(s) \le N$. When $n(s) = N$, we will denote the function by $\phi(\mathcal{U})$, where $\mathcal{U}$ is the finite population under consideration.

Definition 2.5. A function $\phi(s)$ is said to be a *sample free function* (sff) with respect to a sampling design d provided

$$E_p(\phi(s)) = \phi(\mathcal{U}) . \tag{2.10}$$

This shows that an sff $\phi(s)$ serves as an unbiased estimator of the quantity $\phi(\mathcal{U})$. We take an illustrative example.

Example 2.1. Let $d \equiv \mathrm{SRS}(N, n)$ in its reduced form. Fix a value of $k < n$. Let $\psi(s)$ be a function defined on the subsets of $\mathcal{U}$ and let

$$\phi(s) = \sum_{\substack{s' \subset s \\ |s'| = k}} \psi(s') / \binom{n}{k}$$

where the summation extends over all $\binom{n}{k}$ subsamples $s' \subset s$. Then

$$\phi(\mathcal{U}) = \sum_{\substack{s' \\ |s'|=k}} \psi(s')/\binom{N}{k}$$

and

$$\begin{aligned}
E_p(\phi(s)) &= \frac{1}{\binom{N}{n}} \sum_{\substack{s \\ |s|=n}} \sum_{\substack{s' \subset s \\ |s'|=k}} \psi(s')/\binom{n}{k} \\
&= \frac{1}{\binom{N}{n}} \sum_{\substack{s' \subset \mathcal{U} \\ |s'|=k}} \psi(s') \sum_{\substack{s \\ |s|=n \\ s \supset s'}} \frac{1}{\binom{n}{k}} \\
&= \frac{1}{\binom{N}{n}} \binom{N-k}{n-k} \left\{ \sum_{\substack{s' \subset \mathcal{U} \\ |s'|=k}} \psi(s')/\binom{n}{k} \right\} \\
&= \sum_{\substack{s' \subset \mathcal{U} \\ |s'|=k}} \psi(s')/\binom{N}{k} = \phi(\mathcal{U}),
\end{aligned}$$

as expected.

A general procedure for constructing an sff is given below.

Theorem 2.5. *For an arbitrary but fixed k, $1 \le k \le N$, let $\psi(s)$ be a function defined on the subsets of $\mathcal{U}$ and let*

$$\phi(s) = \sum_{\substack{s' \subset s \\ |s'|=k}} \psi(s')\lambda(s', s) \tag{2.11}$$

for every ordered sample $s = (i_1, i_2, \ldots, i_{n(s)})$, $1 \le i_1 < i_2 < \cdots < i_{n(s)} \le N$, $k \le n(s) \le N$. If the functions $\lambda(s', s)$ satisfy

$$\lambda(s', \mathcal{U}) = \sum_{s \supset s'} \lambda(s', s)P(s) \tag{2.12}$$

for every s' of size k, then $\phi(s)$ is an sff with respect to the underlying sampling design.

Proof. We have

$$\begin{aligned}
E_p(\phi(s)) &= \sum_s \phi(s)P(s) \\
&= \sum_s P(s) \sum_{\substack{s' \subset s \\ |s'|=k}} \psi(s')\lambda(s', s)
\end{aligned}$$

$$= \sum_{s' \subset \mathcal{U}} \psi(s') \sum_{s \supset s'} \lambda(s', s) P(s)$$

$$= \sum_{\substack{s' \subset \mathcal{U} \\ |s'|=k}} \psi(s') \lambda(s', \mathcal{U}) = \phi(\mathcal{U}) .$$

Hence the result is proved. □

Various interesting applications of the above theorem have been discussed in Srivastava (1985). We now specialize to the particular case of $k = 1$ and deduce the following result.

Corollary 2.6. *The function*

$$\phi(s) = \sum_{i \in s} \psi(i) \lambda(i, s) \tag{2.13}$$

is an unbiased estimator of

$$\phi(\mathcal{U}) = \sum_{i=1}^{N} \psi(i) \lambda(i, \mathcal{U}) \tag{2.14}$$

provided $\lambda(i, \mathcal{U})$ *is defined as*

$$\lambda(i, \mathcal{U}) = \sum_{s \ni i} \lambda(i, s) P(s) , \qquad 1 \le i \le N . \tag{2.15}$$

An important special choice of $\lambda(i, s)$ in Corollary 2.6 now leads to the following result.

Theorem 2.7. *An unbiased estimator of the population total* $T(\mathbf{Y})$ *is given by*

$$\phi(s, \mathbf{Y}) = \sum_{i \in s} Y_i c_i(s) / \pi_c(i) \tag{2.16}$$

where

$$\pi_c(i) = \sum_{s \ni i} c_i(s) P(s) , \qquad 1 \le i \le N . \tag{2.17}$$

Proof. We set $\psi(i) = Y_i$ and $\lambda(i, s) = c_i(s) / \pi_c(i)$ in (2.13). Then, in view of (2.17), we obtain, using (2.15),

$$\lambda(i, \mathcal{U}) = \sum_{s \ni i} c_i(s) P(s) / \pi_c(i) = 1 , \qquad 1 \le i \le N .$$

Hence, (2.14) reduces to $\phi(\mathcal{U}) = T(\mathbf{Y})$. □

The following corollary is now immediate.

Corollary 2.8. *An unbiased estimator of $T(\mathbf{Y})$ is given by*

$$\phi(s, \mathbf{Y}) = c(s) \sum_{i \in s} Y_i / \pi_c(i) \tag{2.18}$$

where

$$\pi_c(i) = \sum_{s \ni i} c(s) P(s) , \qquad 1 \leq i \leq N . \tag{2.19}$$

In particular, taking $c(s) = 1$, $\phi(s, \mathbf{Y})$ in (2.18) reduces to the HTE given in (2.7). The estimator in (2.18) thus generalizes the HTE in an obvious way and we obtain a class of HT-type estimators. We refer to Murthy (1963) and Srivastava (1985) for various other results in this context.

(Design-) unbiasedness alone cannot be a deciding factor for the choice of a "good design." It is further proposed that the (design-) mean squared error (mse) of the estimator should be minimized. In comparing two given estimators $e_1(s, \mathbf{Y})$ and $e_2(s, \mathbf{Y})$ if it so happens that

$$E_p(e_1(s, \mathbf{Y}) - \theta(\mathbf{Y}))^2 \leq E_p(e_2(s, \mathbf{Y}) - \theta(\mathbf{Y}))^2$$

uniformly in $\mathbf{Y} \in R^N$, then the estimator e_1 is said to be *at least as good* as the estimator e_2 for estimating $\theta(\mathbf{Y})$ under squared error loss. If, in the above inequality, *strict inequality* holds for at least one point $\mathbf{Y} \in R^N$, we say that the estimator e_1 is *better* than the estimator e_2 and in such a case, e_2 is said to be *inadmissible*. Clearly, any inadmissible estimator can be discarded from further consideration. For unbiased estimators, (design-) mse reduces to (design-) variance.

Below we provide a general result of this nature. The reader may recall the definition of equivalent samples given in Section 1.1.

Theorem 2.9. *The estimator $e^*(s, \mathbf{Y})$ defined as*

$$e^*(s, \mathbf{Y}) = {\sum_s}' e(s, \mathbf{Y}) P_d(s) \Big/ {\sum_s}' P_d(s) \tag{2.20}$$

is at least as good as or better than the estimator $e(s, \mathbf{Y})$, where the summation Σ_s' extends over all samples which are equivalent to s in the sampling design under consideration, unless $e(s, \mathbf{Y})$ is the same for all equivalent samples in which case e and e^* coincide.

Proof. It can be easily verified that

$$E_p[e(s, \mathbf{Y})] = E_p(e^*(s, \mathbf{Y})) .$$

Further,

$$\begin{aligned} E_p[e^2(s, \mathbf{Y})] &= \sum_s P_d(s) e^2(s, \mathbf{Y}) \\ &= \sum \sum{}' P_d(s) e^2(s, \mathbf{Y}) \\ &\geq \sum \Big(\sum{}' P_d(s) e(s, \mathbf{Y}) \Big)^2 \Big(\sum{}' P_d(s) \Big)^{-1} \\ &= \sum P_d(s) e^{*2}(s, \mathbf{Y}) \,. \end{aligned}$$

In the above, we have used Cauchy–Schwartz inequality to deduce that

$$\Big(\sum{}' P_d(s)\Big)\Big(\sum{}' e^2(s, \mathbf{Y}) P_d(s)\Big) \geq \Big(\sum{}' P_d(s) e(s, \mathbf{Y})\Big)^2 \,.$$

It is now evident that

$$E_p(e^*(s, \mathbf{Y}) - \theta(\mathbf{Y}))^2 \leq E_p(e(s, \mathbf{Y}) - \theta(\mathbf{Y}))^2$$

uniformly in $\mathbf{Y} \in R^N$ with strict inequality for some $\mathbf{Y}$ unless $e(s, \mathbf{Y}) = e^*(s, \mathbf{Y})$ for every s with $P_d(s) > 0$. Hence the result is established. □

Remark 2.2. The estimator e^* is referred to in the literature as the symmetrized form of the estimator e. This terminology is due to Murthy (1957). Sometimes $e(s, \mathbf{Y})$ is referred to as an *ordered* estimator while $e^*(s, \mathbf{Y})$ is referred to as an *unordered* estimator. It may be seen that if d^* corresponds to the reduced form of the sampling design d, then $E_p(e^*(s, \mathbf{Y})) = E_{p^*}(e^*(s^*, \mathbf{Y}))$ and $E_p(e^{*2}(s, \mathbf{Y})) = E_{p^*}(e^{*2}(s^*, \mathbf{Y}))$ uniformly in $\mathbf{Y} \in R^N$. This is why essentially only the reduced forms of sampling designs are relevant to inference problems.

Turning back to the question of comparison of estimators for a given sampling design d, ideally we put forward the requirement that the choice of the estimator e should minimize $E_p(e(s, \mathbf{Y}) - \theta(\mathbf{Y}))^2$ uniformly in $\mathbf{Y} \in R^N$. This, along with the condition of unbiasedness, forms the criterion of *uniformly minimum variance unbiased estimation*.

Definition 2.6. An estimator $e(s, \mathbf{Y})$ based on a sampling design d, is said to be a uniformly minimum variance unbiased estimator (umvue), also called *best* estimator, for $\theta(\mathbf{Y})$ if and only if

$$E_p(e(s, \mathbf{Y})) = \theta(\mathbf{Y}) \text{ uniformly in } \mathbf{Y} \in R^N$$

and

$$E_p(e(s, \mathbf{Y}) - \theta(\mathbf{Y}))^2 \leq E_p(e'(s, \mathbf{Y}) - \theta(\mathbf{Y}))^2 \text{ uniformly in } \mathbf{Y} \in R^N \,,$$

where $e'(s, \mathbf{Y})$ is *any* other estimator satisfying $E_p(e'(s, \mathbf{Y})) = \theta(\mathbf{Y})$ uniformly in $\mathbf{Y} \in R^N$. As we shall see later, umvue's *seldom* exist.

Remark 2.3. At this stage, we want to emphasize that there exists a fundamental difference between a finite population setup and the usual inferential setup. Unless a finite population has for its units some sort of identification, concepts like sample, sampling design and its implementation become *ill-defined.* This calls for the introduction of *labels* which bear 1:1 correspondence with the units so that the units become identifiable and a sampling design can be meaningfully implemented.

Kempthorne (1978) discussed a small but illuminating example of how ignorance of labels may lead to possibly incomplete knowledge about the population under study. Suppose a population of two units is surveyed by two persons, each one independently choosing just one unit and simply noting the value of the survey variable and ignoring the label. If the two values thus gathered happen to be identical, it is impossible to decide whether or not the population has been exhaustively surveyed! The possibility that the values on the two units are in fact the same, remains undecided! See Kempthorne (1978) for further discussion.

In view of the above discussion, it is recommended that one should fully record the data set $\{(i_1, Y_{i_1}), (i_2, Y_{i_2}), \ldots, (i_{n(s)}, Y_{i_{n(s)}})\}$, utilizing *any information* that is carried by the labels of the sampled units. Under this method of formulation, however, there is a striking general result on the nonexistence of any umvue or best estimator for most sampling designs. We will discuss this in full detail below. We postpone the discussion on the scope of the utilization of unlabelled data to Section 4.11.

2.3. SOME EXISTENCE AND NONEXISTENCE RESULTS

First we present a very general nonexistence result. The proof of this result is quite simple but elegant. It is due to Basu (1971).

Theorem 2.10. *For a given noncensus design and for a general parametric function $\theta(\mathbf{Y})$ which depends effectively on the Y-values, there does not exist any umvue.*

Proof. Let $e(s, \mathbf{Y})$ be an unbiased estimator of $\theta(\mathbf{Y})$ for a given noncensus sampling design d. Fix a point $\mathbf{Y}_0$ in R^N. Next, define the estimator $e^*(s, \mathbf{Y}) = e(s, \mathbf{Y}) - e(s, \mathbf{Y}_0) + \theta(\mathbf{Y}_0)$. It is evident that $E_p[e(s, \mathbf{Y})] = E_p[e^*(s, \mathbf{Y})] = \theta(\mathbf{Y})$ for all $\mathbf{Y} \in R^N$. Further, $e^*(s, \mathbf{Y}_0) = \theta(\mathbf{Y}_0)$ for every s with $P(s) > 0$. Therefore, $V_p(e^*(s, \mathbf{Y})) = 0$ whenever $\mathbf{Y} = \mathbf{Y}_0$. Hence, at any given point in R^N, the variance of a suitably constructed unbiased estimator can be made equal to zero. This clearly demonstrates that there does not exist any umvue of $\theta(\mathbf{Y})$. □

For estimation of a finite population total $T(\mathbf{Y})$, it is therefore evident that there does not exist any best estimator in the class of all unbiased estimators. What happens if we restrict ourselves to a suitable subclass of the large class? It can be shown that the same non-existence result continues to hold even if we restrict to the subclass of linear homogeneous *and* nonhomogeneous unbiased estimators. The above nonexistence results are valid for any noncensus sampling design.

With respect to a further subclass $\mathscr{C}_L$ of all *homogeneous linear unbiased estimators* (hlue's) of $T(\mathbf{Y})$, in our search for the umvue, we have the following mixed results enunciated in Theorem 2.11 and Theorem 2.12, first observed by Godambe (1955) and further clarified by Hanurav (1966).

To formalize these results, we introduce the following definitions.

Definition 2.7. Two samples s_1 and s_2 are said to be *disjoint* if they have no common unit. We will use the notation $s_1 \cap s_2 = \emptyset$ in such a case.

Definition 2.8. A sampling design $d = (S_d, P_d)$ is said to be a *unicluster* sampling design if for *any* two samples $s_1, s_2 \in S_d$, either s_1 and s_2 are disjoint or they are equivalent. Otherwise, a sampling design is called a nonunicluster sampling design.

The sampling designs d_1, d_2, and d_5, listed in Example 1.4, are all examples of nonunicluster sampling designs. Below is an example of a unicluster design in its unreduced form.

Example 2.2. Take $\mathscr{U} = \{1, 2, 3, 4, 5\}$ and

$$d: \quad P_d((1,2)) = 0.2,\ P_d((2,1)) = 0.3,\ P_d((3,5,4)) = 0.4,\ P_d((5,3,4)) = 0.1\,.$$

In the following, we will utilize some characteristic properties of unicluster and nonunicluster sampling designs. These properties may be of independent interest as well. For a unicluster sampling design, either $\pi_{ij} = 0$ or $\pi_{ij} = \pi_i = \pi_j$ for *any* pair (i, j), $1 \le i \ne j \le N$ (Exercise 2.3). Again, define the following subsets of S:

$$S_{ij} = \{s \mid s \in S,\ s \ni i,\ j\}\,,$$
$$S_{i\bar{j}} = \{s \mid s \in S,\ s \ni i,\ s \not\ni j\}\,,$$
$$S_{\bar{i}j} = \{s \mid s \in S,\ s \ni j,\ s \not\ni i\}\,, \qquad 1 \le i \ne j \le N\,.$$

Then in a unicluster sampling design, for any pair (i, j), either $S_{ij} = \emptyset$ or $S_{i\bar{j}} = S_{\bar{i}j} = \emptyset$. On the otherhand, for a nonunicluster sampling design, for *some* pair (i, j), $1 \le i \ne j \le N$, S_{ij} is *necessarily* nonnull and so is *at least* one of the subsets $S_{i\bar{j}}$ and $S_{\bar{i}j}$ (Exercise 2.4).

As to the existence of a hlue of $T(\mathbf{Y})$ based on a nonunicluster sampling design, it is possible to construct an hlue of $T(\mathbf{Y})$ of the type

$$l(s, \mathbf{Y}) = \sum_{i \in s} l_{si} Y_i , \qquad s \in S ,$$

such that, for a *given* pair (i, j), $i \ne j$, for which S_{ij} is *nonnull*,

$$l_{si} = \begin{cases} l_{sj} , & s \ni i, j \\ 0 , & \text{otherwise.} \end{cases}$$

We leave this as an exercise for the reader (Exercise 2.5). We are now ready to state and prove the following results on existence (Theorem 2.11) and nonexistence (Theorem 2.12) of best estimators of $T(\mathbf{Y})$ in the class $\mathscr{C}_L$.

Theorem 2.11. *For a unicluster sampling design, the HTE is the unique best estimator of $T(\mathbf{Y})$ in the class $\mathscr{C}_L$.*

Proof. We start with an hlue $l(s, \mathbf{Y})$ for $T(\mathbf{Y})$ and consider its symmetrized form $l^*(s, \mathbf{Y})$. To be specific, we write $l^*(s, \mathbf{Y}) = \Sigma_{i \in s} l^*_{si} Y_i$.

In view of (2.20), we have $l^*_{si} = l^*_{s^*i}$ for any $i \in s$, s^*, $s \approx s^*$. Since for the design under consideration, for any two samples s' and s'' with positive probabilities, either $s' \cap s'' = \emptyset$ or $s' \approx s''$, we have, in view of the unbiasedness of l^*,

$$\begin{aligned} E_p(l^*(s, \mathbf{Y})) &= \sum_s \Big(\sum_{i \in s} l^*_{si} Y_i \Big) P_d(s) \\ &= \sum_i Y_i \Big(\sum_{s \ni i} l^*_{si} P_d(s) \Big) \\ &= \sum_i Y_i l^*_{s^*i} \Big(\sum_{s \ni i} P_d(s) \Big) = \sum_i Y_i l^*_{s^*i} \pi_i \qquad \text{(recall (1.3))} \\ &= T(\mathbf{Y}) \text{ uniformly in } \mathbf{Y} \in R^N , \end{aligned}$$

which yields $l^*_{si} = \pi_i^{-1}$ so that $l^*(s, \mathbf{Y}) = \Sigma_{i \in s} Y_i / \pi_i$, $s \in S$.

Thus the symmetrized (and, hence, improved) form of *any* hlue $l(s, \mathbf{Y})$ of $T(\mathbf{Y})$ is the unique hlue $l^*(s, \mathbf{Y}) = \Sigma_{i \in s} Y_i / \pi_i$ which is, therefore, the *unique best* estimator of $T(\mathbf{Y})$ inside $\mathscr{C}_L$. Clearly, $l^*(s, \mathbf{Y})$ is the HTE. □

Remark 2.4. Indeed it must be noted that even for such designs, the HTE is not the best among *all* unbiased estimators (Exercise 2.6).

Theorem 2.12. *For a nonunicluster sampling design, there does not exist any best estimator of $T(\mathbf{Y})$ in $\mathscr{C}_L$.*

Proof. The following proof might be more illustrative to the reader than the original proof given in Godambe (1955) and Hanurav (1966). Our proof consists of three steps. First, for a selected subset of N points $\mathbf{Y}$ in R^N we characterize the structure of hlue for $T(\mathbf{Y})$ attaining minimum variance at each of these points simultaneously. Secondly, we provide an hlue of $T(\mathbf{Y})$ attaining minimum variance at a suitably selected point $\mathbf{Y}$ in R^N, different from those considered above. Finally, we show that the estimator in Step I, which happens to be HTE, has a larger variance at this point because the sampling design in nonunicluster. This is how we arrive at a contradiction, thereby establishing the claim. We shall now provide the details of the proof.

Step I. The only hlue of $T(\mathbf{Y})$ which attains minimum variance simultaneously at each of the following N points $\mathbf{Y}_1 = (1, 0, \ldots, 0)$, $\mathbf{Y}_2 = (0, 1, 0, \ldots, 0)$, $\mathbf{Y}_N = (0, 0, \ldots, 0, 1)$ is the HTE. This is a property of the HTE which will be studied in detail in Chapter Three. We defer the proof to Chapter Three to avoid repetition.

Step II. Given a nonunicluster sampling design, we consider a pair of units (i, j), $i \neq j$, such that S_{ij} as well as one of $S_{i\bar{j}}$ and $S_{\bar{i}j}$ are nonnull. It is then possible to construct an hlue of $T(\mathbf{Y})$ of the form $l(s, \mathbf{Y}) = \Sigma_{k \in s} l_{sk} Y_k$, $s \in S$ such that $l_{si} = l_{sj}$, $s \ni i, j$, $l_{si} = l_{sj} = 0$, otherwise. We now select the point $\mathbf{Y}$ with $(Y_i = 1, Y_j = -1, Y_k = 0, 1 \leq k \leq N, k \neq i \neq j)$. Clearly, for this point, $T(\mathbf{Y}) = 0$. Further, at this point $E(l^2(s, \mathbf{Y}))$ assumes the value $\Sigma_{s \ni i} l_{si}^2 P_d(s) + \Sigma_{s \ni j} l_{sj}^2 P_d(s) - 2\,\Sigma_{s \ni i,j} l_{si} l_{sj} P_d(s)$ which is zero by our construction of $l(s, \mathbf{Y})$. Thus, the above estimator has zero variance at this point.

Step III. We have already mentioned that the HTE is the best for the collection of points in Step I. Now we evaluate its performance at the point $\mathbf{Y}$ mentioned in Step II. Referring to the expression for $E(l^2(s, \mathbf{Y}))$ evaluated at this point, we obtain for the HTE the corresponding expression as $(1/\pi_i) + (1/\pi_j) - (2\pi_{ij}/\pi_i\pi_j)$ which is strictly positive. This is because at least one of $S_{i\bar{j}}$ and $S_{\bar{i}j}$ is nonnull and, consequently, either $\pi_i > \pi_{ij}$ or $\pi_j > \pi_{ij}$ or both hold.

Thus there is *no* estimator which attains minimum variance at *all* points in R^N. This completes the proof. □

The notion of unicluster sampling designs generalizes the well-known concept of cluster sampling which formally in its simplest form corresponds to: $N = mk$, $s_1 = (1, \ldots, k)$, $s_2 = (k+1, \ldots, 2k), \ldots, s_m = (N-k+1, \ldots, N)$, $P(s_1) = P(s_2) = \cdots = P(s_m) = 1/m$. See Chapter Seven for more details. Theorem 2.11 points to the fact that a proper choice of the sampling design (namely, unicluster sampling design) certainly leads to the best estimator of the population total among all of its hlue's based on the chosen sampling design. The HTE based on such a design serves as the best estimator. Further, the strategy (d, HTE(d)) is equivalent to its symmetrized version (d^*, HTE(d^*)). Such a design is *not*, however, *unique*. If d_1

and d_2 are two such designs, then, of the two strategies $(d_i, \text{HTE}(d_i))$, $i = 1, 2$, no one is uniformly better than the other in the sense of $E_{p_1}(e_1 - T)^2 \le$ or $\ge E_{p_2}(e_2 - T)^2$, uniformly in $\mathbf{Y} \in R^N$ (Exercise 2.13). This clearly brings out the complexity in the choice of one sampling strategy to work with.

Example 2.3. Take $\mathcal{U} = \{1, 2, \ldots, 10\}$. The following are examples of unicluster designs:

$$d_1\text{:}\quad P((1,2)) = 0.2,\ P((3,4,7)) = 0.3,\ P((5,6,9)) = 0.4,\ P((8,10)) = 0.1\,.$$

$$d_2\text{:}\quad P((2,4,7)) = 0.2,\ P((9,6,3,1)) = 0.5,\ P((10,5,8)) = 0.1\,,$$
$$P((10,8,5)) = 0.1,\ P((1,6,3,9)) = 0.1\,.$$

The reduced form of d_2 is d_2^*:

$$d_2^*\text{:}\quad P^*((2,4,7)) = 0.2,\ P^*((1,3,6,9)) = 0.6\,,$$
$$P^*((5,8,10)) = 1.2\,.$$

Table 2.3 gives the best estimators of $T(\mathbf{Y})$ based on the above two designs.

Table 2.3. Best Estimators of the Population Total

Designs	Best Estimator (HTE) of $T(\mathbf{Y})$
d_1	$\text{HTE}((1,2),\mathbf{Y}) = 5(Y_1 + Y_2)$, $\text{HTE}((3,4,7),\mathbf{Y}) = \frac{10}{3}(Y_3 + Y_4 + Y_7)$, $\text{HTE}((5,6,9),\mathbf{Y}) = \frac{5}{2}(Y_5 + Y_6 + Y_9)$, $\text{HTE}((8,10),\mathbf{Y}) = 10(Y_8 + Y_{10})$
d_2^*	$\text{HTE}((1,3,6,9),\mathbf{Y}) = \frac{5}{3}(Y_1 + Y_3 + Y_6 + Y_9)$, $\text{HTE}((2,4,7),\mathbf{Y}) = 5(Y_2 + Y_4 + Y_7)$ $\text{HTE}((5,8,10),\mathbf{Y}) = 5(Y_5 + Y_8 + Y_{10})$

In concluding this section, we may reiterate that in as much as the behaviors of various estimators are judged with reference to it, the sampling design plays an important role in the present approach. The twin pair (p, e) is said to form a sampling strategy. So far, we have addressed the problem of choice of e (the estimator) for a given p (the sampling design). Ideally, we would also like to have some guidance as to the choice of p. In a similar way, we could formalize the comparison of two different strategies (p_1, e_1) and (p_2, e_2) for inference on the same parametric function under the same loss function. Thus, for example, we would say the strategy (p_1, e_1) is at least as good as (or better than) the strategy (p_2, e_2) for inference on $\theta(\mathbf{Y})$, if it so happens that

$$E_{p_1}(e_1 - \theta)^2 \le E_{p_2}(e_2 - \theta)^2$$

uniformly in $\mathbf{Y} \in R^N$, with strictly inequality for some $\mathbf{Y} \in R^N$.

Further, if the above holds for all sampling strategies of the type (p_2, e_2) in a given class of strategies including (p_1, e_1), then we declare (p_1, e_1) as the *best* in the class. As one may expect, nonexistence result continues to hold in any meaningful subclass of strategies. As seen above, this is true for the subclass comprising of unicluster designs and the HTE.

2.4. DISCUSSION ON UNICLUSTER SAMPLING DESIGNS

According to Theorem 2.12 the nonunicluster sampling designs do *not* provide best estimators of the population total within the class $\mathscr{C}_L$ of *all* hlue's. Koop (1963) and Prabhu-Ajgaonkar (1965) observed that the *same* nonexistence result holds even within certain subclasses of $\mathscr{C}_L$. Godambe and Joshi (1965) and Basu (1971) also observed the same phenomenon to hold in the wide class of all unbiased estimators, removing the restriction of linearity. On the other hand, the unicluster sampling designs are observed to play an affirmative role in the fruitful search for best estimators (Theorem 2.11). However, comparison of strategies based on unicluster sampling designs does not lead to any conclusive findings. Further, the study to be undertaken below will *reveal* a serious drawback of the unicluster designs (excepting a census) from the practical point of view.

The simple fact, though certainly important, that the HTE, based on a unicluster sampling design, is the best estimator of the population total *seldom* satisfies a survey sampling practitioner, for whom it is no less important to know the *extent* of error to be expected in the estimate proposed on any given occasion. A statistical measure of the error is furnished by the square root of the mean squared error (mse). For an unbiased estimator this quantity is the square root of its variance, also called *standard error* of the estimator. In survey sampling practice, it is most often required of the practitioner to be able to provide a numerical assessment of this measure, based on the actual survey data. As we see, this would become possible only if we could estimate the mse from the data. Then we might use {estimate/estimated mse} as a measure of the *relative* error. For some fixed-size (n) designs, this quantity simplifies to $\sqrt{n}\,\bar{\mathbf{y}}/\mathbf{s}_Y$. It may be noted that the quantity $\mathbf{s}_Y/\bar{\mathbf{y}}$ is the sample *coefficient of variation* with $\boldsymbol{\sigma}_Y/\bar{\mathbf{Y}}$ as the corresponding population value.

Mathematically speaking, what is generally needed is an estimator, say ψ, of $E_p(e-\theta)^2$ where e refers to an estimator, usually unbiased for the parameter θ. Moreover, we put forward the requirement of unbiasedness of ψ (Definition 2.4), particularly when e is known to be unbiased for θ, in which case $E_p(e-\theta)^2$ is called the variance of e and is usually denoted by $V_p(e)$ or simply by $V(e)$. It is also desired that ψ be nonnegative since it proposes to estimate a nonnegative quantity. This poses the so-called problem of *nonnegative variance estimation*. See Chapter Three for details.

With regard to unicluster sampling designs and the use of the HTE's, the following unfortunate fact is now readily observed.

Theorem 2.13. *For any unicluster sampling design (not a census), there does not exist any unbiased estimator of $V(HTE)$, the variance of the HTE.*

The proof of this result is deferred to the next chapter where a more general result concerning homogeneous quadratic functions and their unbiased estimators will be discussed.

A survey statistician, therefore, gains very little from such (unicluster) sampling designs which are seen to be important from theoretical considerations! And, hence, the necessity to look for suitable nonunicluster sampling designs remains. Even though this latter class of designs does *not* admit *best* estimators, one may hope that at least some such designs may produce reasonably good estimators (in some sense), providing, further, unbiased variance estimators of the estimators of the population total which retain their nonnegativity as well!

To start with, let $d = (S_d, P_d)$ be a nonunicluster sampling design. Admitting the fact that no best estimator, based on d, exists for $T(\mathbf{Y})$, in our *search* for a reasonably good estimator, what should we do now? Which estimator should we check to see if it is good or not, in some sense? Fortunately, we have available at least the HTE to start with. It seems quite reasonable then to *examine critically* the *status* of the HTE in combination with a *nonunicluster* sampling design. This we do in the next chapter which is devoted mainly to a study of various properties of the HTE. As we shall see later, this study will also lead us to *admissible strategies* on some occasions.

2.5. CONCLUDING REMARKS

We mentioned towards the end of Section 2.2 that identifiability of the units is a distinctive feature of finite population sampling. Accordingly, any *complete* data set, realized on a given occasion, should consist of pairs of the form (unit chosen, study variable on the chosen unit). If, however, we *ignore* the labels of the units and keep a record only of the values of the survey variable, it will *seldom* be possible to draw valid inferences on the population parameters of interest (unless, of course, the sample size is large enough compared to the population size so that useful inferences can still be drawn). To explain this further, let us consider the problem of estimating the population total $T(\mathbf{Y})$. Of the hle's classified in Section 2.2, only e_1 and e_3 are functionally independent of the actual units drawn. Therefore, denoting the observed values by, say $(y_1, y_2, \ldots, y_n)$, an estimator of the form $\Sigma\, a_i y_i$ might be employed to estimate $T(\mathbf{Y})$. But then the criterion of (design-) unbiasedness may be difficult to meet unless a careful choice is made of the underlying sampling design. We will discuss this further in Section 4.1 with reference to a particular sampling design.

So far most of our discussion has been confined to the estimation of only the finite population total $T(\mathbf{Y})$. However, the reader may have noted that all these results are equally valid for estimation of the finite population mean $\bar{\mathbf{Y}} = (1/N)T(\mathbf{Y})$ or the finite population proportion ϕ of an attribute. Here ϕ is defined as $N\phi$ = number of units in the population possessing the attribute. Note that by defining

$$Y_i = \begin{cases} 1, & \text{if the } i\text{th unit possesses the attribute} \\ 0, & \text{otherwise, } 1 \leq i \leq N, \end{cases}$$

we can identify ϕ as the finite population mean.

It must be noted that estimation of the finite population variance (σ_Y^2) is also of interest on some occasions. The general problem of estimation of a quadratic form will be discussed in Chapter Three. Other parameters of interest could be the population median and appropriate percentile points. However, *not* much theoretical work has been done on this latter topic and, as such, it will *not* be discussed in detail in this book. In Section 4.7 we present some results in this direction.

EXERCISES

2.1. Give a formal proof of the following: Of all the forms e_1 to e_5 of homogeneous linear estimators given in Section 2.1, the most general form is provided by e_5.

2.2. Consider the following designs:

$N=2$, d_1:

s	(1)	(1, 2)
$P(s)$	0.3	0.7

$N=3$, d_2:

s	(1, 2)	(1, 3)	(1, 2, 3)
$P(s)$	0.2	0.3	0.5

$N=5$, d_3:

s	(1)	(2, 3)	(2, 3, 4, 5)
$P(s)$	0.3	0.4	0.3

(a) Based on each of these designs, write down the possible forms e_1 to e_5 of the hle's using the coefficients given in Tables 2.A–2.E.

Table 2.A

Order of draw (t)	1	2	3	4
l_t	1	10/7	8/7	2

Table 2.B. Table of Coefficients $l_{i_t t}$

Unit i_t	Order of draw (t)			
	1	2	3	4
1	1	3	2	1
2	7	10/7	3	2
3	8	2	8	1
4	2	2	2	2
5	−7	3	2	−1

Table 2.C

s	(1)	(1, 2)	(1,3)	(2, 3)	(1, 2, 3)	(2, 3, 4, 5)
l_s	10/3	15	11	0	6	10/3

Table 2.D

Unit i_t	1	2	3	4	5
l_{i_t}	1	10/7	5/4	2	3

Table 2.E. Table of Coefficients l_{si_t}

Samples s	Units				
	1	2	3	4	5
(1)	5.0	–	–	–	–
(1, 2)	1.0	4.0	–	–	–
(1, 3)	0.4	–	2.0	–	–
(2, 3)	–	2.0	3.0	–	–
(1, 2, 3)	−0.8	0.4	0.8	–	–
(2, 3, 4, 5)	–	0.8	0.7	2.0	2.0

(b) Verify which of the estimators in (a) are unbiased for the respective population totals.

2.3. Show that a sampling design is unicluster if and only if for *any* pair of units (i, j), either $\pi_{ij} = 0$ or $\pi_{ij} = \pi_i = \pi_j$.

2.4. Recall the definitions of $S_{i\bar{j}}$, $S_{\bar{i}j}$ and S_{ij} for $i \neq j$. For a nonunicluster sampling design, show that for at least one pair (i, j), $1 \leq i \neq j \leq N$, S_{ij} as well as at least one of the subsets $S_{i\bar{j}}$ and $S_{\bar{i}j}$ are necessarily nonnull.

2.5. For a nonunicluster sampling design, construct an hlue of $T(\mathbf{Y})$ of the type $l(s, \mathbf{Y}) = \Sigma_{i \in s} l_{si} Y_i$, $s \in S$ such that for a given pair (i, j), $i \neq j$, for which S_{ij} is nonnull, $l_{si} = l_{sj}$, $s \ni i, j$, $l_{si} = l_{sj} = 0$, otherwise.

2.6. Consider the sampling design d: $P((1,2)) = 0.3$, $P((3,4,5)) = 0.7$.

(a) Suggest an unbiased estimator of $T(\mathbf{Y})$ with smaller variance than the HTE at some point $\mathbf{Y}_0 \in R^5$. Show the point $\mathbf{Y}_0$.

(b) Can you suggest an unbiased estimator of $T(\mathbf{Y})$ with smaller variance than the HTE at the point $\mathbf{Y} = (1,2,3,4,5)$?

2.7. For the design d: $P((1)) = 0.2$, $P((2)) = 0.1$, $P((3)) = 0.3$, $P((1,2,3)) = 0.4$ write down some sets of values of the coefficients involved in each of the hle's e_1 to e_5 so that each one of these estimators becomes unbiased for the population mean.

2.8. Consider the design d: $P((1,2)) = 0.1$, $P((2,3)) = 0.2$, $P((3,4)) = 0.3$, $P((1,2,3,4)) = 0.4$ and the coefficients l_{si} in Table 2.F.

Table 2.F

Samples	Units			
s	1	2	3	4
(1, 2)	a	c	–	–
(2, 3)	–	d	f	–
(3, 4)	–	–	g	i
(1, 2, 3, 4)	b	e	h	j

Find sets of values of the above coefficients satisfying

(i) $a = b + 3$, $d + f = 5$, $h + j = 7$,

(ii) $a + c = 1$, $d + f = 0$, $i + j = 5$,

(iii) $b + e + h + j = 0$, $d + f = -2$,

such that in each case the resulting estimator becomes unbiased for the population total.

2.9. For the unicluster sampling design d: $P((1,2)) = 0.1$, $P((2,1)) = 0.2$, $P((3)) = 0.1$, $P((4,5)) = 0.2$, $P((5,4)) = 0.4$ consider the two hlue's e_1 and e_2 of $T(\mathbf{Y}) = \Sigma_1^5 Y_i$ in Table 2.G.

Table 2.G

	Coefficients for									
	e_1					e_2				
Samples	1	2	3	4	5	1	2	3	4	5
(1, 2)	a	b	–	–	–	10/3	10/3	– –	–	
(2, 1)	$5 - \frac{a}{2}$	$5 - \frac{b}{2}$	–	–	–	10/3	10/3	– –	–	
(3)	–	–	10	–	–	–	–	10	–	–
(4, 5)	–	–	–	$5 - 2c$	$5 - 2d$	–	–	–	5/3	5/3
(5, 4)	–	–	–	c	d	–	–	–	5/3	5/3

(i) Show that $E_p(e_1^2) \geq E_p(e_2^2)$ *uniformly* in the values of the Y_i's.
(ii) Study the case of equality.
(iii) Comment on the result.

2.10. For the nonunicluster sampling design d: $P((1)) = 0.1$, $P((2)) = 0.2$, $P((1, 2)) = 0.7$ consider the two hlue's e_1 and e_2 of $T(\mathbf{Y}) = Y_1 + Y_2$ in Table 2.H.

Table 2.H

$e_1((1), \mathbf{Y}) = (10 - 7a)Y_1$	$e_2((1)), \mathbf{Y}) = \frac{5}{4}Y_1$
$e_1((2), \mathbf{Y}) = (5 - \frac{7}{2}b)Y_2$	$e_2((2), \mathbf{Y}) = \frac{10}{9}Y_2$
$e_1((1, 2), \mathbf{Y}) = aY_1 + bY_2$	$e_2((1, 2), \mathbf{Y}) = \frac{5}{4}Y_1 + \frac{10}{9}Y_2$

(i) Do you think that $E_p(e_1^2) \geq E_p(e_2^2)$ or the reverse, *uniformly* in Y_1 and Y_2?
(ii) Comment on your result.

2.11. For the nonunicluster sampling design d: $P((1, 2)) = 0.1$, $P((2, 1)) = 0.2$, $P((1, 2, 3)) = 0.4$, $P((1, 2, 3, 4)) = 0.1$, $P((3, 2, 4, 1)) = 0.2$ consider the following hlue of $T(\mathbf{Y}) = Y_1 + Y_2 + Y_3 + Y_4$:

$$e((1, 2), \mathbf{Y}) = 4Y_1,\ e((2, 1), \mathbf{Y}) = Y_1,\ e((1, 2, 3), \mathbf{Y}) = Y_1 + Y_2 + Y_3\,,$$
$$e((1, 2, 3, 4), \mathbf{Y}) = 4Y_2 + 4Y_3,\ e((3, 2, 4, 1), \mathbf{Y}) = Y_2 + Y_3 + 5Y_4\,.$$

(i) Construct another hlue of $T(\mathbf{Y})$, say, e^* such that $E_p(e^{*2}) \leq E_p(e^2)$ *uniformly* in the Y_i-values.
(ii) Comment on the result

2.12. Prove that *any* strategy (p, e) can *always* be *uniformly* improved upon by its *symmetrized* version (p^*, e^*).

2.13. Take $N = 4$ and consider the following designs:

$$d_1\colon\ P_{d_1}((1, 2)) = P_{d_1}((3, 4)) = 0.5\,;$$
$$d_2\colon\ P_{d_2}((1, 3)) = P_{d_2}((2, 4)) = 0.5\,.$$

Write e_i for the HTE based on the design d_i $(i = 1, 2)$. Compare $E_{p_1}(e_1^2)$ and $E_{p_2}(e_2^2)$ and comment on the result.

2.14. (a) For the sampling design d_3 in Exercise 2.2, construct an HT-type estimator of the form (2.18) using the following values of $c(s)$:

s	(1)	(2, 3)	(2, 3, 4, 5)
$c(s)$	1	2	3

(b) Compare the variance of the above estimator with that of the HTE at some points to show that there are population values for which the above estimator possesses smaller variance than the HTE.

2.15. Consider the following sampling design: $P((1,2)) = P((1,3)) = 0.5$. Let e be an hlue of $T(\mathbf{Y})$ and define e^* as

$$e^*((1,2),\mathbf{Y}) = e((1,2),\mathbf{Y}) + Y_1^2$$

and

$$e^*((1,3),\mathbf{Y}) = e((1,3),\mathbf{Y}) - Y_1^2\,.$$

Show that e does not uniformly dominate e^* in terms of variance. [This example due to Godambe and Joshi (1965) illustrates that in general it is not enough to confine attention only to the hlue's.]

REFERENCES

Basu, D. (1958). On sampling with and without replacement. *Sankhyā*, **20**, 287–294.

Basu, D. (1971). An essay on the logical foundations of survey sampling I. In: *Foundations of Statistical Inference* (Godambe, V. P., and Sprott, D. S., Eds.), 203–242. Hold, Rinehart and Winston, Toronto.

Cochran, W. G. (1977). *Sampling Techniques*. Third Edition. Wiley, New York.

Deming, W. E. (1960). *Sample Design in Business Research*. Wiley, New York.

Dillman, D. A. (1978). *Mail and Telephone Surveys: The Total Design Method*. Wiley, New York.

Godambe, V. P. (1955). A unified theory of sampling from finite populations. *J. Roy. Statist. Soc.*, **B17**, 269–278.

Godambe, V. P. and Joshi, V. M. (1965). Admissibility and Bayes estimation in sampling finite populations I. *Ann. Math. Statist.*, **36**, 1707–1722.

Hanurav, T. V. (1966). Some aspects of unified sampling theory. *Sankhyā*, **A28**, 175–203.

Horvitz, D. G., and Thompson, D. J. (1952). A generalization of sampling without replacement from a finite universe. *J. Amer. Statist. Assoc.*, **47**, 663–685.

Kempthorne, O. (1978). Some aspects of statistics, sampling and randomization. In: *Contributions to Survey Sampling and Applied Statistics* (David, H. A., Ed.), 11–28. Academic Press, New York.

Koop, J. C. (1963). On the axioms of sample formation and their bearing on the construction of linear estimators in sampling theory for finite universes I, II, III. *Metrika*, **7**, 81–114, 165–204.

Levy, P. S., and Lemeshow, S. (1980). *Sampling for Health Professionals*. Lifetime Learning, Belmont, California.

Murthy, M. N. (1957). Ordered and unordered estimators in sampling without replacement. *Sankhyā*, **18**, 379–390.

Murthy. M. N. (1963). Generalized unbiased estimation in sampling from finite populations. *Sankhyā*, **B25**, 242–262.

Oppenheim, A. N. (1966). *Questionnaire Design and Attitude Measurement*. Basic Books, New York.

Prabhu-Ajgaonkar, S.G. (1965). On a class of linear estimators in sampling with varying probabilities without replacement. *J. Amer. Statist. Assoc.*, **60**, 637–642.

Srivastava, J. (1985). On a general theory of sampling, using experimental design. Concepts I: Estimation. *Bull. Internat. Statist. Inst.*, **51**, Book 2, I.S.I. Centenary Session, Amsterdam.

Sudman, S., and Bradburn, N. M. (1982). *Asking Questions: A Practical Guide to Questionnaire Design*. Jossey-Bass, San Francisco, California.

CHAPTER THREE

The Horvitz–Thompson Estimator

We devote this chapter primarily to a study of the Horvitz–Thompson Estimator (HTE). We deduce an expression for its variance and present some useful and convenient estimators. Next we pose the problem of nonnegative variance estimation and present the relevant available results. We then establish some properties of the HTE and state a few others. In addition, we derive a set of necessary and sufficient conditions for unbiased estimation of quadratic forms and study its implications for the choice of sampling strategies.

3.1. SOME BASIC RESULTS CONCERNING THE HORVITZ–THOMPSON ESTIMATOR (HTE)

Throughout the chapter we will regard the finite population total $T(\mathbf{Y}) = \Sigma_{i=1}^{N} Y_i$ as the parametric function of interest. Here Y is the survey variable related to a finite population of N distinct and labelled units. We take recourse to a sampling design $d = (S_d, P_d)$ yielding π_i's and π_{ij}'s as the first- and second-order inclusion probabilities, respectively. The Horvitz–Thompson estimator of the population total is defined as (see (2.7))

$$\text{HTE}(s, \mathbf{Y}) = \sum_{i \in s} Y_i/\pi_i = \sum_{1}^{N} f_i(s) Y_i/\pi_i\,, \qquad s \in S_d \tag{3.1}$$

where the $f_i(s)$'s are the indicator random variables defined in Section 1.3. It is known from Corollary 2.3 that the HTE in (3.1) is a hlue of $T(\mathbf{Y})$. It would have been appropriate to use a more explicit symbol like HTE $(T(\mathbf{Y})|(s, \mathbf{Y}))$ for HTE to indicate that it is meant to estimate the population total $T(\mathbf{Y})$. However, we will simply use HTE unless there is a danger of confusion.

We will deduce below the expression for $V(\text{HTE})$, the variance of the HTE in (3.1). Indeed, we can also deduce a general expression for the variance of any arbitrary hlue $l(s, \mathbf{Y})$ of $T(\mathbf{Y})$ and this we do first. The estimator $l(s, \mathbf{Y})$ is given by $l(s, \mathbf{Y}) = \Sigma_{i \in s} l_{si} Y_i$, $s \in S_d$. According to Theorem 2.4, the conditions of unbiasedness of $l(s, \mathbf{Y})$ for $T(\mathbf{Y})$ are given by $\Sigma_{s \ni i} l_{si} P_d(s) = 1$, $1 \leq i \leq N$. The expression for $V(l(s, \mathbf{Y}))$, the variance of $l(s, \mathbf{Y})$, is deduced as follows.

$$\begin{aligned} V(l(s, \mathbf{Y})) &= E(l^2(s, \mathbf{Y})) - T^2(\mathbf{Y}) \\ &= \sum_{s \in S_d} P_d(s) l^2(s, \mathbf{Y}) - T^2(\mathbf{Y}) \\ &= \sum_{s \in S_d} P_d(s) \Big\{ \sum_{i \in s} l_{si} Y_i \Big\}^2 - \Big(\sum Y_i \Big)^2 \\ &= \sum_i Y_i^2 \Big\{ \sum_{s \ni i} l_{si}^2 P_d(s) - 1 \Big\} + \sum_{i \neq j} \sum Y_i Y_j \Big\{ \sum_{s \ni i, j} l_{si} l_{sj} P_d(s) - 1 \Big\} . \end{aligned} \tag{3.2}$$

Specializing now on the HTE, we set $l_{si} = 1/\pi_i$, $i \in s$, $s \in S_d$ and obtain, using (1.3) and (1.4),

$$V(\text{HTE}) = \sum_i Y_i^2 \Big\{ \frac{1}{\pi_i} - 1 \Big\} + \sum_{i \neq j} \sum Y_i Y_j \Big\{ \frac{\pi_{ij}}{\pi_i \pi_j} - 1 \Big\} . \tag{3.3}$$

The above expression for V(HTE) was given by Horvitz and Thompson (1952). We may denote it by V_{HT}(HTE). One could also deduce (3.3) directly from the representation of the HTE in (3.1), using (1.13).

For a fixed-size (n) design (see Definition 1.4), the relations (1.7), (1.8), and (1.10) may be used to deduce an alternate expression for V(HTE). Towards this, we first simplify

$$\sum_{i \neq j} \sum (\pi_i \pi_j - \pi_{ij}) \Big\{ \frac{Y_i^2}{\pi_i^2} + \frac{Y_j^2}{\pi_j^2} \Big\}$$

as

$$\begin{aligned} &= 2 \sum_i \Big[Y_i^2 / \pi_i^2 \Big\{ \sum_{j(\neq i)} (\pi_i \pi_j - \pi_{ij}) \Big\} \Big] \\ &= 2 \sum_i [Y_i^2 / \pi_i^2 \{ (n - \pi_i) \pi_i - (n-1) \pi_i \}] \qquad \text{(using (1.7) and (1.10))} \\ &= 2 \sum_i Y_i^2 (1 - \pi_i) / \pi_i , \end{aligned}$$

and hence we write (3.3) as

$$\begin{aligned} V(\text{HTE}) = \frac{1}{2} \sum_{i \neq j} \sum (\pi_i \pi_j - \pi_{ij}) \Big[\frac{Y_i^2}{\pi_i^2} + \frac{Y_j^2}{\pi_j^2} \Big] \\ + \sum_{i \neq j} \sum Y_i Y_j \Big[\frac{\pi_{ij}}{\pi_i \pi_j} - 1 \Big] \end{aligned}$$

$$= \frac{1}{2} \sum_{i \neq j} \sum (\pi_i \pi_j - \pi_{ij}) \left[\frac{Y_i}{\pi_i} - \frac{Y_j}{\pi_j} \right]^2$$

$$= \sum_{i<j} \sum (\pi_i \pi_j - \pi_{ij}) \left[\frac{Y_i}{\pi_i} - \frac{Y_j}{\pi_j} \right]^2 . \tag{3.4}$$

This latter expression for $V(\text{HTE})$ was furnished by Sen (1953) and also by Yates and Grundy (1953). We may call it the Sen–Yates–Grundy variance estimator and denote it as $V_{\text{SYG}}(\text{HTE})$.

The representation (3.4) of $V(\text{HTE})$ has an interesting *implication* in terms of *choice* of a sampling design for its *use* with the HTE. Since, in any case, our aim is always to make use of an estimator with least variance, it would be highly desirable to employ a fixed-size (n) design which satisfies $\pi_i \propto Y_i$ for all i. This would then produce $V(\text{HTE}) = 0$ and, consequently, $T(\mathbf{Y})$ could be estimated with no error at all! However, this is certainly not a realistic situation as the values of the survey variable are never known a priori! On the other hand, if X is an auxiliary variable related to the survey variable Y, then we might think of employing a fixed-size (n) design which at least satisfies $\pi_i \propto X_i$ for all i whenever the values X_i's on X are available beforehand and we have strong reasons to believe that the values of Y and X are "nearly" proportional. For such a situation, we might expect $V(\text{HTE})$ to be appreciably small. It may be said that this, more or less, has been the basis for a good deal of work found in the literature concerning the use of auxiliary information in finite population studies. We will take up this study in considerable detail in Chapter Five.

Our next concern is to obtain an unbiased estimator of $V(\text{HTE})$ as any survey practitioner will inevitably seek to provide an estimate of the error in the estimate (HTE) of the population total (see Section 2.4). We will deduce below a general result concerning unbiased estimation of an arbitrary function of Y say $Q(\mathbf{Y})$ of the form $Q(\mathbf{Y}) = \Sigma\,\Sigma_{1 \le i \le j \le N}\, q_{ij}(Y_i, Y_j)$ where with no loss of generality we assume that each component $q_{ij}(Y_i, Y_j)$ is not identically equal to either zero or a constant. Clearly, a sufficient condition for unbiased estimation of $Q(\mathbf{Y})$ is that each component $q_{ij}(Y_i, Y_j)$ can be estimated separately. A feature of $q_{ij}(Y_i, Y_j)$ which is important in this context is whether it has the representation $q_{ij}(Y_i, Y_j) = h_i(Y_i) + f_j(Y_j)$ in which case it will be referred to as being *separable*. In the following, we present results concerning unbiased estimation of $q_{ij}(Y_i, Y_j)$, utilizing the tool developed in Theorem 2.1.

Theorem 3.1. *If $\pi_{ij} > 0$, then $q_{ij}(Y_i, Y_j)$ has an unbiased estimator given by*

$$\hat{q}_{ij}(s, \mathbf{Y}) = \begin{cases} q_{ij}(Y_i, Y_j)/\pi_{ij}\,, & \textit{if } s \ni i, j \\ 0\,, & \textit{otherwise}\,. \end{cases} \tag{3.5}$$

In the case $q_{ij}(Y_i, Y_j)$ is not separable, $\pi_{ij} > 0$ is also necessary.

Proof. If $\pi_{ij} > 0$, there exists at least one sample with positive probability containing the units i and j. Therefore, using (3.5) and (1.4),

$$E(\hat{q}_{ij}(s, \mathbf{Y})) = \sum_{s \ni i,j} P(s) q_{ij}(Y_i, Y_j)/\pi_{ij} = q_{ij}(Y_i, Y_j)\,.$$

This establishes the first part. Next, assume that $\pi_{ij} = 0$. This means that no sample in the support of the design will contain both units i and j. Suppose now that $q_{ij}(Y_i, Y_j)$ is not separable and there exists an unbiased estimator of it, say $g(s, \mathbf{Y})$. Then we must have $\Sigma_s P(s)g(s, \mathbf{Y}) = q_{ij}(Y_i, Y_j)$ uniformly in $\mathbf{Y} \in R^N$. Clearly, the left hand side in this identity does not contain any term involving both Y_i and Y_j as $\pi_{ij} = 0$. However, the right hand side involves Y_i and Y_j together as $q_{ij}(Y_i, Y_j)$ is not separable. This is a contradiction.

In the case $q_{ij}(Y_i, Y_j)$ is separable in the form $q_{ij}(Y_i, Y_j) = h_i(Y_i) + f_j(Y_j)$, a necessary and sufficient condition for its unbiased estimator to exist is that $\pi_i > 0$ and $\pi_j > 0$ in which case one such estimator is given by

$$\hat{q}_{ij}(s, \mathbf{Y}) = \begin{cases} 0\,, & \text{if } s \not\ni i, j \\ h_i(Y_i)/\pi_i\,, & \text{if } s \ni i,\, s \not\ni j \\ f_j(Y_j)/\pi_j\,, & \text{if } s \ni j,\, s \not\ni i \\ h_i(Y_i)/\pi_i + f_j(Y_j)/\pi_j\,, & \text{if } s \ni i, j\,. \end{cases}$$

We have also deduced that a necessary and sufficient condition for an unbiased estimator of $q_{ii}(Y_i, Y_i)$ to exist is that $\pi_i > 0$, in which case one such estimator is given by

$$\hat{q}_{ii}(s, \mathbf{Y}) = \begin{cases} 0\,, & \text{if } s \not\ni i \\ q_{ii}(Y_i, Y_i)/\pi_i\,, & \text{if } s \ni i\,. \end{cases}$$

Having estimated all relevant components $q_{ij}(Y_i, Y_j)$ in $Q(\mathbf{Y})$, an unbiased estimator of $Q(\mathbf{Y})$ would be given by

$$\hat{Q}(s, \mathbf{Y}) = \sum\sum_{1 \le i \le j \le N} \hat{q}_{ij}(s, \mathbf{Y})\,. \tag{3.6}$$

By now it should be clear that in the representation of $Q(\mathbf{Y}) = \Sigma\,\Sigma_{1 \le i \le j \le N}\, q_{ij}(Y_i, Y_j)$, if no component $q_{ij}(Y_i, Y_j)$ is separable, then the conditions $\pi_{ij} > 0$, $1 \le i \ne j \le N$ are both necessary and sufficient. □

Specializing to quadratic forms, the following useful results can be stated as a direct consequence of Theorem 3.1. Results of this kind have been discussed by Hanurav (1966). J. N. K. Rao (1979) has provided further results in this direction.

Corollary 3.2. *An unbiased estimator of*

$$Q(\mathbf{Y}) = \sum\sum_{1 \le i \le j \le N} a_{ij} Y_i Y_j \tag{3.7}$$

exists if and only if $\pi_{ij} > 0$ *whenever* $a_{ij} \neq 0$, *in which case one such estimator is given by*

$$\hat{Q}(s, \mathbf{Y}) = \sum_{i \in s} a_i Y_i^2 / \pi_i + \sum_{\substack{i<j \\ i,j \in s}} \sum a_{ij} Y_i Y_j / \pi_{ij} . \tag{3.8}$$

As an important application of Corollary 3.2, we now suggest two generally different unbiased estimators of the population variance σ_Y^2. Towards this we first note that σ_Y^2 can be expressed in the following forms:

$$\sigma_Y^2 = \frac{1}{N} \sum_{i=1}^{N} (Y_i - \bar{\mathbf{Y}})^2 = \frac{N-1}{N^2} \sum_{i=1}^{N} Y_i^2 - \frac{1}{N^2} \sum_{i \neq j} \sum Y_i Y_j , \tag{3.9}$$

and

$$\sigma_Y^2 = \frac{1}{2N^2} \sum_{i \neq j} \sum (Y_i - Y_j)^2 . \tag{3.10}$$

From the above it is clear that in the notation of (3.7), $a_{ij} \neq 0$, $1 \leq i \neq j \leq N$. Therefore, a necessary and sufficient condition for the existence of an unbiased estimator of σ_Y^2 is that $\pi_{ij} > 0$ for all pairs (i, j). When these conditions are satisfied in a sampling design, utilizing (3.8) we can derive the following unbiased estimators of σ_Y^2 based on the representations in (3.9) and (3.10), respectively.

$$\hat{\sigma}_Y^2(1) = \frac{N-1}{N^2} \sum_{i \in s} Y_i^2 / \pi_i - \frac{1}{N^2} \sum_{\substack{i \neq j \\ i,j \in s}} \sum Y_i Y_j / \pi_{ij} . \tag{3.11}$$

$$\hat{\sigma}_Y^2(2) = \frac{1}{2N^2} \sum_{\substack{i \neq j \\ i,j \in s}} \sum (Y_i - Y_j)^2 / \pi_{ij} . \tag{3.12}$$

As we mentioned earlier, these two estimators of σ_Y^2 are, in general, different. For some sampling designs they do coincide. One such example is a fixed-size (n) design with constant first- and second-order inclusion probabilities for $n \geq 2$. It is interesting to observe that the estimator in (3.12) is always nonnegative. In other words, we can say that the population variance has at least one nonnegative unbiased estimator if $\pi_{ij} > 0$ for all pairs (i, j).

In the literature on survey sampling the term *variance estimation* usually refers to estimation of the variance of an estimator. In the following we address this problem when, for the purpose of estimation of a finite population total, the estimator employed is homogeneous linear unbiased.

Corollary 3.3. *An unbiased estimator of the variance of the homogeneous linear unbiased estimator* $l(s, \mathbf{Y}) = \Sigma_{i \in s} l_{si} Y_i$ *of the population total* *exists if and only if* $\pi_{ij} > 0$ *for every pair* (i, j), *in which case one such estimator is given by*

$$\hat{V}(l(s, \mathbf{Y})) = \sum_{i \in s} b_i Y_i^2 / \pi_i + \sum_{\substack{i<j \\ i,j \in s}} \sum b_{ij} Y_i Y_j / \pi_{ij} , \qquad (3.13)$$

where, as in (3.2)

$$b_i = \sum_{s' \ni i} l_{s'i}^2 P(s') - 1 \quad \text{and} \quad b_{ij} = 2\Big[\sum_{s' \ni i,j} l_{s'i} l_{s'j} P(s') - 1\Big] . \qquad (3.14)$$

Proof. Looking at the expression for $V(l(s, \mathbf{Y}))$ in (3.2), we observe that in the set up of Corollary 3.2, $a_i = b_i$ and $a_{ij} = b_{ij}$. If for a pair (i, j), $b_{ij} \neq 0$, then we appeal to Corollary 3.2 and conclude that π_{ij} must be positive. On the other hand, if for a pair (i, j), $b_{ij} = 0$ then we have

$$\sum_{s' \ni i,j} l_{s'i} l_{s'j} P(s') = 1$$

which in its turn implies that $\pi_{ij} > 0$, since otherwise there is no sample s' with $P(s') > 0$ containing the units i and j, thereby reducing the left hand side to zero. Therefore, in either case $\pi_{ij} > 0$ for every pair (i, j), is necessary. The sufficiency part is obvious in view of the estimator in (3.13). □

Remark 3.1. In (3.13) we gave *one* unbiased variance estimator based on explicit use of the π_i's and π_{ij}'s. Unbiased variance estimator is not, however, generally unique and this brings us to emphasize the following crucial point regarding unbiased variance estimation. While the verification of positivity of π_{ij}'s is a must for the existence of unbiased variance estimators, the explicit knowledge and use of the π_{ij}'s can be bypassed in some cases. It is very important to be aware of this since there are many useful and interesting survey designs for which an explicit derivation of π_{ij}'s is extremely tedious, if not impossible. Some such designs will be presented in Sections 5.5 and 5.6. In each case we will provide an unbiased estimator of the population total and its unbiased variance estimator, bypassing the use of π_i's and π_{ij}'s. An example of this is given in Exercise 3.17.

Turning back to the use of the HTE, it is evident that $V(\text{HTE})$ is unbiasedly estimable if and only if $\pi_{ij} > 0$ for every pair (i, j).

Whenever a sampling design provides $\pi_{ij} > 0$ for *all* (i, j), we may write, following (3.3) and (3.13),

$$\hat{V}_{\text{HT}}(\text{HTE}) = \sum_{i \in s} \frac{Y_i^2}{\pi_i} \left\{ \frac{1}{\pi_i} - 1 \right\} + \sum_{\substack{i \neq j \\ i,j \in s}} \sum \frac{Y_i Y_j}{\pi_{ij}} \left\{ \frac{\pi_{ij}}{\pi_i \pi_j} - 1 \right\} . \qquad (3.15)$$

Also, following (3.4) and (3.13), for a fixed-size sampling design with $\pi_{ij} > 0$ for all (i, j),

$$\hat{V}_{\text{SYG}}(\text{HTE}) = \sum_{i<j,i,j \in s}\sum \left(\frac{Y_i}{\pi_i} - \frac{Y_j}{\pi_j}\right)^2 \left\{\frac{\pi_i \pi_j - \pi_{ij}}{\pi_{ij}}\right\}. \tag{3.16}$$

We will refer to $\hat{V}_{\text{HT}}(\text{HTE})$ as the Horvitz–Thompson variance estimator and to $\hat{V}_{\text{SYG}}(\text{HTE})$ as the Sen–Yates–Grundy variance estimator. For a unicluster sampling design (see Definition 2.8), unless it is a census, the conditions for existence of an unbiased estimator of $V(\text{HTE})$ are *not* all satisfied. This provides the proof of Theorem 2.13 enunciated before and, as mentioned in Section 2.5, thereby brings out a major drawback of the whole class of unicluster sampling designs from practical considerations.

Even though a vast majority of (*but not all*) nonunicluster sampling designs provide means to unbiasedly estimate $V(\text{HTE})$, an unbiased estimate of the form of (3.15) or (3.16) does *not always* turn out to be nonnegative, which is unfortunate. This simply means that for a given sampling design and a proposed estimator, the matrix of the underlying quadratic form may not remain *nonnegative definite*. Occasionally these designs would lead to negative estimates of variance. We will study this problem more closely in the next section. We close this section with an example.

Example 3.1. Let $U = \{1, 2, 3, 4, 5\}$ and adopt the following sampling design:

$$d: \quad P((1,3,5)) = \frac{4}{10}, \quad P((2,3,5)) = \frac{3}{10}, \quad P((1,2,4)) = \frac{1}{10},$$
$$P((3,4,5)) = \frac{2}{10}.$$

Suppose the population values are $Y_1 = 5$, $Y_2 = 10$, $Y_3 = 6$, $Y_4 = Y_5 = 0$. As an estimator of $V(\text{HTE})$, if we use $\hat{V}_{\text{SYG}}(\text{HTE})$, then we end up with a negative value of -15 for the sample $(1, 3, 5)$. On the other hand, if we use $\hat{V}_{\text{HT}}(\text{HTE})$, we end up with a negative value of -75 for the sample $(1, 2, 4)$. See exercise 3.16 in this context.

3.2. NONNEGATIVE UNBIASED VARIANCE ESTIMATION

In this section we study the problem of nonnegative unbiased variance estimation with reference to the Horvitz–Thompson Estimator (HTE). The expressions (3.15) and (3.16) represent two well-known estimators of $V(\text{HTE})$. Both these estimators may occasionally take negative values and this is considered highly undesirable. A careful choice of the underlying sampling design might result in assuring nonnegativity of these estimators.

Also, other nonnegative unbiased variance estimators (nnuve's) might be available in some cases. In the discussion below, we will restrict ourselves to only fixed-size (n) designs and study the possible forms of nnuve's. Moreover, we will assume throughout that $\pi_{ij} > 0$, $1 \le i \ne j \le N$ as, otherwise, we do not have unbiased variance estimators. This implies $n \ge 2$.

If we set $Z_i = Y_i/\pi_i$, $1 \le i \le N$, then (3.4) and (3.16) reduce, respectively, to

$$V_{\text{SYG}}(\text{HTE}) = \sum_{i<j}\sum (\pi_i \pi_j - \pi_{ij})(Z_i - Z_j)^2 \tag{3.17}$$

and

$$\hat{V}_{\text{SYG}}(\text{HTE}) = \sum_{\substack{i<j \\ i,j \in s}}\sum \{(\pi_i \pi_j - \pi_{ij})/\pi_{ij}\}(Z_i - Z_j)^2 \, . \tag{3.18}$$

As regards *nonnegative* unbiased estimation of $V_{\text{SYG}}(\text{HTE})$ in (3.17), the estimator (3.18) certainly serves the purpose whenever the underlying sampling design provides $0 < \pi_{ij} \le \pi_i \pi_j$, $1 \le i \ne j \le N$. In the literature attempts have been made to construct designs satisfying such conditions. For example, it is clear that if our sampling design is a fixed-size design with uniform first- and second-order inclusion probabilities, then the condition $0 < \pi_{ij} \le \pi_i \pi_j$, $1 \le i \ne j \le N$ is trivially satisfied. Using some results in the theory of design of experiments as well as trade-off (Hedayat, 1990), Hedayat and Majumdar (1990), established that for every population of size N and sample of size n, there exists a survey design with at least two distinct π_{ij}'s, and with all π_i's equal, which provides a nonnegative unbiased estimator for the variance of the Horvitz–Thompson estimator of the population total. Three such survey designs are presented in Table 3.1 for the case $N = 7$, $n = 3$. Satisfying the conditions $0 < \pi_{ij} \le \pi_i \pi_j$, $1 \le i \ne j \le N$ may *not* be possible by use of arbitrary designs. Moreover, a different nnuve might exist in some cases. It, therefore, seems reasonable first to characterize the class of all possible nnuve's of the HTE for any arbitrary sampling design. Then, on any given occasion, we would essentially look inside that class. It may be noted that our results do *not* guarantee the existence of an nnuve for an arbitrary sampling design.

We will restrict ourselves only to the class of homogeneous *quadratic* estimators of (3.17). Vijayan (1975) has initiated this type of study and established the results to be discussed below.

Theorem 3.4. *For a fixed-size* (n) *design*

(i) *every nonnegative homogeneous quadratic unbiased variance estimator of the* HTE *is necessarily of the form*

$$\hat{V}(s, \mathbf{Y}) = \sum_{\substack{i<j \\ i,j \in s}}\sum b_{ij}(s)(Z_i - Z_j)^2 \tag{3.19}$$

Table 3.1. Examples of Sampling Designs of Size 3 Based on 7 Units Having Unequal Second-Order Inclusion Probabilities While Guaranteeing the Sen–Yates–Grundy Variance Estimator to Be Estimated Nonnegatively

Samples s	d_1 $28P(s)$	d_2 $28P(s)$	d_3 $28P(s)$
1 2 4	3	2	1
2 3 5	3	3	2
3 4 6	4	4	4
4 5 7	4	4	4
1 5 6	4	4	4
2 6 7	4	3	3
1 3 7	4	4	4
2 3 4	1	1	1
1 2 5	1	1	1
2 4 6	0	1	1
1 2 7	0	1	1
2 4 5	0	0	1
1 2 3	0	0	1

where b_{ij}'s satisfy the conditions

$$\sum_{s \ni i,j} b_{ij}(s) P_d(s) = \pi_i \pi_j - \pi_{ij}\,, \qquad 1 \le i \ne j \le N\,, \tag{3.20}$$

and

(ii) *with*

$$b_{ii}(s) = -\sum_{j(\ne i)} b_{ij}(s)\,, \qquad 1 \le i \le N \tag{3.21}$$

and

$$B(s) = (b_{ij}(s))$$

the estimator (3.19) *is nonnegative if for every sample s with $P_d(s) >$ 0, the $n \times n$ matrix $B(s)$ is negative semidefinite.*

Proof. Let $Q(s, \mathbf{Y})$ be an arbitrary nonnegative homogeneous quadratic unbaised estimator of V(HTE). Utilizing (2.2), we may rewrite $Q(s, \mathbf{Y})$ in the form

$$Q(s, \mathbf{Y}) = \sum_{i,j \in s}\sum q_{ij}(s) Y_i Y_j = \sum_{i,j \in s}\sum a_{ij}(s) Z_i Z_j\,. \tag{3.22}$$

Using (3.17) we observe that $V(\text{HTE})$ vanishes at the point $Z_1 = Z_2 = \cdots = Z_N$. This means that $Q(s, \mathbf{Y})$ must also vanish at this point for all samples. Referring to (3.22), this implies that

$$\sum_{\substack{i,j \\ i,j \in s}} \sum a_{ij}(s) = 0 \text{ for every sample } s \text{ with } P_d(s) > 0 \tag{3.23}$$

and consequently

$$\sum_{j \in s} a_{ij}(s) = 0 \text{ for every } i \in s \text{ and for every } s\,. \tag{3.24}$$

We now write $Q(s, \mathbf{Y})$ as follows:

$$\begin{aligned}
Q(s, \mathbf{Y}) &= \sum_{i,j \in s} \sum a_{ij}(s) Z_i Z_j = \sum_{i \in s} a_{ii}(s) Z_i^2 + \sum_{\substack{i \neq j \\ i,j \in s}} \sum a_{ij}(s) Z_i Z_j \\
&= -\sum_{\substack{i \neq j \\ i,j \in s}} \sum a_{ij}(s) Z_i^2 + \sum_{\substack{i \neq j \\ i,j \in s}} \sum a_{ij}(s) Z_i Z_j \\
&= -\sum_{\substack{i < j \\ i,j \in s}} \sum a_{ij}(s)(Z_i - Z_j)^2 = \sum_{\substack{i < j \\ i,j \in s}} \sum b_{ij}(s)(Z_i - Z_j)^2
\end{aligned}$$

writing $b_{ij}(s) = -a_{ij}(s)$, $i, j \in s$. Finally we observe that by the unbiasedness of $Q(s, \mathbf{Y})$ for $V(\text{HTE})$

$$\begin{aligned}
E_d Q(s, \mathbf{Y}) &= \sum_{s \in S_d} \left[\sum_{i<j} \sum b_{ij}(s)(Z_i - Z_j)^2 \right] P_d(s) \\
&= \sum_{i<j} \sum \left[\sum_{s \ni i,j} b_{ij}(s) P_d(s) \right] (Z_i - Z_j)^2 \\
&= V(\text{HTE}) = \sum_{i<j} \sum (\pi_i \pi_j - \pi_{ij})(Z_i - Z_j)^2 \\
&\qquad \text{uniformly in all } Z_1, Z_2, \ldots, Z_N\,,
\end{aligned}$$

which implies (3.20). To prove (ii), we need to show that $B(s)$ has to be negative semidefinite for every s with $P_d(s) > 0$ for $\hat{V}(s, \mathbf{Y})$ to be nonnegative for all $Z_1, Z_2, \ldots, Z_N$. Since $b_{ij}(s) = -a_{ij}(s)$, then

$$b_{ii}(s) = -\sum_{j(\neq i)} b_{ij}(s) = \sum_{j(\neq i)} a_{ij}(s) = -a_{ii}(s)\,.$$

Thus $B(s) = -A(s) = (-a_{ij}(s))$. Now $\hat{V}(s, \mathbf{Y})$ in (3.19) is ≥ 0 if $A(s)$ corresponding to $Q(s, \mathbf{Y})$ in (3.22) is positive semidefinite or equivalently if $B(s)$ is negative semidefinite. □

The Sen–Yates–Grundy variance estimator $\hat{V}_{\text{SYG}}(\text{HTE})$ in (3.18) can be easily identified as a member of the class given by (3.19) by setting $b_{ij}(s) = \{(\pi_i \pi_j - \pi_{ij})/\pi_{ij}\}$, $i \neq j$, $i, j \in s$, $P_d(s) > 0$. These readily satisfy (3.20). This estimator would thus turn out to be nnuve whenever the matrix $((b_{ij}(s)))$ with $b_{ii}(s) = -\Sigma_{j \in s, j(\neq i)} b_{ij}(s)$, $i \in s$ would be nnd for every s with $P_d(s) > 0$. On the other hand, the Horvitz–Thompson variance estimator $\hat{V}_{\text{HT}}(\text{HTE})$ in (3.15) is *not* a member of the above class and hence it cannot serve as a nnuve of the HTE for a fixed-size (n) design, unless, of course, it coincides with (3.16) uniformly in the Y-values for some such design.

In the particular case of $n = 2$, the result is even stronger.

Corollary 3.5. *For $n = 2$, homogeneous quadratic nnuve's exist if and only if $0 < \pi_{ij} \leq \pi_i \pi_j$, $1 \leq i \neq j \leq N$ and the only nnuve, when it exists, is that of Sen–Yates–Grundy, namely the* $\hat{V}_{\text{SYG}}(\text{HTE})$ in (3.18).

Proof. Clearly (3.19) reduces, for $s = (i, j)$ with $i < j$, to $(Z_i - Z_j)^2 b_{ij}(s)$ and for this to be nonnegative, it is necessary and sufficient that $b_{ij}(s) \geq 0$. However, (3.20) simplifies to the single term $b_{ij}(s)\pi_{ij} = \pi_i \pi_j - \pi_{ij}$ yielding $b_{ij}(s) = \{(\pi_i \pi_j - \pi_{ij})\}/\pi_{ij}$. The rest is now clear. □

Remark 3.2. Lanke (1975) has demonstrated the validity of the result of Corollary 3.5 within the general class of *all nnuve*'s.

Note that the above result holds within the framework of *reduced* sampling designs. For an unreduced sampling design, of course, the uniqueness of the Sen–Yates–Grundy variance estimator is *not* claimed here. One might set, for example,

$$\hat{V}(\text{HTE}) = \begin{cases} \pi_i \pi_j - \pi_{ij})(Z_i - Z_j)^2 b(s) & \text{if } s = (i, j) \\ (\pi_i \pi_j - \pi_{ij})(Z_i - Z_j)^2 b(s') & \text{if } s' = (j, i) \end{cases}$$

with $b(s)$, $b(s') > 0$ and $b(s)P_d(s) + b(s')P_d(s') = 1$. The case of $b(s) = b(s') = 1/\pi_{ij}$, $1 \leq i \neq j \leq N$ would then correspond to the Sen–Yates–Grundy variance estimator.

For any arbitrary fixed-size (n) design, therefore, so long as the conditions $\{0 < \pi_{ij} \leq \pi_i \pi_j,\ 1 \leq i \neq j \leq N\}$ are satisfied, there exists at least one nnuve of the HTE and one could just take $\hat{V}_{\text{SYG}}(\text{HTE})$ in (3.19) as one such estimator. For $n = 2$, however, these conditions are necessary and (3.19) is the only nnuve. What then can be said for $n > 2$? Supposing that the above conditions are not satisfied, is $\hat{V}_{\text{SYG}}(\text{HTE})$ still nonnegative and, if not, does there exist any other nnuve? Below we cite two examples dealing with these important questions. The first example provides a sampling design in which the sufficient conditions $\pi_{ij} \leq \pi_i \pi_j$ are not satisfied and yet $\hat{V}_{\text{SYG}}(\text{HTE})$ is nonnegative. The second example provides a sampling design in which the sufficient conditions $\pi_{ij} \leq \pi_i \pi_j$ are violated and $\hat{V}_{SYG}$(HTE) cannot be guaranteed to be nonnegative. However, we are able to construct another estimator of the variance of HTE which is unbiased and nonnegative. We will not pursue these problems further here. The interested reader is referred to results in J.N.K. Rao (1979, 1988) in which further references may be found.

Example 3.2. $N = 5$, $n = 3$. Consider the sampling design given in Table 3.2.

Table 3.2. A Fixed-Size (3) Sampling Design

s	(1, 2, 3)	(1, 4, 5)	(2, 4, 5)	(3, 4, 5)
$P(s)$	$\frac{1}{4}$	$\frac{1}{4}$	$\frac{1}{4}$	$\frac{1}{4}$

For this design, $\pi_{45} > \pi_4 \pi_5$. However, it can be verified that the estimator in (3.18) is nnuve.

Again, with reference to the Example 3.1, the reader can verify that the choice of $b_{ij}(s)$'s exhibited in Table 3.3 leads to an unbiased nonnegative estimator of V(HTE) in (3.22).

3.3. SOME PROPERTIES OF THE HTE

In this section we want to investigate the extent to which the HTE possesses properties of a good estimator. We recall that for unicluster sampling designs, the HTE is known to be the uniformly minimum variance unbiased estimator within the class of homogeneous linear unbiased estimators of the population total (see Theorem 2.11). Throughout this section, the underlying sampling design will be taken to be arbitrary. Let d be any design and let e_1 and e_2 be any two unbiased estimators of the parametric function $\theta(\mathbf{Y})$. Suppose it so happens that

$$E_p(e_1 - \theta)^2 \leq E_p(e_2 - \theta)^2 \quad \text{for all } \mathbf{Y} \in R^N, \text{ with strict inequality for at least one } \mathbf{Y} \in R^N. \tag{3.25}$$

Table 3.3. Table of Coefficients for Unbiased Nonnegative Estimation of V(HTE)

	Coefficients $b_{ij}(s)$, $1 \leq i < j \leq N$, $(i, j) \in s$									
Samples (s)	(1, 2)	(1, 3)	(1, 4)	(1, 5)	(2, 3)	(2, 4)	(2, 5)	(3, 4)	(3, 5)	(4, 5)
(1, 3, 5)	0	1/8	0	1/8	0	0	0	0	−1/16	0
(2, 3, 5)	0	0	0	0	1/5	0	1/5	0	−1/10	0
(1, 2, 4)	1	0	1/2	0	0	1/5	0	0	0	0
(3, 4, 5)	0	0	0	0	0	0	0	7/20	−7/40	7/20

Clearly, it makes no sense to use the estimator $e_2(s, \mathbf{Y})$ since a *uniformly* better estimator $e_1(s, \mathbf{Y})$ is available. This gives rise to the following concept. (Recall the discussion preceding Theorem 2.9).

Definition 3.1. An estimator $e_2(s, \mathbf{Y})$ is defined as *inadmissible* within the class of all unbiased estimators of $\theta(\mathbf{Y})$ whenever there exists another estimator $e_1(s, \mathbf{Y})$ satisfying (3.25).

Carrying this further, we say that an estimator is *admissible* if it cannot be *uniformly* improved upon by another estimator. Mathematically, this would mean that an estimator $e(s, \mathbf{Y})$ is admissible if and only if with respect to *any* other estimator $e'(s, \mathbf{Y})$, either $E_p(e - \theta)^2 = E_p(e' - \theta)^2$ for all $\mathbf{Y} \in R^N$ or, there exists a nonempty subset of R^N, depending possibly on $e(s, \mathbf{Y})$ and $e'(s, \mathbf{Y})$, say $W(e, e', \mathbf{Y})$, such that

$$E_p(e - \theta)^2 < E_p(e' - \theta)^2 \qquad \text{for all } \mathbf{Y} \in W(e, e', \mathbf{Y}) .$$

This would simply mean that the estimator $e(s, \mathbf{Y})$ cannot be *uniformly* improved upon by any other estimator and hence we may not be totally unjustified in using $e(s, \mathbf{Y})$ for inferring on $\theta(\mathbf{Y})$. This is why admissibility may be regarded as a minimal requirement to be expected of a good estimator.

Specializing to the problem of estimating the population total $T(\mathbf{Y})$, we now raise the following questions. Does there exist an admissible estimator? If so, is it unique? What is the status of the HTE with regard to admissibility? For a nonunicluster sampling design, we have noted in Chapter Two (immediately after Definition 2.4) that one can construct various error functions to produce plenty of unbiased estimators of the population total. It can be argued that at least one of these estimators must be admissible as, otherwise, inadmissibility of *all* conceivable estimators would ultimately lead to the existence of an estimator with zero variance uniformly in $\mathbf{Y} \in R^N$. This is a contradiction. On the other hand, such an admissible estimator is also clearly not unique. In our search for one such admissible estimator, it may be worth examining the status of the HTE. We present the following results due to Godambe (1960) and Joshi (1965).

Theorem 3.6. *The HTE is admissible within the class $\mathscr{C}_L$ of all hlue's of $T(\mathbf{Y})$.*

Theorem 3.7. *The HTE is admissible within the wider class of all unbiased estimators of $T(\mathbf{Y})$.*

We will discuss below only the proof of Theorem 3.6. We will simply write $P(s)$ for $P_d(s)$. Let $l(s, \mathbf{Y}) = \Sigma_{i \in s} l_{si} Y_i$, $P(s) > 0$ be an arbitrary hlue of $T(\mathbf{Y})$. The coefficients l_{si}'s satisfy $\Sigma_{s \ni i} l_{si} P(s) = 1$, $1 \le i \le N$. (See (2.1) and

(2.8)). The expressions (3.2) and (3.3), respectively, represent $V(l(s, \mathbf{Y}))$ and $V(\text{HTE})$. Let $\mathbf{Y}(i) = (0, 0, \ldots, Y_i, \ldots, 0)$, $1 \leq i \leq N$, $Y_i \neq 0$. Then

$$V(l(s, \mathbf{Y}))|_{\mathbf{Y}=\mathbf{Y}(i)} = Y_i^2\Big(\sum_{s \ni i} l_{si}^2 P(s) - 1\Big) \tag{3.26}$$

while

$$V(\text{HTE})|_{\mathbf{Y}=\mathbf{Y}(i)} = Y_i^2\Big(\frac{1}{\pi_i} - 1\Big)\,. \tag{3.27}$$

Now note that by the Cauchy–Schwartz inequality,

$$\Big[\sum_{s \ni i} l_{si}^2 P(s)\Big]\Big[\sum_{s \ni i} P(s)\Big] \geq \Big[\sum_{s \ni i} l_{si} P(s)\Big]^2 ,$$

that is, $[\Sigma_{s \ni i} l_{si}^2 P(s)] \geq 1/\pi_i$ since $\Sigma_{s \ni i} l_{si} P(s) = 1$, and $\Sigma_{s \ni i} P(s) = \pi_i$. This yields (3.26) $\geq$ (3.27) with "=" if and only if $l_{si} = 1/\pi_i$ for every $i \in s$ with $P(s) > 0$. Hence, taking $W = \bigcup_{i=1}^N \{(0, 0, \ldots, Y_i, 0, \ldots, 0) \mid Y_i \neq 0\}$, we see that $V(l(s, \mathbf{Y})) \geq V(\text{HTE})$ for all $\mathbf{Y} \in W \subset R^N$ with strict inequality for at least one $\mathbf{Y} \in W$, as, otherwise, $l_{si} = 1/\pi_i$ for every $i \in s$, with $P(s) > 0$ in which case $l(s, \mathbf{Y})$ coincides with the HTE. This establishes admissibility of the HTE inside the class $\mathscr{C}_L$. □

Remark 3.3. Suppose the set of possible Y-values is a subset K of R^N. Even then the proof would carry through so long as $W \subseteq K$. However in the situation where $W \not\subseteq K$, it may be possible to uniformly improve over the HTE by suggesting suitable unbiased estimators for some sampling designs. Further, unbiasedness is also important above. If this requirement is relaxed, then the HTE still remains admissible within the class of all estimators for a fixed-size sampling design. However, such results may not be generally true for variable-size designs even when the estimators are linear homogeneous. See Godambe and Joshi (1965) in this context.

The proof of Theorem 3.7 is complicated since the HTE has to be judged against *all* unbiased estimators of $T(\mathbf{Y})$, not merely against the linear ones. The reader is referred to Joshi (1965). Also see Exercise 3.5.

Other properties of the HTE have been discussed in the literature. We will not pursue those topics here. The interested reader is referred to Godambe (1955) and Godambe and Joshi (1965).

The above technique of establishing admissibility of the HTE is *not* widely applicable. It depends largely on our feeling about the performance of the HTE at a specified set of points, namely the points $\mathbf{Y}(i)$, $1 \leq i \leq N$. Failing to *discover* such points, the technique would *not* be applicable. Indeed, other useful methods have been suggested from time to time. For example, Patel and Dharmadhikari (1977, 1978) developed entirely different tools to produce a large class of admissible estimators. See Sinha and

Pantula (1986) for additional results in this direction. Earlier, Roy and Chakravarti (1960) presented some results, again from a different point of view. See Sinha (1976) in this context. Recently, Meeden and Ghosh (1983) presented yet another interesting approach in constructing admissible estimators for the finite population total. Sengupta (1980, 1982, 1983) and Meeden and Ghosh (1984) established admissibility of some commonly used estimators. Undoubtedly, admissibility in finite population sampling continues to be a fascinating and fruitful area of research.

We conclude this chapter by citing some results on admissible strategies. We recall that a sampling strategy is a combination (d, e) of a sampling design d and an estimator e. It is also expressed as (p, e) where p stands for the sampling plan generated by $\{P_d(s) \mid s \in S_d\}$. For the purpose of drawing inference on a given parametric function $\theta(\mathbf{Y})$, we may decide to use various sampling strategies $(p_1, e_1), (p_2, e_2), \ldots$. Quite naturally, then, we have to suggest methods for comparing different sampling strategies arising in a given context. From cost considerations, the sampling designs may be said to be comparable provided the expected sample size is the same for all the competing designs. In particular, each one could be a fixed-size (n) design for a specified n. Then the criterion for comparing the strategies may be taken as one of minimizing the mean square error $E_p(e - \theta(\mathbf{Y}))^2$. A strategy (p, e) is said to be inadmissible if there exists another competing strategy (p', e') such that $E_{p'}(e' - \theta)^2 \leq E_p(e - \theta)^2$ for all $\mathbf{Y} \in R^N$ with strict inequality for at least one point in R^N. The strategy (p_0, e_0) is said to be admissible if it is *not* inadmissible. Admissible strategies might be sought by restricting the estimators (based on any competing design) to suitable subclasses such as unbiased estimators. In such a case, all competing strategies are also termed unbiased strategies.

It is true that in most cases there is *no* unique admissible strategy available for estimating the finite population mean. As regards the strategies based on the HTE, the following result was established by Ramakrishnan (1975).

Theorem 3.8. *For estimating the finite population total or mean, among all unbiased strategies based on sampling designs having the same expected sample size n (integer), any fixed-size (n) design coupled with the HTE is admissible.*

In particular, we may derive the following corollary.

Corollary 3.9. *The strategy* (SRS(N, n), *sample mean*) *serves as an admissible unbiased strategy for estimation of the finite population total or the mean among all unbiased strategies based on sampling designs having expected sample size n.*

We refer the reader to Cassel et al. (1977) and also to Chaudhuri and Vos (1988) and Chaudhuri (1988) for an excellent discussion on this topic

EXERCISES

3.1. For each of the sampling designs in Table 3.A, write down the expression for the HTE of the population total and an estimate of its variance for all the samples involved in the design. Verify directly $E(\text{HTE}) = T(\mathbf{Y})$ and $E(\hat{V}_{\text{HT}}(\text{HTE})) = V(\text{HTE})$. Is your estimator $\hat{V}$ nonnegative?

Table 3.A

d_1:

s	(1)	(2)	$\cdots$	(N)	$(1, 2, 3, \ldots, N)$
$P(s)$	p_1	p_2	$\cdots$	p_N	q

d_2:

s	(1, 2)	(2, 3)		$(N-1, N)$	$(N, 1)$	$(1, 2, 3, \ldots, N)$
$P(s)$	p_1	p_2	$\cdots$	p_{N-1}	p_N	q

d_3:

s	(1)	(1, 2)	(3)	(3, 4)	$\cdots$	$(N-1)$	$(N-1, N)$	$(1, 2, 3, \ldots, N)$
$P(s)$	p_1	p_2	p_3	p_4	$\cdots$	p_{N-1}	p_N	q

$\left(0 < p_i, q < 1, \sum p_i + q = 1\right)$

d_4:

s	(1)	(1, 2)	(1, 2, 3)	$\cdots$	$(1, 2, 3, \ldots, N)$
$P(s)$	p_1	p_2	p_3	$\cdots$	p_N

$\left(0 < p_i < 1, \sum p_i = 1\right)$

3.2. For which of the sampling designs in Table 3.B would you recommend use of the HTE? Why?

Table 3.B

d_1:

s	(1)	(2)	(3)	$\cdots$	(N)
$P(s)$	p_1	p_2	p_3	$\cdots$	p_N

d_2:

s	(1, 2)	(2, 3)	$\cdots$	$(N-1, N)$	$(N, 1)$
$P(s)$	p_1	p_2	$\cdots$	p_{N-1}	p_N

d_3:

s	$(1, 2, \ldots, N_1)$	$(N_1+1, N_1+2, \ldots, N_1+N_2)$	$\cdots$	$(N-N_k+1, \ldots, N)$
$P(s)$	p_1	p_2	$\cdots$	p_k

d_4:

s	(1)	(1, 2)	(1, 2, 3)	$\cdots$	$(1, 2, 3, \ldots, N)$
$P(s)$	p_1	p_2	p_3	$\cdots$	p_N

$\left(0 < p_i < 1, \sum p_i = 1\right)$

3.3. Table 3.C displays two pairs of fixed-size (n) sampling designs. Examine, for each pair, whether there exists any difference between the two sampling designs so far as the use of the HTE (for estimating the population total) is concerned.

Table 3.C

d_1: SRS$(N = n^2, n)$

d_2: $P((i, i+1, i+2, \ldots, i+n-1)) = 1/(n^2+n)$; $i = 1, n+1, 2n+1, \ldots, n^2-n+1$;

$$P((i_1, i_2, \ldots, i_n)) = 1/(n^n + n^{n-1}), \quad 1 \le i_1 \le n$$
$$n+1 \le i_2 \le 2n$$
$$\vdots$$
$$n^2 - n + 1 \le i_n \le n^2 (= N)$$

d_1: SRS$(N = mn, n)$ with $m \ge n$

d_2: $P((i_1, i_2, \ldots, i_n)) = n(n-1)/N(N-1)C^{m-2}_{n-2}$,

$$im + 1 \le i_1 < \cdots < i_n \le (i+1)m \text{ for } i = 0, 1, 2, \ldots, n-1$$
$$= n(n-1)/N(N-1)m^{n-2}$$
$$1 \le i_1 \le m,\ m+1 \le i_2 \le 2m, \ldots,$$
$$N - m + 1 \le i_n \le N\ (=mn)$$

Comment on the result.

3.4. Let d_i represent SRS(N, n_i) design, $i = 1, 2$. Here n_1 and n_2 are two fixed integers satisfying $1 \le n_1 < n_2 \le N$. Define a sampling design d as $d = pd_1 + qd_2$, $0 \le p$, $q \le 1$, $p + q = 1$ in the sense that $P_d(s) = pP_{d_1}(s) + qP_{d_2}(s)$ for all s.

(i) Deduce expressions for π_i's and π_{ij}'s for the designs d_1, d_2, and d.

(ii) Find expressions for V(HTE) and $\hat{V}_{\text{HT}}$(HTE) based on the sampling designs d_1, d_2, and d.

(iii) For the design d, considering V(HTE) as a function of p, say, V_p(HTE), determine p_0 such that $V_{p_0}(\text{HTE}) = \min_p V_p(\text{HTE})$. Interpret your result.

3.5. Let d be a uniform sampling design on $\mathcal{U} = \{1, 2, 3\}$ with support $s_1 = (1)$, $s_2 = (2)$, and $s_3 = (2, 3)$. Prove that the HTE of $T(\mathbf{Y})$ is admissible in the class of all unbiased estimators of $T(\mathbf{Y})$.

3.6. For the sampling design in Exercise 3.5 prove that the estimator $e(s, \mathbf{Y})$ defined by

$$e(s, \mathbf{Y}) = \begin{cases} 3Y_1, & \text{if } s = (1) \\ 3Y_2, & \text{if } s = (2) \\ 3Y_3, & \text{if } s = (2,3) \end{cases}$$

is an admissible estimator in the class of all unbiased (linear or nonlinear) estimators of $T(\mathbf{Y})$.

3.7. While the estimator in Exercise 3.6 is admissible it does not take full advantage of the information provided by the data (see case $s = s_3$). Explain this phenomenon.

3.8. For a population of size 4, give an example of a sampling design with an admissible estimator, different from the HTE, for the population total in the class of hlue's.

3.9. Fill in the blanks in the following statements:

(i) The Horvitz–Thompson estimator for the population mean, based on any sampling design, is given by HTE$(s, \mathbf{Y}) =$ _______.

(ii) The expression due to Sen–Yates–Grundy for the variance of the HTE is derived in the context of a _______ sampling design and is given by _______.

(iii) An example of a sampling design, other than SRS design, where the nonnegative variance estimation problem can be handled satisfactorily is given by _______.

(iv) An example of a sampling design where the nonnegativity problem cannot be resolved is given by _______.

3.10. Construct a fixed-size sampling design and an estimator of V(HTE) satisfying the conditions of Theorem 3.4 such that for some i and j, $\pi_{ij} > \pi_i \pi_j$ and yet the estimator is nonnegative for all choices of $\mathbf{Y}$. Show, therefore, that, while $\pi_{ij} \leq \pi_i \pi_j$, $\forall i \neq j$, is sufficient for the existence of a nnuve of V(HTE) for a fixed size design, it is not a necessary one.

3.11. For the following sampling design, show that both $\hat{V}_{\text{HT}}$(HTE) and $\hat{V}_{\text{SYG}}$(HTE) may assume negative values: $P((1,2)) = P((3,4)) = 6/16$, $P((1,3)) = P((1,4)) = P((2,3)) = P((2,4)) = 1/16$.

3.12. Prove that for a fixed-size (2) design, $\hat{V}_{\text{HT}}(\text{HTE}) = \hat{V}_{\text{SYG}}(\text{HTE})$ if and only if the reduced sampling design is SRS$(N, 2)$.

3.13. Construct a sampling design in which V(HTE) can be unbiasedly estimated, but every unbiased estimator of it can assume negative values.

3.14. Prove that for a fixed-size (2) sampling design $\hat{V}_{\text{HT}}$(HTE) is nonnegative if and only if the reduced sampling design is SRS$(N, 2)$.

3.15. Use the identity

$$T(\mathbf{Y}) = \sum_{i=1}^{N} Y_i = \frac{1}{2(N-1)} \sum\sum_{1 \leq i \neq j \leq N} (Y_i + Y_j)$$

in combininination with (3.5) to form an hlue of $T(\mathbf{Y})$ different from the HTE based on a sampling design for which $\pi_{ij} > 0$, $1 \leq i \neq j \leq N$. Deduce the variance expression for your estimator and provide an unbiased estimator of the variance. Compare your estimator with the HTE for a hypothetical population.

3.16. Give an example of a sampling design and a choice of $\mathbf{Y} \in R^N$ such that for one and the same sample, both $\hat{V}_{\text{HT}}(\text{HTE})$ and $\hat{V}_{\text{SYG}}(\text{HTE})$ assume negative values. What would you do in such a situation?

3.17. For the sampling design in Exercise 1.26, consider the following estimator. If $G_1 = \{a, b\}$ and $G_2 = \{c, d, e\}$ is the selected partition and $s = (i_1, i_2)$ is the selected sample with $i_1 \in G_1$, $i_2 \in G_2$, then construct the estimator $\hat{T}(\mathbf{Y})$ as

$$\hat{T}(\mathbf{Y}) = (a + b)Y_{i_1}/i_1 + (c + d + e)Y_{i_2}/i_2 .$$

(a) Verify directly that $\hat{T}(\mathbf{Y})$ is an unbiased estimator of the population total.

(b) Show that the above estimator has variance given by

$$V(\hat{T}(\mathbf{Y})) = 6 \sum_{i=1}^{5} (Y_i^2/i) - 2T^2(\mathbf{Y})/5 .$$

(c) Verify directly that $\hat{V}(\hat{T}(\mathbf{Y}))$ defined below serves as a nonnegative unbiased estimator of $V(\hat{T}(\mathbf{Y}))$.

$$\hat{V}(\hat{T}(\mathbf{Y})) = \frac{2}{3}\left[\left(\frac{15Y_{i_1}}{i_1} - \hat{T}(\mathbf{Y})\right)^2\left(\frac{i_1}{a+b}\right) + \left(\frac{15Y_{i_2}}{i_2} - \hat{T}(\mathbf{Y})\right)^2\left(\frac{i_2}{c+d+e}\right)\right].$$

Note that no explicit use of the π_i's and π_{ij}'s have been made above.

3.18. Consider the following sampling design for a sample of size 2.

$$d: \quad P_d((i, j)) = \frac{1}{(N-1)^2}, \qquad 1 \leq i < j \leq N-1 ;$$

$$P_d((i, N)) = \frac{N}{2(N-1)^2}, \qquad 1 \leq i \leq N-1 .$$

Now consider the HTE and also an HT-type estimator of the form (2.18) with

$$c(s) = \begin{cases} 2, & \text{if } s = (i, j), 1 \leq i < j \leq N-1 ; \\ 1, & \text{if } s = (i, N), 1 \leq i \leq N-1 . \end{cases}$$

Compare the variances of the two estimators at all points of the form $(Y_1 = Y_2 = \cdots = Y_{N-1} = 1,\ Y_N = u)$ by varying u over both positive and negative values. Comment on the result.

3.19. With reference to an HT-type estimator of the form (2.18), define

$$\bar{c} = \sum_s c(s)P(s)\,,$$

$$\pi_c(i) = \sum_{s \ni i} c(s)P(s)\,, \qquad 1 \le i \le N\,,$$

$$\pi_c(i, j) = \sum_{s \ni i,j} c(s)P(s), \qquad 1 \le i < j \le N\,.$$

(a) Show that for a fixed-size (n) design,
 (i) $\Sigma_{i=1}^{N} \pi_c(i) = n\bar{c}$,
 (ii) $\Sigma_{j(\neq i)} \pi_c(i, j) = (n-1)\pi_c(i)$,
 (iii) $\Sigma\,\Sigma_{i \neq j} \pi_c(i, j) = n(n-1)\bar{c}$.

(b) Write down the coefficient of Y_i^2 in the expression for $V(\phi(s, \mathbf{Y}))$ where $\phi(s, \mathbf{Y})$ is the HT-type estimator in (2.18).

(c) Using Cauchy–Schwartz inequality, show that the coefficient of Y_i^2 in $V(\phi(s, \mathbf{Y})$ is at least equal to the corresponding coefficient of Y_i^2 in V(HTE). Comment on this result.

REFERENCES

Cassel, C. Särndal, C., and Wretman, J. H. (1977). *Foundations of Inference in Survey Sampling*. Wiley, New York.

Chaudhuri, A. (1988). Optimality of sampling strategies. In: *Handbook of Statistics, Vol. 6, Sampling* (Krishnaiah, P. R., and Rao, C. R., Eds.), 47–96. North-Holland, Amsterdam.

Chaudhuri, A., and Vos, J. W. E. (1988). *Unified Theory and Strategies of Survey Sampling*. North-Holland, Amsterdam.

Godambe, V. P. (1955). A unified theory of sampling from finite populations. *J. Roy. Statist. Soc.*, **B17**, 269–278.

Godambe, V. P. (1960). An admissible estimate for any sampling design. *Sankhyā*, **22**, 285–288.

Godambe, V. P., and Joshi, V. M. (1965). Admissibility and Bayes estimation in sampling finite populations I. *Ann. Math. Statist.*, **36**, 1707–1722.

Hanurav, T. V. (1966). Some aspects of unified sampling theory. *Sankhyā*, **A28**, 175–203.

Hedayat, A. (1990). The theory of trade-off for t-designs. In: *The IMA volume in Math. and Its Applications*, *Coding Theory and Design Theory*, *Part II*, *Design Theory*, Vol. 21, (Dijen Ray-Chaudhuri, Ed.), 101–126. Springer-Verlag, Berlin.

Hedayat, H., and Majumdar, D. (1990). On fixed size sampling designs providing nonnegative Sen-Yates-Grundy Variance Estimator. Stat. Lab. Tech. Report, Dept. of Math., Stat. & Comp. Sci., Univ. of Illinois, Chicago.

Horvitz, D. G., and Thompson, D. J. (1952). A generalization of sampling without replacement from a finite universe. *J. Amer. Statist. Assoc.*, **47**, 663–685.

Joshi, V. M. (1965). Admissibility and Bayes estimation in sampling finite populations II, III. *Ann. Math. Statist.*, **36**, 1723–1729, 1730–1742.

Lanke, J. (1975). Some contributions to the theory of survey sampling. Ph.D. Thesis, University of Lund.

Meeden, G., and Ghosh, M. (1983). Choosing between experiments: application to finite population sampling. *Ann. Statist.* **11**, 296–305.

Meeden, G., and Ghosh, M. (1984). On the admissibility and uniform admissibility of ratio type estimators. In: *Statistics: Applications and New Directions. Proceedings ISI Golden Jubilee Internat. Conference* (Ghosh, J. K., and Roy, J., Eds.), 378–387. Indian Statistical Institute, Calcutta.

Patel, H. C., and Dharmadhikari, S. W. (1977). On linear invariant unbiased estimators in survey sampling. *Sankhyā*, **C39**, 21–27.

Patel, H.C., and Dharmadhikari, S. W. (1978). Admissibility of Murthy's and Midzuno's estimators within the class of linear unbiased estimators of finite population totals. *Sankhyā*, **C40**, 21–28.

Ramakrishnan, M. K. (1975). Choice of an optimum sampling strategy I. *Ann. Statist.*, **2**, 669–679.

Rao, J. N. K. (1979). On deriving mean square errors and their non-negative unbiased estimators in finite population sampling. *J. Indian Statist. Assoc.*, **17**, 125–136.

Rao, J. N. K. (1988). Variance estimation in sample surveys. In: *Handbook of Statistics*, *Vol. 6*, *Sampling* (Krishnaiah, P. R., and Rao, C. R., Eds.), 427–447. North-Holland, Amsterdam.

Roy, J., and Chakravarti, I. M. (1960). Estimating the mean of a finite population. *Ann. Math. Statist.*, **31**, 392–398.

Sen, A. R. (1953). On the estimate of the variance in sampling with varying probabilities. *J. Indian Soc. Argic. Statist.*, **5**, 119–127.

Sengupta, S. (1980). On the admissibility of symmetrized Des-Raj estimator for PPSWOR samples of size two. *Cal. Statist. Assoc. Bull.*, **29**, 35–44.

Sengupta, S. (1982). Admissibility of the symmetrized Des-Raj estimator for fixed size sampling designs of size two. *Cal. Statist. Assoc. Bull.*, **31**, 201–205.

Sengupta, S. (1983). On the admissibility of Hartley-Ross estimator. *Sankhyā*, **B45**, 471–474.

Sinha, B. K. (1976). On balanced sampling schemes. *Cal. Statist. Assoc. Bull.*, **25**, 129–138.

Sinha, B. K., and Pantula, S. G. (1986). Linear invariance and admissibility in sampling finite populations. *Sankhyā*, **B48**, 246–257.

Vijayan, K. (1975). On estimating the variance in unequal probability sampling. *J. Amer. Statist. Assoc.*, **70**, 713–716.

Yates, F., and Grundy, P. M. (1953). Selections without replacement from within strata with probability proportional to size., **B15**, 253–261.

CHAPTER FOUR

Simple Random and Allied Sampling Designs

We devote this chapter entirely to a discussion of finite population inference problems based on the most commonly used and very basic sampling design, namely the simple random sampling design and its allied forms. We discuss at length the problems of estimation of the mean, proportion, variance, and quantiles. Next we address the question of determination of sample size for a predetermined margin of error. We also examine the question of best linear unbiased estimation of a finite population mean using an unlabelled data set in this framework.

4.1. SIMPLE RANDOM SAMPLING DESIGNS

Simple Random Sampling (SRS) designs are very special members of the vast class of *probability sampling* designs. These are well known to the samplers for their ready and simple applicability. Moreover, they form the basis of many other sampling designs encountered in theory and practice. For a population of size N, an SRS(N, n) sampling design is a uniform fixed-size (n) sampling design defined as

$$P((i_1, i_2, \ldots, i_n)) = \{N(N-1)(N-2)\cdots(N-n+1)\}^{-1},$$
$$1 \le i_1 \ne i_2 \ne \cdots \ne i_n \le N.$$

In its reduced form, it is given by

$$P^*((j_1, j_2, \ldots, j_n)) = \binom{N}{n}^{-1}, \qquad 1 \le j_1 < j_2 < \cdots < j_n \le N.$$

For simplicity of notation, however, throughout this chapter the probability associated with the reduced form of the sample s will be denoted by $P(s)$ instead of $P^*(s)$. Clearly, it is a sampling design where every sample of the same size (n) has the same chance of being selected. The underlying scheme of sampling has been discussed earlier in Example 1.8. As suggested by that scheme, the operational steps in obtaining such a sample of size n consists of selecting the units one by one with equal probability and without replacement until n units have been selected.

A sample selected by an SRS(N, n) design is commonly referred to as a *random* (*sr*) sample. Unfortunately, to nonexperts the term random sample frequently connotes a miniature of the population. For example, in studying the income level of a community consisting of various income groups, a random sample may *not* contain representatives from all economic levels. Indeed, a random sample may well contain members of only one or two income groups. What makes the sample random is the underlying random process of selection and not the composition of the sample actually drawn on a given occasion. We should not judge a selected random sample as extreme or unrepresentative of the population based on its composition. The discomfort about the possibility of getting a random sample, which may not be representative in the common sense of the term, can be eliminated to a large extent by adopting sampling designs other than SRS designs. Some such designs will be introduced and discussed in the subsequent sections of this and other chapters. We would like to emphasize that the motivation for using an SRS design rests, apart from its simplicity and ready applicability, on the understanding that the survey variable has a more or less uniform distribution on the units of the population under consideration. Finally, we stress again that a random sample should *not* be interpreted as a *representative* sample. This latter terminology has been investigated in detail in a series of excellent review papers by Kruskal and Mosteller (1979a–c, 1980).

4.2. COMPOSITION AND DECOMPOSITION OF A SIMPLE RANDOM SAMPLE

Suppose we want to draw an SRS(N, n) sample of size n from a given population and it is observed that an SRS(N, n^*) sample of size $n^* > n$ is already available. We might then think of drawing an SRS(n^*, n) subsample of size n from the available sample of size n^*. It is interesting to investigate the properties of the resulting sampling design. In particular, we might be interested to know whether the subsample of size n thus arrived at is, in fact, a simple random (*sr*) sample from the original population. Again, if $n^* < n$, we might think of using the n^* units already available and combining them with another simple random sample of $(n - n^*)$ units drawn from the remaining $(N - n^*)$ units of the population. Then the ultimate sample would be a composition of the two separate samples of sizes n^* and $(n - n^*)$. We would like to know if such a composition would result in a simple random sample of the original population. We resolve these issues in the following theorems.

Theorem 4.1. *Simple random subsampling of n units from a simple random sample of $n*$ $(>n)$ units results in simple random sampling of n units from the original population.*

Proof. Let $s^* = (i_1, i_2, \ldots, i_{n^*})$ be an *sr* sample of size $n^*(>n)$ from a given population of size N. Let $s = (j_1, j_2, \ldots, j_n)$ be an *sr* subsample of size n from the sample s^*. Then the unconditional distribution of samples of the type s is given by

$$P((j_1, j_2, \ldots, j_n)) = \sum P(s^*)P((j_1, j_2, \ldots, j_n)|s^*)$$

where Σ extends over all $s^* \supset (j_1, j_2, \ldots, j_n)$. Cleary, $P(s^*) = \binom{N}{n^*}^{-1}$ and $P((j_1, j_2, \ldots, j_n)|s^*) = 0$ if $(j_1, j_2, \ldots, j_n) \not\subset s^*$ while $P((j_1, j_2, \ldots, j_n)|s^*) = \binom{n^*}{n}^{-1}$ if $(j_1, j_2, \ldots, j_n) \subset s^*$. Hence, $P((j_1, j_2, \ldots, j_n)) = \binom{N}{n^*}^{-1}\binom{n^*}{n}^{-1}$ {# of s^* containing $(j_1, j_2, \ldots, j_n)$ as a subset} $= \binom{N}{n^*}^{-1}\binom{n^*}{n}^{-1}\binom{N-n}{n^*-n} = \binom{N}{n}^{-1}$. This completes the proof. □

Theorem 4.2. *A composition of two simple random samples, the first of size n_1 drawn from the original population and the second of size n_2 drawn from the subpopulation of $(N - n_1)$ units not covered by the first sample, constitutes a simple random sample of size $n_1 + n_2$ from the original population.*

The proof follows essentially along the same lines as above and is left as an exercise for the reader (Exercise 4.1).

We conclude this section by citing yet another method of combining two simple random samples. Suppose the two samples are drawn independently each from the entire population. If they are combined into a new sample by counting the repeated units only once, then the resulting sampling design will no longer be a fixed-size design. Of course, it will be a probability sampling design but of a special type. For such designs, the first- and second-order inclusion probabilities may be worked out without any difficulty and these may be used in deducing the mean and the variance of the number of distinct units contained in the composite samples. Inference based on such sampling designs will not be treated separately in this book.

4.3. THE POPULATION PARAMETERS

As before, let Y denote the survey variable of interest. We assume that Y is of a quantitative character and it takes values $Y_1, Y_2, \ldots, Y_N$ on the units $1, 2, \ldots, N$, respectively. The following notations are standard in the literature.

$T(\mathbf{Y}) = \Sigma_{i=1}^{N} Y_i$ = population total of the Y-values

$\bar{\mathbf{Y}} = \Sigma_{i=1}^{N} Y_i/N$ = population mean of the Y-values

$\sigma_Y^2 = \Sigma_{i=1}^{N} (Y_i - \bar{\mathbf{Y}})^2/N = N^{-1} \Sigma Y_i^2 - \bar{\mathbf{Y}}^2$ = population variance of the Y-values with divisor N

$\mathbf{S}_Y^2 = \Sigma_{i=1}^{N} (Y_i - \bar{\mathbf{Y}})^2/(N-1)$ = population variance of the Y-values with divisor $(N-1)$.

Clearly, $N\sigma_Y^2 = (N-1)\mathbf{S}_Y^2$. When the survey variable is understood from the context, we will simply write σ^2 or $\mathbf{S}^2$ for the population variance.

Our primary interest is to estimate the population total or the population mean and its standard error. (We will discuss other parameters of interest in Section 4.7). It will be seen that the estimation of the population variance σ^2 is incidental to this study. Of course, a study of the estimation of σ^2 as a measure of dispersion of the survey variables in the population could be of independent interest. For example, suppose the governing board of a large township is planning to impose a monthly fee to the residential household for rendering a particular community service such as garbage disposal. It would be convenient to advocate and implement a suggestion that for all households this fee be uniform. On the other hand, the board might be concerned about the possibility of complaints regarding unfairness of this proposal on the ground that not all households are equally benefitted by this service. To avoid such a situation and to justify the proposal, the board might be advised to study the nature of variation in the benefits to be derived by various households and accordingly decide on uniformity of the monthly fee if such a variation is found to be reasonably small.

4.4. ESTIMATION OF POPULATION PARAMETERS

We now turn to the estimation problems based on SRS(N, n) designs, assuming $2 \leq n < N$. As we have noted earlier, an SRS(N, n) design with $2 \leq n < N$ is a nonunicluster sampling design of fixed size n. Therefore, there exists *no* best unbiased estimator of the population total or mean. However, as noted earlier in Chapters Two and Three, admissible unbiased estimators are indeed available. Of course, the HTE is one of them and it is traditionally used for SRS designs because of its simplicity and simple statistical interpretation. Here we shall write $\hat{\theta}(\mathbf{Y})$ to indicate an unbiased estimator for the parameter $\theta(\mathbf{Y})$.

Below we first determine the values of π_i's and π_{ij}'s for such a design.

Lemma 4.3. *For an* SRS(N, n) *design,*

$$\pi_i = n/N\,, \qquad 1 \leq i \leq N\,, \tag{4.1}$$

$$\pi_{ij} = n(n-1)/N(N-1)\,, \qquad 1 \leq i \neq j \leq N\,. \tag{4.2}$$

Proof. We have, by definition,

$$\pi_i = \sum_{s \ni i} P(s) = \binom{N}{n}^{-1} [\text{\# of samples containing the unit i}]$$
$$= \binom{N}{n}^{-1}\binom{N-1}{n-1} = n/N .$$

Also,

$$\pi_{ij} = \sum_{s \ni i,j} P(s) = \binom{N}{n}^{-1} [\text{\# of samples containing the units } i \text{ and } j]$$
$$= \binom{N}{n}^{-1}\binom{N-2}{n-2} = n(n-1)/N(N-1) .$$

Now we state the following results: □

Theorem 4.4. *Under* SRS(N, n) *design, the sample mean*

$$\bar{\mathbf{y}} = \sum_{i \in s} Y_i/n \tag{4.3}$$

and the sample variance

$$\mathbf{s}_Y^2 = \sum_{i \in s} (Y_i - \bar{\mathbf{y}})^2/(n-1) \tag{4.4}$$

are unbiased estimators of the population mean $\bar{\mathbf{Y}}$ *and the population variance* $\mathbf{S}_Y^2$, *respectively. Further, the variance of* $\bar{\mathbf{y}}$ *and an unbiased estimator of this variance are respectively given by*

$$V(\bar{\mathbf{y}}) = (n^{-1} - N^{-1})\mathbf{S}_Y^2 \tag{4.5}$$

and

$$\hat{V}(\bar{\mathbf{y}}) = (n^{-1} - N^{-1})\mathbf{s}_Y^2 . \tag{4.6}$$

Proof. The proof is based on (3.1), (3.4), and (3.18) which refer, respectively, to the expressions for the HTE, its variance $V_{\mathrm{SYG}}(\mathrm{HTE})$, and unbiased estimator $\hat{V}_{\mathrm{SYG}}(\mathrm{HTE})$ of the variance, for estimating a finite population total based on a fixed size sampling design. We suitably modify these expressions for estimating a finite population mean and substitute the values of π_i's and π_{ij}'s from (4.1) and (4.2). We deduce the following results:

$$\hat{\bar{\mathbf{Y}}} = N^{-1} \sum_{i \in s} Y_i/\pi_i = n^{-1} \sum_{i \in s} Y_i = \text{sample mean} = \bar{\mathbf{y}} ,$$

$$V(\bar{\mathbf{y}}) = N^{-2} \sum_{i<j}\sum (\pi_i\pi_j - \pi_{ij})\{(Y_i/\pi_i) - (Y_j/\pi_j)\}^2$$
$$= N^{-2}(N^2/n^2)[(n^2/N^2) - \{n(n-1)/N(N-1)\}] \sum_{i<j}\sum (Y_i - Y_j)^2$$

$$= N^{-2}(N-n)n^{-1}(N-1)^{-1} \sum_{i<j}\sum (Y_i - Y_j)^2$$

$$= N^{-1}(N-n)n^{-1}\left\{\sum_{i=1}^{N} (Y_i - \bar{\mathbf{Y}})^2/(N-1)\right\}$$

$$= N^{-1}(N-n)n^{-1}\mathbf{S}_Y^2 = (n^{-1} - N^{-1})\mathbf{S}_Y^2 .$$

Similarly, we can show that

$$\hat{V}(\bar{\mathbf{y}}) = N^{-1}(N-n)n^{-1}\left\{\sum_{i\in s} (Y_i - \bar{\mathbf{y}})^2/(n-1)\right\} = N^{-1}(N-n)n^{-1}\mathbf{s}_Y^2$$

$$= (n^{-1} - N^{-1})\mathbf{s}_Y^2 = n^{-1}\mathbf{s}_Y^2(1 - nN^{-1}) .$$

The factor $(1 - nN^{-1})$ is known as the finite population correction (fpc). Since $E[\hat{V}(\bar{\mathbf{y}})] = V(\bar{\mathbf{y}})$, it is also clear that $E(\mathbf{s}_Y^2) = \mathbf{S}_Y^2$. □

The standard error (s.e.) of the estimate of the mean $\bar{\mathbf{Y}}$ is given by $(n^{-1} - N^{-1})^{1/2}\mathbf{S}_Y$. As an approximation, we may write

$$V(\bar{\mathbf{y}}) \simeq \mathbf{S}_Y^2/n \qquad \text{and} \qquad \hat{V}(\bar{\mathbf{y}}) \simeq \mathbf{s}_Y^2/n .$$

This works out fairly well when the sampling fraction n/N is small (usually less than 5%).

We have noted above that for SRS(N, n) designs with $n \geq 2$, the sample variance $\mathbf{s}_Y^2$ serves as an unbiased estimator of the population variance $\mathbf{S}_Y^2$. Thus we have a nonnegative estimator of the population variance as well. Another point to be noted is that the above simple forms of estimators are based primarily on the fact that π_i's and π_{ij}'s are the same for all units and for all pairs of units, respectively. In other words, whenever a fixed-size (n) sampling design has constant first- and second-order inclusion probabilities, we obtain $\bar{\mathbf{y}}$ and $\mathbf{s}_Y^2$ as unbiased estimators for $\bar{\mathbf{Y}}$ and $\mathbf{S}_Y^2$, respectively. As we shall see in Section 4.9, there are sampling designs other than SRS(N, n) designs having this feature for the first- and second-order inclusion probabilities. It may further be added that for SRS(N, n) designs, $\hat{V}_{\mathrm{HT}}$(HTE) coincides with $\hat{V}_{\mathrm{SYG}}$(HTE). (Exercise 4.2).

Example 4.1. It is required to estimate the total volume of timber in a forest. There are a total of 97 trees numbered serially as $1, 2, \ldots, 97$. An SRS of 20 trees is located and the volume of timber (denoted as V) in each of these trees is subsequently determined. The following data are available:

ΣV_i = sum of the observations = 45,000 c.ft.,
ΣV_i^2 = sum of squares of the observations = 1.2×10^8 (c.ft.)^2.

Denote by $T(\mathbf{V})$ the total volume of timber in the forest. Then we might use $\hat{T}(\mathbf{V}) = N\bar{\mathbf{v}} = 218250$ c.ft. The estimated s.e. of this estimate is given by $N\{n^{-1} - N^{-1}\}^{1/2}\mathbf{s}_v$ where $\mathbf{s}_v$ is the sample s.d. with divisor $(n-1)$. We have $\mathbf{s}_v^2 = 0.99337 \times 10^8$ (c.ft.)^2 so that estimated s.e. $= 1.9176 \times 10^4$ c.ft.

Remark 4.1. If it is believed that the population contains a few units with large or extreme values, then a simple random sample mean may deliberately overestimate or underestimate the true population mean, depending on whether or not the sample contains some of those units. In such cases the sample mean may be suitably modified to deal with the situation. We refer to Hidiroglou and Srinath (1981) for a discussion on this topic. Other related references are also to be found there.

4.5. ESTIMATION OF A POPULATION PROPORTION

In some enquiries the survey variable may be qualitative in nature. In such a case our interest may focus on possession of the attribute and the parameter to be estimated may naturally be ϕ, the true proportion of population units possessing the attribute. We may relate ϕ to $\bar{\mathbf{Y}}$ by defining another survey variable Y for which

$$\begin{aligned} Y_i &= 1 \quad \text{if the } i\text{th unit possesses the attribute,} \\ &= 0 \quad \text{if the } i\text{th unit does not possess the attribute.} \end{aligned}$$

We may then use the results of Theorem 4.4 to deduce the following results. The reader may easily derive the proofs.

Theorem 4.5. *Under* SRS(N, n) *design, based on the HTE, an unbiased estimator of the population proportion ϕ is given by the sample proportion $\hat{\phi}$ of units possessing the attribute. Further,*

$$V(\hat{\phi}) = (n^{-1} - N^{-1})\{N\phi(1-\phi)/(N-1)\} \tag{4.7}$$

$$\hat{V}(\hat{\phi}) = (n^{-1} - N^{-1})\{n\hat{\phi}(1-\hat{\phi})/(n-1)\}\,. \tag{4.8}$$

As approximations to (4.7) and (4.8) which work well when the population size N is large and the sampling fraction is small,

$$V(\hat{\phi}) \simeq \phi(1-\phi)/n\,,$$
$$\hat{V}(\hat{\phi}) \simeq \hat{\phi}(1-\hat{\phi})/(n-1)\,.$$

Example 4.2. We wish to estimate the proportion of households in Oak Park, IL, living in apartments/houses which are rented as opposed to being owned. Suppose a complete list of the households is available and it shows a total of 1700 households. A random sample of 50 households is selected and the following information obtained: rented, 27; owned, 23. Then the sample proportion of rented households is 0.54 and this serves as an estimate of the population proportion with estimated s.e. (approx.) given by $\{\hat{\phi}(1-\hat{\phi})/(n-1)\}^{1/2} = 0.0712$. The adequacy of the sample size will be judged later immediately following Example 4.3.

4.6. DETERMINATION OF SAMPLE SIZE

The expression (4.5) suggests that the variance of the estimate $\bar{\mathbf{y}}$ of the population mean $\bar{\mathbf{Y}}$ in SRS(N, n) design is a decreasing function of the sample size n. As the reciprocal of the variance is generally regarded as a measure of precision of an estimate, it turns out that the precision will be increased by increasing the sample size. However, an increase in the sample size inevitably increases the cost of operation of the survey. In a survey the cost incurred is usually composed of an overhead cost C_0 plus the cost of data collection and analysis. This latter component depends on the sample size and is usually taken as being directly proportional to the sample size. In other words, the total cost incurred in a survey involving a sample of size n is represented by the *linear cost structure* $C(n) = C_0 + cn$.

It is clear that we have to strike a balance between $C(n)$ and $V(\bar{\mathbf{y}})$ in (4.5). Usually our choice of sample size is decided by a budgetary constraint. Thus, if B is the given budget, then the sample size n is taken as the integral part of $(B - C_0)/c$. The precision attained by using this sample size is clearly given by the reciprocal of (4.5) with n as given above. On the other hand, sometimes the sample size has to be determined in such a way that the precision of the estimate attains a desired level with the minimum budget. Thus, if V_0 is the stated upper limit to the variance of an estimate, then the sample size n_0 must be so determined that

$$(n_0^{-1} - N^{-1})\mathbf{S}_Y^2 \le V_0 < ((n_0 - 1)^{-1} - N^{-1})\mathbf{S}_Y^2 . \tag{4.9}$$

The optimal sample size n_0 satisfying (4.9) cannot be obtained unless, of course, we assume that $V_0/\mathbf{S}_Y^2$ is a known quantity. However, we must note that this assumption is seldom met in practice. Thus an exact solution to n_0 is impossible to arrive at in most situations. The following approximate solution is suggested in the literature. We take a preliminary simple random sample of size n_1 and estimate $\mathbf{S}_Y^2$ by $\mathbf{s}_1^2$, which is analogous to $\mathbf{s}_Y^2$ in (4.4) but based on the preliminary sample with divisor $(n_1 - 1)$. Then the overall sample size n_0 is determined from the approximate formula

$$n_0 = \mathbf{s}_1^2\left(1 + \frac{2}{n_1}\right)\Big/ V_0 . \tag{4.10}$$

A second sample of size $(n_0 - n_1)$ is selected from the remaining $(N - n_1)$ units in the population, using simple random sampling procedure. As to the choice of n_1, there is no hard and fast rule. We have to be guided by the extent and nature of variation of the survey variables in the population. After the second sample is selected, the two samples are combined together to form an ultimate sample of size n_0. Then the estimate of $\bar{Y}$ is taken as the mean of all the n_0 observations in the combined sample.

We now turn to the problem of determining the sample size for estimating a population proportion ϕ. With a budgetary constraint, the sample size is determined as before. On the other hand, with a given bound V_0 to the variance of the estimate $\hat{\phi}$, the sample size n_0 has to satisfy

$$\begin{aligned}(n_0^{-1} - N^{-1})N\phi(1-\phi)(N-1)^{-1} &\leq V_0 \\ &< ((n_0 - 1)^{-1} - N^{-1})N\phi(1-\phi)(N-1)^{-1} .\end{aligned}$$

Unlike the other situation, here we can bound the quantity $\phi(1-\phi)$ by 0.25. Hence, an upper bound to n_0 is obtained by using the equation

$$(n_0^{-1} - N^{-1})N/4(N-1) = V_0 . \tag{4.11}$$

When N is large, this yields, ignoring the fpc,

$$n_0 = (4V_0)^{-1} . \tag{4.12}$$

In addition, the knowledge of an upper limit to ϕ in case $\phi < 0.5$ or of a lower limit to ϕ if $\phi > 0.5$ may be utilized to reduce the sample size from the above upper bound to a more reasonable number.

Again the sample size may have to be determined so that the sample estimate $\hat{\phi}$ of the true population proportion ϕ does *not* differ by more than a stated *margin of error d* in more than $100\alpha\%$ of the cases. This means that we have to set up a probability inequality of the form

$$\Pr\{|\hat{\phi} - \phi| \geq d\} \leq \alpha \tag{4.13}$$

and solve for n for given d and α. It is well known that the distribution of $n\hat{\phi}$ is hypergeometric. Using the normal approximation, we may take

$$\{\hat{\phi} - \phi\}/\{\phi(1-\phi)(n^{-1} - N^{-1})\}^{1/2}$$

as a standard normal deviate. Using $z_{\alpha/2}$ for the upper $100\alpha\%$ point of a standard normal distribution, we may derive, from (4.13), the estimating equation

$$d = z_{\alpha/2}\{\phi(1-\phi)(n^{-1} - N^{-1})\}^{1/2} , \tag{4.14}$$

which gives

$$n = n_0/\{1 + (n_0/N)\}\,, \qquad n_0 = d^{-2} z^2_{\alpha/2} \phi(1 - \phi)\,. \tag{4.15}$$

Again, a knowledge as to a suitable upper or lower bound to ϕ is helpful. Otherwise, we use $n_0 = z^2_{\alpha/2}/4d^2$ and evaluate the value of n.

See Exercise 4.4 for a similar result relating to estimation of a finite population mean with bounded margin of relative error.

Example 4.3. Suppose, in the setup of Example 4.2, it is known in advance that at most 60% of the households reside in apartments/houses which are rented. We wish to determine the minimum sample size (necessary under SRS) to ensure that our estimate of the true population proportion lies within a limit of 0.10 with confidence coefficient of 0.95. In using (4.15), we substitute 0.50 for ϕ and obtain $n_0 = d^{-2} z^2_{\alpha/2}(0.5)^2 = 96$ and hence, $n = n_0/(1 + n_0/N) = 91$. Thus a random sample of 91 households would be adequate for this purpose.

We see that a sample of size 50 is inadequate to ensure the stated margin of error. We can determine the margin of error in the estimate provided by the data based on the sample of size 50 by using the formula (4.14). We take $n = 50$, $z_{\alpha/2} = 1.96$, $N = 1700$, and $\phi = 0.54$ (the estimate obtained through our data). Calculations yield $d = 0.138$. Using $\phi = 0.5$, we would obtain $d = 0.139$. This shows the extent of difference in the margin of error is about 0.001. If this could be tolerated, additional observations need not be taken; otherwise, we would take an additional 41 observations using an SRS design on the remaining units of the population.

4.7. INFERENCE ON ORDERED VALUES FROM SRS DESIGNS

Besides the population mean, variance, and the proportion, there are other parameters of the finite population which might be of interest. Some such examples are the population median, range, quantiles, or percentiles all of which are functions of the ordered observations in the population. Unfortunately, very little research has been done in this direction and, as yet, we do not have satisfactory results available. Specializing on simple random sampling, however, some studies have been made for possible inference on the ordered observations. In this section we shall present some such results. The reader may consult Konijn (1973) in this context. Other useful references are Sedransk (1969), Sedransk and Meyer (1978), Sedransk and Smith (1988), Meyer (1987), and Francisco and Fuller (1991).

For the sake of simplicity of presentation, we shall assume that the values $Y_1, \ldots, Y_N$ of the survey variable are all distinct and we will denote by $Y_{(1)} < \cdots < Y_{(N)}$ the ordered values. Further, it will be assumed that the labels $1, \ldots, N$ contain no information as to the ordering of the values $Y_1, \ldots, Y_N$. For example, it is not necessarily true that $Y_i < Y_j$ for $i < j$. In such a situation, it is not clear how the information on labels of units included in a sample would be useful for inference problems relating to ordered values $Y_{(1)}, \ldots, Y_{(N)}$. We may, therefore, ignore the labels after collection of data and, instead, denote the sample observations in the ordered from by $y_{(1)} < \cdots < y_{(n)}$. It is true that these observations represent a random collection of the ordered values in the population. However, the correspondence between r and k for which $Y_{(r)} = y_{(k)}$ is assumed to be unknown for any pair (r, k). Under such a setup, we intend to present some results relating to inference on parameters of the type $Y_{(a)}$, $Y_{(b)} - Y_{(a)}$, $[Y_{(a)}, Y_{(b)}]$ for specified values of a, b, $1 \le a < b \le N$. We start with the following definition.

Definition 4.1. If I_1 and I_2 are two intervals on the real line with $I_1 \subset I_2$, then I_2 is said to be an outer interval for I_1 and I_1 is said to be an inner interval for I_2.

Suppose we want to estimate $Y_{(a)}$ based on simple random sample data $y_{(1)}, y_{(2)}, \ldots, y_{(n)}$. It is clear that $Y_{(1)} \le y_{(1)} < y_{(n)} \le Y_{(N)}$ so that no unbiased estimator of $Y_{(1)}$ or of $Y_{(N)}$ exists. As regards $Y_{(a)}$ for $1 < a < N$, it is equally true that no unbiased estimator based on $y_{(1)}, y_{(2)}, \ldots, y_{(n)}$ exists. We may, therefore, try to provide suitable one-sided or two-sided confidence intervals for $Y_{(a)}$ with a specified confidence coefficient close to $(1 - \alpha)$.

Theorem 4.6. *Under* SRS(N, n) *and strict ordering of Y values, an outer confidence interval for $Y_{(a)}$ with confidence coefficient at least as much as (and close to)* $(1 - \alpha)$ *is provided by* $[y_{(r)}, y_{(s)}]$ *where r and s satisfy*

$$\left\{\sum_{t=r}^{s-1} \binom{a}{t}\binom{N-a}{n-t} + \binom{a-1}{s-1}\binom{N-a}{n-s}\right\} \Big/ \binom{N}{n} \ge 1 - \alpha\,, \tag{4.16}$$

$$\left\{\sum_{t=r+1}^{s-1} \binom{a}{t}\binom{N-a}{n-t} + \binom{a-1}{s-1}\binom{N-a}{n-s}\right\} \Big/ \binom{N}{n} < 1 - \alpha\,, \tag{4.17}$$

$$\left\{\sum_{t=r}^{s-2} \binom{a}{t}\binom{N-a}{n-t} + \binom{a-1}{s-2}\binom{N-a}{n-s+1}\right\} \Big/ \binom{N}{n} < 1 - \alpha\,. \tag{4.18}$$

Proof. Clearly, our choice of (r, s) must lead to $P\{y_{(r)} \le Y_{(a)} \le y_{(s)}\} \ge 1 - \alpha$ while $P\{y_{(r+1)} \le Y_{(a)} \le y_{(s)}\} < 1 - \alpha$ as well as $P\{y_{(r)} \le Y_{(a)} \le y_{(s-1)}\} < 1 - \alpha$. To evaluate $P\{y_{(r)} \le Y_{(a)} \le y_{(s)}\}$, for example, we set $y_{(t)} \le Y_{(a)}$ and $y_{(t+1)} \ge Y_{(a+1)}$ for some t, $r \le t \le s - 1$. This means we must have t sample units chosen out of $\{Y_{(1)}, Y_{(2)}, \ldots, Y_{(a)}\}$ and $(n - t)$ sample units chosen out of $\{Y_{(a+1)}, \ldots, Y_{(N)}\}$. The probability of this last event, under simple random sampling, is given by $\binom{a}{t}\binom{N-a}{n-t}/\binom{N}{n}$, based on the hypergeometric probability formula. There is still another possibility left out. This corresponds to $y_{(s)} = Y_{(a)}$ for which the probability is $\binom{a-1}{s-1}\binom{N-a}{n-s}/\binom{N}{n}$. The rest is now clear. □

Next, suppose we are interested in the specified population quantile interval $[Y_{(a)}, Y_{(b)}]$ and we want to provide inner or outer confidence interval estimate for it. We may employ a sample interval $[y_{(r)}, y_{(s)}]$ for this purpose. This time the choice of (r, s) will be different. We state and prove the following result for an outer confidence interval. A similar result for an inner interval can be established without any difficulty.

Theorem 4.7. *Under* SRS(N, n) *and strict ordering of the Y values, the sample interval* $[y_{(r)}, y_{(s)}]$ *provides an outer confidence interval for the specified population quantile interval* $[Y_{(a)}, Y_{(b)}]$ *with confidence coefficient at least as much as* (*and close to*) $(1-\alpha)$ *provided r and s satisfy*

$$\sum\sum_{r\le u<v\le s-1} \binom{a}{u}\binom{N-b+1}{n-v+1}\binom{b-a-1}{v-u-1}\Big/\binom{N}{n} \ge (1-\alpha)\,, \quad (4.19)$$

$$\sum\sum_{r+1\le u<v\le s-1} \binom{a}{u}\binom{N-b+1}{n-v+1}\binom{b-a-1}{v-u-1}\Big/\binom{N}{n} < (1-\alpha)\,, \quad (4.20)$$

$$\sum\sum_{r\le u<v\le s-2} \binom{a}{u}\binom{N-b+1}{n-v+1}\binom{b-a-1}{v-u-1}\Big/\binom{N}{n} < (1-\alpha)\,. \quad (4.21)$$

Proof. Clearly, our choice of r and s must be such that $P\{y_{(r)} \le Y_{(a)} < Y_{(b)} \le y_{(s)}\} \ge 1-\alpha$ while $P\{y_{(r+1)} \le Y_{(a)} < Y_{(b)} \le y_{(s)}\} < 1-\alpha$ as well as $P\{y_{(r)} \le Y_{(a)} < Y_{(b)} \le y_{(s-1)}\} < 1-\alpha$. To evaluate $P\{y_{(r)} \le Y_{(a)} < Y_{(b)} \le y_{(s)}\}$, for example, we pick up the sample order statistics $y_{(v)}$ and $y_{(u)}$ satisfying $y_{(r)} \le y_{(u)} < y_{(v)} \le y_{(s)}$. Then, whenever we have $y_{(u)} \le Y_{(a)}$ and $y_{(v)} \ge Y_{(b)}$, the requirement $y_{(r)} \le Y_{(a)} < Y_{(b)} \le y_{(s)}$ is satisfied for $r \le u < v \le s-1$. Now $P\{y_{(u)} \le Y_{(a)}, y_{(v)} \ge Y_{(b)}\}$ is given by $\binom{a}{u}\binom{N-b+1}{n-v+1}\binom{b-a-1}{v-u-1}/\binom{N}{n}$, based on the hypergeometirc probability formula. The rest clearly follows. □

We might be interested in drawing suitable inference about the length of a specified quantile interval $[Y_{(a)}, Y_{(b)}]$. For this problem, no satisfactory result is known in terms of inner or outer confidence interval.

Another interesting problem would be to specify a sample interval $[y_{(r)}, y_{(s)}]$ which contains at least a specified number t of the population values with a specified coverage probability.

Theorem 4.8. *Under* SRS(N, n) *and strict ordering of the Y values, the random interval* $[y_{(r)}, y_{(s)}]$ *includes at least t population values with coverage* *probability of at least as much as (and close to)* $(1-\alpha)$ *provided r and s satisfy*

$$\sum_{l\geq t}\sum_{j=r}^{N-n+s-t+1}\binom{j-1}{r-1}\binom{l-2}{s-r-1}\binom{N-j-l+1}{n-s}\Big/\binom{N}{n}\geq 1-\alpha\,, \tag{4.22}$$

$$\sum_{l\geq t}\sum_{j=r}^{N-n+s-t}\binom{j-1}{r-1}\binom{l-2}{s-r-2}\binom{N-j-l+1}{n-s+1}\Big/\binom{N}{n}< 1-\alpha\,, \tag{4.23}$$

$$\sum_{l\geq t}\sum_{j=r+1}^{N-n+s-t+1}\binom{j-1}{r}\binom{l-2}{s-r-2}\binom{N-j-l+1}{n-s}\Big/\binom{N}{n}< 1-\alpha\,. \tag{4.24}$$

We leave the proof to the reader (Exercise 4.5).

Remark 4.2. We observe that in each of the above cases, the exact solution depends heavily on numerical computation. Some such computations are reported in Meyer (1987). Moreover, in the above, we explicitly assumed $Y_{(1)} < Y_{(2)} < \cdots < Y_{(N)}$. In case this assumption is not tenable, we may derive probability inequalities analogous to those presented here. We refer to Meyer (1987) for this and related matters.

Example 4.4. The purpose of this example is to illustrate an application of Theorem 4.8. Suppose there are 10 different population values and we want to know what sample interval in a random sample of size 4 would contain at least 5 population values with a coverage probability of at least 95%. We set $t = 5$ and $\alpha = 0.05$ and solve for r and s satisfying (4.22)–(4.24) subject to $1 \leq r < s \leq 4$. We come up with the following facts. The sample range $[y_{(1)}, y_{(4)}]$ contains at least 5 population values with confidence coefficient 97% while each of the intervals $[y_{(1)}, y_{(3)}]$ and $[y_{(2)}, y_{(4)}]$ has coverage probability only 67%. Therefore, the sample range $[y_{(1)}, y_{(4)}]$ is the required interval.

In Exercise 4.6 the reader is asked to solve similar problems.

Remark 4.3. We emphasize that because we are adopting an SRS sampling design the above computations were made possible without *any* knowledge as to the true ordering of the values of the survey variable. Without an *explicit* knowledge of the true orderings, however, we may *not* be in a position to carry out such computations for an *arbitrary* sampling design. This problem will be reconsidered in Section 4.10.

4.8. MIXTURES OF SIMPLE RANDOM SAMPLING (SRS) DESIGNS

Suppose d_1 and d_2 are two different sampling designs. Then a mixture design d is formed of d_1 and d_2 based on the mixing proportions p and q, $0 < p$,

$q<1$, $p+q=1$, in the following way: $P_d(s)=pP_{d_1}(s)+qP_{d_2}(s)$, $s\in S$. Explicitly written,

$$\begin{aligned} P_d(s) &= pP_{d_1}(s)\,, && \text{if } P_{d_2}(s)=0 \\ &= qP_{d_2}(s)\,, && \text{if } P_{d_1}(s)=0 \\ &= pP_{d_1}(s)+qP_{d_2}(s)\,, && \text{if } P_{d_1}(s)>0,\ P_{d_2}(s)>0\,. \end{aligned} \tag{4.25}$$

It is clear that d is a sampling design. When the components d_1 and d_2 are fixed size sampling designs of sizes n_1 and n_2, $n_1\neq n_2$, we get a mixture design where every sample is of size n_1 or n_2. To implement such a design in practice, it is enough to perform a Bernoulli trial with success probability p and failure probability q. Then, if success occurs, we implement d_1; otherwise, we implement d_2. Similarly, a mixture of $k(\geq 2)$ sampling designs can be defined.

A special type of mixture SRS designs is of particular interest to us. This is mainly because of its simplicity and ready applicability. Following Basu (1958), we discuss this in considerable detail below.

Fix a value of $n<N$. Consider (reduced) SRS designs of sizes $1,2,\ldots,n$. Call them $d_1,d_2,\ldots,d_n$ in that order. Let Δ stand for the usual *difference operator* defined by the relation $\Delta f(x)=f(x+1)-f(x)$. Then $\Delta^2 f(x) = \Delta[\Delta f(x)] = \Delta[f(x+1)-f(x)] = \Delta f(x+1)-\Delta f(x) = f(x+2)-2f(x+1)+f(x)$. Likewise, for an integer k, $\Delta^k f(x) = \Sigma_{j=0}^{k}(-1)^{k-j}\binom{k}{j}f(x+j)$. Taking $f(x)=x^n$ we can evaluate $\Delta^\nu x^n$ in a similar fashion. Denote by $\Delta^\nu O^n$ the value of $\Delta^\nu x^n$ evaluated at $x=0$. It can be shown that $\Delta^\nu O^n=\Sigma_{j=0}^{\nu}(-1)^j\binom{\nu}{j}(\nu-j)^n$. We will use the identity $N^n=\Sigma_{\nu=1}^{n}\binom{N}{\nu}\Delta^\nu O^n$. We now take the mixing proportions as $p_1,p_2,\ldots,p_n$ for the designs $d_1,\ldots,d_n$, respectively, where $p_\nu=N^{-n}\binom{N}{\nu}\Delta^\nu O^n$, $1\leq\nu\leq n$. The sampling design $d=\Sigma\, p_\nu d_\nu$ thus obtained will be a variable size design with sizes $n(s)=1,2,\ldots,n$ and $P(n(s)=\nu)=\binom{N}{\nu}\Delta^\nu O^n/N^n=p_\nu$, $1\leq\nu\leq n$. There are exactly $\binom{N}{\nu}$ samples of size ν in the mixture design d. A typical sample $s=(i_1,i_2,\ldots,i_\nu)$ of size ν has probability

$$P_d(s)=p_\nu/\binom{N}{\nu}=\frac{\Delta^\nu O^n}{N^n}\,,\qquad 1\leq\nu\leq n, 1\leq i_1<i_2<\cdots<i_\nu\leq N\,. \tag{4.26}$$

Recognizing that probability can be interpreted as relative frequency, we can give the following frequency-based interpretation of the sampling design d. There are all together N^n n-ples, a typical n-ple with repetitions ignored and resulting in a sample of the type $s=(i_1,i_2,\ldots,i_\nu)$ has $\Delta^\nu O^n$ copies, $1\leq i_1<i_2<\cdots<i_\nu\leq N$, $1\leq\nu\leq n$. Interpreted this way, the sampling scheme corresponding to the design d turns out to be extremely simple and readily applicable. We draw n units at random, one at a time and *with replacement*. We call the collection $(j_1,j_2,\ldots,j_n)$. Clearly, there may be some repeated units in this collection. We count each repeated unit only once and reorder the distinct units, ν in number, as $(i_1,i_2,\ldots,i_\nu)$, $1\leq i_1<$

$i_2 < \cdots < i_\nu \le N$. These then constitute a sample from the resulting sampling design d. This is because there are exactly N^n such collections of the form $(j_1, j_2, \ldots, j_n)$ of which $\Delta^\nu O^n$ involve the typical ν units $(i_1, i_2, \ldots, i_\nu)$ and this is true for all $1 \le i_1 < i_2 < \cdots < i_\nu \le N$, $1 \le \nu \le n$. Similar results hold for the case of $n \ge N$. See Basu (1958).

Remark 4.4. The particular mixture design discussed above is commonly termed in the literature a *simple random sampling with replacement* (SRSWR) sampling design. It is generally defined as a sampling design with *equal* probability of all the N^n n-ples of the type $(i_1, i_2, \ldots, i_n)$. However, this does *not* fit our definition of a sampling design (Definition 1.1) which assigns probabilities to samples consisting only of ordered or unordered *distinct* units. We, therefore, regard SRSWR sampling design as a mixture sampling design in the sense discussed above. We recognize the SRSWR scheme of sampling as a convenient and readily applicable means of generating the particular mixture design.

For the above mixture sampling design as also for a general mixture design, the first- and second-order inclusion probabilities can be immediately written down. The general formula $\pi_d(i, j, \ldots) = p\pi_{d_1}(i, j, \ldots) + q\pi_{d_2}(i, j, \ldots)$ applies. Of course, for the above mixture design, we can directly verify that $\pi_d(i) = 1 - (1 - N^{-1})^n$ and $\pi_d(i, j) = 1 - 2(1 - N^{-1})^n + (1 - 2N^{-1})^n$. Accordingly, we may further deduce, using (1.11) and (1.12), that

$$E_d[n(s)] = \sum \pi_d(i) = N\{1 - (1 - N^{-1})^n\}, \tag{4.27}$$

$$V_d[n(s)] = N\{(1 - N^{-1})^n - (1 - 2N^{-1})^n\} + N^2\{(1 - 2N^{-1})^n - (1 - N^{-1})^{2n}\}. \tag{4.28}$$

Further, it can be shown that (Exercise 4.7)

$$E\{[n(s)]^{-1}\} = \sum_{j=1}^{N} j^{n-1}/N^n. \tag{4.29}$$

For inference on the population total or mean based on mixture designs, the general approach would be to use the HTE again. However, for a mixture design, the computations of HTE and its variance estimator are complicated in many cases. Also the nonnegative variance estimation problem is not easily resolved. Thus, alternative estimators might be sought. For the particular mixture d of SRS designs under consideration, one such alternative estimator assumes a very simple form and this will be discussed below in detail.

Recall Theorem 4.4 in this context. We know that for the sampling design d_ν, $E(\bar{\mathbf{y}}_\nu) = \bar{\mathbf{Y}}$ and $V(\bar{\mathbf{y}}_\nu) = (\nu^{-1} - N^{-1})\mathbf{S}_Y^2$ where $\bar{\mathbf{y}}_\nu$ is the mean of the ν

observations obtained under d_ν. Suppose for the mixture design d as well we decide to use the sample mean $\bar{\mathbf{y}}_{n(s)}$ of the $n(s)$ observations available. Then, since $d = \Sigma\, p_\nu d_\nu$,

$$E(\bar{\mathbf{y}}_{n(s)}) = E_1 E_2(\bar{\mathbf{y}}_{n(s)}) = \Sigma\, p_\nu E(\bar{\mathbf{y}}_\nu) = \bar{\mathbf{Y}} \tag{4.30}$$

$$\begin{aligned} V(\bar{\mathbf{y}}_{n(s)}) &= V_1\{E_2(\bar{\mathbf{y}}_{n(s)})\} + E_1\{V_2(\bar{\mathbf{y}}_{n(s)})\} \\ &= E_1\{V_2(\bar{\mathbf{y}}_{n(s)})\} \qquad \text{since } E_2(\bar{\mathbf{y}}_\nu) = \bar{\mathbf{Y}} \text{ for all } \nu \\ &= \Sigma\, p_\nu(\nu^{-1} - N^{-1})\mathbf{S}_Y^2 = \{E(n(s))^{-1} - N^{-1}\}\mathbf{S}_Y^2 . \end{aligned} \tag{4.31}$$

Further, for $\nu \geq 2$, $E(\mathbf{s}_\nu^2) = \mathbf{S}_Y^2$ where $\mathbf{s}_\nu^2 = (\nu - 1)^{-1} \Sigma_{i \in s} (Y_i - \bar{\mathbf{y}}_\nu)^2$ is the variance of the sample observations with divisor $(\nu - 1)$. This follows from Theorem 4.4. Thus we may propose

$$\hat{\mathbf{S}}_Y^2 = \begin{cases} 0 & \text{if } \nu = 1 \\ \mathbf{s}_\nu^2[1 - N^{-n+1}]^{-1} , & \text{if } \nu \geq 2 \end{cases} \tag{4.32}$$

as an unbiased estimator of $\mathbf{S}_Y^2$. Further, it is nonnegative. Thus,

$$\hat{V}(\bar{\mathbf{y}}_{n(s)}) = \{E(n(s))^{-1} - N^{-1}\}\hat{\mathbf{S}}_Y^2 \tag{4.33}$$

can be used to provide an estimate of the variance of the estimate of the population mean. Other alternative estimators are also available (Exercise 4.8).

It is clear that $\nu \leq n$ so that $E(\nu^{-1}) \geq n^{-1}$. Hence, use of the mixture design d yields an estimator $\bar{\mathbf{y}}_{n(s)}$ which performs no better than $\bar{\mathbf{y}}_n$, the sample mean based on observations under d_n alone. In other words, the component d_n used alone would provide a uniformly better estimator than the use of *any* combination of $d_1, \ldots, d_n$. However, it must be realized that this sort of comparison is *not* totally fair as the two competitors do not entail the same cost of the survey. It is clear that d_n is based on a much higher cost than the mixture design d as generally the design d would result in a sample of size less than n (i.e., a sample consisting of less than n *distinct* units). Basu (1958) has discussed this aspect of comparison of the two designs d and d_n.

In the literature, the sample mean based on all the units $j_1, \ldots, j_n$ in the collection under simple random sampling *with* replacement has also been considered and studied; see Cochran (1977). It has been demonstrated, as expected, that such an estimator in *inadmissible* and $\bar{\mathbf{y}}_{n(s)}$ is uniformly better. See Basu (1958), Korwar and Serfling (1970) and Asok (1980) for a comparison of the two estimators. For further results in this direction, we refer to Pathak (1962, 1988) and to Sinha and Sen (1989).

Example 4.5. The purpose of this example is to illustrate the use of the formulae presented above. Consider a population of 20 units and take $n = 5$.

Then the mixing proportions $p_\nu = \binom{20}{\nu}\Delta^\nu O^5/20^5$, $\nu = 1, 2, \ldots, 5$ assume the values (up to four decimal places) 0, 0.0018, 0.0534, 0.3634, and 0.5814, respectively. This yields $E((n(s))^{-1}) = 0.22583$. Also, using (4.29), we would obtain $E(\nu^{-1}) = \Sigma\, j^{n-1}/N^n = 0.22583$. Again, suppose we come up with the collection (12, 2, 17, 5, 12) following the application of the scheme of sampling so that the actual sample drawn is $s = (2, 5, 12, 17)$ with $n(s) = 4$. Assume for example, $Y_2 = 12$, $Y_5 = 25$, $Y_{12} = 17$, and $Y_{17} = 20$ so that $\bar{\mathbf{y}}_{n(s)} = 18.5$ and $\mathbf{s}^2_{n(s)} = 29.6667$. Then, according to (4.30), (4.32), and (4.33), $\hat{\bar{\mathbf{Y}}} = \bar{\mathbf{y}}_{n(s)} = 18.5$, $\hat{\mathbf{S}}^2_Y = \mathbf{s}^2_\nu(1 - N^{-n+1})^{-1} = 29.6669$, and $\hat{V}(\hat{\bar{\mathbf{Y}}}) = \{E(n(s))^{-1} - N^{-1}\}\, \hat{\mathbf{S}}^2_Y = 5.2163$. Therefore, an estimate of the population mean is given by 18.5 along with the estimated s.e of 2.28.

4.9. SOME VARIANTS OF SRS DESIGNS

So far as the use of the HTE is concerned, it is evident that the properties of the HTE depend solely on the π_i's and π_{ij}'s. It is interesting to investigate the extent to which these inclusion probabilities characterize the sampling designs. For example, $\pi_i = n/N$ and $\pi_{ij} = n(n-1)/N(N-1)$ for all $1 \le i < j \le N$, if the sampling design is an SRS(N, n). We might ask if SRS(N, n) is the only sampling design with the above stated values for the first- and second-order inclusion probabilities. The answer is in the negative unless $n = 2$. Indeed sampling designs having this feature exist which are not uniform on all samples of size n.

To formalize the concepts and problems in a general setup, we introduce the following definition.

Definition 4.2. Two sampling designs d_1 and d_2 based on the same population of size N are said to be *equivalent with respect to the first-order and second-order inclusion probabilities* if

$$\pi_{d_1}(i) = \pi_{d_2}(i) \quad \text{and} \quad \pi_{d_1}(i, j) = \pi_{d_2}(i, j) \quad \text{for all } 1 \le i < j \le N. \tag{4.34}$$

Hereafter, for simplicity, we may say that the two sampling designs d_1 and d_2 are *equivalent* (written as $d_1 \approx d_2$) if they are equivalent in the sense of Definition 4.2. We note that the condition $\pi_{d_1}(i) = \pi_{d_2}(i)$ for all $1 \le i \le N$ along with the relation $E(n(s)) = \Sigma\, \pi_i$ implies that the expected costs (under a linear cost structure) should be equal if $d_1 \approx d_2$, a natural demand for the concept to be practically meaningful.

We recall that the support of an SRS(N, n) design in its reduced form consists of all $\binom{N}{n}$ samples of size n from the N units in the population. Also, the probability distribution over the support is uniform, that is, $P(s) = \binom{N}{n}^{-1}$ for all samples in the support. The following problems have been raised and satisfactorily resolved. Is there any fixed size (n) sampling design which is equivalent to an SRS(N, n) design having one of the following possibilities:

(i) its support size is fewer than $\binom{N}{n}$ but the probability distribution over the support is uniform;

(ii) its support size is less than $\binom{N}{n}$ and the probability distribution over the support is not uniform;

(iii) the support size is $\binom{N}{n}$ but the probability distribution over the support is not uniform?

The answers to both the problems (ii) and (iii) are in the affirmative. However, there are values of N and n for which the answer to (i) may be in the negative. Before we state some general results, we present an illustrative example.

Example 4.6. Let $N = 6$, $n = 3$. Each of the sampling designs in Table 4.1 is equivalent to SRS(6, 3) which has support $\binom{6}{3} = 20$. The first one is an example of (i), the second one is an example of (ii), while the third one is an example of (iii).

Table 4.1. Sampling Designs Equivalent to SRS(6, 3) Design

d_1: $P((1,2,5)) = P((1,2,6)) = P((1,3,4)) = P((1,3,5)) = P((1,4,6))$
$= P((2,3,4)) = P((2,3,6)) = P((2,4,5)) = P((3,5,6)) = P((4,5,6)) = 0.1$

d_2: $P((1,2,3)) = P((1,2,4)) = P((1,2,5)) = P((1,3,6)) = P((1,4,6))$
$= P((1,5,6)) = P((2,3,6)) = P((2,4,6)) = P((2,5,6)) = 1/30$; $P((1,3,4))$
$= P((1,3,5)) = P((1,4,5)) = P((2,3,4)) = P((2,3,5)) = P((2,4,5))$
$= P((3,4,6)) = P((3,5,6)) = P((4,5,6)) = 2/30$; $P((1,2,6)) = 3/30$

d_3: $P((1,2,3)) = P((1,2,4)) = P((1,3,6)) = P((1,4,5)) = P((1,5,6))$
$= P((2,3,5)) = P((2,4,6)) = P((2,5,6)) = P((3,4,5)) = P((3,4,6)) = 1/30$
$P((1,2,5)) = P((1,2,6)) = P((1,3,4)) = P((1,3,5)) = P((1,4,6))$
$= P((2,3,4)) = P((2,3,6)) = P((2,4,5)) = P((3,5,6)) = P((4,5,6)) = 2/30$

For each of these sampling designs we can readily verify that $\pi_i = 3/6$ and $\pi_{ij} = 6/30$ for all $1 \leq i < j \leq 6$.

There are pairs (N, n) for which there is no sampling design equivalent to SRS(N, n) with the support size smaller than $\binom{N}{n}$ and with uniform distribution of probability over the support. One such example is the pair (8, 3). The proof for this particular case is not difficult and is left as an exercise (Exercise 4.9).

An important problem in this area is the following. Suppose we would like to construct a sampling design d^* equivalent to SRS(N, n) and with minimum support size. What can be said a priori about the support size of

d^*? Since $\pi_{ij} > 0$ for all i, j, every pair of units must show up in some samples. There are altogether $\binom{N}{2}$ pairs and each sample of size n yields $\binom{n}{2}$ pairs. Hence we need at least $\binom{N}{2}/\binom{n}{2}$ samples in the support. This essentially yields the lower bound in the following inequality

$$\{(N/n)\{(N-1)/(n-1)\}\} \leq \text{support size of } d^* \leq \binom{N}{2} \tag{4.35}$$

where $\{x\}$ denotes the integer part of the real number x. Arguments leading to the upper bound $\binom{N}{2}$ are more technical and we suggest the interested reader consults Foody and Hedayat (1977), Wynn (1977), and Hedayat (1979).

The following results have also been discussed in the literature.

Theorem 4.9. *For each pair (N, n) there is a sampling design equivalent to* SRS(N, n) *design whose support size is less than $\binom{N}{n}$ and the probability distribution over the support is not uniform.*

Theorem 4.10. *For each pair (N, n) there is a sampling design equivalent to* SRS(N, n) *design whose support size is $\binom{N}{n}$ but the probability distribution over the support is not uniform.*

Examples of sampling designs satisfying the conditions of both the above theorems were exhibited earlier in the Example 4.6. We omit the proofs of the above theorems. The interested reader may consult Sinha (1976), Foody and Hedayat (1977), Hedayat (1979), and Sengupta (1979). Apart from their theoretical interest, sampling designs which are equivalent to SRS designs have useful practical applications in controlled survey sampling. To save cost and time, we may want certain combinations of units to be excluded from the support or to have small probability attributed to them. We may wish to put as much probability as possible on those samples (i.e., combinations of units) which, for some reason or other, are considered to be highly preferred. At the same time, we want to enjoy the aspect of simplicity of data analysis as evidenced by simple random sampling. This has put forward the concept of controlled sampling. Chakrabarti (1963), Avadhani and Sukhatme (1973), Foody and Hedayat (1977), Wynn (1977), Hedayat (1979), and Sengupta (1982) discuss problems related to controlled sampling. We will not get into these details. Instead, we will present two examples.

Example 4.7. Suppose we want a sampling design based on samples of size 3 for a population of size 8 such that the design is equivalent to an SRS(8, 3) design but has the added property that a particular sample, e.g., (1, 2, 3) has the maximum chance of being selected while any sample containing 2 of these units has zero chance of selection. The following sampling design serves this purpose. $P((1, 2, 3)) = 6/56$,

$P((2,5,8)) = P((3,7,8)) = 5/56$, $P((2,6,7)) = P((3,4,5)) = 4/56$, $P((1,4,7)) = P((1,4,8)) = P((1,5,6)) = P((1,5,7)) = P((1,6,8)) = 3/56$, $P((2,4,6)) = P((2,4,7)) = P((3,4,6)) = P((4,6,8)) = P((3,5,6)) = 2/56$, $P((2,4,5)) = P((2,4,8)) = P((4,5,8)) = P((3,6,7)) = P((3,6,8)) = P((5,6,7)) = P((5,7,8)) = 1/56$.

Example 4.8. Suppose there are 16 units in a population arranged in the form of a 4×4 array as shown in Table 4.2. We want a sampling design of size 4 which will be equivalent to SRS(16, 4) design but will have *no* sample involving more than 2 units from *any* row or column.

Table 4.2

1	2	3	4
5	6	7	8
9	10	11	12
13	14	15	16

The following scheme of sampling results in such a sampling design.

(i) Choose independently one row at random and one column at random. The union of the two contains 7 elements. Delete them and consider the resulting configuration which is a 3×3 array of the form shown in Table 4.3.

Table 4.3

a	b	c
d	e	f
g	h	i

(ii) Delete all the 3 units in one transversal chosen at random in the array in Table 4.3. Note that there are altogether six transversals, namely (a, e, i), (g, e, c), (a, f, h), (d, b, i), (g, b, f), and (d, h, c).

(iii) Choose any 4 of the remaining 6 units via SRS(6, 4). Declare these units as forming the ultimate sample.

We leave the verification to the reader (Exercise 4.16).

In the context of controlled sampling, the manipulation of probabilities over possible samples can often be translated in the language of linear and mathematical programming as originally suggested by Foody and Hedayat (1977). Thus, we can lexicographically order the samples and attach an unknown probability, say x_i, to the ith sample. If we can translate our sampling requirements in the form of minimization (or maximization) of an objective function of the x's subject to a set of constraints, then we are ready to use one of the available softwares such as LINDO or OSL and

Table 4.4

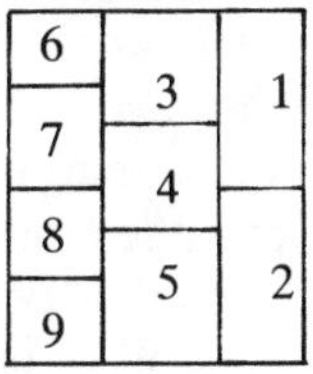

obtain a solution (if it exists) to our survey problem. Let us demonstrate this idea by an example. Suppose we want to choose three states from the nine neighboring states: Wisconsin (1), Illinois (2), Minnesota (3), Iowa (4), Missouri (5), North Dakota (6), South Dakota (7), Nebraska (8), and Kansas (9). We desire to build a survey design which minimizes the total probability over those states which are mutually adjacent. Further, we would like to have the first and the second ordered inclusion probabilities be identical to those in SRS(9, 3). If we look at the map of the United States, we see that these nine states can be depicted as in Table 4.4, and thus we desire to minimize the total probability over 8 samples

$$(1,2,4), (1,3,4), (2,4,5), (3,4,7), (3,6,7), (4,5,8), (4,7,8) \text{ and } (5,8,9)\,.$$

We used LINDO, and in a fraction of a second the computer produced two survey designs. In one design 72 samples including the above 8 samples were excluded from the support. Further, the design is uniform on its support. In another design 62 samples including the above 8 samples were excluded from the support. However, the design is not uniform on its support. In both designs $\pi_i = 3/9$ and $\pi_{ij} = 1/12$ as we desired. The support of the first design consists of the following 12 samples:

$$(1,2,8), (1,3,6), (1,4,5), (1,7,9), (2,3,7), (2,4,9)$$
$$(2,5,6), (3,4,8), (3,5,9), (4,6,7), (5,7,8), (6,8,9)$$

and each sample was assigned the probability 1/12. The second design had the following samples in the support:

$$\begin{gathered}(1,2,9), (1,3,6), (1,3,7), (1,4,5), (1,4,8), (1,5,8)\,,\\ (1,6,7), (2,3,5), (2,3,8), (2,4,7), (2,5,6), (2,6,8)\\ (3,4,6), (3,4,9), (3,5,9), (3,7,8), (4,5,6), (4,8,9)\,,\\ (5,7,8), (5,7,9), (6,7,9), (6,8,9)\end{gathered}$$

and the probability over each sample was 1/24 except for the two samples 1, 2, 9 and 2, 4, 7, each having 2/24. J. N. K. Rao and Nigam (1990) have followed this line of thinking for producing controlled sampling designs with other objective functions and constraints.

Fienberg and Tanur (1987) point out an interesting application of variants of SRS designs where some specified samples would be desired to receive high probability of selection. They report as follows "... The practical value of such sampling plans is especially apparent when government statistical agencies designing elaborate surveys need to ensure the presence of sample units in selected political jurisdictions that are the constituencies of elected representatives who must approve the budgets for the surveys. In the United States, such considerations have been formally built into the decennial redesign of the Current Population Survey." The Current Population Survey (CPS) is the largest continuing personal sample of households in the United States and it is conducted monthly by the U.S. Bureau of the Census. The data from the CPS are the only comprehensive source of information on personal and socioeconomic characteristics of the total population between decennial censuses. It is compiled with utmost care and is based on several thousands of households located in hundreds of counties.

4.10. INFERENCE FROM ORDER STATISTICS BASED ON OTHER DESIGNS

We will follow the notations and the definitions given in Section 4.7. Suppose we wish to provide an outer confidence interval for $Y_{(k)}$ for a specified k. Let d be a fixed-size (n) sampling design adopted for this purpose. Assume, momentarily, that $Y_{i_1} < Y_{i_2} < \cdots < Y_{i_N}$ for a specific permutation $(i_1, i_2, \ldots, i_N)$ of the labels $(1, 2, \ldots, N)$. The following results can be easily derived.

$$\begin{aligned}
P[y_{(1)} \le Y_{(1)} \le y_{(n)}] &= \pi_{i1}\,, \\
P[y_{(1)} \le Y_{(2)} \le y_{(n)}] &= \pi_{i_1} + \pi_{i_2} - \pi_{i_1 i_2}\,, \\
P[y_{(1)} \le Y_{(3)} \le y_{(n)}] &= (\pi_{i_1} + \pi_{i_2} + \pi_{i_3}) - (\pi_{i_1 i_2} + \pi_{i_1 i_3} + \pi_{i_2 i_3}) \\
&\quad + \pi_{i_1 i_2 i_3} \qquad \text{for } n \ge 3 \\
&= (\pi_{i_1} + \pi_{i_2} + \pi_{i_3}) - (\pi_{i_1 i_2} + \pi_{i_2 i_3}) \qquad \text{for } n = 2\,.
\end{aligned} \tag{4.36}$$

We now discuss the complexity in the computation of these probabilities in situations where the true ordering $i_1, i_2, \ldots, i_N$ is *not* known. We consider the following simple example of a sampling design.

Example 4.9. $N = 7$, $n = 3$, $P((1, 2, 3)) = 0.1$, $P((1, 4, 6)) = 0.2$, $P((2, 5, 7)) = 0.3$, $P((3, 4, 7)) = 0.4$. It is easily verified that $P[y_{(1)} \le Y_{(1)} \le y_{(3)}] = 0.3$ if $Y_{(1)} = Y_1$ and 0.7 if $Y_{(1)} = Y_7$. Similarly, $P[y_{(1)} \le Y_{(7)} \le y_{(3)}] = 0.4$ if $Y_{(7)} = Y_2$ and 0.3 if $Y_{(7)} = Y_5$.

This shows that in general, for an arbitrary sampling design the exact coverage probability $P[y_{(1)} \leq Y_{(k)} \leq y_{(n)}]$ would be difficult to compute without any additional information on the true ordering of the values of the survey variable. Of course, a lower bound to this probability can be computed for small values of N by direct enumeration.

We now specialize to $k = 1$ and 2 which provide computational simplicity in the study of the nature of designs. From (4.36), it is clear that for $k = 1$, we need constancy of the π_i's and for $k = 2$ we need constancy of $(\pi_i + \pi_j - \pi_{ij})$for *all* pairs (i, j). This latter condition is equivalent to constancy of the π_{ij}'s for fixed size designs. It is thus evident that variants of SRS designs provide natural extensions of SRS designs for which $P[y_{(1)} \leq Y_{(k)} \leq y_{(n)}]$ can be computed without any difficulty for $k = 1$ or 2 as also for $k = N - 1$ or N. It may be seen that for $3 \leq k \leq N - 2$, such designs do *not* necessarily provide computational simplicity. Again, any mixture of SRS designs is *indeed* suitable for the computation of $P[y_{(1)} \leq Y_{(k)} \leq y_{(n)}]$ for *any* k, $1 \leq k \leq N$. This is because the required probability will be the corresponding mixture of component probabilities arising out of the mixing SRS designs. Readers familiar with t-designs (e.g., a balanced incomplete block design is a 2-design) can readily verify that a t-design allows computational simplicity for $k \leq t$ or $k \geq N - t + 1$. There is an extensive literature on t-designs; see for example, Hedayat and Kageyama (1980).

We conclude this section with an interesting observation relating to a comparison between SRS with and without replacement sampling schemes. We first make a comparison between d_0: SRS(N, n_0) versus d: a mixture design of the type pSRS$(N, n_1) + q$SRS(N, n_2) where $n_0 = pn_1 + qn_2$ and $0 < p < 1$, $q = 1 - p$. Consider the problem of providing an outer interval for the population range $[Y_{(1)}, Y_{(N)}]$. Clearly, the coverage probability attains its maximum when we use $[y_{(1)}, y_{(n)}]$ as an outer interval. Computations readily yield

$$P_{d_0}[y_{(1)} \leq Y_{(1)} < Y_{(N)} \leq y_{(n_0)}] = n_0(n_0 - 1)/N(N - 1) = p_0 ,$$

$$P_d[y_{(1)} \leq Y_{(1)} < Y_{(N)} \leq y_{(n)}] = \{pn_1(n_1 - 1) + qn_2(n_2 - 1)\}/N(N - 1) = p^* .$$

It is straightforward to verify that $p_0 < p^*$. Specializing to a general mixture design d^*: $\Sigma_\nu\, q_\nu$SRS(N, ν) where $\Sigma\, \nu q_\nu = n_0$, $0 < q_\nu < 1$, $\Sigma\, q_\nu = 1$, it is again seen that

$$\sum_\nu q_\nu P_{d_\nu}[y_{(1)} \leq Y_{(1)} < Y_{(N)} \leq y_{(\nu)}] > P_{d_0}[y_{(1)} \leq Y_{(1)} < Y_{(N)} \leq y_{(n_0)}] .$$

In particular, it then follows that the with replacement SRS scheme with the same average sample size as a without replacement SRS scheme provides higher coverage probability than the latter. This is interesting since in customary inference the conclusion is completely *opposite*.

4.11. SOME ASPECTS OF UNLABELLED DATA

The fact that π_i's and π_{ij}'s are constants for an SRS(N, n) design have a very useful implication in the recording of data. For such designs we need not record data in the explicit form $\{(i_1, Y_{i_1}), (i_2, Y_{i_2}), \ldots, (i_n, Y_{i_n})\}$. It is enough to collect only the values of the survey variable for the units covered by the sample without any regard as to their identification in terms of the labels of the sampled units. Such a collection of observations will be referred to as *unlabelled* data. This would still validate the computation of the HTE and its variance estimate. On the other hand, for any other sampling design (excluding those which are equivalent to SRS designs) not having this feature, it is not enough to record only the values of the survey variable without the underlying labels, as we may *not* be able to provide unbiased estimators of the population parameters of interest.

It should, however, be emphasized that by ignoring the labels of the selected units in the recorded data set, we are limiting our choice of an unbiased estimator of the population mean or the population total only to the HTE alone. Otherwise, we could possibly use estimators other than the HTE. See Exercise 4.17 in this context.

Turning back to unlabelled data arising out of simple random sampling, we want to present some properties of such data sets. As is traditionally done, we will conveniently denote by $y_1, y_2, \ldots, y_n$ a set of unlabelled observations obtained on the survey variable Y under simple random sampling. It is then evident that $y_1, y_2, \ldots, y_n$ represent a rearrangement of the sample observations. Therefore, these unlabelled observations may be regarded as being generated under two probability distributions (randomization), the first one corresponding to an SRS design and the next one corresponding to a *random* arrangement of the data obtained in a sample. As random variables, $y_1, y_2, \ldots, y_n$ possess certain properties which are explained below. We assume $n \geq 2$.

Theorem 4.11. *Under an* SRS(N, n) *design and a random arrangement of data yielding the unlabelled data set* $(y_1, y_2, \ldots, y_n)$,

$$P(y_i = Y_k) = 1/N, \qquad 1 \leq i \leq n,\ 1 \leq k \leq N, \tag{4.37}$$

$$P(y_i = Y_k, y_j = Y_l) = 1/N(N-1), \qquad 1 \leq i \neq j \leq n,\ 1 \leq k \neq l \leq N. \tag{4.38}$$

Hence,

$$E(y_i) = \bar{\mathbf{Y}}, \qquad V(y_i) = \sigma_Y^2, \qquad \mathrm{Cov}(y_i, y_j) = -\sigma_Y^2/(N-1). \tag{4.39}$$

Further, denoting the sample mean and variance by $\bar{\mathbf{y}}$ *and* $\mathbf{s}_Y^2$, *respectively*,

$$E(\bar{\mathbf{y}}) = \bar{\mathbf{Y}}, \qquad V(\bar{\mathbf{y}}) = \left(\frac{1}{n} - \frac{1}{N}\right)\mathbf{S}_Y^2. \tag{4.40}$$

Moreover, an unbiased estimator of $V(\bar{\mathbf{y}})$ is given by

$$\hat{V}(\bar{\mathbf{y}}) = \left(\frac{1}{n} - \frac{1}{N}\right)\mathbf{s}_Y^2 . \tag{4.41}$$

Proof. Recalling that two types of randomizations are acting on the y_i's, we have, for $1 \le i \le n$, $1 \le k \le N$,

$$\begin{aligned} P(y_i = Y_k) &= \sum_{s \ni k} P_d(s) P_R(y_i = Y_k \mid s \ni k) \\ &= \frac{(n-1)!}{n!} \sum_{s \ni k} P_d(s) = \pi_k / n = n / nN = 1/N . \end{aligned}$$

In the above the suffix d refers, as usual, to the sampling design and the suffix R refers to random permutation of the sample observations $\{Y_k \mid k \in s\}$, on units in the sample s, yielding the unlabelled data set $y_1, y_2, \ldots, y_n$. For $1 \le i \ne j \le n$ and $1 \le k \ne l \le N$,

$$\begin{aligned} P(y_i = Y_k, y_j = Y_l) &= \sum_{s \ni k,l} P_d(s) P_R(y_i = Y_k, y_j = Y_l \mid s \ni k, l) \\ &= \frac{(n-2)!}{n!} \sum_{s \ni k,l} P_d(s) \\ &= \pi_{kl} / n(n-1) = 1/N(N-1) . \end{aligned}$$

Therefore,

$$\begin{aligned} E(y_i) &= \bar{\mathbf{Y}} , \qquad E(y_i^2) = \sum_k Y_k^2 / N \\ E(y_i y_j) &= \sum\sum_{k \ne l} Y_k Y_l / N(N-1) = \bar{\mathbf{Y}}^2 - \sigma_Y^2/(N-1) . \end{aligned}$$

Hence,

$$V(y_i) = \sigma_Y^2 \quad \text{and} \quad \operatorname{Cov}(y_i, y_j) = -\sigma_Y^2/(N-1) .$$

These lead to

$$E(\bar{\mathbf{y}}) = \bar{\mathbf{Y}} \quad \text{and} \quad V(\bar{\mathbf{y}}) = \left(\frac{1}{n} - \frac{1}{N}\right)\mathbf{S}_Y^2 .$$

It also follows quite readily that $E(\mathbf{s}_Y^2) = \mathbf{S}_Y^2$ and hence all the results are established. □

Remark 4.5. In the above, we have shown that the covariance between any two observations in an unlabelled data set arising out of a simple random sample of size $n(\ge 1)$ is $-\sigma_Y^2/(N-1)$. This value is independent of the sample size n. An alternative interesting proof of this result has been given by Bondy and Zlot (1976). Appealing to symmetry, they first argue that $y_1, y_2, \ldots, y_n$ have the same variance and the same covariance between any two of them. Therefore,

$$V(\bar{y}) = V\Big(\sum_i y_i/n\Big) = \Big[\sum_i V(y_i) + \sum\sum_{i \neq j} \text{Cov}(y_i, y_j)\Big]\Big/ n^2$$

$$= [nV(y_1) + n(n-1)\text{Cov}(y_1, y_2)]/n^2 .$$

When $n = N$, $\bar{y}$ coincides with $\bar{Y}$ which is a fixed quantity so that $V(\bar{y})$ reduces to zero. This yields

$$NV(y_1) + N(N-1)\text{Cov}(y_1, y_2) = 0$$

and, hence, the covariance between any two observations turns out to be $-\sigma_Y^2/(N-1)$.

Remark 4.6. The expressions in (4.39) provide what may be termed a *linear estimation model* for estimation of the population mean $\bar{Y}$ using the unlabelled observations $y_1, y_2, \ldots, y_n$. This has an interesting implication in terms of *best linear unbiased estimate* (blue) for $\bar{Y}$. As was first pointed out by Neyman (1934), the mean $\bar{y}$ is the unique unbiased estimator with minimum variance among all unbiased estimators of $\bar{Y}$ which are linear in the observations $y_1, y_2, \ldots, y_n$. The reader will note that this result stands in *sharp contrast* to the nonexistence result established in Chapter Two. It must be clearly understood that the difference between the two results lies in the fundamental concept and utilization of labelling. For labelled data, we have compared estimators based on $\{Y_i \mid i \in s\}$, where s is a chosen sample, and for unlabelled data we are comparing estimators based on $y_1, \ldots, y_n$, where these observations undergo the model in (4.39). It is true that the values of $\Sigma_{i\in s} Y_i/n$ and $\bar{y}$ do agree under all circumstances but the two are to be regarded as estimators arising in two different contexts. See Exercise 4.17 in this matter.

We conclude this section by giving a proof of the fact that $\bar{y}$ is the blue of $\bar{Y}$ under the model in (4.39).

Theorem 4.12. *In simple random sampling resulting in unlabelled observations $y_1, y_2, \ldots, y_n$, $\bar{y}$ is the blue of $\bar{Y}$.*

Proof. Several proofs can be given. We give the simplest one based on an elementary inequality. First note that $y_1, \ldots, y_n$ obey the model in (4.39). Let $\Sigma_{i=1}^n c_i y_i$ be a linear unbiased estimator of $\bar{Y}$. Then the coefficients $c_1, c_2, \ldots, c_n$ must satisfy the condition $\Sigma_{i=1}^n c_i = 1$. Next, using (4.39), we can write

$$V\Big(\sum_{i=1}^n c_i y_i\Big) = \sigma_Y^2\Big[\sum_{i=1}^n c_i^2 - \frac{1}{N-1}\sum\sum_{i\neq j} c_i c_j\Big]$$

$$= \sigma_Y^2 \left[\frac{N \sum_{i=1}^{n} c_i^2}{N-1} - \frac{1}{N-1} \left(\sum_{i=1}^{n} c_i \right)^2 \right]$$

$$= \sigma_Y^2 \left[\left(N \sum_{i=1}^{n} c_i^2 / N - 1 \right) - (1/(N-1)) \right].$$

To minimize the above expression, we have to minimize $\Sigma_{i=1}^n c_i^2$ subject to the restriction $\Sigma_{i=1}^n c_i = 1$. Clearly, the solution is $c_1 = c_2 = \cdots = c_n = 1/n$ as is evident by an application of Cauchy–Schwartz inequality. The resulting minimum variance is $((1/n) - (1/N))N\sigma_Y^2/(N-1)$, which is the same as $((1/n) - (1/N))\mathbf{S}_Y^2$ as in (4.40). □

EXERCISES

4.1. Prove Theorem 4.2.

4.2. Show that for SRS(N, n) design with $n \geq 2$,

$$\hat{V}_{\text{HT}}(\bar{\mathbf{y}}) = \hat{V}_{\text{SYG}}(\bar{\mathbf{y}}) = (n^{-1} - N^{-1})\mathbf{s}_Y^2 .$$

Comment on the result.

4.3. (a) Show that for a fixed-size (n) design, the sample mean $\bar{\mathbf{y}}$ is an unbiased estimator of the population mean $\bar{\mathbf{Y}}$ if and only if $\pi_i = n/N$, $1 \leq i \leq N$. Show further that the variance of $\bar{\mathbf{y}}$ is proportional to $\mathbf{S}_Y^2$ if and only if $\pi_{ij} = n(n-1)/N(N-1)$, $1 \leq i \neq j \leq N$.

(b) Deduce the corresponding results for variable size sampling designs.

4.4. Suppose it is required to estimate the population mean $\bar{\mathbf{Y}}$ with the relative absolute error not exceeding a given quantity d. Prove that under an SRS(N, n) design this will be achieved provided the sample size is determined through the approximate relation $n_0 = z_{\alpha/2}^2 V_0^2/d^2$, where $V_0 = \mathbf{S}_Y/\bar{\mathbf{Y}}$ (population coefficient of variation) and the coverage probability is at least $1 - \alpha$.

4.5. Prove Theorem 4.8.

4.6. Consider an SRS(15, 5) design.

(a) What would be the coverage probability of the sample range $[y_{(1)}, y_{(5)}]$ as an outer confidence interval for the population quantile range $[Y_{(5)}, Y_{(10)}]$?

(b) What is the probability that the sample range contains at least 50% of the population values?

(c) Find the largest value of t so that the probability that the sample range contains at least t of the population values is nearly equal to 95%.

4.7. Deduce (4.29).
Hint: Define $\varphi(N, n) = E\{(n(s))^{-1}|(N, n)\}$. Show that $\varphi(N, n) = \{(N-1)/N\}^n \varphi(N-1, n) + 1/N$. Use a frequency approach.

4.8. Consider the expression for $V(\bar{\mathbf{y}}_{n(s)})$ in (4.31). Starting with $\{(n(s))^{-1} - N^{-1}\}\hat{\mathbf{S}}_Y^2$, suggest an alternative unbiased estimator for $V(\bar{\mathbf{y}}_{n(s)})$.

4.9. Show that there is no uniform fixed-size (3) sampling design other than an SRS(8, 3) design available for a population of 8 units with constant first- and second-order inclusion probabilities for these units.

4.10. Prove the left hand side inequality in (4.35).

4.11. Prove the right hand side inequality in (4.35).

4.12. A population consists of N units of which N_1 units are peculiar in nature. It is decided to use the following scheme of sampling.

(a) Select the N_1 peculiar units outright and include them in the sample;

(b) Select $(n - N_1)$ additional units using SRS design on the remaining $(N - N_1)$ units of the population.

Consider the problem of estimation of the population mean $\bar{\mathbf{Y}}$. Suggest an unbiased estimator and its variance estimator. Is the latter nonnegative?

4.13. Consider the following k sampling designs:

$$d_1\text{: SRS}(N, n_1),\ d_2\text{: SRS}(N, n_2), \ldots, d_k\text{: SRS}(N, n_k)$$

where $1 \le n_1 < n_2 < \cdots < n_k \le N$. Let $p_1, p_2, \ldots, p_k$ be the mixing proportions in the mixture sampling design $d = \Sigma\, p_j d_j$. Denote by $\bar{\mathbf{y}}_j$ the sample mean based on n_j *observations arising out of* d_j, $1 \le j \le k$, and by $\bar{\mathbf{y}}_d$ the mean based on the mixture sampling design.

(a) Show that $V(\bar{\mathbf{y}}_d) > V(\bar{\mathbf{y}}_k)$ irrespective of the choice of the mixing proportions $p_1, p_2, \ldots, p_k$.

(b) If $p_1', p_2', \ldots, p_k'$ is another set of mixing proportions leading to another mixture design $d' = \Sigma\, p_j' d_j$, how would you compare the two mixture designs d and d'?

(c) Can you construct interesting examples of such mixture designs and compare them?

(d) Give some examples of such mixtures which are equivalent in some sense.

4.14. A market research analyst selected n households via SRS from a community of N households, all listed in a telephone directory. Her purpose is to estimate the proportion of households (in that community) favoring the brand A of a certain product to the brand B. She decides to collect data via telephone inquiry and immediately record the response in her personal computer file opened for this purpose. Upon completion of the job, she notices that information from one household is missing and has not been recorded by mistake. She then tries to reach that household for the second time but in vain. How would you analyze the available data? Justify your procedure.

4.15. Suppose d is a sampling design of fixed size n based on a population of size N for which $\pi_i = n/N$ and $\pi_{ij} = n(n-1)/N(N-1)$ for $1 \leq i \neq j \leq N$. No other information is available on the design d.

(a) What is the lower bound on $P(s)$ for any sample s of size n?

(b) Show that $P(s) \leq 1/N$ for any sample under d.

4.16. Work out the details in connection with Example 4.8.

4.17. For SRS(N, n) design, consider the following estimator of the population mean $\bar{\mathbf{Y}}$:

$$\begin{aligned} e^*(s, \mathbf{Y}) &= (Y_1/n) + (1 - N^{-1})\bar{\mathbf{y}}'(s), \qquad && s \ni 1, \\ &= (1 - N^{-1})\bar{\mathbf{y}}(s), && s \not\ni 1, \end{aligned}$$

where

$$\bar{\mathbf{y}}'(s) = \sum_{i \in s, i \neq 1} Y_i/(n-1) \qquad \text{and} \qquad \bar{\mathbf{y}}(s) = \sum_{i \in s} Y_i/n .$$

(a) Show that $e^*(s, \mathbf{Y})$ is an unbiased estimator of $\bar{\mathbf{Y}}$.

(b) Compute an expression for the variance of $e^*(s, \mathbf{Y})$.

(c) Show that among all hlue's of $\bar{\mathbf{Y}}$, $e^*(s, \mathbf{Y})$ attains minimum variance at the point $(1, 0, 0, \ldots, 0)$ and zero variance at the point $(0, 1, 1, \ldots, 1)$.

(d) Show that among all hlue's of $\bar{\mathbf{Y}}$, $e^*(s, \mathbf{Y})$ also minimizes the sum of the variances at the points $(0, 1, 0, 0, \ldots, 0)$, $(0, 0, 1, 0, \ldots, 0), \ldots, (0, 0, 0, \ldots, 0, 1)$.

(e) Justify that $e^*(s, \mathbf{Y})$ is an admissible estimator. Comment on your result.

4.18. Consider the population $\mathcal{U}$ of size N. Draw a sample s_1 using an SRS(N, n_1) design on $\mathcal{U}$. Next, draw another sample s_2 using an SRS($N - n_1, n_2$) design on $\mathcal{U} - s_1$. Let $\bar{\mathbf{y}}(s_1)$ and $\bar{\mathbf{y}}(s_2)$, respectively, be the sample means based on observations in s_1 and s_2.

(a) Show that both are unbiased for the population mean.
(b) Show that $V(\bar{y}(s_1)) \gtreqless V(\bar{y}(s_2))$ according as $n_1 \lesseqgtr n_2$.
(c) Denote by s the combined sample and by $\bar{y}(s)$ the combined sample mean. Compute $V(\bar{y}(s))$ and, hence or otherwise, $\text{Cov}(\bar{y}(s_1), \bar{y}(s_2))$.
(d) Is $\bar{y}(s)$ the best linear combination of $\bar{y}(s_1)$ and $\bar{y}(s_2)$ for estimating the population mean?

4.19. In the above exercise, suppose the second sample of size n_2 is also drawn from the entire population independently of the first sample s_1. Then a combined sampling design can be defined as follows: $P(s) = f_s / \binom{N}{n_1}\binom{N}{n_2}$ where f_s = number of ways a sample s can be formed using s_1 and s_2 together.

(a) Taking $N = 6$ and $n_1 = n_2 = 3$, show that the combined design is of the form:

$$\begin{aligned} d:\quad & P((i, j, k)) = 1/400\,, && 1 \le i < j < k \le 6\,, \\ & P((i, j, k, l)) = 12/400\,, && 1 \le i < j < k < l \le 6\,, \\ & P((i, j, k, l, m)) = 30/400\,, && 1 \le i < j < k < l < m \le 6\,, \\ & P((1, 2, 3, 4, 5, 6)) = 20/400\,. \end{aligned}$$

(b) For the combined design with parameters N, n_1, n_2, compute π_i's and π_{ij}'s.
(c) Suggest a suitable inference procedure based on the combined data for the estimation of the population mean and variance.
(d) Show that *any* estimator suggested for this sampling design can be improved when the design is changed to SRS$(N, n_1 + n_2)$. Comment on this result.

4.20. From a simple random sample of n units a simple random subsample of n_1 units is taken and added to the original sample. Show that the mean based on the $(n + n_1)$ observations is an unbiased estimator of the population mean but its variance is greater than the variance of the mean based on the original n units by the approximate factor $(1 + (3n_1/n))(1 + (n/n_1))^{-2}$.

Hint. Represent the mean as the weighted mean and proceed.

4.21. Consider a linear cost structure so that $C(s) = C_0 + cn(s)$ is the cost incurred in data collection and analysis for a sample s of size $n(s)$. For a sampling design d, we may then define the average cost of implementation of d as $C(d) = C_0 + cE(n(s))$.

(a) For the mixture design $d = \Sigma\, p_\nu d_\nu$, $p_\nu = \binom{N}{\nu}\Delta^\nu O^n / N^n$, $\nu = 1, 2, \ldots, n$ give an expression for the average cost.

(b) Suggest an SRS design d^* with the same average cost as d above.

(c) Compare the performance of the two designs d and d^* for estimation of the population mean.

4.22. In the same setup as above, make a comparison of two mixture designs, based on a given set of component designs, having the same average cost.

4.23. Below is a part of a sampling design which can be made to a sampling design by one additional sample. Can you develop it into one which could be equivalent to an SRS(7, 3) design?

$$P((1,2,4)) = 3/21,\ P((1,3,5)) = P((3,4,7)) = P((4,5,6)) = 2/21\ ,$$
$$P((1,3,6)) = P(1,5,7)) = P((2,3,5)) = P((2,3,6))$$
$$= P((2,3,7)) = P((2,5,6)) = P((2,5,7))$$
$$= P((2,6,7)) = P((3,4,6)) = P(4,5,7)) = 1/21\ .$$

4.24. It is desired to have a sampling design equivalent to an SRS (6, 3) design with the property that all samples except (1, 2, 3) receive positive probability. Is there any such sampling design? If so, construct one of them.

4.25. A certain population $\mathcal{U}$ is to be studied at two different time periods t_1 and t_2. Denote by $Y_1, Y_2, \ldots, Y_N$ the values of the survey variable at time t_1 and by $Z_1, Z_2, \ldots, Z_N$ those at time t_2. The survey variables Y and Z usually refer to the same characteristic in the population. The purpose is to estimate $\bar{\mathbf{Y}} - \bar{\mathbf{Z}}$ using one of the following sampling strategies:

I. At time t_1, adopt SRS(N, n_1) and select a sample s_1. At time t_2, draw a sample s_0 of size n_0 adopting SRS(n_1, n_0) on s_1. Also draw a sample s_0' of size n_0' adopting SRS($N - n_1$, n_0') on $\mathcal{U} - s_1$. Then combine the two samples to form a sample s_2, of size $n_2 = n_0 + n_0'$. Estimate $\bar{\mathbf{Y}}$ and $\bar{\mathbf{Z}}$ using the sample means based on data collected on Y and Z, respectively, from s_1 and s_2.

II. Proceed as above at time t_1. At time t_2, s_0 is the same as before. Draw a sample s_{00}' of size n_0' adopting SRS($N - n_0$, n_0') on $\mathcal{U} - s_0$. Form then s_2' as the compositon of s_0 and s_{00}'. Estimate $\bar{\mathbf{Y}}$ as before and $\bar{\mathbf{Z}}$ using the mean of $n_2 (= n_0 + n_0')$ observations collected on Z covered by s_2'.

III. Proceed as above at time t_1. At time t_2. adopt SRS(N, n_2) on the entire population $\mathcal{U}$ and draw a sample s_2''. Estimate $\bar{\mathbf{Y}}$ and $\bar{\mathbf{Z}}$ using, respectively, the sample means.

(a) Prove that

$$V(\bar{\mathbf{y}}-\bar{\mathbf{z}}) = \left(1-\frac{n_1}{N}\right)\frac{\mathbf{S}_Y^2}{n_1} + \left(1-\frac{n_2}{N}\right)\frac{\mathbf{S}_Z^2}{n_2} - 2(1-f)\frac{n_0}{n_1 n_2}\mathbf{S}_{YZ}$$

where

$$\begin{aligned} f &= \frac{n_1 n_2}{N n_0}\,, && \text{for sampling strategy I,} \\ &= \frac{n_1 n_2}{N n_0} - \frac{(n_1 - n_0)(n_2 - n_0)}{(N - n_0) n_0}\,, && \text{for sampling strategy II ,} \\ &= 1\,, && \text{for sampling strategy III.} \end{aligned}$$

(b) With respect to the sign of $\mathbf{S}_{YZ}$, which sampling strategy would minimize $V(\bar{\mathbf{y}}-\bar{\mathbf{z}})$? Interpret your result.

4.26. Denote by $\bar{\mathbf{y}}(m)$ the sample mean based on a sample of size m drawn according to SRS(N, m).

(a) Show that for $1 \le n \le N-1$,

$$P\{\bar{\mathbf{y}}(n) \le \bar{\mathbf{Y}} - \epsilon\} = P\left\{\bar{\mathbf{y}}(N-n) \ge \bar{\mathbf{Y}} + \frac{n}{N-n}\,\varepsilon\right\},$$

$$P\{|\bar{\mathbf{y}}(n) - \bar{\mathbf{Y}}| \le t\} = P\left\{|\bar{\mathbf{y}}(N-n) - \bar{\mathbf{Y}}| \le \frac{n}{N-n}\,t\right\}.$$

(b) Observe that $\bar{\mathbf{y}}_{(n)}$ and $\bar{\mathbf{y}}_{(N-n)}$ are linearly related to each other. Comment on the applicability of the Central Limit Theorem on $\bar{\mathbf{y}}_{(n)}$ as $n \to N$.

4.27. There have been repeated complaints that the emergency room of a hospital does *not* properly handle the emergency cases related to a certain minority group in the population. A government agency decides to audit the files of the emergency room, accumulated during a 6-month period, for the purpose of assessing the situation. It is decided to estimate the proportion of files of the minority group, which have *not* been properly handled.

(a) How would you sample the files, given that altogether 5329 files have been accumulated during the period under study?

(b) Based on your sampling design, how would you estimate the parameter in question?

(c) Provide a method for estimating the variance of the estimator you propose.

4.28. Denote by E_R and V_R the operators of mean and variance under rearrangement of sample data yielding the unlabelled observations $y_1, y_2, \ldots, y_n$. Show that

$$E_R(y_i|s) = \bar{\mathbf{y}}\,, \qquad V_R(y_i|s) = (n-1)\mathbf{s}_Y^2/n\,, \qquad \mathrm{Cov}_R(y_i, y_j|s) = -\mathbf{s}_Y^2/n\,.$$

Hence, establish (4.39) using the following facts: $E = E_pE_R$, $V = V_pE_R + E_pV_R$, and $\mathrm{Cov} = E_p\,\mathrm{Cov}_R + \mathrm{Cov}_pE_R$. The reader may consult Section 2b.3 in C. R. Rao (1965) regarding such combined operators.

REFERENCES

Asok, C. (1980). A note on the comparison between simple mean and mean based on distinct units in sampling with replacement. *Amer. Statistician*, **34**, 158.

Avadhani, M. S., and Sukhatme, B. V. (1973). Controlled sampling with equal probabilities and without replacement. *Internat. Statist. Rev.*, **41**, 175–182.

Basu, D. (1958). On sampling with and without replacement. *Sankhyā*, **20**, 287–294.

Bondy, W. H., and Zlot, W. (1976). The standard error of the mean and the difference between means for finite populations. *Amer. Statistician*, **30**, 96–97.

Chakrabarti, M. C. (1963). On the use of incidence matrices of designs in sampling from finite populations. *J. Indian Statist. Assoc.*, **1**, 78–85.

Cochran, W. G. (1977). *Sampling Techniques*. Third Edition. Wiley, New York.

Fienberg, S. E., and Tanur, J. M. (1987). Experimental and sampling structures: Parallels diverging and meeting. *Internat. Statist. Rev.*, **55**, 75–96.

Foody, W., and Hedayat, A. (1977). On theory and applications of BIB designs with repeated blocks. *Ann. Statist.*, **5**, 932–945. (1979). Corrigendum, *Ann. Statist.*, **7**, 925.

Francisco, C. A., and Fuller, W. A. (1991). Quantile estimation with a complex Survey design. *Ann. Statist.*, **19**, 454–469.

Hedayat, A. (1979). Sampling designs with reduced support sizes. In: *Optimizing Methods in Statistics* (Rustagi, J., Ed.), 273–288. Academic Press, New York.

Hedayat, A., and Kageyama, S. (1980). The family of *t*-designs—Part I. *J. Statist. Plann. Infer.*, **4**, 173–212.

Hidiroglou, M. A., and Srinath, K. P. (1981). Some estimators of a population total from simple random samples containing large units. *J. Amer. Statist. Assoc.*, **76**, 690–695.

Korwar, R. M., and Serfling, R. J. (1970). On averaging over distinct units in sampling with replacement. *Ann. Math. Statist.*, **41**, 2132–2134.

Kruskal, W., and Mosteller, F. (1979a). Representative sampling, I: Nonscientific literature. *Internat. Statist. Rev.*, **47**, 13–24.

Kruskal, W., and Mosteller, F. (1979b). Representative sampling, II: Scientific literature, excluding statistics. *Internat. Statist. Rev.*, **47**, 111–127.

Kruskal, W., and Mosteller, F. (1979c). Representative sampling, III: The current statistical literature. *Internat. Statist. Rev.*, **47**, 245–265.

Kruskal, W., and Mosteller, F. (1980). Representative sampling, IV: The history of the concepts in statistics, 1895–1939. *Internat. Statist. Rev.*, **48**, 169–195.

Konijn, H. S. (1973). *Statistical Theory of Sample Survey Design and Analysis*. North-Holland, New York.

Meyer, J. S. (1987). Outer and inner confidence intervals for finite population quantile interval. *J. Amer. Statist. Assoc.*, **82**, 201–204.

Neyman, J. (1934). On the two different aspects of the representative method: The method of stratified sampling and the method of purposive selection. *J. Roy. Statist. Soc.*, **97**, 558–625.

Pathak, P. K. (1962). On simple random sampling with replacement. *Sankhyā*, **A24**, 287–302.

Pathak, P. K. (1988). Simple random sampling. In: *Handbook of Statistics, Vol. 6, Sampling* (Krishnaiah, P. R., and Rao, C. R., Eds.), 97–109. North-Holland, Amsterdam.

Rao, C. R. (1965). *Linear Statistical Inference and its Applications*. Wiley, New York.

Rao, J. N. K., and Nigam, A. K. (1990). Optimal controlled sampling designs. *Biometrika* **77**, in press.

Sedransk, J. (1969). Some elementary properties of systematic sampling. *Skand. Aktuar.*, **52**, 39–47.

Sedransk, J., and Meyer, J. S. (1978). Confidence intervals for the quantiles of a finite population: simple random and stratified simple random sampling. *J. Roy. Statist. Soc.*, **B40**, 239–252.

Sedransk, J., and Smith, P. J. (1988). Inference for finite population quantiles. In: *Handbook of Statistics, Vol. 6, Sampling* (Krishnaiah. P. R., and Rao, C. R., Eds.), 267–289. North-Holland, Amsterdam.

Sengupta, S. (1979). On the construction of non-invariant balanced sampling designs. *Cal. Statist. Assoc. Bull.*, **28**, 9–24.

Sengupta, S. (1982). Construction of some non-invariant balanced sampling designs. *Cal. Statist. Assoc. Bull.*, **31**, 165–185.

Sinha, B. K. (1976). On balanced sampling schemes. *Cal. Statist. Assoc. Bull.*, **25**, 129–138.

Sinha, B. K. and Sen, P. K. (1989). On averaging over distinct units in sampling with replacement. *Sankhyā*, **B51**, 65–83.

Wynn, H. P. (1977). Convex sets of finite population plans. *Ann. Statist.*, **5**, 414–418.

CHAPTER FIVE

Uses of Auxiliary Size Measures in Survey Sampling: Strategies Based on Probability Proportional to Size Schemes of Sampling

In many survey sampling situations, either during the course of data collection on the survey variable or even beforehand, auxiliary information on other variable(s) may be made available. Usually auxiliary information from one or more sources can be suitably blended together to yield an auxiliary size measure. Various sampling strategies which utilize such measures at the design stage and/or at the estimation stage find extensive use in practice. In this chapter we provide a thorough treatment of some of the existing sampling strategies which utilize auxiliary size measures primarily at the design stage.

5.1. NATURE AND SCOPE OF AUXILIARY SIZE MEASURES

In Chapter Two we learned that a survey sampling procedure deals with the twin problem of suggesting a sampling design and providing suitable estimators of relevant parametric functions. It is imperative that good survey statisticians make every attempt to improve the quality of these two ingredients with the hope of producing optimal results for the intended survey. We also learned in Chapter Three that the first- and second-order inclusion probabilities in a sampling design play a key role in the inference stage, particularly if we limit our estimator of the population total to the HTE. In Chapter Four we introduced sampling designs of fixed and variable sizes which provide, among other things, constant first- and second-order inclusion probabilities for the units in a given population. As we saw, this led to a significant simplification of the data analysis step. Notice that this was achieved without taking advantage of any additional available information present at the various stages of the survey.

The purpose of this and the next chapter is to familiarize the reader with various types of sampling strategies suggested in the literature based on utilization of auxiliary information. Some such strategies are based on utilization of auxiliary information at the estimation stage (keeping the sampling design simple) while others utilize the auxiliary information at the design stage or at both stages. Knowledge of the different methods of data collection and analysis utilizing auxiliary information should prove helpful to a survey sampling practitioner.

Useful auxiliary information could come from a variety of sources. For example, previous census data on agricultural production might be useful in developing a sampling methodology for studying the current year's production. Again, a knowledge of acreage under an agricultural commodity, via analysis of satellite pictures, might be useful in estimating the production of that particular commodity. As a third example, a quick eye-estimate by an expert might be useful in suggesting a suitable sampling methodology for estimating the volume of timber in a forest. In practice such auxiliary information is usually made available to the survey statistician (possibly at the stage of preparation of the frame) at no or very little extra cost compared to the cost of data collection on the survey variable. We would like to emphasize, however, that use of such additional information may not always lead to increased precision. If no suitable justification can be provided, such additional information may and should be ignored.

For the sake of presentation of the sampling strategies, we will assume that additional or auxiliary information relates to the values $X_1, X_2, \ldots, X_N$ of an auxiliary variable X on the population units $1, 2, \ldots, N$, respectively. For the type of application we discuss, it is assumed that $X_1, X_2, \ldots, X_N$ are strictly positive (rational) numbers. These are often referred to as *size measures* or *auxiliary size measures* since they relate to the magnitude of an auxiliary variable. What we call here an auxiliary variable X could, in reality, arise out of an integration of several distinct sources of information which are suitably combined by the survey practitioner for the problem at hand. Thus, for example, X could be a known real-valued function of t different auxiliary variables $Z_1, Z_2, \ldots, Z_t$, where Z_i is related to the ith source of information, $1 \leq i \leq t$. We will, therefore, start with the following setup.

A population $\mathcal{U}$ of N units has to be surveyed and the population total $T(\mathbf{Y})$ relating to a survey variable Y has to be efficiently estimated. The availability of an auxiliary variable X is pointed out and the auxiliary size measures $X_1, X_2, \ldots, X_N$ are presented beforehand to the survey statistician. The objective is to suggest a suitable sampling strategy, possibly making use of available auxiliary information. At this stage, it is almost imperative to postulate an explicit relationship between the survey variable Y and the auxiliary variable X. A vague notion of possible relation between

Y and X will not lead to an improved design. On the other hand, perfect knowledge of a concrete mathematical relation between Y and X is certainly too much to expect. We have to be guided by experience in postulating any meaningful relation between the two variables.

The implementation of the sampling strategies to be presented in this chapter relies on an understanding of a possible relation among the values of the survey variable and those of the auxiliary variable. Roughly stated, it is assumed that the points (X_i, Y_i), $1 \le i \le N$ do not vary substantially from a line passing through the origin. At first sight this appears to be too demanding. However, we must emphasize that the survey statistician is not limited to only one functional form of $Z_1, Z_2, \ldots, Z_t$, the auxiliary variables available at hand. Indeed, the practitioner can utilize knowledge of the population under study and search for a function $g(\cdot)$ so that

$$X = g(Z_1, Z_2, \ldots, Z_t)$$

assumes positive values and approximately satisfies the requirement stated above.

Before leaving this section we must emphasize that if the survey practitioner is unable to find such a function $g(\cdot)$, the designs to be presented in this chapter cannot be expected to provide satisfactory results. Failing to effectively utilize auxiliary information at the design stage, the survey statistician might still be able to benefit from such information at the estimation stage. This will be treated in considerable detail in Chapter Six.

5.2. USE OF AUXILIARY SIZE MEASURES AT DESIGN STAGE: ΠPS SAMPLING DESIGNS

In developing the sampling design, we can be guided by the following considerations. Suppose we decide to adopt a fixed-size (n) sampling design and subsequently use the HTE. Then the variance function of the HTE will assume the form given in (3.4). Since our aim is to control this variance, one way to achieve this would be to select a sampling design for which the quantities $\{(Y_i/\pi_i) - (Y_j/\pi_j)\}$, $1 \le i \ne j \le N$ are appreciably small. A version of this requirement which is feasible is that $\{(X_i/\pi_i) - (X_j/\pi_j)\}$, $1 \le i \ne j \le N$ be made as small as possible. This would hopefully reduce the variance in (3.4), at least in situations where values on the survey variable and the auxiliary variable are *nearly* proportional. This point was discussed earlier in Chapter Three in the paragraph following (3.4). It thus seems reasonable to study the constructional aspects of sampling designs utilizing auxiliary size measures with the following motivation.

Suppose a population of N units is to be sampled to produce a sample of fixed-size n. A priori we know the values of $X_1, X_2, \ldots, X_N$ on an auxiliary variable X, associated with the units $1, 2, \ldots, N$, respectively. The fixed-size (n) sampling design is supposed to provide each π_i proportional to the corresponding X_i value, that is,

$$\pi_i = \lambda X_i\,, \qquad 1 \le i \le N \tag{5.1}$$

so that the quantities $\{(X_i/\pi_i) - (X_j/\pi_j)\}$, $1 \le i \ne j \le N$ are, indeed, all zeroes. Clearly this is *not* possible *unless* the X_i's are of the same sign. We shall, therefore, assume that all the X_i's are strictly positive. Since the resulting design is to be of fixed-size (n), we must have, according to (1.7), $\Sigma_{i=1}^{N} \pi_i = n$ while $0 < \pi_i \le 1$ for each i, $1 \le i \le N$. Consequently, the constant of proportionality in (5.1) equals $n/T(\mathbf{X})$ where $T(\mathbf{X}) = \Sigma_{i=1}^{N} X_i$. This suggests the requirement

$$\pi_i = nX_i/T(\mathbf{X})\,, \qquad 1 \le i \le N \tag{5.2}$$

where the X_i's must satisfy

$$0 < nX_i \le T(\mathbf{X})\,, \qquad 1 \le i \le N\,. \tag{5.3}$$

Hereafter, we shall denote $X_i/T(\mathbf{X})$ by p_i, to be called the *normed* size measure of unit i, so that

$$0 < np_i \le 1,\ 1 \le i \le N\,; \qquad \sum_{1}^{N} p_i = 1\,. \tag{5.4}$$

It is useful to introduce the following concept.

Definition 5.1. A sampling design whose first-order inclusion probabilities satisfy $\pi_i \propto X_i$, $1 \le i \le N$, is called a *ΠPS sampling design*.

In the present case we will work exclusively with fixed-size (n) ΠPS sampling designs. We use the notation ΠPS($N, n, \mathbf{p}$) to designate a fixed-size (n) ΠPS sampling design with respect to the normed size measure vector $\mathbf{p} = (p_1, p_2, \ldots, p_N)$. In such a case the components of $\mathbf{p}$ have to satisfy (5.4). Such designs are also known as IPPS designs in the literature. Here IPPS is as an abbreviation for *Inclusion Probability Proportional to Size*. It must be noted that for a given $\mathbf{p}$, a ΠPS($N, n, \mathbf{p}$) design (even in its reduced form) is *not*, however, unique unless $n = 1$ or $N - 1$.

In the beginning of this section we gave a justification for using a fixed size ΠPS sampling design. Before proceeding further we present an example of a ΠPS sampling design.

Example 5.1. For $N = 6$, $n = 3$ and $\mathbf{p} = (2/17,\ 3/17,\ 4/17,\ 1/17,\ 2/17,\ 5/17)$, the design with $P((2,3,6)) = 6/17$, $P((3,5,6)) = 4/17$, $P((1,3,4)) = 2/17$, $P((1,2,6)) = 3/17$, $P((1,5,6)) = P((4,5,6)) = 1/17$ is a ΠPS($6, 3, \mathbf{p}$) design.

The sampling design given in the above example reveals one undesirable feature of some ΠPS sampling designs, which can arise unless we demand additional properties. In the above design not all the π_{ij}'s are positive so that unbiased variance estimation would be a problem. To overcome this difficulty, we must impose the requirement $\pi_{ij} > 0$ for each pair (i, j), $i \neq j$, $i, j = 1, 2, \ldots, N$. Another desirable requirement on such a sampling design is that for every pair (i, j), $\pi_i \pi_j \geq \pi_{ij}$. Then the Sen–Yates–Grundy variance estimator will be nonnegative regardless of the values of the survey variable. This is not an unreasonable requirement on a sampling design which is supposed to be used in combination with the HTE. Unfortunately, in the class of ΠPS sampling designs, not every design satisfying $\pi_{ij} > 0$ for each pair (i, j) enjoys this additional desirable feature. This can be verified from the following example of a ΠPS sampling design with the same values of N, n, and $\mathbf{p}$ as above: $P((2,3,6)) = 5/17$, $P((3,5,6)) = 3/17$, $P((1,3,6)) = 2/17$, $P((4,5,6)) = P((1,2,6)) = P((1,5,6)) = P((3,4,6)) = P((1,2,4)) = P((2,5,6)) = P((1,2,3)) = 1/17$. In this example, $\pi_{36} > \pi_3 \pi_6$.

It thus appears that in order for a ΠPS sampling design to be practical, we must also require

$$0 < \pi_{ij} \leq \pi_i \pi_j \,, \qquad 1 \leq i \neq j \leq N \,. \tag{5.5}$$

Hanurav (1967) has carried the arguments still further. We shall, however, be satisfied with these requirements and now discuss some methods for constructing such sampling designs. Various methods for this and allied problems have been discussed in the literature. The general problem seems quite difficult to solve and so far only very few workable solutions are available. The case of $n = 2$ received considerable attention from early researchers in the area and extension of these methods to the general case have been discussed by some researchers. Below we first take up the case of $n = 2$ and present some basic solutions which are applicable under all circumstances. We follow this by presenting further simple solutions for the case of $n = 2$, which, however, require additional conditions to be satisfied by the size measures. The general case is presented at the end. In what follows, we shall assume, with no loss of generality, that $np_i < 1$ for all i. See Exercise 5.32 for the case of $np_i = 1$ for at least one i. There is a vast literature on sampling schemes which meet various specific requirements. The interested reader may in particular consult Brewer and Hanif (1983) and Chaudhuri and Vos (1988).

5.3. ΠPS SAMPLING DESIGNS OF SIZE 2

We will present a few popular ΠPS sampling designs of size 2. Some of these will be presented in terms of the unit drawing algorithm $A(q_1, q_2, q_3)$ described in Section 1.2. Recall that both $q_1(\cdot)$ and $q_3(\cdot\,|s)$ describe probabilities on the elements of $\mathcal{U}$. In other words, $0 \leq q_1((i)) \leq 1$, $1 \leq i \leq N$, $\Sigma_{i=1}^{N} q_1((i)) = 1$, and $0 \leq q_3((i)|s) \leq 1$, $1 \leq i \leq N$, $\Sigma_{i=1}^{N} q_3((i)|s) = 1$, $0 \leq q_2(s) \leq 1$. Further, recall that according to the algorithm $A(q_1, q_2, q_3)$, the probability of generating a sample $s = (i_1, i_2, \ldots, i_{n(s)})$ is given by (1.1).

Brewer's Method

Brewer (1963) suggested a ΠPS sampling scheme. This is described in the following theorem.

Theorem 5.1. *The algorithm $A(q_1, q_2, q_3)$ with the choice*

$$q_1((i)) = \{p_i(1-p_i)(1-2p_i)^{-1}\}\left[\sum_{k=1}^{N} \{p_k(1-p_k)(1-2p_k)^{-1}\}\right]^{-1},$$

$$q_2((i)) = 1\,, \qquad q_2(s) = 0 \text{ otherwise},$$

$$q_3((j)|(i)) = p_j(1-p_i)^{-1}\,, \qquad j \neq i,\ 1 \leq i \leq N\,,$$

results in a ΠPS *sampling design satisfying* $0 < \pi_{ij} \leq \pi_i \pi_j$, $1 \leq i \neq j \leq N$.

Proof. It is easy to verify that

$$P((i, j)) = \{p_i p_j (1-2p_i)^{-1}\}\left[\sum_{k=1}^{N} \{p_k(1-p_k)(1-2p_k)^{-1}\}\right]^{-1} \tag{5.6}$$

for each pair (i, j), $1 \leq i \neq j \leq N$. This gives

$$\begin{aligned}\pi_{ij} &= P((i, j)) + P((j, i)) \\ &= p_i p_j \{(1-2p_i)^{-1} + (1-2p_j)^{-1}\}\left[1 + \sum_{k=1}^{N} p_k^2 (1-2p_k)^{-1}\right]^{-1},\end{aligned} \tag{5.7}$$

for each pair (i, j), $1 \leq i < j \leq N$. From this it follows that $\pi_i = 2p_i$, $1 \leq i \leq N$. Further,

$$\pi_i \pi_j - \pi_{ij} = 4 p_i p_j A_{ij} \tag{5.8}$$

where

$$A_{ij} = \left\{\sum_{\substack{k \neq i, \\ \neq j}} p_k (1-2p_k)^{-1}\right\}\left[1 + \sum_{k=1}^{N} p_k (1-2p_k)^{-1}\right]^{-1}, \tag{5.9}$$

which is strictly positive for any pair (i, j), $1 \leq i < j \leq N$. □

Durbin's Method

Following Durbin (1967), we can establish the following result.

Theorem 5.2. *The algorithm* $A(q_1, q_2, q_3)$ *with the choice*

$$\begin{aligned} q_1((i)) &= p_i ,\\ q_2((i)) &= 1 , \qquad q_2(s) = 0 \textit{ otherwise},\\ q_3((j)|(i)) &= p_j\{(1-2p_i)^{-1} + (1-2p_j)^{-1}\}\\ &\quad \times \left[\sum_{k(\neq i)} p_k\{(1-2p_i)^{-1} + (1-2p_k)^{-1}\} \right]^{-1} ,\\ &\qquad j \neq i, \quad 1 \leq i \leq N , \end{aligned}$$

results in a ΠPS *sampling design satisfying* $0 < \pi_{ij} \leq \pi_i \pi_j$, $1 \leq i \neq j \leq N$.

Proof. It is clear that

$$P((i, j)) = p_i p_j \{(1-2p_i)^{-1} + (1-2p_j)^{-1}\} \left[1 + \sum_{k=1}^{N} p_k(1-2p_k)^{-1} \right]^{-1} \tag{5.10}$$

for $1 \leq i \neq j \leq N$. The rest of the verification is left to the reader (Exercise 5.1). □

Remark 5.1. The two preceding methods lead to the same expressions for the π_{ij}'s. It is thus possible that different sampling schemes result in the same reduced sampling design, even though the underlying *unreduced* sampling designs are different. This is consistent with the one-to-one correspondence established in Theorem 1.1 and Theorem 1.2.

J. N. K. Rao's Method

J. N. K. Rao (1965) suggested a *rejective* sampling procedure which also leads to the *same* reduced sampling design as those of Brewer and Durbin. According to this method, we draw the first unit of the sample using $P((i)) = p_i$, $1 \leq i \leq N$. Next, the second unit is drawn using $P((i)) \propto p_i/(1-2p_i)$, $1 \leq i \leq N$, without any regard to the unit drawn on the first occasion. If now the two units thus selected turn out to be *distinct*, the sampling operation is stopped and the units are taken to form the sample. Otherwise, the entire operation is discarded and the whole process is started afresh. This is continued until we come up with two distinct units. These units are then taken to form the ultimate sample.

We now present the design aspect of this sampling method.

Theorem 5.3. *The rejective sampling procedure of Rao produces a fixed-size* (2) ΠPS *sampling design satisfying* $0 < \pi_{ij} \le \pi_i \pi_j$, $1 \le i \ne j \le N$.

The proof rests on a verification of

$$P((i, j)) = p_i p_j (1 - 2p_j)^{-1} \left[\sum_{k=1}^{N} p_k (1 - p_k)(1 - 2p_k)^{-1} \right]^{-1} \quad (5.11)$$

for any pair (i, j), $1 \le i \ne j \le N$. We leave it to the reader (Exercise 5.2). This yields $\pi_{ij} = P((i, j)) + P((j, i))$ in the same form as that obtained under the two methods described earlier. It is thus evident that this method shares all the properties of the other two methods. Some interesting relations connecting the three methods are presented in Exercise 5.3.

Example 5.2. The purpose of this example is to show the sampling designs which arise out of the different methods discussed so far. Consider a population with $N = 5$ units and the normed size measures given by $\mathbf{p} = (0.4, 0.2, 0.2, 0.1, 0.1)$. In Table 5.1, we show the various sampling designs derived according to the methods presented above.

Table 5.1. Unreduced Sampling Designs: Values[a] of $P(i, j)$, $1 \le i \ne j \le 5$

i/j	1	2	3	4	5
1	X	48/235 16/235 32/235	48/235 16/235 32/235	24/235 6/235 15/235	24/235 6/235 15/235
2	16/235 48/235 32/235	X	8/235 8/235 8/235	4/235 3/235 7/470	4/235 3/235 7/470
3	16/235 48/235 32/235	8/235 8/235 8/235	X	4/235 3/235 7/470	4/235 3/235 7/470
4	6/235 24/235 15/235	3/235 4/235 7/470	3/235 4/235 7/470	X	3/470 3/470 3/470
5	6/235 24/235 15/235	3/235 4/235 7/470	3/235 4/235 7/470	3/470 3/470 3/470	X

[a]The top figures correspond to Brewer's method; the middle figures correspond to Rao's method; the bottom figures correspond to Durbin's method.

Sunter's Method

Following Sunter (1986), we consider the units of the population one by one and decide on their inclusion in the ultimate sample. In doing so, we consider the following framework.

The units in the population are so arranged that

$$2p_1 < 1,\ 2p_2 < \sum_{2}^{N} p_i = 1 - p_1,$$

$$2p_3 < \sum_{3}^{N} p_i = 1 - p_1 - p_2, \ldots, 2p_{N-2} < \sum_{N-2}^{N} p_i . \tag{5.12}$$

This can be achieved, for example, by arranging the units in increasing order of the p_i-values. Let us define

$$p_i^* = \sum_{k=i}^{N} p_k , \qquad 1 \le i \le N , \tag{5.13}$$

so that (5.12) is equivalent to $2p_i < p_i^*$, $1 \le i \le N-2$.

We are now in a position to describe Sunter's method. It consists of three steps.

Step I. Consider the first unit and perform a Bernoulli trial with success probability $p = 2p_1$. Decide on inclusion of the first unit in the sample in case success occurs. In any case, pass on to the next step.

Step II. Denote by n_i the number of units yet to be sampled when the ith unit is reached. Clearly, $n_2 = 1$ or 2 depending on the result of the first trial in the previous step. In either case, perform a Bernoulli trial with $p = n_2 p_2 / p_2^*$ to decide on inclusion of the second unit in the sample in case a success occurs.

In general, if $n_i > 0$, perform a Bernoulli trial with $p = n_i p_i / p_i^*$ at the ith stage to decide on inclusion of the ith unit in the ultimate sample.

Continuing this procedure as needed up to the stage $(N-2)$, we may end up with a sample of size 2 in which case we terminate sampling. On the other hand, we may have obtained one or no unit at all for inclusion in the sample. In the latter case, we proceed to the following step.

Step III. First consider the situation when $n_{N-1} = 1$. This means that only one unit has so far been selected from among the first $(N-2)$ units and one more unit has yet to be selected. Then perform a Bernoulli trial with success probability p'_{N-1}/p^*_{N-1} where p'_{N-1} is a number to be suitably chosen shortly. If success occurs, we select the $(N-1)$th unit in the sample. Otherwise, we select the Nth unit in the sample. In the situation when $n_{N-1} = 2$, no unit has so far been selected from among the first $(N-2)$ units and, therefore, we select the last two units in the sample.

This method results in a fixed-size (2) design with $\pi_i = 2p_i$, $1 \le i \le N-2$. Further, for $i = N-1$ or N, $\pi_i = 2p_i$ holds if and only if the choice of p'_{N-1} is based on the formula

$$p'_{N-1}/p^*_{N-1} = \{2p_{N-1} - P(n_{N-1} = 2)\}/2\{p_{N-1} + p_N - P(n_{N-1} = 2)\} . \tag{5.14}$$

Clearly such a choice is feasible provided $P(n_{N-1} = 2) < 2\min(p_{N-1}, p_N)$. It can be shown that for such a sampling design, $0 < \pi_{ij} < \pi_i \pi_j$ holds for each pair (i, j), $1 \le i \ne j \le N$, provided $P(n_{N-1} = 2) < 4p_{N-1}p_N$ holds. We note that this last inequality implies the former one. In *most* situations there are some arrangements of the units satisfying this last inequality along with those in (5.12). Exercise 5.4 is concerned with verification of this inequality.

With reference to the population studied in Example 5.2, in order to apply Sunter's method the units need to be rearranged. For example, the ordering 1, 2, 4, 3, 5 would suffice. The design arising out of this method is the following: $P((1,2)) = 128/480$, $P((1,3)) = 132/480$, $P((1,4)) = 64/480$, $P((1,5)) = 60/480$, $P((2,3)) = 33/480$, $P((2,4)) = 16/480$, $P((2,5)) = 15/480$, $P((3,4)) = 11/480$, $P((3,5)) = 16/480$, $P((4,5)) = 5/480$. Further, $\pi_{12} = \pi_{13} = 0.32$, $\pi_{14} = \pi_{15} = 0.16$, $\pi_{23} = 0.16$, $\pi_{24} = \pi_{25} = \pi_{34} = \pi_{35} = 0.08$, and $\pi_{45} = 0.04$. We readily verify $0 < \pi_{ij} < \pi_i \pi_j$, for each pair (i, j).

In case $p_{N-1} = p_N$, we note that Sunter's method is quite straightforward to apply, since in (5.14), p'_{N-1}/p^*_{N-1} reduces to 1/2. When p_{N-1} is different from p_N, we can suggest the following combination of Sunter's method and a method of Hanurav (1967) as an alternative scheme.

Suppose without loss of generality $p_1 \le p_2 \le \cdots \le p_{N-1} < p_N$. We perform a Bernoulli trial with success probability

$$\delta = 2(1 - p_N)(p_N - p_{N-1})/(1 - p_N - p_{N-1}) .$$

If it is a success we select u_N and another unit u_i with probability $p_i/(1 - p_N)$, $1 \le i \le N-1$. If it is a failure, we apply Sunter's method with p_i's replaced by

$$p^*_i = p_i\left\{1 - \frac{\delta}{2(1 - p_N)}\right\} \Big/ (1 - \delta) , \qquad 1 \le i \le N-2$$

$$p^*_{N-1} = p^*_N = (p_N - \delta/2)/(1 - \delta) .$$

See Exercise 5.5 in this context.

Emptying Box Principle

This is a fairly general method of producing *all* conceivable ΠPS sampling designs. This method was developed by Hedayat and Lin (1980) and later expanded by Hedayat et al. (1989). We shall discuss this method for a general sample size in the next section.

Lahiri–Midzuno Method

Lahiri (1951) and Midzuno (1952) independently described a very simple sampling scheme for selecting n units from a population of N units when auxiliary size measures are available beforehand. See Exercise 5.29 for a description of their original scheme. The scheme can be suitably modified to yield a ΠPS sampling scheme. This method of construction of a ΠPS sampling scheme may be regarded as the simplest of all methods, even in a general setup. However, it is *not* always applicable. We shall discuss this modified method for a general sample size in the next section.

5.4. ΠPS SAMPLING DESIGNS OF SIZE n

We will present two specific methods.

Lahiri–Midzuno Method.

Following Lahiri (1951) and Midzuno (1952), we establish the following result.

Theorem 5.4. *Whenever $np_i \geq (n-1)(N-1)^{-1}$ for every i, $1 \leq i \leq N$, the algorithm $A(q_1, q_2, q_3)$ with the choice*

$$q_1((i)) = [(N-1)/(N-n)][np_i - (n-1)(N-1)^{-1}] = p'_i, \quad \text{say } 1 \leq i \leq N \tag{5.15}$$

$$q_2((i)) = 1, \; q_2(s) = \begin{cases} 1, & \text{for all } s \text{ with } n(s) \leq n-1 \\ 0, & \text{otherwise}; \end{cases}$$

$$q_3((j)|(i_1, i_2, \ldots, i_k)) = (N-k)^{-1}, \qquad 1 \leq j \neq i_1 \neq i_2 \neq \cdots \neq i_k \leq N,$$
$$k = 1, 2, \ldots, n-1$$

results in a ΠPS fixed-size (n) sampling design satisfying $0 < \pi_{ij} \leq \pi_i \pi_j$, $1 \leq i \neq j \leq N$.

The proof readily follows from the fact that

$$\pi_i = p'_i + (1-p'_i)[(n-1)/(N-1)] = np_i, \qquad 1 \leq i \leq N \tag{5.16}$$

and

$$\pi_{ij} = \{(p'_i + p'_j)(n-1)/(N-1)\} + \{(1 - p'_i - p'_j)(n-1)(n-2)/(N-1)(N-2)\}. \tag{5.17}$$

In case $n = 2$, the first unit of the sample is selected using

$$P((i)) = [(N-1)/(N-2)][2p_i - (N-1)^{-1}] = p'_i, \qquad \text{say } 1 \leq i \leq N \tag{5.18}$$

while the second unit is selected at random from the remaining population units. This then results in

$$\pi_i = 2p_i , \tag{5.19}$$

$$\pi_{ij} = (p'_i + p'_j)/(N-1) . \tag{5.20}$$

See Exercise 5.6 for a relaxation of the condition for applicability of this method.

Example 5.3. Consider a population with $N = 10$ units and with the following X-values: 78, 83, 87, 93, 96, 102, 108, 111, 117, 125. Then $T(\mathbf{X}) = 1000$. The least p_i value is 0.078. For a sample of size $n = 3$, the stated condition for applicability of Midzuno's scheme is: $p_i > (n-1)/n(N-1) = 2/27 = 0.0741$ for each $i, 1 \le i \le 10$. This condition is satisfied here and so we can apply Midzuno's scheme. We now present p'_i-values using (5.15): 0.015, 0.034, 0.050, 0.073, 0.085, 0.108, 0.131, 0.142, 0.166, 0.196. Next we may apply the cumulative total method to draw the first unit of the sample. We select a 3-digit random number, say 305. Then this corresponds to the 6th unit so that $i_1 = 6$. We now select the other two units of the sample using SRS from the remaining 9 units of the population. Suppose we get $i_2 = 2$ and $i_3 = 8$. Then the sample generated by this procedure is $s = (6, 2, 8)$. We have $P((6, 2, 8)) = 0.0015$ and the probability of the unordered sample is $P^*(s^*) = 0.0079$. Moreover, by (5.16) and (5.17), $\pi_2 = 0.249$, $\pi_6 = 0.306$, $\pi_8 = 0.333$, $\pi_{26} = 0.0554$, $\pi_{28} = 0.062$, $\pi_{68} = 0.0764$.

Sampford's Method

This is a generalization of J. N. K. Rao's method described earlier. According to Sampford (1967), we draw the first unit of the population using $P((i)) = p_i$, $1 \le i \le N$. Next, we draw $(n-1)$ additional units, one at a time, from the entire population, each time using $P((i)) \propto p_i/(1 - np_i)$, $1 \le i \le N$. We now examine the entire collection of the units drawn so far. If *all* of them happen to be *distinct*, we accept them and use them to form a sample. Otherwise, we *reject* the *entire* collection and the process is started all over again *until* we come up with a set of *n distinct* units. Quite naturally, then, this procedure has been termed a *rejective* sampling procedure.

We now investigate the nature of the resulting sampling design.

Theorem 5.5. *The rejective sampling procedure of Sampford results in a fixed-size* (n) ΠPS *sampling design satisfying* $0 < \pi_{ij} \le \pi_i \pi_j$, $1 \le i \ne j \le N$.

Proof. It is clear that the method results in a fixed-size (n) sampling design. We now show that it also yields

$$\pi_i = np_i\,, \qquad 0 < \pi_{ij} < \pi_i\pi_j,\ 1 \le i \ne j \le N\,. \tag{5.21}$$

Let $s = (i_1, i_2, \ldots, i_n)$ be a typical sample generated by this method. We can easily see that for $1 \le i_1 \ne i_2 \ne \cdots \ne i_n \le N$,

$$P(s) \propto (p_{i_1}p_{i_2} \cdots p_{i_n})/\{(1 - np_{i_2})(1 - np_{i_3}) \ldots (1 - np_{i_n})\}\,. \tag{5.22}$$

Moreover, in its reduced form,

$$P^*(s^*) = \delta_{i_1 i_2 \ldots i_n} \Big/ \Big[\sum_{1 \le k_1 < k_2 < \cdots < k_n \le N} \cdots \sum \delta_{k_1 k_2 \ldots k_n} \Big] \tag{5.23}$$

$$\text{for } 1 \le i_1 < i_2 < \cdots < i_n \le N$$

$$= 0\,, \qquad \text{Otherwise}$$

where

$$\delta_{ijk\ldots} = \frac{(p_i p_j p_k \ldots)(1 - p_i - p_j - p_k - \cdots)}{(1 - np_i)(1 - np_j)(1 - np_k)\cdots}\,. \tag{5.24}$$

We now compute π_i's and π_{ij}'s for this sampling design in the following manner.

Let

$$A_1 = \sum_{s \ni 1} P(s) \qquad \text{and} \qquad A_{\bar{1}} = \sum_{s \not\ni 1} P(s)$$

where $P(s)$ is as in (5.22). First note that the numerator of $A_{\bar{1}}$ is given by

$$\sum_{2 \le i_1 \ne i_2 \ne \cdots \ne i_n \le N} \cdots \sum \{(p_{i_1} p_{i_2} \cdots p_{i_n})/(1 - np_{i_2}) \ldots (1 - np_{i_n})\}$$

$$= \sum_{2 \le i_2 \ne i_3 \ne \cdots \ne i_n \le N} \cdots \sum \frac{\{(1 - p_1 - p_{i_2} - p_{i_3} - \cdots - p_{i_n})(p_{i_2} p_{i_3} \cdots p_{i_n})\}}{(1 - np_{i_2})(1 - np_{i_3}) \ldots (1 - np_{i_n})}\,. \tag{5.25}$$

Next, the numerator of A_1 can be obtained from (5.22) by putting in turn every $i_j = 1$, $1 \le j \le n$ and summing over the resulting expressions. The first subtotal of $(n-1)!$ terms is given by

$$p_1 \sum_{2 \le i_2 \ne \cdots \ne i_n \le N} \cdots \sum p_{i_2} p_{i_3} \cdots p_{i_n} / \{(1 - np_{i_2}) \ldots (1 - np_{i_n})\}\,.$$

Further, the jth subtotal for $j \ge 2$ is given by

$$p_1(1 - np_1)^{-1} \sum_{\substack{2 \le i_1 \ne \cdots \ne i_n \le N \\ \text{excluding } i_j}} \cdots \sum \frac{\{p_{i_1} p_{i_2} \cdots p_{i_{j-1}} p_{i_{j+1}} \cdots p_{i_n}\}}{(1 - np_{i_2}) \ldots (1 - np_{i_{j-1}})(1 - np_{i_{j+1}}) \ldots (1 - np_{i_n})}$$

which can be rewritten as

$$p_1(1-np_1)^{-1} \sum_{2 \le i_2 \ne \cdots \ne i_n \le N} \cdots \sum \{(1-np_{i_j})p_{i_2}p_{i_3}\cdots p_{i_n}/(1-np_{i_2})\cdots(1-np_{i_n})\}\,.$$

Summing all these terms, we get the numerator of A_1

$$= \sum_{2 \le i_2 \ne \cdots \ne i_n \le N} \cdots \sum \frac{p_1(1-np_1)^{-1}(p_{i_2}\cdots p_{i_n})(1-np_1+1-np_{i_2}+\cdots+1-np_{i_n})}{(1-np_{i_2})\cdots(1-np_{i_n})}\,.$$

$$= np_1(1-np_1)^{-1} \sum_{2 \le i_2 \ne \cdots \ne i_n \le N} \cdots \sum \frac{(1-p_1-p_{i_2}-\cdots-p_{i_n})p_{i_2}p_{i_3}\cdots p_{i_n}}{(1-np_{i_2})\cdots(1-np_{i_n})}\,.$$

Hence, referring to (5.25), we have

$$\text{Numerator of } A_1 = np_1(1-np_1)^{-1} \times \text{numerator of } A_{\bar{1}}\,. \tag{5.26}$$

This gives $\pi_1/(1-\pi_1) = np_1/(1-np_1)$ so that $\pi_1 = np_1$. It is now evident from the symmetry of the sampling design that $\pi_i = np_i$ for each i, $1 \le i \le N$. Next, we compute (for example) π_{12}. To that end, we define the following quantities.

$$\alpha = \sum_{3 \le i_2 \ne \cdots \ne i_n \le N} \cdots \sum \{(p_{i_2}p_{i_3}\cdots p_{i_n})/(1-np_{i_2})\cdots(1-np_{i_n})\} \tag{5.27}$$

$$\beta = \sum_{3 \le i_2 \ne \cdots \ne i_n \le N} \cdots \sum \{(p_{i_2}\cdots p_{i_n})(1-p_{i_2}-\cdots p_{i_n})/(1-np_{i_2})\cdots(1-np_{i_n})\}\,. \tag{5.28}$$

Further, in addition to A_1 and $A_{\bar{1}}$, let

$$A_2 = \sum_{s \ni 2} P(s)\,, \qquad A_{\bar{2}} = \sum_{s \not\ni 2} P(s)\,, \qquad A_{12} = \sum_{s \ni 1,2} P(s)\,,$$

$$A_{1\bar{2}} = \sum_{s \ni 1, \not\ni 2} P(s)\,, \qquad A_{\bar{1}2} = \sum_{s \not\ni 1, \ni 2} P(s)\,, \qquad A_{\bar{1}\bar{2}} = \sum_{s \not\ni 1, \not\ni 2} P(s)\,.$$

Then it is evident that

$$\pi_1 = A_1\,, \qquad \pi_2 = A_2\,, \qquad \pi_{12} = A_{12}\,, \qquad A_{\bar{1}\bar{2}} = 1 - \pi_1 - \pi_2 + \pi_{12}\,.$$

We have seen above that the numerator of A_1 is equal to $\{(np_1/(1-np_1))\}$ multiplied by the numerator of $A_{\bar{1}}$. Now we want to deduce a similar relation connecting the numerators of $A_{1\bar{2}}$, $A_{\bar{1}2}$, and $A_{\bar{1}\bar{2}}$ using the quantities α and β.

Numerator of $A_{\bar{1}\bar{2}}$

$$= \sum_{2 \le i_1 \ne i_2 \ne \cdots \ne i_n \le N} \cdots \sum \{(p_{i_1} p_{i_2} \cdots p_{i_n})/(1 - np_{i_2}) \cdots (1 - np_{i_n})\}$$

$$= \sum_{3 \le i_2 \ne \cdots \ne i_n \le N} \cdots \sum \frac{\{(p_{i_2} p_{i_3} \cdots p_{i_n})(1 - p_1 - p_2 - p_{i_2} - \cdots - p_{i_n})\}}{(1 - np_{i_2})(1 - np_{i_3}) \cdots (1 - np_{i_n})}$$

$$= \beta - (p_1 + p_2)\alpha . \tag{5.29}$$

Again,

Numerator of $A_{\bar{1}2}$

$$= \sum_{3 \le i_2 \ne \cdots \ne i_n \le N} \cdots \sum [\{(p_2 p_{i_2} \cdots p_{i_n})/(1 - np_{i_2}) \cdots (1 - np_{i_n})\}$$

$$+ \{(p_2 p_{i_2} \cdots p_{i_n})/(1 - np_2)(1 - np_{i_3}) \cdots (1 - np_{i_n})\} + \cdots$$

$$+ \{(p_2 p_{i_2} p_{i_3} \cdots p_{i_n})/(1 - np_2)(1 - np_{i_2}) \cdots (1 - np_{i_{n-1}})\}]$$

$$= n \sum_{3 \le i_2 \ne \cdots \ne i_n \le N} \cdots \sum \frac{\{(p_2 p_{i_2} \cdots p_{in})(1 - p_2 - p_{i_2} - \cdots - p_{i_n})\}}{(1 - np_2)(1 - np_{i_2}) \cdots (1 - np_{i_n})}$$

$$= \{np_2/(1 - np_2)\} \sum_{3 \le i_2 \ne \cdots \ne i_n \le N} \cdots \sum \frac{(p_{i_2} \cdots p_{i_n})(1 - p_2 - p_{i_2} - \cdots - p_{i_n})}{(1 - np_{i_2}) \cdots (1 - np_{i_n})}$$

$$= np_2(\beta - p_2\alpha)/(1 - np_2) . \tag{5.30}$$

Similarly, the numerator of $A_{1\bar{2}} = np_1(\beta - p_1\alpha)/(1 - np_1)$. These now yield in an obvious way

$$\pi_1 = \pi_{12} + \pi_{1\bar{2}} = \pi_{12} + np_1(\beta - p_1\alpha)/(1 - np_1)c ,$$

$$\pi_2 = \pi_{12} + \pi_{\bar{1}2} = \pi_{12} + np_2(\beta - p_2\alpha)/(1 - np_2)c , \tag{5.31}$$

and also

$$1 - \pi_1 - \pi_2 + \pi_{12} = \pi_{\bar{1}\bar{2}} = \{\beta - (p_1 + p_2)\alpha\}/c , \tag{5.32}$$

where

$$c = n \sum_{1 \le i_1 < i_2 < \cdots < i_n \le N} \sum \cdots \sum \delta_{i_1 i_2 \ldots i_n} \tag{5.33}$$

with $\delta_{ijk\ldots}$ as in (5.24). We now solve for π_{12} from these equations, using $\pi_1 = np_1$ and $\pi_2 = np_2$. Two cases arise depending on whether or not $p_1 = p_2$. In either case, it turns out that $\pi_{12} = \pi_1\pi_2(1 - c^{-1}(\alpha/n))$, which is less than $\pi_1\pi_2$ since $\alpha > 0$. This illustrates the fact that the sampling design is a ΠPS design of fixed size n with the desired properties of the first- and second-order inclusion probabilities. □

It must however be noted that in general the exact expressions of π_{ij}'s are quite involved. Asok and Sukhatme (1976) have evaluated approximate expressions for π_{ij} correct to $O(N^{-4})$.

Example 5.4. In this example we apply Sampford's technique to generate a sampling design of size $n = 3$ for a population of $N = 5$ with the p_i-values: 0.10, 0.15, 0.20, 0.25, 0.30. We calculate $z_i = p_i/(1 - 3p_i)$: 0.1428, 0.2727, 0.5000, 1.0000, 3.0000 with a total of 4.9155. Next we calculate the normed z_i's and call them q_i's: 0.0291, 0.0555, 0.1017, 0.2034, 0.6103. Applying (5.22) and (5.23) we end up with the following form of the resulting sampling design: $P((1, 2, 3)) = 0.0091$, $P((1, 2, 4)) = 0.0167$, $P((1, 2, 5)) = 0.0450$, $P((1, 3, 4)) = 0.0275$, $P((1, 3, 5)) = 0.0734$, $P((1, 4, 5)) = 0.1284$, $P((2, 3, 4)) = 0.0467$, $P((2, 3, 5)) = 0.1224$, $P((2, 4, 5)) = 0.2100$, $P((3, 4, 5)) = 0.3208$. The values of the π_i's and the π_{ij}'s are given in Table 5.2.

Table 5.2. Inclusion Probabilities of First- and Second-Order

	1	2	3	4	5
1	0.3001	0.0708	0.1100	0.1726	0.2468
2		0.4499	0.1782	0.2734	0.3774
3			0.5999	0.3950	0.5166
4				0.7501	0.6592
5					0.9000

Hedayat–Lin Method

The method to be discussed now may be regarded as a universal method in the sense that all conceivable sampling designs, no matter how they are devised, can be generated by this method. It is able to produce ΠPS sampling designs with support size less than $\binom{N}{n}$ and thus can be used in controlled sampling situations as well. It works with no further condition except the basic necessary conditions $np_i < 1$, $i = 1, 2, \ldots, N$. For given N, n, and the normed size vector $\mathbf{p}$, the method produces more than one ΠPS($N, n, \mathbf{p}$) sampling design.

The basic idea is to translate the problem of constructing a ΠPS sampling design into a game of emptying N labelled boxes, each filled with a certain number of indistinguishable objects, in a series of rounds. The associated game aspect of the problem will be discussed first. We have N boxes labelled as $1, 2, \ldots, N$. In all, we have M indistinguishable objects. These M objects are distributed into these N boxes, so that there are $k_i > 0$ objects in the ith box. A round of size $n < N$ is defined to be the selection of any n distinct boxes and the removal of one object from each. We say we have a successful game of size n if we can empty these N boxes in a series of rounds of size n

each. Two rounds of size n are said to be *identical* if the selected n boxes are the same in these two rounds. Otherwise, they are said to be *distinct*.

For our purpose we will keep track of boxes being selected in each round. Two major questions at this stage are: What are the necessary and sufficient conditions under which a successful game can be played? If the setup allows a successful game, then in how many distinct ways can we play the game? The answer to question two is unknown to us and the answer to question one is given below.

Theorem 5.6. *The necessary and sufficient conditions for the existence of a successful game based on a series of rounds of size n played on N boxes containing* $k_1, k_2, \ldots, k_N$ *objects are*

$$\text{(i)} \quad M \equiv 0 (\text{mod } n) , \qquad \text{(ii)} \quad \max_i k_i \le M/n \tag{5.34}$$

where $M = k_1 + k_2 + \cdots + k_N$.

Proof. Since in each round we remove n objects, to have a successful game the total number of objects must be a multiple of n. Also, a successful game has precisely M/n rounds. Clearly, no box can contain more objects than the total number of rounds. This establishes the necessity of conditions (i) and (ii). To prove the sufficiency of these conditions, we shall identify a successful game. In the first round we remove one object from those n boxes which contain the largest number of objects. This is called a *max-round*. Let the second and the remaining rounds also be max-rounds on these N boxes with the residual objects. These M/n max-rounds produce a successful game (Exercise 5.9). □

The following example will elucidate the procedure outlined above.

Example 5.5. Consider the following system: $N = 6$, $n = 3$, $k_1 = 6$, $k_2 = 9$, $k_3 = 12$, $k_4 = 3$, $k_5 = 6$, and $k_6 = 15$. Here $M = 51$ and $M/n = 17$. The necessary conditions are satisfied. The 17 rounds of size 3 exhibited in Table 5.3 will remove all the 51 objects.

Remark 5.2. We would like to make the following important observations in the context of Example 5.5.

(i) Except in rounds 1, 2, 3, 6, 13, and 15 we had more than one choice in selecting $n = 3$ boxes under the max-round principle.
(ii) We could modify our procedure and empty boxes 3, 5, and 6 in rounds 13, 14, and 15 and boxes 1, 2, and 4 in rounds 16 and 17.

The above example clearly demonstrates that there are many options in the formation of rounds in general and our procedure could easily be

Table 5.3

Box No.	1	2	3	4	5	6
No. of objects:	6	9	12	3	6	15
Round 1		1	1			1
Residuals	6	8	11	3	6	14
Round 2		1	1			1
Residuals	6	7	10	3	6	13
Round 3		1	1			1
Residuals	6	6	9	3	6	12
Round 4	1		1			1
Residuals	5	6	8	3	6	11
Round 5			1		1	1
Residuals	5	6	7	3	5	10
Round 6		1	1			1
Residuals	5	5	6	3	5	9
Round 7	1		1			1
Residuals	4	5	5	3	5	8
Round 8		1			1	1
Residuals	4	4	5	3	4	7
Round 9	1	1			1	
Residuals	3	4	4	3	4	6
Round 10		1	1			1
Residuals	3	3	3	3	4	5
Round 11	1				1	1
Residuals	2	3	3	3	3	4
Round 12		1		1		1
Residuals	2	2	3	2	3	3
Round 13			1		1	1
Residuals	2	2	2	2	2	2
Round 14	1			1	1	
Residuals	1	2	2	1	1	2
Round 15		1	1			1
Residuals	1	1	1	1	1	1
Round 16			1	1		1
Residuals	1	1	0	0	1	0
Round 17	1	1			1	
Residuals	0	0	0	0	0	0

Table 5.4

Round	Selected Boxes	Round	Selected boxes
1	2, 3, 6	10	2, 3, 6
2	2, 3, 6	11	1, 5, 6
3	2, 3, 6	12	2, 4, 6
4	1, 3, 6	13	3, 5, 6
5	3, 5, 6	14	1, 4, 5
6	2, 3, 6	15	2, 3, 6
7	1, 3, 6	16	3, 4, 6
8	2, 5, 6	17	1, 2, 5
9	1, 3, 6		

modified throughout the process. The properties are important when we apply this procedure in constructing ΠPS sampling designs with additional properties.

To introduce and utilize the next idea we have recorded in Table 5.4 the results of the 17 rounds of the game for Example 5.5. As we can observe from Table 5.4, there are several identical rounds in this game. For example, rounds 1, 2, 3, 6, 10, and 15 are identical. Indeed, this game has precisely 9 distinct rounds. Thus, rather than playing rounds 1, 2, 3, 6, 10, and 15 we could play just round 1 and remove 6 objects from each of the three boxes 2, 3, and 6. A similar simplification applies for other sets of identical rounds. This observation leads us to the notion of a round of *frequency* f and *size* n. A round is said to be of frequency f and size n if n boxes are selected and f objects are removed from each. Thus, by combining rounds 1, 2, 3, 6, 10, and 15 in Example 5.5 we will have a round of frequency 6 and size 3. This notion raises the following useful question. Suppose a setup can have a successful game of size n. In any given round, how large can the frequency be so that a successful game can be devised based on the residual objects? The answer is this: the frequency can be made arbitrarily large so long as the necessary conditions (5.34) are not violated for the residual objects. The following example elucidates this point.

Example 5.6. Let $N = 5$, $n = 3$, $k_1 = k_2 = k_3 = 3$, $k_4 = 4$, and $k_5 = 5$. Then, the frequency of the first round cannot be made bigger than one if box number 5 is not in the round. However, the frequency of the first round can be made as large as three if box numbers 5 and 4 are included in the round. The reason for the former case is this: any round of frequency 2 or more will cause the violation of the necessary condition (ii) of (5.34). For the latter case the necessary conditions (5.34) will be preserved. A successful game starting with a round of frequency three and size three is given in Table 5.5.

Table 5.5

Box No.	1	2	3	4	5
No. of objects:	3	3	3	4	5
Round 1: frequency 3			3	3	3
Residuals	3	3	0	1	2
Round 2: frequency 2	2	2			2
Residuals	1	1	0	1	0
Round 3: frequency 1	1	1		1	
Residuals	0	0	0	0	0

Suppose a setup can have a successful game. What would be the maximum frequency of a max-round so that a successful game on the residuals can be devised? The answer is contained in the following.

Theorem 5.7. *If necessary, renumber the boxes so that* $k_1 \geq k_2 \geq \cdots \geq k_N$. *Then the maximum frequency of the first max-round of size n is equal to* $\min\{k_n, (M/n) - k_{n+1}\}$, *where* $M = k_1 + k_2 + \cdots + k_N$.

Proof. Clearly the frequency of the first max-round cannot exceed k_n. On the other hand if we remove f objects from boxes $1, 2, \ldots, n$, then to be able to devise a successful game on the residuals we must verify the necessary conditions (5.34). Condition (i) of (5.34) is trivially satisfied for any $f \leq k_n$. To preserve condition (ii) of (5.34), it is necessary that the residuals $k_1 - f$, $k_2 - f, \ldots, k_n - f$, $k_{n+1}, \ldots, k_N$ after the removal of f objects from boxes 1 to n should satisfy

$$k_{n+1} \leq \frac{\sum_{i=1}^{N} k_i - nf}{n} = \frac{M}{n} - f\,. \qquad (5.35)$$

This completes the proof. □

As a corollary to Theorem 5.7 we can state that within the process of implementing a successful game we can introduce a max-round with the maximum frequency as long as this maximum frequency is $\min\{k'_n, (M'/n) - k'_{n+1}\}$, where $k'_1 \geq k'_2 \geq \cdots \geq k'_N$ are ordered residuals and $M' = k'_1 + k'_2 + \cdots + k'_N$.

To illustrate the preceding idea we apply a series of max-rounds with maximum frequency at each round on the setup of Example 5.6. We rearrange the boxes so that, to begin with, the contents are in nonincreasing order. Example 5.7 illustrates this feature of the game.

Example 5.7. Let $N = 6$, $n = 3$, and $k_1 = 6$, $k_2 = 9$, $k_3 = 12$, $k_4 = 3$, $k_5 = 6$, $k_6 = 15$. A series of five max-rounds with maximum frequency in each round for a game of emptying boxes with these contents is exhibited in Table 5.6.

Table 5.6

Box No.	6	3	2	1	5	4
No. of objects:	15	12	9	6	6	3
Max-round 1: frequency 9	9	9	9			
Residuals	6	3	0	6	6	3
Max-round 2: frequency 5	5			5	5	
Residuals	1	3	0	1	1	3
max-round 3: frequency 1	1	1				1
Residuals	0	2	0	1	1	2
max-round 4: frequency 1		1		1		1
Residuals	0	1	0	0	1	1
max round 5: frequency 1		1			1	1
Residuals	0	0	0	0	0	0

A successful game whose rounds are all max-rounds with maximum frequency is called a *game with saturated max-rounds*. Example 5.7 is one such game. In Exercise 5.10 the reader is asked to prove that, if a set up with N boxes allows a successful game, then the game with *saturated* max-rounds will empty the boxes in no more than N distinct rounds. However, a game with saturated max-rounds does not necessarily possess the minimum number of distinct rounds (Exercise 5.11).

We have developed enough ideas to be useful for our sampling problems. We shall shortly show that, by approximating the given normed size measures $p_1, p_2, \ldots, p_N$ by rational numbers, we can establish a direct correspondence between ΠPS sampling designs and the games discussed above. It may be noted that implementation of any sampling scheme discussed in the previous sections also tacitly rests on the assumption that the normed size measures are rational numbers. Therefore we shall assume that the normed size measures $p_1, p_2, \ldots, p_N$ are, indeed, rational numbers.

We shall now state and prove three theorems. The first two theorems show that games can be used to construct ΠPS sampling designs with or without additional features. The third theorem shows that for any given ΠPS sampling design we can construct a game which can be used to produce the same ΠPS sampling design. These theorems together establish the fact that all ΠPS sampling designs can be generated via games described here.

Theorem 5.8. *For any N, n, and normed size measures $p_1, p_2, \ldots, p_N$ with $np_i = \pi_i \le 1$, there exists a game that can be converted into a* ΠPS$(N, n, \mathbf{p})$ *sampling design.*

Proof. Let q be a positive integer so that $k_i = np_i q$ is an integer, $i = 1, 2, \ldots, N$. Then, irrespective of the value of q, the positive integers $k_1, k_2, \ldots, k_N$ will satisfy the necessary conditions (5.34) and thus a successful game of size n can be played based on N, n, and $k_1, k_2, \ldots, k_N$. Play any such game (for example, a game with saturated max-rounds). Now construct a sampling design of size n in the following way. Let the support consist of those samples of size n associated with distinct rounds in the game. Let the probability over any sample $(i_1, i_2, \ldots, i_n)$ of size n be the sum of the frequencies of the rounds associated with the boxes labelled $(i_1, i_2, \ldots, i_n)$ emptied in the game, divided by q. Clearly this results in a sampling design (Exercise 5.12) with

$$
\begin{aligned}
\pi_i &= \text{sum of the frequencies of rounds containing the } i\text{th unit}/q \\
&= np_i q/q = np_i\ . \qquad \square
\end{aligned}
$$

Example 5.8. Let $N = 5$, $n = 3$, and $p_1 = 2/16$, $p_2 = 3/16$, $p_3 = 4/16$, $p_4 = 2/16$, and $p_5 = 5/16$. For the choice of $q = 16$, we obtain $k_1 = 6$, $k_2 = 9$, $k_3 = 12$, $k_4 = 6$, and $k_5 = 15$. We first exhibit a series of rounds in Table 5.7 for this emptying box problem.

Next we exhibit the corresponding ΠPS$(5, 3, \mathbf{p})$ sampling design. For example, the sample $(2, 3, 5)$ has frequency 2 in each of the rounds 1, 4, and 7. Therefore, according to Theorem 5.8, $P((2, 3, 5)) = (2 + 2 + 2)/16 = 6/16$. Similarly the rest of the samples and the corresponding probabilities can be determined. The resulting sampling design is: $P((2, 3, 5)) = 6/16$. $P((1, 3, 5)) = P((3, 4, 5)) = 3/16$, $P((1, 2, 4)) = P((1, 2, 5)) = P((2, 4, 5)) = P((1, 4, 5)) = 1/16$.

Let us now look at the procedure outlined in Theorem 5.8 from other viewpoints. The only demand we formally imposed on the procedure of this theorem was that π_i be proportional to p_i as was specified in the definition of ΠPS sampling designs. Otherwise, we left the procedure very flexible so that we can adjust or modify it to produce desirable ΠPS sampling designs. If, for example, we further demand that $\pi_{ij} > 0$ for each pair (i, j), $1 \le i \ne j \le N$, then we should adjust the procedure in light of the following fact. The integer q must be chosen large enough so that

$$
\min_i k_i \ge \left\{ \frac{N-1}{n-1} \right\}, \text{ the smallest integer not less than } (N-1)/(n-1)\,; \tag{5.36}
$$

otherwise, there will not be enough samples to cover *all* pairs (i, j), $i \ne j = 1, 2, \ldots, N$ (Exercise 5.13). Indeed, we shall present a technique

Table 5.7

Box No.	1	2	3	4	5
No. of objects:	6	9	12	6	15
Round 1: frequency 2		2	2		2
Residuals	6	7	10	6	13
Round 2: frequency 2	2		2		2
Residuals	4	7	8	6	11
Round 3: frequency 1	1	1		1	
Residuals	3	6	8	5	11
Round 4: frequency 2		2	2		2
Residuals	3	4	6	5	9
Round 5: frequency 1	1		1		1
Residuals	2	4	5	5	8
Round 6: frequency 2			2	2	2
Residuals	2	4	3	3	6
Round 7: frequency 2		2	2		2
Residuals	2	2	1	3	4
Round 8: frequency 1		1		1	1
Residuals	2	1	1	2	3
Round 9: frequency 1			1	1	1
Residuals	2	1	0	1	2
Round 10: frequency 1	1	1			1
Residuals	1	0	0	1	1
Round 11: frequency 1	1			1	1
Residuals	0	0	0	0	0

much more useful than selecting a large q. We call it a *multiplier technique*. It works like this: we start the game with any q so long as all $k_i = np_iq$ are integers and keep track of all pairs (i, j), $i \neq j$, while going through the rounds. At any stage, if a round is designed to empty a box, we check in advance whether the samples associated with the previous rounds and one to be produced at the moment cover all pairs involving the unit represented by that box. If not, we multiply all the residuals by a large enough integer h so that in future rounds we can include those pairs which were not covered before. In other words, we can inflate the residuals, if necessary, by a multiplier, adopted at any stage of operation of the game.

There is a slight modification in forming the corresponding ΠPS sampling design if one or more multipliers is used. As before, the samples in the support are those determined by the rounds. However, in assigning probabilities over these samples we should take into account the values of those multipliers and the stages at which they were introduced. To make the notation and presentation simple let there be one multiplier only, denoted by h. Suppose s is a sample. Let $r(s)$ be the sum of the frequencies of the rounds which produced s before the introduction of the multiplier. Let $r'(s)$ be the sum of the frequencies of the rounds which produced s after the multiplier was introduced. We now allot the probability $P(s)$ to s as a quantity proportional to $hr(s) + r'(s)$. It is easy to verify that the resulting sampling design is a $\Pi PS(N, n, \mathbf{p})$ sampling design where $\mathbf{p}$ is the original normed size vector (Exercise 5.14). In a similar and parallel fashion we can define the corresponding ΠPS sampling design if more than one multiplier were to be used.

There are cases for which $\pi_{ij} > 0$, $i \neq j = 1, 2, \ldots, N$, without introducing a multiplier. Example 5.8 is one such example. Now we give an example where a single multiplier is used to achieve the positivity of π_{ij} for all $i \neq j = 1, 2, \ldots, N$.

Example 5.9. Let $N = 8$, $n = 3$, and $p_1 = 2/18$, $p_2 = 3/18$, $p_3 = 1/18$, $p_4 = 5/18$, $p_5 = 1/18$, $p_6 = 2/18$, $p_7 = 1/18$, $p_8 = 3/18$. For $q = 18$ we obtain $k_1 = 6$, $k_2 = 9$, $k_3 = 3$, $k_4 = 15$, $k_5 = 3$, $k_6 = 6$, $k_7 = 3$, $k_8 = 9$. For this choice of q

$$\min_i k_i = 3 < \left\{\frac{N-1}{n-1}\right\} = 4$$

and therefore, if we go on and empty 8 boxes with these k_i's, we will end up with a ΠPS sampling design in which some of the π_{ij}'s will be zero. Of course, we could avoid this by selecting a large q, but it will necessitate going through many rounds. In Table 5.8 we will go on with these k_i's and introduce a multiplier so that π_{ij}'s will become positive for all $i \neq j = 1, 2, \ldots, 8$.

Note that in Table 5.8 we increased the values of the residuals at the end of round 4 as otherwise the next round would have emptied boxes prior to achieving $\pi_{ij} > 0$ for all pairs (i, j). In this situation the multiplier $h = 4$ was enough to ensure that we could judiciously select rounds eventually leading to positivity of all π_{ij}'s. We also observe that by the end of round 13 we did achieve our goal. Thus we were at liberty to select any pattern for the last three rounds of the game.

Since in this case $h = 4$, each round before round 5 is counted four times in computing the probabilities over the selected samples. The resulting ΠPS sampling design with all $\pi_{ij} > 0$ is exhibited in Table 5.9. It is interesting to observe that this sampling design has excluded $\binom{8}{3} - 16 = 40$ samples from the support. While reducing the support size was not a goal, this example indicates that we can manipulate the emptying box procedure to exclude some undesirable samples, thereby making it useful in controlled sampling.

We now summarize the preceding ideas in the form of a theorem.

Table 5.8

Box No.	1	2	3	4	5	6	7	8
No. of objects:	6	9	3	15	3	6	3	9
Round 1: frequency 8		8		8				8
Residuals	6	1	3	7	3	6	3	1
Round 2: frequency 5	5			5		5		
Residuals	1	1	3	2	3	1	3	1
Round 3: frequency 1			1		1		1	
Residuals	1	1	2	2	2	1	2	1
Round 4: frequency 1			1	1	1			
Residuals	1	1	1	1	1	1	2	1
Modified Residuals (multiplier = 4)	4	4	4	4	4	4	8	4
Round 5: frequency 1	1				1		1	
Residuals	3	4	4	4	3	4	7	4
Round 6: frequency 1		1				1	1	
Residuals	3	3	4	4	3	3	6	4
Round 7: frequency 1				1			1	1
Residuals	3	3	4	3	3	3	5	3
Round 8: frequency 1	1	1						1
Residuals	2	2	4	3	3	3	5	2
Round 9: frequency 1		1			1		1	
Residuals	2	1	4	3	2	3	4	2
Round 10: frequency 1					1	1		1
Residuals	2	1	4	3	1	2	4	1
Round 11: frequency 1			1			1	1	
Residuals	2	1	3	3	1	1	3	1
Round 12: frequency 1	1		1					1
Residuals	1	1	2	3	1	1	3	0
Round 13: frequency 1	1	1	1					
Residuals	0	0	1	3	1	1	3	0
Round 14: frequency 1			1	1			1	
Residuals	0	0	0	2	1	1	2	0
Round 15: frequency 1				1	1		1	
Residuals	0	0	0	1	0	1	1	0
Round 16: frequency 1				1		1	1	
Residuals	0	0	0	0	0	0	0	0

Table 5.9

Sample	Probability	Sample	Probability	Sample	Probability
(2, 4, 8)	32/72	(1, 4, 6)	20/72	(3, 5, 7)	4/72
(3, 4, 5)	4/72	(1, 5, 7)	1/72	(2, 6, 7)	1/72
(4, 7, 8)	1/72	(1, 2, 8)	1/72	(2, 5, 7)	1/72
(5, 6, 8)	1/72	(3, 6, 7)	1/72	(1, 3, 8)	1/72
(3, 4, 7)	1/72	(1, 2, 3)	1/72	(4, 5, 7)	1/72
(4, 6, 7)	1/72				

Theorem 5.9. *For any given $N, n < N$ and normed size measures $p_1, p_2, \ldots, p_N$ with $np_i \leq 1$, $1 \leq i \leq N$, there are games resulting in sampling designs with various support sizes and varieties of probabilities on each support with $\pi_{ij} > 0$, for each pair (i, j), $1 \leq i \neq j \leq N$.*

The next theorem demonstrates that the emptying box procedure is so general that it will produce *all* ΠPS sampling designs.

Theorem 5.10. *For given N, $n < N$ and normed size vector* $\mathbf{p}$, *the emptying box procedure produces all* ΠPS$(N, n, \mathbf{p})$ *sampling designs.*

Proof. We need to prove that if we are given an arbitrary ΠPS$(N, n, \mathbf{p})$ sampling design generated by a method other than the emptying box procedure, then there exists a q and a series of rounds which can produce the given ΠPS sampling design. Let $s_1, s_2, \ldots, s_m$ be the samples and $P(s_1)$, $P(s_2), \ldots, P(s_m)$ be the corresponding probabilities for the given sampling design. Compute $\pi_i = np_i$, $i = 1, 2, \ldots, N$, for this design. With no loss of generality we shall assume that these probabilities are rational. We select an integer q so that $P(s_j)q$ is an integer for $j = 1, 2, \ldots, m$. Let

$$k_i = np_i q = q\pi_i = q \sum_{s \ni i} P(s)\,, \qquad i = 1, 2, \ldots, N\,. \tag{5.37}$$

Clearly, the k_i's are integers. Now consider the game with these N, n, and $k_1, k_2, \ldots, k_N$. Suppose s_j in the support consists of $j_1, j_2, \ldots, j_n$. Then we remove $P(s_j)q$ objects from boxes $j_1, j_2, \ldots, j_n$ with a round of frequency $P(s_j)q$. We follow this procedure for $j = 1, 2, \ldots, m$. These series of rounds will empty all N boxes since in all we have removed

$$\sum_{j=1}^{m} nP(s_j)q = nq = k_1 + k_2 + \cdots + k_N \tag{5.38}$$

objects. Now if we form the corresponding ΠPS sampling design it will be identical to the one we started with. Clearly they both have the same support. Moreover, the probability which the sampling design generated by the game puts on the sample s_j will be equal to $P(s_j)q/q = P(s_j)$, as we desired. □

It is obvious that, if a ΠPS sampling design possesses any additional properties, these will also be exhibited by the sampling design obtained through the emptying box procedure.

While the emptying box procedure is capable of producing all ΠPS sampling designs, it is not easy to specify a set of predetermined rounds so that the generated ΠPS sampling design enjoys additional properties such as $0 < \pi_{ij} \le \pi_i \pi_j$. For moderate values of N and n we can perhaps generate by computer a sufficiently large number of ΠPS sampling designs and select the one which is desirable to us. However, it would be very desirable to be able to specify a series of predetermined rounds which produces a ΠPS sampling design with $0 < \pi_{ij} \le \pi_i \pi_j$, for all $i \ne j$. Relying, for example, on the scheme of Sampford (1967), and utilizing Theorem 5.10 we know that in all cases there are rounds with the above desirable properties. But we need to specify it in a concrete form. For general N, n, and the normed size vector $\mathbf{p}$ the solution seems complicated. However, for $n = 2$ we will produce below a class of solutions which are readily applicable.

First we apply Theorem 5.9 to reproduce the sampling design arising out of the Brewer–Rao–Durbin (BRD) sampling schemes using the emptying box procedure. According to their schemes, we have from (5.7) for the pair (i, j) the joint inclusion probability given by

$$p_i p_j \{(1-2p_i)^{-1} + (1-2p_j)^{-1}\}\left[1 + \sum_{k=1}^{N} p_k^2 (1-2p_k)^{-1}\right]^{-1} = \pi_{ij}^0 .$$

We now select an integer A so large that $A\pi_{ij}^0 = n_{ij}$ is an integer for *every* pair (i, j). Then we apply the emptying box procedure in the following way. We simply select the pair of boxes (i, j) in one single round of frequency n_{ij}, $1 \le i < j \le N$. It is evident that we end up with the same reduced form as the sampling designs underlying the above schemes of sampling.

We now suggest a modification to this procedure which will generate a class of ΠPS sampling designs each meeting the requirements $0 < \pi_{ij} \le \pi_i \pi_j$, $1 \le i < j \le N$. The modification calls for relaxation in the choice of A. This time we demand that the choice of A, say A^*, is such that

$$n_{ij}^* = A^* \pi_{ij}^0 \ge 1 \qquad \text{for } \textit{every} \text{ pair } (i, j) \tag{5.39}$$

but n_{ij}^*'s *not* necessarily being integers, while

$$k_i = \sum_{j(\ne i)} n_{ij}^* = A^* \sum_{j(\ne i)} \pi_{ij}^0 = A^* \pi_i \tag{5.40}$$

is an integer for every i, $1 \le i \le N$. Condition (5.40) can be easily met in practice by choosing $A^* = tq$ where t is an integer and q is such that $q\pi_i$ is an integer for each i. This point will be illustrated further in an example to be given at the end. Suppose now that we place k_i objects in the ith box, $1 \le i \le N$, and that we start emptying the boxes in pairs. We select each pair

(i, j) in a single round of frequency n_{ij}, where $n_{ij} = [n^*_{ij}]$ = greatest integer not exceeding n^*_{ij}, $1 \le i < j \le N$. After completing this $\binom{N}{2}$ series of rounds, we may be left with a total of $B = \Sigma_{i=1}^{N} m_i$ objects where $m_i = k_i - \Sigma_{j(\neq i)} n_{ij}$ is the number of objects left in the ith box. We now continue the game of emptying the boxes with these residual objects, removing pairs until the end. That we actually end up with a successful game is established in the following lemma.

Lemma 5.1. *With m_i objects in the ith box, $1 \le i \le N$, the boxes can be emptied in pairs.*

Proof. Clearly, our task is to verify the conditions in (5.34). First observe that

$$M = \sum_i m_i = \sum_i \Big\{k_i - \sum_{j(\neq i)} n_{ij}\Big\} = \sum k_i - \sum_{i\neq j}\sum n_{ij} = A^* \sum \pi_i - \sum_{i\neq j}\sum n_{ij}$$

$$= 2A^* - \sum_{i\neq j}\sum n_{ij} = 2\Big\{A^* - \sum_{i<j}\sum n_{ij}\Big\},$$

so that M is an even integer. Next we show that $2m_i \le M$, that is, $m_i \le \Sigma_{j(\neq i)} m_j$ for each i. We distinguish between two cases.

Case I. $m_i \ge 1$ for each $i, 1 \le i \le N$. In this case, $\Sigma_{j(\neq i)} m_j \ge N - 1$. We show that $m_i \le N - 1$ for each $i, 1 \le i \le N$. Note that $n_{ij} = [n^*_{ij}] > n^*_{ij} - 1$ so that

$$m_i = k_i - \sum_{j(\neq i)} n_{ij} \le k_i - \sum_{j(\neq i)} (n^*_{ij} - 1) = k_i - \sum_{j(\neq i)} n^*_{ij} + (N-1)$$

$$= A^*\pi_i - A^* \sum_{j(\neq i)} \pi^0_{ij} + N - 1 = N - 1 .$$

Case II. $m_i = 0$ for some i. Assume $m_i = 0$ for $i = 1, 2, \ldots, l$ for some l, $1 \le l \le N-2$. This means that each $n^*_{ij} (= A^*\pi^0_{ij})$ is an integer for $i = 1, 2, \ldots, l$, $j = 1, 2, \ldots, N$. Therefore, for any $i > l$,

$$m_i = k_i - \sum_{j(\neq i)} n_{ij} = A^*(\pi_i - \pi^0_{i1} - \cdots - \pi^0_{il}) - \sum_{\substack{j=l+1\\(j\neq i)}}^{N} n_{ij} = A^* \sum_{\substack{j=l+1\\(j\neq i)}}^{N} \pi^0_{ij} - \sum_{\substack{j=l+1\\(j\neq i)}}^{N} n_{ij}$$

$$= A^* \sum_{\substack{j=l+1\\(j\neq i)}}^{N} \pi^0_{ji} - \sum_{\substack{j=l+1\\(j\neq i)}}^{N} n_{ji} = \sum_{\substack{j=l+1\\(j\neq i)}}^{N} (A^*\pi^0_{ji} - n_{ji}) .$$

Clearly the quantity $A^*\pi^0_{ji} - n_{ji}$ is involved in the expression for m_j for $l + 1 \le j(\neq i) \le N$. Therefore, $m_i \le \Sigma_{j=l+1,(j\neq i)}^{N} m_j$ for $i > l$. This establishes the claim. □

It may be noted that l cannot exceed $N-2$. In view of the above, we may now conclude that there always exists a successful game, no matter how the second series of rounds is played. It thus turns out that in the process we have produced a ΠPS sampling design of fixed size 2. Moreover, any sampling design so generated also satisfies $0<\pi_{ij}\leq\pi_i\pi_j$, $1\leq i<j\leq N$, if our choice of A^* is such that

$$4p_ip_jA^*\geq(N-1)\left[1+\sum_{k=1}^{n}p_k(1-2p_k)^{-1}\right]\left[\sum_{\substack{k\neq i\\ \neq j}}p_k(1-2p_k)^{-1}\right]^{-1} \tag{5.41}$$

for each pair (i, j). (See (5.8) and (5.9) in this context.) To verify this last claim, we observe that for any such design, d,

$$\pi_d(i, j)=(n_{ij}+\gamma_{ij})/A^*\,, \tag{5.42}$$

where $n_{ij}=[n^*_{ij}]=[A^*\pi^0_{ij}]$ and γ_{ij} is the total frequency of the last set of rounds involving the boxes $((i, j)$, $1\leq i<j\leq N$. Since $\gamma_{ij}\leq\min\{m_i, m_j\}$, we conclude that $\gamma_{ij}\leq N-1$ for each pair (i, j) since $m_i\leq N-1$ as noted earlier in Lemma 5.1. Therefore, from (5.42), we have

$$\pi_d(i, j)\leq(n_{ij}+N-1)/A^*\leq(n^*_{ij}+N-1)/A^*=\pi^0_{ij}+(N-1)/A^*\,, \tag{5.43}$$

so that, if we now choose A^* satisfying $\pi^0_{ij}+(N-1)/A^*\leq\pi_i\pi_j$ for each pair (i, j), we are done. This yields $A^*\geq(N-1)\{\pi_i\pi_j-\pi^0_{ij}\}^{-1}$, which simplifies to (5.41). Hence we have proved the following result.

Theorem 5.11. *For given N and the normed size measures $p_1, p_2, \ldots, p_N$ with $2p_i<1$, $1\leq i\leq N$, there are various games which can be played to produce a family of ΠPS$(N, 2, \mathbf{p})$ sampling designs with properties $0<\pi_{ij}\leq\pi_i\pi_j$, $1\leq i\neq j\leq N$.*

We will now take up an illustrative example.

Example 5.10. Let $N=5$, $n=2$, and $p_1=0.4$, $p_2=p_3=0.2$, $p_4=p_5=0.1$. Based on BRD methods we obtain the following values of π^0_{ij}'s. $\pi^0_{12}=\pi^0_{13}=64/235$, $\pi^0_{14}=\pi^0_{15}=30/235$, $\pi^0_{23}=16/235$, $\pi^0_{24}=\pi^0_{25}=7/235$, $\pi^0_{34}=\pi^0_{35}=7/235$, and $\pi^0_{45}=3/235$. For the choice of $A=235$ we obtain $k_1=188$, $k_2=k_3=94$, $k_4=k_5=47$. Consequently, the game with precisely 10 distinct rounds illustrated in Table 5.10 will produce BRD sampling designs in the reduced form. This has been listed in Table 5.13.

As alternatives to BRD schemes, we will now produce some other sampling designs with a choice $A^*<235$. Subject to satisfying (5.39), (5.40),

Table 5.10

Box No.	1	2	3	4	5
No. of objects:	188	94	94	47	47
Round 1: frequency 64	64	64			
Residuals	124	30	94	47	47
Round 2: frequency 64	64		64		
Residuals	60	30	30	47	47
Round 3: frequency 30	30			30	
Residuals	30	30	30	17	47
Round 4: frequency 30	30				30
Residuals	0	30	30	17	17
Round 5: frequency 16		16	16		
Residuals	0	14	14	17	17
Round 6: frequency 7		7		7	
Residuals	0	7	14	10	17
Round 7: frequency 7		7			7
Residuals	0	0	14	10	10
Round 8: frequency 7			7	7	
Residuals	0	0	7	3	10
Round 9: frequency 7			7		7
Residuals	0	0	0	3	3
Round 10: frequency 3				3	3
Residuals	0	0	0	0	0

and (5.41) we can take $A^* \geq 150$. For the choice of $A^* = 150$ we obtain $k_1 = 120$, $k_2 = k_3 = 60$, $k_4 = k_5 = 30$. This time the first 10 distinct rounds will be uniquely specified following the description mentioned before Lemma 5.1. We have presented this part of the game in Table 5.11. The second part of the game which consists of emptying the residuals after round 10 is now left to the discretion of the sampler. We list in Table 5.12 two such options and in Table 5.13 we display the two resulting sampling designs along with the BRD design.

It is easy to check that any choice of $A^* = 5t$, $t \geq 30$, would serve the purpose. In particular $t = 47$ produces BRD designs. The factor 5 is meant to take care of (5.40). Further, it may be noted that condition (5.41) is not necessary (Exercise 5.15).

Remark 5.3. Among the available procedures only the emptying box procedure can be manipulated to produce ΠPS sampling designs with additional desirable features such as exclusion of some undesirable samples

Table 5.11

Box No.	1	2	3	4	5
No. of objects:	120	60	60	30	30
Round 1: frequency 40	40	40			
Residuals	80	20	60	30	30
Round 2: frequency 40	40		40		
Residuals	40	20	20	30	30
Round 3: frequency 19	19			19	
Residuals	21	20	20	11	30
Round 4: frequency 19	19				19
Residuals	2	20	20	11	11
Round 5: frequency 10		10	10		
Residuals	2	10	10	11	11
Round 6: frequency 4		4		4	
Residuals	2	6	10	7	11
Round 7: frequency 4		4			4
Residuals	2	2	10	7	7
Round 8: frequency 4			4	4	
Residuals	2	2	6	3	7
Round 9: frequency 4			4		4
Residuals	2	2	2	3	3
Round 10: frequency 1				1	1
Residuals	2	2	2	2	2

Table 5.12

Option 1						Option 2					
Box No.	1	2	3	4	5	Box No.	1	2	3	4	5
Residuals (after round 10)	2	2	2	2	2	Residuals (after round 10)	2	2	2	2	2
Round 11: frequency 2	2	2				Round 11: fequency 1	2		2		
Residuals	0	0	2	2	2	Residuals	0	2	0	2	2
Round 12: frequency 1			1	1		Round 12: frequency 1		1		1	
Residuals	0	0	1	1	2	Residuals	0	1	0	1	2
Round 13: fequency 1			1		1	Round 13: frequency 1		1			1
Residuals	0	0	0	1	1	Residuals	0	0	0	1	1
Round 14: frequency 1				1	1	Round 14: frequency 1				1	1
Residuals	0	0	0	0	0	Residuals	0	0	0	0	0

Table 5.13

Samples in the Support	Probabilities		
	BRD Schemes	Option 1	Option 2
(1, 2)	64/235	42/150	40/150
(1, 3)	64/235	40/150	42/150
(1, 4)	30/235	19/150	19/150
(1, 5)	30/235	19/150	19/150
(2, 3)	16/235	10/150	10/150
(2, 4)	7/235	4/150	5/150
(2, 5)	7/235	4/150	5/150
(3, 4)	7/235	5/150	4/150
(3, 5)	7/235	5/150	4/150
(4, 5)	3/235	2/150	2/150

(if $n > 2$) from the support and putting a high or low probability over selected samples which are more or less desired, for example, from an economical point of view. This flexibility, on the other hand, has left us with many unsettled questions. For example, for a given sampling setup with $n \geq 3$, how should we play the associated game, besides the one related to Sampford's design, so that the resulting sampling design enjoys the property

Table 5.14

Samples	Probabilities for Design No. 1	Probabilities for Design No. 2
(1, 2, 3)	–	1/34
(1, 2, 4)	–	–
(1, 2, 5)	2/34	–
(1, 2, 6)	–	3/34
(1, 3, 4)	–	–
(1, 3, 5)	–	–
(1, 3, 6)	6/34	7/34
(1, 4, 5)	2/34	1/34
(1, 4, 6)	–	–
(1, 5, 6)	2/34	–
(2, 3, 4)	–	1/34
(2, 3, 5)	–	1/34
(2, 3, 6)	12/34	7/34
(2, 4, 5)	–	–
(2, 4, 6)	2/34	2/34
(2, 5, 6)	2/34	3/34
(3, 4, 5)	–	–
(3, 4, 6)	2/34	1/34
(3, 5, 6)	4/34	6/34
(4, 5, 6)	–	1/34

$0 < \pi_{ij} \le \pi_i \pi_j$, for all $i \ne j$? This is an interesting and challenging problem for the reader to consider.

Before closing the section we present in Table 5.14 two ΠPS$(6, 3, \mathbf{p})$ sampling designs with $0 < \pi_{ij} \le \pi_i \pi_j$ for all $i \ne j$, where the normed size measures are $p_1 = 2/17$, $p_2 = 3/17$, $p_3 = 4/17$, $p_4 = 1/17$, $p_5 = 2/17$, $p_6 = 5/17$, both of them produced by the emptying box procedure. One of these has support size 9 while the other has support size 12. The one with support size 9 is the same as the one in Example 5.8. In Exercise 5.15, the reader is asked to produce a game which results in the other sampling design.

Additional results on the emptying box procedure are given in Hedayat et al. (1989) and Hedayat (1990).

5.5. PPS SCHEME OF SAMPLING AND RELATED ESTIMATORS

So far our inclination towards using auxiliary information was guided by the consideration of the exclusive use of the HTE for estimating the finite population total or mean. We decided on some desirable features of fixed size sampling designs to support the use of the HTE and described various sampling methods for meeting these objectives.

In the literature some other methods of sampling utilizing auxiliary information have been suggested. These can be implemented in a straightforward manner to quite easily produce fixed-size (n) designs for any value of $n \ge 2$. However, the HTE is generally inapplicable because of algebraic complication in the expressions for the π_i's and π_{ij}'s. Instead, an alternative estimator with some appealing properties is available with each such design. This avoids the use of the π_{ij}'s, an issue emphasized in Remark 3.1. We shall study one such sampling strategy in detail below.

As usual, we consider a labelled population $\mathcal{U}$ of N units and denote by p_i the normed size measure of the ith unit, $1 \le i \le N$. According to this method, a sample $s = (i_1, i_2, \ldots, i_n)$ is drawn utilizing the sampling algorithm $A(q_1, q_2, q_3)$ where

$$q_1((i)) = p_i\,, \qquad 1 \le i \le N\,,$$

$$q_2(s) = \begin{cases} 1\,, & \text{if } n(s) \le n-1 \\ 0\,, & \text{otherwise}\,, \end{cases}$$

$$q_3((i)|s = (i_1, i_2, \ldots, i_k)) = p_i/(1 - p_{i_1} - p_{i_2} - \cdots - p_{i_k})\,, \tag{5.44}$$
$$1 \le i(\ne i_1, \ne i_2 \ne \cdots \ne i_k) \le N,\ 1 \le k \le n-1$$
$$\text{for any set of } (i_1, i_2, \ldots, i_k),\ 1 \le i_1 \ne i_2 \ne \cdots \ne i_k \le N\,.$$

In terms of the original size measures $X_1, \ldots, X_N$, we may write

$$q_1((i)) = X_i/T(\mathbf{X})\,,$$
$$q_3((i)|s = (i_1, i_2, \ldots, i_k)) = X_i/(T(\mathbf{X}) - X_{i_1} - \cdots - X_{i_k})\,. \tag{5.45}$$

The resulting sampling design is given by

$$P(s) = p_{i_1} p_{i_2} \cdots p_{i_n} / (1 - p_{i_1})(1 - p_{i_1} - p_{i_2}) \cdots (1 - p_{i_1} - p_{i_2} - \cdots - p_{i_{n-1}}) \tag{5.46}$$

where $s = (i_1, i_2, \ldots, i_n)$, $1 \leq i_1 \neq i_2 \neq \cdots \neq i_n \leq N$.

The sampling design, in its reduced form, will involve *ordered* samples of the type $s^* = (j_1, j_2, \ldots, j_n)$ where $1 \leq j_1 < \cdots < j_n \leq N$ and the corresponding probability $P^*(s^*)$ would be obtained by summing the probabilities of $n!$ components of the type (5.46).

Lahiri (1951) developed a simple and interesting method to implement $q_1(\cdot)$ and $q_3(\cdot)$ in practice. We may describe the method in fairly general terms in the following manner. Suppose there are N^* labelled units and we wish to select one of them such that $P((i)) = X_i / \Sigma X_i$, $1 \leq i \leq N^*$, where the X_i's are positive integers. We can replace the usual cumulative total method (mentioned in Section 1.2) by the following method.

We select an arbitrary but fixed integer $X^* \geq \max_i X_i$. Then we select a random pair (r, s) where $1 \leq r \leq N^*$ and $1 \leq s \leq X^*$. If $X_r \geq s$, we choose the rth unit as the sampled unit. Otherwise, we discard the pair (r, s) and make a fresh random selection of a pair (r', s') and examine if $X_{r'} \geq s'$ or not. We continue this procedure until, for some random pair (r, s), $X_r \geq s$ for the first time. Then we stop and decide on the rth unit as the sampled unit. It can be easily verified that this method yields

$$P((i)) = (X_i / N^* X^*) \Big/ \left\{ 1 - \sum_{j=1}^{N^*} \left(1 - \frac{X_j}{X^*} \right) / N^* \right\} = X_i \Big/ \sum X_j \tag{5.47}$$

as desired. We leave the verification to the reader (Exercise 5.17).

In application, N^* varies from N to $N - n + 1$ but we can keep the choice of X^* fixed for all sampling operations at different stages. Thus to implement Lahiri's method in the present setup, at the first stage we will work with all the N units ($N^* = N$) and at the subsequent stages we will work with the units remaining after each successful sampling operation (N^* = number of residual units). However, the choice of X^* made in the beginning can be left unaltered at all subsequent stages. This greatly simplifies the sampling operation compared to the cumulative total method.

The sampling scheme described through the algorithm in (5.44) has been termed PPS sampling scheme, in contrast to the ΠPS sampling scheme studied earlier, since at each stage the probability of selecting one particular unit, if it is not already included in the sample, is proportional to its size. The sampling design may be denoted by the symbol PPS$(N, n, \mathbf{p})$.

The following is another alternative convenient method for implementing the PPS sampling scheme of size n described via the algorithm in (5.44).

Select the first unit using probability proportional to size by any convenient method, for example Lahiri's method. Select the next unit from the *entire* population using probability proportional to size. Repeat this procedure until n *distinct* units are obtained. These n distinct units (repetitions ignored) then constitute a PPS sample of size n.

For a PPS$(N, 2, \mathbf{p})$ sampling design,

$$\pi_i = p_i + \sum_{j(\neq i)} p_j p_i/(1-p_j) = p_i\left\{1 + \sum_{j(\neq i)} p_j/(1-p_j)\right\}, \tag{5.48}$$

$$\pi_{ij} = p_i p_j/(1-p_i) + p_j p_i/(1-p_j) = p_i p_j\{(1/(1-p_i) + (1/(1-p_j))\}, \tag{5.49}$$

and these quantities satisfy $0 < \pi_{ij} < \pi_i \pi_j$, $1 \leq i \neq j \leq N$. For $n > 2$, the expressions for the π_i's and π_{ij}'s are quite complicated. However, as we shall see shortly, there are estimators other than the HTE which bypass direct use of these inclusion probabilities.

Des Raj (1956) suggested a simple unbiased estimator of the population total with some desirable features. Murthy (1957) modified and uniformly improved the Des Raj estimator using its symmetrized form. This latter estimator is popularly known as Murthy's estimator or symmetrized Des Raj estimator. Several researchers have studied the admissibility property of Murthy's estimator and other related estimators. We do not elaborate on this. The interested readers are referred to Patel and Dharmardhikari (1978) and Sengupta (1980, 1982). We present below results concerning the Des Raj and Murthy estimators.

Suppose $s = (i_1, i_2, \ldots, i_n)$ is an *ordered* sample generated by the PPS$(N, n, \mathbf{p})$ method of sampling. We wish to estimate the population total $T(\mathbf{Y})$ based on the observations $\{Y_i \mid i \in s\}$. Below $\hat{T}_D(\mathbf{Y})$ and $\hat{T}_M(\mathbf{Y})$ will denote, respectively, the estimators suggested by Des Raj and Murthy.

Theorem 5.12. *For the* PPS$(N, n, \mathbf{p})$ *sampling design, the Des Raj estimator, based on the sample* $s = (i_1, i_2, \ldots, i_n)$, *given by*

$$\hat{T}_D(\mathbf{Y}) = (t_1 + t_2 + \cdots + t_n)/n = \bar{t}((i_1, i_2, \ldots, i_n)), \tag{5.50}$$

where

$$\begin{aligned} t_1 &= Y_{i_1}/p_{i_1}, \\ t_2 &= Y_{i_1} + \{Y_{i_2}(1 - p_{i_1})/p_{i_2}\} \\ &\vdots \\ t_n &= (Y_{i_1} + \cdots + Y_{i_{n-1}}) + \{Y_{i_n}(1 - p_{i_1} - \cdots - p_{i_{n-1}})/p_{i_n}\} \end{aligned} \tag{5.51}$$

is unbiased for the population total $T(\mathbf{Y})$. *The expression for the variance of* $\hat{T}_D(\mathbf{Y})$ *is given by*

$$V(\hat{T}_D(\mathbf{Y})) = \sum_{i=1}^{n} V(t_i)/n^2 . \tag{5.52}$$

Further, an unbiased nonnegative variance estimator is given by

$$\hat{V}(\hat{T}_D(\mathbf{Y})) = \sum_{i=1}^{n} (t_i - \bar{t})^2/n(n-1)\,, \tag{5.53}$$

where $\bar{t}$ is the same as $\bar{t}((i_1, \ldots, i_n))$ in (5.50).

Proof. Based on the first unit i_1, we suggest the estimator

$$t_1 = Y_{i_1}/p_{i_1}\,, \tag{5.54}$$

which is an unbiased estimator of $T(\mathbf{Y})$ since

$$E(t_1) = \sum_{i_1} (Y_{i_1}/p_{i1})p_{i_1} = \sum_{i_1} Y_{i_1} = T(\mathbf{Y})\,.$$

Next, based on the first two units i_1 and i_2, we suggest the estimator

$$t_2 = Y_{i_1} + \{Y_{i_2}(1 - p_{i_1})/p_{i_2}\}\,. \tag{5.55}$$

This estimator is also unbiased for the population total since the conditional expectation of $\{Y_{i_2}(1 - p_{i_1})/p_{i_2}\}$ given (i_1) is

$$\begin{aligned} E_2\{Y_{i_2}(1 - p_{i_1})/p_{i_2}|(i_1)\} &= \sum_{i_2(\neq i_1)} \frac{Y_{i_2}(1 - p_{i_1})}{p_{i_2}} \cdot \frac{p_{i_2}}{1 - p_{i_1}} \\ &= \sum_{i_2(\neq i_1)} Y_{i_2} = T(\mathbf{Y}) - Y_{i_1}\,, \end{aligned}$$

so that

$$E_2(t_2|(i_1)) = Y_{i_1} + E_2\{Y_{i_2}(1 - p_{i_1})/p_{i_2}|(i_1)\} = Y_{i_1} + T(\mathbf{Y}) - Y_{i_1} = T(\mathbf{Y})\,.$$

This gives

$$E(t_2) = E_1E_2(t_2|(i_1)) = T(\mathbf{Y})\,.$$

Moreover,

$$E(t_1t_2) = E_1E_2(t_1t_2) = E_1\{t_1E_2(t_2|(i_1))\} = T^2(\mathbf{Y})\,, \tag{5.56}$$

so that $\text{Cov}(t_1, t_2) = 0$. This argument can be carried further without any difficulty and we can conclude that, based on the units $(i_1, i_2, \ldots, i_n)$, the estimators $t_1, t_2, \ldots, t_n$ given in (5.51) form a set of n uncorrelated unbiased estimators of $T(\mathbf{Y})$. This led Des Raj to suggest the estimator $\hat{T}_D(\mathbf{Y})$ in (5.50) which is clearly unbiased for $T(\mathbf{Y})$. Becasue of zero covariance between any pair of the estimators $t_1, t_2, \ldots, t_n$, we obtain (5.52). Taking expectation and simplifying directly, we can verify (5.53). □

It is evident that the Des Raj estimator $\hat{T}_D(\mathbf{Y})$ in (5.50) is an ordered estimator and, hence, it can be symmetrized to yield a uniformly improved estimator using the symmetrizing procedure developed in Chapter Two. This task was accomplished by Murthy (1957).

Theorem 5.13. *For the* PPS$(N, n, \mathbf{p})$ *sampling design, Murthy's estimator based on a sample s^* is unbiased for the population total and has the representation*

$$\hat{T}_M(\mathbf{Y}) = \sum_{i \in s^*} Y_i P^*(s^*|(i)) \Big/ \sum_{i \in s^*} p_i P^*(s^*|(i)) \qquad (5.57)$$

where $P^*(s^*|(i))$ is the conditional probability of the sample s^* given that it contains the unit i in the first position. Further, the variance of the estimator is given by

$$V(\hat{T}_M(\mathbf{Y})) = \sum_i Y_i^2 \Big\{ \sum_{s^* \ni i} P^{*2}(s^*|(i)) P^{*-1}(s^*) - 1 \Big\}$$
$$+ \sum\sum_{i \neq j} \Big\{ \sum_{s^* \ni i,j} P^*(s^*|(i)) P^*(s^*|(j)) P^{*-1}(s^*) - 1 \Big\} Y_i Y_j \,. \qquad (5.58)$$

An unbiased nonnegative variance estimator is also given by $\hat{V}(\hat{T}_M(\mathbf{Y}))$ shown in (5.59):

$$P^{*-2}(s^*) \sum\sum_{\substack{i,j \in s^* \\ i<j}} p_i p_j \{ P^*(s^*) P^*(s^*|(i, j)) - P^*(s^*|(i)) P^*(s^*|(j)) \}$$
$$\times (Y_i p_i^{-1} - Y_j p_j^{-1})^2 \,. \qquad (5.59)$$

Proof. We refer to Theorem 2.9 and expression (2.20) regarding the method of construction of an unordered (symmetrized) estimator starting with any ordered estimator. Since the Des Raj estimator for $T(\mathbf{Y})$ is unbiased, it is evident that Murthy's estimator is also unbiased. Applied in the present context, it follows that the symmetrized form of the Des Raj estimator in (5.50) would be given, for a sample $s^* = (j_1, j_2, \ldots, j_n)$, by

$$\sum{}' \bar{t}((i_1, i_2, \ldots, i_n)) P((i_1, i_2, \ldots, i_n)) \Big/ \sum{}' P((i_1, i_2, \ldots, i_n)) \,, \qquad (5.60)$$

where Σ' refers to the $n!$ permutations of the indices $j_1, j_2, \ldots, j_n$, $1 \leq j_1 < j_2 < \cdots < j_n \leq N$, resulting in permutations of the form $i_1, i_2, \ldots, i_n$. The expression (5.60) can now be simplified to yield (5.57). The simplification is carried out in detail for the denominator and is only outlined for the numerator.

The denominator in (5.60) represents the probability of getting the unordered sample s^* using the PPS$(N, n, \mathbf{p})$ design. The sum Σ' has $n!$ terms which can be divided into n sets of $(n-1)!$ each. In the rth set we include all those terms for which $i_1 = j_r$ for a specified $j_r \in s^*$. This, however, yields

$$\sum' P((i_1, i_2, \ldots, i_n)) = \sum_{r=1}^{n} p_{j_r} \sum'' P((i_1, i_2, \ldots, i_n) | i_1 = j_r) \quad (5.61)$$

where Σ'' refers to the sum of the conditional probabilities of all PPS samples resulting in the sample s^* with the first position being occupied by j_r. We have denoted $\Sigma'' P((i_1, i_2, \ldots, i_n)|(i_1 = j_r))$ by $P^*(s^*|(j_r))$. Hence, the denominator of (5.60) has the representation $\Sigma_{i \in s^*} p_i P^*(s^*|(i))$ which is indicated in (5.57). As regards the numerator of (5.57), we again start with (5.60) and split Σ' into n subtotals as before. It can be seen that the term involving Y_{j_r} in the rth subtotal corresponding to $i_1 = j_r$ is

$$\frac{1}{n} \{(Y_{j_r}/p_{j_r}) + (n-1) Y_{j_r}\} p_{j_r} P^*(s^*|(j_r)) \,. \quad (5.62)$$

Next the coefficient of Y_{j_r} has to be collected from each of the other subtotals corresponding to $i_2 = j_r, i_3 = j_r, \ldots, i_n = j_r$. This can be done in a routine manner and the terms can be added to (5.62). The resulting expression will, on simplification, yield $P^*(s^*|(j_r))$ as the coefficient of Y_{j_r}, thereby producing the numerator of (5.57). The case of $n = 2$ is elaborated at the end (see p. 141).

We now deduce below the expression for $V(\hat{T}_M(\mathbf{Y}))$. Using (5.57), we have

$$\begin{aligned} V(\hat{T}_M(\mathbf{Y})) &= \sum_{s^*} \hat{T}_M^2(\mathbf{Y}) P^*(s^*) - T^2(\mathbf{Y}) \\ &= \sum_{s^*} P^{*-1}(s^*) \Big\{ \sum_{i \in s^*} Y_i P^*(s^*|(i)) \Big\}^2 - T^2(\mathbf{Y}) \\ &= \sum_{s^*} P^{*-1}(s^*) \Big\{ \sum_{i \in s^*} Y_i^2 P^{*2}(s^*|(i)) \\ &\quad + \sum_{\substack{i \neq j \\ i, j \in s^*}} \sum Y_i Y_j P^*(s^*|(i)) P^*(s^*|(j)) \Big\} - T^2(\mathbf{Y}) \end{aligned} \quad (5.63)$$

which yields, on simplification, (5.58).

Finally, we proceed to establish (5.59) which gives an expression for an unbiased estimator of $V(\hat{T}_M(\mathbf{Y}))$ in (5.63). We note that the second representation of $V(\hat{T}_M(\mathbf{Y}))$ in (5.63) has two components. The first component can be estimated unbiasedly by

$$P^{*-2}(s^*) \Big\{ \sum_{i \in s^*} Y_i P^*(s^*|(i)) \Big\}^2 , \quad (5.64)$$

utilizing the tool developed in Theorem 2.1. Therefore, the task of estimating $V(\hat{T}_M(\mathbf{Y}))$ unbiasedly is reduced to the problem of estimation of $T^2(\mathbf{Y})$ unbiasedly. By an application of Corollary 3.2, it is clear that we can unbiasedly estimate this quadratic form since for the PPS($N, n, \mathbf{p}$) sampling design, $\pi_{ij} > 0$, $1 \le i \ne j \le N$. Unfortunately, however, we face many computational complexities in evaluating the expressions for π_i's and π_{ij}'s for PPS($N, n, \mathbf{p}$) sampling designs unless the sample size is small enough.

We shall try below to rewrite $T^2(\mathbf{Y})$ in a form which lends itself to an immediate application of Theorem 2.1, bypassing an explicit use of π_i's and π_{ij}'s. Towards this, we first observe that

$$\sum_{s^* \ni i} P^*(s^* | (i)) = 1 , \qquad 1 \le i \le N ,$$

and

$$\sum_{s^* \ni i, j} P^*(s^* | (i, j)) = 1 , \qquad 1 \le i \ne j \le N .$$

Here $P^*(s^*|(i, j))$ refers to the conditional probability of the sample s^* given that it contains the unit i in the first position and the unit j in the second position. Utilizing the above facts, we express $T^2(\mathbf{Y})$ as

$$\begin{aligned}
T^2(\mathbf{Y}) &= \sum_{i=1}^{N} Y_i^2 + \sum_{\substack{i \ne j \\ 1}}^{N}\sum Y_i Y_j \\
&= \sum_{i=1}^{N} Y_i^2 \Big(\sum_{s^* \ni i} P^*(s^*|(i)) \Big) + \sum_{\substack{i \ne j \\ 1}}^{N}\sum Y_i Y_j \Big(\sum_{s^* \ni i, j} P^*(s^*|(i, j)) \Big) \\
&= \sum_{s^*} \sum_{i \in s^*} P^*(s^*|(i)) Y_i^2 + \sum_{s^*} \sum_{\substack{i \ne j \\ i, j \in s^*}} \sum P^*(s^*|(i, j)) Y_i Y_j \\
&= \sum_{s^*} P^*(s^*) \Big[P^{*-1}(s^*) \sum_{i \in s^*} Y_i^2 P^*(s^*|(i)) \\
&\qquad + P^{*-1}(s^*) \sum_{\substack{i \ne j \\ i, j \in s^*}} \sum P^*(s^*|(i, j)) Y_i Y_j \Big] .
\end{aligned}$$

This yields, by an application of Theorem 2.1,

$$\Big[\sum_{i \in s^*} P^{*-1}(s^*) P^*(s^*|(i)) Y_i^2 + \sum_{\substack{i \ne j \\ i, j \in s^*}} \sum P^{*-1}(s^*) P^*(s^*|(i, j)) Y_i Y_j \Big] \tag{5.65}$$

as an unbiased estimator of $T^2(\mathbf{Y})$. Combining this with (5.64), we end up with the following unbiased variance estimator.

$$\hat{V}(\hat{T}_M(\mathbf{Y})) = P^{*-2}(s^*) \sum_{i\in s^*} P^*(s^*|(i))\{P^*(s^*|(i)) - P^*(s^*)\} Y_i^2$$

$$+ P^{*-2}(s^*)\left\{\sum\sum_{\substack{i,j\in s^*\\ i\neq j}} [P^*(s^*|(i))P^*(s^*|(j)) - P^*(s^*)P^*(s^*|(i,j))]Y_iY_j\right\}. \tag{5.66}$$

We now claim that the variance estimator in (5.66) coincides with that in (5.59). To establish this claim, one might evaluate and compare the coefficient of Y_i^2 and Y_iY_j in both the expressions. We leave the details of this verification to the reader. Instead, we follow another method which is more instructive. The expression in (5.59) reminds us of the Sen–Yates–Grundy variance estimator discussed in Chapter Three. We will try to rewrite the variance estimator in (5.66) by utilizing the technique of formation of Sen–Yates–Grundy variance estimator. We start with the expression

$$\sum\sum_{\substack{i,j\in s^*\\ i\neq j}} p_ip_j\{P^*(s^*)P^*(s^*|(i,j)) - P^*(s^*|(i))P^*(s^*|(j))\}(Y_ip_i^{-1} - Y_jp_j^{-1})^2 \tag{5.67}$$

and simplify the square term in (5.67) as

$$2\sum\sum_{\substack{i,j\in s^*\\ i\neq j}} p_jp_i^{-1}\{P^*(s^*)P^*(s^*|(i,j)) - P^*(s^*|(i))P^*(s^*|(j))\}Y_i^2$$

$$= 2\sum_{i\in s^*} Y_i^2p_i^{-1}\left[P^*(s^*)\sum_{\substack{j\in s^*\\ j\neq i}} p_jP^*(s^*|(i,j)) - P^*(s^*|(i))\sum_{\substack{j\in s^*\\ j\neq i}} p_jP^*(s^*|(j))\right]$$

$$= 2\sum_{i\in s^*} Y_i^2p_i^{-1}\left[P^*(s^*)(1-p_i)\sum_{\substack{j\in s^*\\ j\neq i}} p_j(1-p_i)^{-1}P^*(s^*|(i,j))\right.$$

$$\left. - P^*(s^*|(i))\sum_{\substack{j\in s^*\\ j\neq i}} p_jP^*(s^*|(j))\right]$$

$$= 2\sum_{i\in s^*} Y_i^2p_i^{-1}\left[P^*(s^*)(1-p_i)P^*(s^*|(i)) - P^*(s^*|(i))\right.$$

$$\left. \times\{P^*(s^*) - p_iP^*(s^*|(i))\}\right]$$

$$= 2\sum_{i\in s^*} Y_i^2p_i^{-1}[P^{*2}(s^*|(i)) - P^*(s^*)P^*(s^*|(i))]p_i$$

$$= 2\sum_{i\in s^*} Y_i^2[P^{*2}(s^*|(i)) - P^*(s^*)P^*(s^*|(i))].$$

On the other hand, the cross-product term in (5.67) is seen to be equal to

$$(-2)\left\{\sum_{\substack{i,j\in s^* \\ i\neq j}}\sum \{P^*(s^*)P^*(s^*|(i,j)) - P^*(s^*|(i))P^*(s^*|(j))\}Y_iY_j\right\}. \quad (5.68)$$

Therefore, $\hat{V}(\hat{T}_M(\mathbf{Y}))$ in (5.66) simplifies to (5.59). Pathak and Shukla (1966) and Subrahmanya (1967) have demonstrated the nonnegativity of (5.59) by establishing that for every $s^* \ni i, j$,

$$P^*(s^*)P^*(s^*|(i,j)) \geq P^*(s^*|(i))P^*(s^*|(j)). \qquad \square$$

Andreatta and Kaufman (1986) and T. J. Rao et al. (1988) have given alternative simpler proofs of the above inequality. See Exercise 5.33 in this context.

We now specialize to the case of $n = 2$ and give simple explicit expressions for the estimators and their variance estimators in terms of the observed values and the normed size measures.

$$\text{Ordered sample} = (i, j), \qquad P((i,j)) = p_ip_j/(1-p_i),$$

$$t_1 = Y_i/p_i, \qquad t_2 = Y_i + Y_j(1-p_i)/p_j,$$

$$\hat{T}_D(\mathbf{Y}) = \bar{t}((i,j)) = (t_1+t_2)/2 = [Y_i/p_i + Y_i + Y_j(1-p_i)/p_j]/2.$$

$$\text{Permuted sample} = (j, i), \qquad P((j,i)) = p_jp_i/(1-p_j),$$

$$\bar{t}((j,i)) = [Y_j/p_j + Y_j + Y_i(1-p_j)/p_i]/2,$$

$$\hat{T}_M(\mathbf{Y}) = \frac{\bar{t}((i,j))P((i,j)) + \bar{t}((j,i))P((j,i))}{P((i,j)) + P((j,i))} \quad (5.69)$$

$$= (2-p_i-p_j)^{-1}[\{Y_i(1-p_j)/p_i\} + \{Y_j(1-p_i)/p_j\}],$$

$$P^*(s^*) = P((i,j)) + P((j,i)) = p_ip_j\{(1-p_i)^{-1} + (1-p_j)^{-1}\},$$

$$P^*(s^*|(i)) = p_j/(1-p_i),\ P^*(s^*|(j)) = p_i/(1-p_j),\ P^*(s^*|(i,j)) = 1,$$

$$\hat{V}(\hat{T}_D(\mathbf{Y})) = \{\bar{t}((i,j)) - \bar{t}((j,i))\}^2/2.$$

$$\hat{V}(\hat{T}_M(\mathbf{Y})) = (1-p_i)(1-p_j)(1-p_i-p_j)(2-p_i-p_j)^{-2}\{Y_ip_i^{-1} - Y_jp_j^{-1}\}^2. \quad (5.70)$$

Clearly, (5.70) is recognized to be nonnegative.

Now we give an illustrative example.

Example 5.11. A population consists of 15 units and the following values are available for an auxiliary variable X.

Units	1	2	3	4	5	6	7	8	9	10	11	12	13	14	15
X_i	2.3	3.0	4.1	2.6	5.3	6.0	2.8	5.2	11.3	7.2	8.0	3.5	4.2	3.8	5.2

Suppose we want to adopt a PPS sampling scheme and generate a sample of size 3. Then we will proceed through the following steps, using Lahiri's method.

Step I. Modify the X_i-values to $X_i^* = 10X_i$, $1 \le i \le 15$ so that X_i^*'s are all integers. We will work with X_i^*'s now.

Step II. Select a value of $X^* > \max_i(X_i^*) = 113$. We choose $X^* = 120$.

Step III. (a) Select a random pair (r_1, r_2) where $1 \le r_1 \le 15$ and $1 \le r_2 \le 120$. Consulting a random number table, suppose we obtain $r_1 = 7$ and $r_2 = 92$. Then $X_{r_1}^* = X_7^* = 28 < r_2 = 92$ and so we discard this pair and make a fresh selection. Suppose this time we obtain $r_1 = 5$ and $r_2 = 38$. Then $X_5^* = 53 > r_2 = 38$ and so we take $i_1 = 5$ as the first unit of our sample. Next we go to (b).

(b) Choose again a random pair (r_1, r_2), $1 \le r_1 \le 15$, $r_1 \ne 5$, and $1 \le r_2 \le X^* = 120$. Note that the range of r_2 is left unaltered and only the range of values of r_1 excludes the integer 5. Suppose we select $r_1 = 12$ and $r_2 = 28$. Then since $X_{12}^* = 35 > r_2 = 28$, we take $i_2 = 12$. Next we go to (c).

(c) Exclude 5 and 12 from the values of r_1 and make a random choice of the pair (r_1, r_2). Suppose this time we select $r_1 = 4$ and $r_2 = 59$. Then $X_4^* = 26 < 59$, we discard this pair and make a fresh selection of $r_1 = 7$ and $r_2 = 25$. This time $X_7^* = 28 > 25$ so that we take $i_3 = 7$.

Thus, at the end, we come up with the *ordered* sample $s = (5, 12, 7)$ and proceed to collect data on these units with respect to the survey variable. Suppose specifically that $Y_5 = 50$, $Y_7 = 30$, and $Y_{12} = 40$. Then we compute, using $T(\mathbf{X}) = \Sigma X_i = 74.5$ and (5.51),

$$t_1 = Y_5/p_5 = 50 \times 74.5/5.3 = 702.83\ ,$$

$$t_2 = Y_5 + Y_{12}(1 - p_5)/p_{12} = 50 + (40 \times 69.2/3.5) = 840.86\ ,$$

$$t_3 = (Y_5 + Y_{12}) + Y_7(1 - p_5 - p_{12})/p_7$$

$$= (90) + 30 \times 65.7/2.8 = 793.93\ ,$$

$$\bar{t} = (t_1 + t_2 + t_3)/3 = 779.21\ ,$$

$$\sum (t_i - \bar{t})^2 = 11837.056\ .$$

Therefore, using (5.50), the Des Raj estimator of the population total is given by

$$\bar{T}_D(\mathbf{Y}) = (t_1 + t_2 + t_3)/3 = 779.21 = \bar{t}\ .$$

Table 5.15

(1) Ordered Sample s	(2) $P(s)$	(3) t_1	(4) t_2	(5) t_3	(6) $\bar{t}(s)$	(7) $P(s)\bar{t}(s)$
(5, 7, 12)	152×10^{-6}	702.83	791.43	838.86	777.71	118211.92×10^{-6}
(5, 12, 7)	153×10^{-6}	702.83	840.86	793.93	779.21	119219.13×10^{-6}
(7, 5, 12)	146×10^{-6}	798.21	706.41	838.86	781.16	114049.36×10^{-6}
(7, 12, 5)	143×10^{-6}	798.21	849.43	713.40	787.01	112542.43×10^{-6}
(12, 5, 7)	149×10^{-6}	851.43	709.81	793.93	785.06	116973.94×10^{-6}
(12, 7, 5)	144×10^{-6}	851.43	800.71	713.40	788.51	113545.44×10^{-6}

The estimated s.e. is obtained from (5.53) as

$$\left\{\sum (t_i - \bar{t})^2/6\right\}^{1/2} = 44.42\,.$$

To compute the symmetrized estimator of Murthy, we prepare Table 5.15. Then, using (5.57), we get Murthy's estimator of the population total as

$$\begin{aligned}\hat{T}_M(\mathbf{Y}) &= \sum' \bar{t}(s)P(s) \Big/ \sum' P(s)\\ &= \frac{\text{total of column (7)}}{\text{total of column (2)}}\\ &= 783.02\,.\end{aligned}$$

The estimated s.e. of the estimate can be calculated using (5.59). We leave the details of the calculation to the reader (Exercise 5.19).

5.6. RAO–HARTLEY–COCHRAN SCHEME OF SAMPLING AND RELATED ESTIMATOR

In concluding this chapter, we present another convenient and well-known sampling strategy based on the utilization of auxilliary information. This is suggested in J.N.K. Rao et al. (1962). We start with the concept of *random grouping* of the population under study.

Given a population $\mathcal{U}$ of size N and a fixed set of integers $N_1, N_2, \ldots, N_n$ satisfying $N_i \geq 1$, $1 \leq i \leq n$, $\Sigma_{i=1}^n N_i = N$, we say that $\{G_1, G_2, \ldots, G_n\}$ forms an $(N, n; N_1, N_2, \ldots, N_n)$ random grouping of $\mathcal{U}$ if $\mathcal{U} = \bigcup_{i=1}^n G_i$ and $|G_i| = N_i$, $1 \leq i \leq n$ and, moreover, for every i, $1 \leq i \leq n$, in the formation of G_i, every N_i-ple of $\mathcal{U}$ has an equal chance of being selected. One way of forming such a random grouping is to let G_1 be a random sample from $\mathcal{U}$ drawn according to an SRS(N, N_1) design, G_2 be a random sample from $\mathcal{U} - G_1$ drawn according to an SRS($N - N_1, N_2$) design, and so on. At the $(n-1)$th stage, G_{n-1} is taken to be a random sample drawn from $\mathcal{U} - (G_1 \cup G_2 \cup \cdots \cup G_{n-2})$ according to an SRS($N - \Sigma_{i=1}^{n-2} N_i, N_{n-1}$) design, and, finally, we form G_n with the remaining N_n units of $\mathcal{U}$.

We now describe the Rao–Hartley–Cochran (RHC) scheme of sampling for a population $\mathcal{U}$ of size N with the normed size measures $p_1, p_2, \ldots, p_N$ for the units in the population. This method consists of two steps. Below we regard $N_1, N_2, \ldots, N_n$ as arbitrary but fixed, satisfying $N_i \geq 1$, $1 \leq i \leq n$, $\Sigma_{i=1}^{n} N_i = N$.

RHC Sampling Scheme

Step I. Form an $(N, n; N_1, N_2, \ldots, N_n)$ random grouping of $\mathcal{U}$, say $\{G_1, G_2, \ldots, G_n\}$.

Step II. Draw one unit per group using PPS method of sampling independently within each group.

According to the PPS method, given the random grouping, the chance of selecting the ith unit in the sample is p_i/P_k if $i \in G_k$, where $P_k = \Sigma_{j\in G_k} p_j$, $1 \leq k \leq n$, $1 \leq i \leq N$.

The estimator of the population total suggested under the RHC scheme of sampling is given by

$$\hat{T}_{\text{RHC}}(\mathbf{T}) = \sum_{k=1}^{n} Y_{i_k} P_k / p_{i_k} \tag{5.71}$$

where $s = (i_1, i_2, \ldots, i_n)$ is the selected sample with the unit i_k drawn from G_k, $1 \leq k \leq n$.

Below we state and prove results relating to the properties of this estimator.

Theorem 5.14. *Under the RHC scheme of sampling generating the grouping $\{G_1, G_2, \ldots, G_n\}$ and resulting in the sample $s = (i_1, i_2, \ldots, i_n)$, the RHC estimator $\hat{T}_{\text{RHC}}(\mathbf{Y})$ in (5.71) is unbiased for the population total. The variance of the estimator is given by*

$$V(\hat{T}_{\text{RHC}}(\mathbf{Y})) = \left\{\left(\sum_{k=1}^{n} N_k^2 - N\right)\Big/(N^2 - N)\right\}\left\{\sum_{i=1}^{N}\left(\frac{Y_i}{p_i} - T(\mathbf{Y})\right)^2 p_i\right\}. \tag{5.72}$$

Further, a nonnegative unbiased variance estimator is given by

$$\hat{V}(\hat{T}_{\text{RHC}}(\mathbf{Y})) = \left\{\left(\sum_{k=1}^{n} N_k^2 - N\right)\Big/\left(N^2 - \sum_{k=1}^{n} N_k^2\right)\right\}\left\{\sum_{k=1}^{n}\left(\frac{Y_{i_k}}{p_{i_k}} - \hat{T}_{\text{RHC}}(\mathbf{Y})\right)^2 P_k\right\}. \tag{5.73}$$

Proof. We use the conditional expectation argument. Given the random grouping $\{G_1, G_2, \ldots, G_n\}$,

$$E(Y_{i_k}P_k/p_{i_k}|G_k) = \sum_{i \in G_k} Y_i \tag{5.74}$$

and, hence,

$$E(\hat{T}_{\text{RHC}}(\mathbf{Y})|\{G_1, G_2, \ldots, G_n\}) = \sum_{k=1}^{n} \sum_{i \in G_k} Y_i = T(\mathbf{Y}) \tag{5.75}$$

for every possible random grouping. Therefore,

$$E(\hat{T}_{\text{RHC}}(\mathbf{Y})) = E_R[E(\hat{T}_{\text{RHC}}(\mathbf{Y})|\{G_1, G_2, \ldots, G_n\})] = T(\mathbf{Y}), \tag{5.76}$$

thereby establishing the unbiasedness of $\hat{T}_{\text{RHC}}(\mathbf{Y})$ for $T(\mathbf{Y})$.

As regards $V(\hat{T}_{\text{RHC}}(\mathbf{Y}))$, we can use the formula

$$V(\cdot) = E_R[V(\cdot|\{G_1, G_2, \ldots, G_n\})] + V_R[E(\cdot|\{G_1, G_2, \ldots, G_n\})]. \tag{5.77}$$

Note that, because selection is done independently in each group,

$$V(\hat{T}_{\text{RHC}}(\mathbf{Y})|\{G_1, G_2, \ldots, G_n\}) = \sum_{k=1}^{n} \left\{ \sum_{i \in G_k} \left(\frac{Y_i P_k}{p_i} - \sum_{i \in G_k} Y_i \right)^2 p_i/P_k \right\}. \tag{5.78}$$

Using (5.75) and (5.78), we have, from (5.77),

$$V(\hat{T}_{\text{RHC}}(\mathbf{Y})) = E_R \sum_{k=1}^{n} \left\{ \sum_{i \in G_k} \left(\frac{Y_i P_k}{p_i} - \sum_{i \in G_k} Y_i \right)^2 p_i/P_k \right\}. \tag{5.79}$$

We now follow the approach in J. N. K. Rao (1979) to deduce that (5.79) simplifies to (5.72) and also that (5.73) provides a nonnegative unbiased estimator of (5.72). Towards this, we first note that the quantity inside $\{\cdots\}$ in (5.79) can be expressed as

$$\sum_{i \in G_k} \left(\frac{Y_i P_k}{p_i} - \sum_{i \in G_k} Y_i \right)^2 p_i/P_k = \sum_{\substack{i<j \\ i,j \in G_k}}\sum \left(\frac{Y_i}{p_i} - \frac{Y_j}{p_j} \right)^2 p_i p_j . \tag{5.80}$$

Therefore, using an indicator function $I((i, j)|k)$ defined as

$$I((i, j)|k) = \begin{cases} 1, & \text{if } i, j \in G_k \\ 0, & \text{otherwise}, \end{cases} \tag{5.81}$$

we deduce that

$$
\begin{aligned}
V(\hat{T}_{\mathrm{RHC}}(\mathbf{Y})) &= E_R\Big[\sum_{k=1}^{n} \mathop{\sum\sum}_{\substack{i<j\\ i,j\in G_k}} \Big(\frac{Y_i}{p_i} - \frac{Y_j}{p_j}\Big)^2 p_i p_j\Big] \\
&= E_R\Big[\mathop{\sum\sum}_{\substack{i<j\\1}}^{N} \Big(\frac{Y_i}{p_i} - \frac{Y_j}{p_j}\Big)^2 p_i p_j \Big\{\sum_{k=1}^{n} I((i,\, j)|k)\Big\}\Big]. \qquad (5.82)
\end{aligned}
$$

It now remains to evaluate the expression for $E_R(I((i,\, j)|k))$ which is quite straightforward. Recalling that random grouping can be effectively implemented using simple random sampling, we obtain

$$
E_R(I((i,\, j)|k)) = \frac{N_k(N_k - 1)}{N(N-1)} \qquad (5.83)
$$

since G_k is of size N_k, $1 \le k \le n$. The expression (5.82) now simplifies to (5.72) which thus gives the variance of the RHC estimator of the population total.

Finally we wish to show that $\hat{V}(\hat{T}_{\mathrm{RHC}}(\mathbf{Y}))$ in (5.73) serves as a nonnegative unbiased estimator of $V(\hat{T}_{\mathrm{RHC}}(\mathbf{Y}))$ in (5.72). Towards this, we first rewrite the expression in (5.73) as

$$
\begin{aligned}
&\Big(\sum_{k=1}^{n} N_k^2 - N\Big)\Big/\Big(N^2 - \sum_{k=1}^{n} N_k^2\Big)\Big\{\sum_{k=1}^{n}\Big(\frac{Y_{i_k}}{p_{i_k}} - \hat{T}_{\mathrm{RHC}}(\mathbf{Y})\Big)^2 P_k\Big\} \\
&= \Big(\sum_{k=1}^{n} N_k^2 - N\Big)\Big/\Big(N^2 - \sum_{k=1}^{n} N_k^2\Big)\Big\{\mathop{\sum\sum}_{\substack{k\neq l\\1}}^{n}\Big(\frac{Y_{i_k}}{p_{i_k}} - \frac{Y_{i_l}}{p_{i_l}}\Big)^2 P_k P_l\Big\}.
\end{aligned}
$$

In the above, we now compute conditional expectation of the term inside $\{\ldots\}$, given the random grouping $\{G_1, G_2, \ldots, G_n\}$, and obtain, in view of independent implementation of PPS designs within each group,

$$
\begin{aligned}
&E\Big[\mathop{\sum\sum}_{\substack{k\neq l\\1}}^{n}\Big(\frac{Y_{i_k}}{p_{i_k}} - \frac{Y_{i_l}}{p_{i_l}}\Big)^2 P_k P_l \big| \{G_1, G_2, \ldots, G_n\}\Big] \\
&= \mathop{\sum\sum}_{\substack{k\neq l\\1}}^{n} \sum_{i\in G_k}\sum_{j\in G_l}\Big(\frac{Y_i}{p_i} - \frac{Y_j}{p_j}\Big)^2 P_k P_l \frac{p_i}{P_k}\frac{p_j}{P_l} \\
&= \mathop{\sum\sum}_{\substack{k\neq l\\1}}^{n} \sum_{i\in G_k}\sum_{j\in G_l}\Big(\frac{Y_i}{p_i} - \frac{Y_j}{p_j}\Big)^2 p_i p_j\,.
\end{aligned}
$$

Therefore,

$$
\begin{aligned}
&E\{\hat{V}(\hat{T}_{\mathrm{RHC}}(\mathbf{Y}))\}\\
&\quad = E_R[E[\hat{V}(\hat{T}_{\mathrm{RHC}}(\mathbf{Y}))|\{G_1, G_2, \ldots, G_n\}]]\\
&\quad = \frac{\Sigma N_k^2 - N}{N^2 - \Sigma N_k^2} E_R\Big[\sum\sum_{\substack{k\neq l\\1}}^{n} \sum_{i\in G_k}\sum_{j\in G_l}\Big(\frac{Y_i}{p_i} - \frac{Y_j}{p_j}\Big)^2 p_i p_j\Big]\\
&\quad = \frac{\Sigma N_k^2 - N}{N^2 - \Sigma N_k^2} \sum\sum_{\substack{k\neq l\\1}}^{n} \sum\sum_{\substack{i<j\\1}}^{N}\Big(\frac{Y_i}{p_i} - \frac{Y_j}{p_j}\Big)^2 p_i p_j E_R\{I(i)|G_k)I((j)|G_l)\}
\end{aligned}
$$

where

$$
I((i)|G_k) = \begin{cases} 1\,, & \text{if } i \in G_k \\ 0\,, & \text{otherwise}\,. \end{cases}
$$

It is evident that

$$
E_R\{I(i)|G_k)I((j)|G_l)\} = \frac{N_k N_l}{N(N-1)}\,, \qquad 1 \le k \neq l \le n
$$

and, hence, finally

$$
\begin{aligned}
E\{\hat{V}(\hat{T}_{\mathrm{RHC}}(\mathbf{Y}))\} &= \frac{\Sigma N_k^2 - N}{N^2 - \Sigma N_k^2} \sum\sum_{\substack{i<j\\1}}^{N}\Big(\frac{Y_i}{p_i} - \frac{Y_j}{p_j}\Big)^2 p_i p_j \sum\sum_{\substack{k\neq l\\1}}^{n} N_k N_l / N(N-1)\\
&= \frac{\Sigma N_k^2 - N}{N(N-1)} \sum\sum_{\substack{i<j\\1}}^{N}\Big(\frac{Y_i}{p_i} - \frac{Y_j}{p_j}\Big)^2 p_i p_j\,,
\end{aligned}
$$

which coincides with (5.72). Clearly, $\hat{V}(\hat{T}_{\mathrm{RHC}}(\mathbf{Y}))$ is nonnegative. □

Remark 5.4. So far, in the above, we have assumed that the different group sizes $N_1, N_2, \ldots, N_n$ are arbitrary but fixed integers satisfying $N_k \ge 1$, $1 \le k \le n$, $\Sigma_{k=1}^n N_k = N$. By examining the variance expression in (5.72), we can conclude that the choice of $N_1, N_2, \ldots, N_n$ which would minimize $V(\hat{T}_{\mathrm{RHC}}(\mathbf{Y}))$ corresponds to

$$
N_1 = N_2 = \cdots = N_n = N/n \qquad \text{if } N \text{ is divisible by } n
$$

and

$$
\begin{aligned}
N_1 = N_2 = \cdots = N_u &= m + 1\\
N_{u+1} = N_{u+2} = \cdots = \cdots = N_n &= m \qquad \text{if } N = mn + u,\ 1 \le u \le n-1\,.
\end{aligned}
$$

In other words, to make the most profitable use of the RHC sampling strategy, we should use random grouping with group sizes as nearly equal as possible.

Cheng and Li (1983, 1987) and Dasgupta and Sinha (1983) have made some studies regarding properties, such as minimaxity, of the RHC sampling strategy with the above optimum choice of group sizes.

We now take an illustrative example.

Example 5.12. Suppose we want to estimate the population total based on a sample of size $n = 3$ taken from the population specified in Example 5.11. We will use the RHC sampling strategy. We take $N_1 = N_2 = N_3 = 5$ as this corresponds to the optimum choice of N_i's.

Step I. We consult a table of random numbers. Suppose we obtain the following random grouping:

$$G_1 = \{7, 3, 12, 8, 1\},\ G_2 = \{9, 14, 6, 2, 10\},\ G_3 = \{4, 5, 11, 13, 15\}\ .$$

Step II. We now utilize the size measures given in Example 5.11 and apply a PPS method in each group for selecting one unit independently from each. Suppose this results in the following three units:

$$i_1 = 7,\ i_2 = 2,\ i_3 = 13\ .$$

We now carry out the following computations, assuming the Y-values in the sample to be

$$Y_7 = 30,\ Y_2 = 32,\ Y_{13} = 45\ ,$$

$$P_1 = \sum_{i \in G_1} p_i = 0.2403,\ P_2 = \sum_{i \in G_2} p_i = 0.4201,\ P_3 = \sum_{i \in G_3} p_i = 0.3396\ ,$$

$$p_{i_1}/P_1 = 0.1564,\ p_{i_2}/P_2 = 0.1008,\ p_{i_3}/P_3 = 0.1700\ .$$

From (5.71),

$$\hat{T}(\mathbf{Y}) = 773.98\ .$$

From (5.73),

$$\hat{V}(\hat{T}(\mathbf{Y})) = 208.1410\ .$$

Hence, the estimate of the population total is 773.98 and the estimated s.e. is $\sqrt{208.1410} \doteq 14.43$.

EXERCISES

5.1. Give a detailed proof of Theorem 5.2

5.2. With reference to J. N. K. Rao's method, give a proof of (5.11).

5.3. Prove the following results:

(i) $P((i, j))$ under Brewer's method $= P((j, i))$ under J. N. K. Rao's method.

(ii) $2P((i, j))$ under Durbin's method $= P((i, j))$ under Brewer's method $+ P((i, j))$ under J. N. K. Rao's method.

5.4. (a) Consider $N = 4$ and let $0 < 2p_i < 1$, $\Sigma_{i=1}^4 p_i = 1$. The purpose is to show that there is an arrangement of these p_i's, say $p_1^*, p_2^*, p_3^*, p_4^*$, satisfying $0 < 2p_1^* < 1$, $0 < 2p_2^* < 1 - p_1^*$, $(1 - 2p_1^*)(1 - 2p_2^*(1 - p_1^*)^{-1}) < 4p_3^* p_4^*$. Establish the following results.

(i) There is at least one pair (i, j) for which $p_i + 2p_j < 1$. Further, there is at most one pair (i, j) for which $p_i + 2p_j \geq 1$, $p_j + 2p_i \geq 1$.

(ii) If $p_i + 2p_j < 1$ for *all* ordered pairs (i, j), there exists an arrangement satisfying the above inequalities (argue by contradiction).

(iii) If $p_i + 2p_2 < 1 < p_2 + 2p_1$, the arrangement $(p_1, \min(p_3, p_4), p_2, \max(p_3, p_4))$ satisfies the above inequalities whenever $2(p_1 + p_2)(1 - p_2) \geq 1$; otherwise, the arrangement $(p_2, \min(p_3, p_4), p_1, \max(p_3, p_4))$ satisfies the above inequalities.

(iv) If $p_1 + 2p_2 \geq 1$, $p_2 + 2p_1 \geq 1$, the arrangement $(p_1, \min(p_3, p_4), p_2, \max(p_3, p_4))$ satisfies the inequalities whenever $2(1 - p_1)(p_1 + p_2) < 1$. If $2(1 - p_1)(p_1 + p_2) \geq 1$, in which case $2(1 - p_2)(p_1 + p_2) < 1$, the arrangement $(p_2, \min(p_3, p_4), p_1, \max(p_3, p_4))$ satisfies the inequalities.

(b) Consider $N \geq 5$ and let $0 < 2p_i < 1$, $\Sigma_{i=1}^N p_i = 1$. Suppose there is an arrangement of the p_i's, say, $p_1^*, p_2^*, \ldots, p_N^*$, satisfying $0 < 2p_1^* < 1$, $0 < 2p_2^* < 1 - p_1^*, \ldots, 0 < 2p_{N-2}^* < 1 - p_1^* - \cdots - p_{N-3}^*$. If, further, $\max(2p_{N-1}^*, 2p_N^*) < p_{N-3}^* + p_{N-2}^* + p_{N-1}^* + p_N^*$, show that the arrangement $(p_1^*, \ldots, p_N^*)$ satisfies the inequality

$$(1 - 2p_1^*)(1 - 2p_2^*(1 - p_1^*)^{-1}) \ldots (1 - 2p_{N-2}^*$$
$$\times (1 - p_1^* - \cdots - p_{N-3}^*)^{-1}) < 4p_{N-1}^* p_N^*.$$

Hint: Use induction on N.

(c) Show that in Sunter's method, $\pi_{ij} < \pi_i \pi_j$, $1 \le i < j \le N$.

5.5. (a) Illustrate Sunter's method by taking an example with $N = 10$ and $n = 2$.

(b) Show that the sampling scheme based on the combination of Sunter's method and Hanurav's technique produces a ΠPS sampling design with the properties $0 < \pi_{ij} \le \pi_i \pi_j$, $1 \le i \ne j \le N$.

5.6. (a) For the Lahiri–Midzuno scheme, show that the reduced form of the sampling design is given by

$$P((i_1, i_2, \ldots, i_n)) = (p'_{i_1} + p'_{i_2} + \cdots + p'_{i_n})\binom{N-1}{n-1}^{-1},$$
$$1 \le i_1 < i_2 < \cdots < i_n \le N .$$

Hence show that for $n = 2$, it reduces to π_{ij} in (5.20).

(b) Show that for a general sample size n, Lahiri–Midzuno scheme provides π_{ij}'s satisfying $0 < \pi_{ij} < \pi_i \pi_j$, $1 \le i < j \le N$.

(c) Using the technique of Exercise 1.25, show that Lahiri–Midzuno scheme is applicable when

$$(p_{i_1} + p_{i_2} + \cdots + p_{in}) \ge \frac{n-1}{N-1}$$

for all possible combinations $(i_1, i_2, \ldots, i_n)$ of $(1, 2, \ldots, N)$. Is this an improvement over the condition cited in Theorem 5.4?

5.7. Give an example of a situation where Lahiri–Midzuno scheme applies for a sample of size $n = 4$. Implement the sampling scheme and produce a sample.

5.8. Illustrate Sampford's method by taking an example with $n = 3$. Compute joint inclusion probabilities for pairs of units covered in your sample.

5.9. (a) With reference to the Hedayat–Lin method, show that the max-round rule produces a successful game.

(b) Show that the following procedure also produces a successful game whenever (5.34) is satisfied.

Write $b = M/n$ and consider a $b \times n$ rectangular array of bn empty cells. Next consider an ordered sequence of length M involving the integers $1, 2, \ldots, N$ formed as follows.

$$\begin{array}{c} 1\ 1 \cdots\cdots 1\ 2\ 2 \cdots\cdots 2 \cdots\cdots N\ N \cdots\cdots N \\ \leftarrow k_1 \text{ times} \rightarrow \leftarrow k_2 \text{ times} \rightarrow \cdots \leftarrow k_N \text{ times} \rightarrow . \end{array}$$

Now start filling up the cells of the $b \times n$ array beginning at the north-west corner and going vertically down the cells, using b elements from the left of the sequence in succession. After the first column of the array is over, move to the top of the next column and start filling up the cells, again going vertically down, using the next set of b elements from the sequence. Continue this procedure until all the cells are filled up. The rows of the array then form the rounds of a successful game.
Illustrate the above method with reference to Example 5.5.

5.10. Suppose a setup with N boxes allows a successful game. Show that the game with saturated max-round will empty the boxes in no more than N distinct rounds (ignoring the frequency factor). Further show that the same phenomenon holds for games whose rounds are saturated but not necessarily max-rounds.

5.11. Consider the following setup:

Boxes	1	2	3	4	5
No. of objects	7	5	4	2	2

(a) Display the game with saturated max-rounds for $n = 2$.
(b) Show that there is another game available with fewer distinct rounds than the one with saturated max-rounds.

5.12. Complete the proof of Theorem 5.8.

5.13. Justify (5.36).

5.14. Show that the multiplier technique illustrated before Example 5.9 yields a ΠPS sampling design.

5.15. (a) With reference to Example 5.10, find an alternative choice of A^* which also serves the purpose. Comment on the result.
(b) Produce a game which results in design no. 2 of Table 5.14.

5.16. (a) Give an example of a fixed-size ($n = 3$) ΠPS sampling design, for a population of 7 units, also satisfying $0 < \pi_{ij}$ for all pairs (i, j).
(b) Reproduce the sampling design playing a suitable emptying box game.
(c) Did you play a max-round game in (b)? If not, play it now and display the corresponding sampling design. Does this sampling design yield $0 < \pi_{ij}$ for all pairs (i, j)?

5.17. For Lahiri's method, establish (5.47).

5.18. Show that the expression (5.60) for $\hat{T}_M(\mathbf{Y})$ simplifies to (5.57). Compute $V(\hat{T}_M(\mathbf{Y}))$ for a sample of size 2 and show directly that an unbiased variance estimator is given by (5.66). Also simplify (5.59) for $n = 2$ and verify that it reduces to (5.66).

5.19. For Example 5.11, compute the estimated s.e. of the estimate of the population total.

5.20. For PPS fixed-size (n) sampling design, consider the following estimators based on a sample $s = (i_1, i_2, \ldots, i_n)$:

$$t'_1 = Y_{i_1}/p_{i_1},\ t'_2 = Y_{i_2}(1 - p_{i_1})/p_{i_1} p_{i_2}(N-1), \ldots,$$

$$t'_n = \frac{Y_{i_n}(1 - p_{i_1})(1 - p_{i_2}) \ldots (1 - p_{i_{n-1}})}{p_{i_1} p_{i_2}, \ldots, p_{i_n}(N-1)(N-2) \ldots (N-n+1)} .$$

(a) Show that $E(t'_k) = T(\mathbf{Y})$, $1 \le k \le N$.
(b) Derive expressions for $V(t'_k)$, $\mathrm{Cov}(t'_k, t'_r)$, $k \ne r$.
(c) Define $\bar{t}' = \Sigma\, t'_k/n$. Show that $\bar{t}'$ is an unbiased estimator of $T(\mathbf{Y})$.
(d) Deduce an expression for the symmetrized form of $\bar{t}'$.

5.21. Assume the X_i's to be positive integers satisfying $nX_i < T(\mathbf{X})$ for each i. Assume further that $k = T(\mathbf{X})/n$ is an integer. Arrange the units $1, 2, \ldots, N$ in some order and calculate the cumulative totals $T_1 = X_1$, $T_2 = X_1 + X_2, \ldots, T_l = X_1 + \cdots + X_l, \ldots, T_N = T(\mathbf{X})$. The following *PPS linear systematic sampling* procedure was introduced by Madow (1949): (i) Draw one random integer r in the interval $[1, k]$. This gives the random start. (ii) If $T_{i_{1-1}} < r \le T_{i_1}$, $T_{i_{2-1}} < r + k \le T_{i_2}, \ldots, T_{i_{n-1}} < r + (n-1)k \le T_{i_n}$, form a sample of size n as $s = (i_1, i_2, \ldots, i_n)$.
(a) Show that this results in a fixed-size (n) ΠPS sampling design.
(b) Does the resulting sampling design ensure $\pi_{ij} > 0$ for all pairs (i, j)? If so, why? If not, what additional conditions would be needed on the size measure X_i's?

5.22. In the setup of the previous exercise, suppose $T(\mathbf{X})$ is not divisible by n. Define $k = [T(\mathbf{X})/n]$. Consider the following revised sampling plan, known as circular systematic sampling procedure. (i) Draw a random integer z from 1 to $T(\mathbf{X})$. (ii) If $T_{i_1 - 1} < z \le T_{i_1}$, $T_{i_{2-1}} < z + k \le T_{i_2}, \ldots, T_{i_n - 1} < z + (n-1)k \le T_{i_n}$, where the integers $z + k, \ldots, z + (n-1)k$, are reduced mod $T(\mathbf{X})$, form a sample of size n as $s = (i_1, i_2, \ldots, i_n)$.
(a) Show that this also results in a fixed-size (n) ΠPS sampling design.
(b) What happens to the positivity of π_{ij} for each pair (i, j)?

5.23. Let d_n be a fixed-size (n) ΠPS($N, n, \mathbf{p}$) sampling design. Suppose the design d_n has been implemented and a sample s_n has been drawn. At this stage we are told that there has been a sudden budget cut so that we can only sample a set of $n_0 < n$ units. The survey statistician then recommends one to take a simple random subsample of size n_0 from the selected sample s_n and claims that this subsample will generate yet another ΠPS($N, n_0, \mathbf{p}$) sampling design, with the resulting sample size n_0.

(i) Do you think that the statistician is making a fool of himself?

(ii) If the original sampling design d_n has additional features like $0 < \pi_{ij} \leq \pi_i \pi_j$ for $1 \leq i < j \leq N$, how much of this is retained by the reduced size sampling design?

(iii) If instead of simple random subsampling, one follows some other sampling design to generate a sample of size $n_0 < n$ from any given sample s_n of size n, what would you expect to be the resulting reduced size sampling design in terms of ΠPS sampling?

5.24. (a) It was initially decided to have a PPS sample of size 5 from a population of size 15. An ordered sample $(9, 3, 12, 7, 1)$ was drawn. However, due to sudden change of plan, data could be collected only on the first three ordered units $(9, 3, 12)$ of the sample. Show that it would still be possible to provide an unbiased estimator of the population mean along with its standard error. Would the precision be reduced?

(b) If, in the above situation, data were collected on *any* three units of the sample, would it be possible to provide an unbiased estimator of the population total?

5.25. In the notation of Section 5.5, show that $V(t_i) \leq V(t_j)$ for any $i > j$. Using this or other facts, show that $V(\bar{t}) \leq V(t_1)/n$. Show further that strict inequality holds unless Y_i is proportional to p_i.

5.26. Show that $\hat{V}(\hat{T}_M(\mathbf{Y}))$ in (5.59) is the unique nnuve of $V(\hat{T}_M(\mathbf{Y}))$ when $n = 2$.

5.27. (a) Show that for a PPS($N, 2, \mathbf{p}$) sampling design, $0 < \pi_{ij} < \pi_i \pi_j$, $1 \leq i \neq j \leq N$.

(b) It is decided to draw a sample of size n utilizing the normed size measures $p_1, p_2, \ldots, p_N$. The following sampling design is suggested. Draw 2 units using PPS($N, 2, \mathbf{p}$) design. Next, draw additional $(n-2)$ units from the remaining $(N-2)$ population units using simple random sampling. The combination of the two samples results in a sample of size n. Denote by $\pi_i(n)$ and $\pi_{ij}(n)$ the inclusion probability of unit i and the joint inclusion probability of units i and j in the ultimate sample of size n. Denote by $\pi_i(2)$

and $\pi_{ij}(2)$ the corresponding quantities in the PPS$(N, 2, \mathbf{p})$ sampling design so that $\pi_i(2)$ and $\pi_{ij}(2)$ are given in (5.48) and (5.49), respectively. Show that $0 < \pi_{ij}(n) < \pi_i(n)\pi_j(n)$, $1 \le i \ne j \le N$.

5.28. Consider the population of 15 units specified in Example 5.11. Suppose it is decided to adopt the RHC scheme of sampling for a sample of size $n = 4$. Show all the operations and produce a sample based on an optimum choice of the N_i's. Taking the Y-values for the selected units in different groups as shown below, provide an estimate of the population total along with its estimated s.e.

$$Y_{i_1} = 30,\ Y_{i_2} = 40,\ Y_{i_3} = 50,\ Y_{i_4} = 60\,.$$

5.29. Lahiri (1951) and Midzuno (1952) suggested the following method, known as the Probability Proportional to Aggregate Size (PPAS) method, of drawing a sample of size n utilizing auxiliary size measures. Draw one unit using a PPS method. Draw the remaining $(n-1)$ units using SRS$(N-1, n-1)$ design on the remaining $(N-1)$ units of the population.

(a) Show that the resulting sampling design, in its reduced form, is given by

$$P((i_1, i_2, \ldots, i_n)) = \sum_{r=1}^{n} X_{i_r} \Big/ \binom{N-1}{n-1} T(\mathbf{X})$$

for $1 \le i_1 < i_2 < \cdots < i_n \le N$.

(b) Show that for this scheme of sampling, $0 < \pi_{ij} < \pi_i \pi_j$ for all pairs (i, j), whenever $n \ge 2$.

(c) Suggest a suitable estimator, other than the HTE, to couple with this scheme, so that the estimator is unbiased and has a very simple form.
Hint: Use an estimator of the type $e_3(s, Y)$ shown in Table 2.1.

(d) What are the similarities and dissimilarities of this scheme to the corresponding ΠPS scheme (based on this method)?

5.30. (a) For each of the following schemes of sampling, verify if $\pi_i(n) \ge \pi_j(n) \leftrightarrow p_i \ge p_j$. Treat the case of $n = 2$ separately. (i) PPS, (ii) PPAS, (iii) RHC with random grouping, and (iv) RHC with PPAS grouping.

(b) For each of the above schemes, verify if $\pi_i(n)$ is increasing in n for every i, $1 \le i \le N$.

5.31. When two or more separate surveys are to be conducted on the same population, it will occasionally be advisable to optimally integrate these surveys in such a way that the expected number of distinct units in the combined sample is the least. This could save time and effort in the overall collection of data. We explain the idea for two separate surveys d_1 and d_2, each of size one on a population $\mathscr{U}$ of size N. See Keyfitz (1951), Mitra and Pathak (1984), Kabadi et al. (1988) for related references and further results on this topic.

An optimal integration of d_1 and d_2 calls for the combination of a sampling design (joint probability distribution) d on $\mathscr{U} \times \mathscr{U}$ so that d_1 and d_2 are captured as its marginals and in addition, if we implement d instead of d_1 and d_2, we obtain a sample s of the type (i, j) with the expected number of distinct units in the sample $(n(s))$ a minimum. Note that $n(s) = 1$ or 2 according as $i = j$ or $i \neq j$.

Let

$$P_{d_1}((i)) = P_{1i}\,, \qquad P_{d_2}((i)) = p_{2i}\,, \qquad P_d((i, j)) = P_{ij}$$

so that

$$p_{1i} = \sum_{j=1}^{N} P_{ij}\,, \qquad 1 \le i \le N\,; \qquad p_{2j} = \sum_{i=1}^{N} P_{ij}\,, \qquad 1 \le j \le N\,.$$

(a) Show that $E_p(n(s)) = 2 - \Sigma_{i=1}^{N} P_{ii}$.

(b) Show that a design d_0 for which

$$P_{d_0}((i, i)) = \min\{p_{1i}, p_{2i}\}\,, \qquad 1 \le i \le N$$

is optimal in the sense that

$$E_{p_0}(n(s)) \le E_p(n(s))$$

where p corresponds to any other integrated survey d.

(c) Show that the following sampling algorithm leads to a solution to d_0.

	Survey 1				Survey 2	
d_1:	1	2 . . . N		d_2:	1	2 . . . N
	p_{11}	$p_{12} \cdots p_{1N}$			p_{21}	$p_{22} \cdots p_{2N}$

Implement d_1 for survey 1. If the jth unit is selected, then select the jth unit for survey 2 as well provided $p_{2j} > p_{1j}$. If $p_{2j} < p_{1j}$, perform a Bernoulli experiment with success probability p_{2j}/p_{1j}. If this results in a success, select jth unit for survey 2. Otherwise, for survey 2, select one unit from those units w in the population for which $p_{2w} > p_{1w}$, assuming a probability of selection proportional to $(p_{2w} - p_{1w})$.

5.32. Suppose that $p_1 = 1/n$. Show that the following two-step procedure yields a $\Pi PS(N, n, \mathbf{p})$ sampling design.

Step I. Construct a $\Pi PS(N-1, n-1, \mathbf{p}^*)$ design where $p_i^* = p_i/(1-p_1)$, $i \neq 1$. Implement it to generate a sample.
Step II. Add unit 1 to the sample selected in Step I.

5.33. For the $\Pi PS(N, n, \mathbf{p})$ sampling design,

(a) Show that the probability of selection of the ordered estimator $s^* = (1, 2, \ldots, n)$ is given by

$$P^*(s^*) = (T(\mathbf{X}) - X_1 - \cdots - X_n) \int_0^\infty e^{-\lambda T(\mathbf{X})} \prod_{j=1}^{n} (e^{\lambda X_j} - 1)\, d\lambda$$
$$= (1 - p_1 - p_2 - \cdots - p_n) \int_0^1 t^{-(p_1 + p_2 + \cdots + p_n)} \prod_{k=1}^{n} (1 - t^{p_k})\, dt\,.$$

Hint: You may use induction on n.

(b) Show further that

$$0 < P^*(s^*) < P^*(s^* | (1)) < P^*(s^* | (1, 2)) < \cdots$$
$$< P^*(s^* | (1, 2, \ldots, n)) = 1\,.$$

Hint: You may again use induction on n.

(c) Noting that $(1 - t^p)^{-1}$ is increasing in t, show that for every pair (i, j), $1 \leq i \neq j \leq n$,

$$\left\{\int_0^1 (1 - t^{p_i})^{-1} \prod_{k=1}^{n} (t^{-p_k} - 1)\, dt\right\} \left\{\int_0^1 (1 - t^{p_j})^{-1} \prod_{k=1}^{n} (t^{-p_k} - 1)\, dt\right\}$$
$$< \left\{\int_0^1 (1 - t^{p_i})^{-1} (1 - t^{p_j})^{-1} \prod_{k=1}^{n} (t^{-p_k} - 1)\, dt\right\}$$
$$\times \left\{\int_0^1 \prod_{k=1}^{n} (t^{-p_k} - 1)\, dt\right\}$$

and, hence or, otherwise, deduce that

$$P^*(s^*) P^*(s^* | (i, j)) > P^*(s^* | (i)) P^*(s^* | (j)) \text{for every pair} (i, j),$$
$$1 \leq i \neq j \leq n\,.$$

Hint: Apply Cauchy–Schwartz inequality-integral version.

(d) Generalizing (c), show further that

(i) $P^*(s^* | (1, i)) P^*(s^* | (1, j)) < P^*(s^* | (1)) P^*(s^* | (1, i, j))$for every pair$(i, j)$, $2 \leq i \neq j \leq n$;

(ii) $P^*(s^*|(1))P^*(s^*|(2)) \ldots P^*(s^*|(k))$
$< P^*(s^*)P^*(s^*|(1, 2, \ldots, k)),\ 2 \le k \le n.$

(e) Show that for every pair (i, j), $1 \le i \ne j \le n$,

$$P^*(s^*)P^*(s^*|(i, j)) - P^*(s^*|(i))P^*(s^*|(j)) > \frac{p_i p_j}{(1 - p_i - p_j)} \{P^*(s^*|(i, j)) - P^*(s^*|(i))\} \times \{P^*(s^*|(i, j)) - P^*(s^*|(j))\}.$$

5.34. For the PPS$(N, n, \mathbf{p})$ sampling design, show that

(a) $\pi_i < \pi_j \Leftrightarrow p_i < p_j$ and $\pi_i > \pi_j \Leftrightarrow p_i > p_j$;

(b) for $0 < p_1 \le p_2 \le \cdots \le p_N < 1$, $np_1 \le \pi_1 \le n/N \le \pi_N \le np_N$;

(c) for $p_i = 1/N$, $\pi_i \ge n/N$ and for $p_i \le 1/N$, $\pi_i \ge np_i$ while for $p_i > 1/N$, $\pi_i \not< np_i$.

REFERENCES

Andreatta, G., and Kaufman, G. M. (1986). Estimation of finite population properties when sampling is without replacement and proportional to magnitude. *J. Amer. Statist. Assoc.*, **81**, 657–666.

Asok, C., and Sukhatme, B. V. (1976). On Sampford's procedures of unequal probability sampling without replacement. *J. Amer. Statist. Assoc.*, **71**, 912–918.

Brewer, K. R. W. (1963). A model of systematic sampling with unequal probabilities. *Austral. J. Statist.*, **5**, 5–13.

Brewer, K. R. W., and Hanif, M. (1983). *Sampling with Unequal Probabilities*. Lecture Notes in Statistics, No. 15. Springer-Verlag, New York.

Chaudhury, A., and Vos, J. W. E. (1988). *Unified Theory and Strategies of Survey Sampling*. North-Holland, Amsterdam.

Cheng, C.-S., and Li, K.-C. (1983). A minimax approach to sample surveys. *Ann. Statist.*, **11**, 552–563.

Cheng, C.-S., and Li, K.-C. (1987). Optimality criteria in survey sampling. *Biometrika*, **74**, 337–345.

Dasgupta, A., and Sinha, B. K. (1983). A decision-theoretic analysis of superpopulation models in finite population sampling. Tech. Report #23, Indian Statistical Institute, Calcutta.

Des Raj (1956). Some estimators in sampling with varying probabilities without replacement. *J. Amer. Statist. Assoc.*, **51**, 269–284.

Durbin, J. (1967). Design of multistage surveys for the estimation of sampling errors. *Appl. Statist.*, **16**, 152–164.

Hanurav, T. V. (1967). Optimum utilization of auxiliary information: sampling of two units for a stratum. *J. Roy. Statist. Soc.*, **B29**, 374–391.

Hedayat, A. (1990). The game of emptying boxes: Theory and application. Texas

Tech. University, Mathematics Series Visiting Scholars Lectures, 1988–1989, Vol. 16, 49–71.

Hedayat, A., and Lin, B. Y. (1980). A complete class theorem for probability proportional to size sampling designs. *Bull. Inst. Math. Statist.*, **9**, 297 (abstract).

Hedayat, A., Lin, B. Y., and Stufken, J. (1989). The construction of ΠPS sampling designs through a method of emptying boxes. *Ann. Statist.*, **17**, 1886–1905.

Kabadi, S. N., Chandrasekaran, R., and Nair, K. P. K. (1988). Optimal integration of several surveys. *Sankhyā*, **B50**, 153–156.

Keyfitz, N. (1951). Sampling with probability proportional to size: adjustment for changes in probabilities. *J. Amer. Statist. Assoc.*, **46**, 105–109.

Lahiri, D. B. (1951). A method of sample selection providing unbiased ratio estimates. *Bull. Internat. Statist. Inst.*, **33**, Book 2, 133–140.

Madow, W. G. (1949). On the theory of systematic sampling II. *Ann. Math. Statist.*, **20**, 333–354.

Midzuno, H. (1952). On the sampling system with probability proportionate to sum of sizes. *Ann. Inst. Statist. Math.*, **3**, 99–107.

Mitra, S. K., and Pathak, P. K. (1984). Algorithms for optimal integration of two or three surveys. *Scand. J. Statist.*, **11**, 257–263.

Murthy, M. N. (1957). Ordered and unordered estimators in sampling without replacement. *Sankhyā*, **18**, 379–390.

Patel, H. C., and Dharmadhikari, S. W. (1978). Admissibility of Murthy's and Midzuno's estimator within the class of linear unbiased estimators of finite population totals. *Sankhyā*, **C40**, 21–28.

Pathak, P. K., and Shukla, N. D. (1966). Non-negativity of a variance estimator. *Sankhyā*, **A28**, 41–46.

Rao, J. N. K. (1965). On two simple schemes of unequal probability sampling without replacement. *J. Indian Statist. Assoc.*, **3**, 173–180.

Rao, J. N. K. (1979). On deriving mean square errors and their non-negative unbiased estimators in finite population sampling. *J. Indian Statist. Assoc.*, **17**, 125–136.

Rao, J. N. K., Hartley, H. O., and Cochran, W. G. (1962). A simple procedure of unequal probability sampling without replacement. *J. Roy. Statist. Soc.*, **B24**, 482–491.

Rao, T. J., Sengupta, S., and Sinha, B. K. (1988). Some probability inequalities for PPSWOR sampling scheme. Tech. Report No. 14/88, Stat-Math Division, Indian Statistical Institute, Calcutta. To appear in *Metrika* (1990–91).

Sampford, M. R. (1967). On sampling without replacement with unequal probabilities of selection. *Biometrika*, **54**, 499–513.

Sengupta, S. (1980). On the admissibility of the symmetrized Des Raj estimator for PPSWOR samples of size two. *Cal. Statist. Assoc. Bull.*, **29**, 35–44.

Sengupta, S. (1982). Admissibility of the symmetrized Des Raj estimator for fixed size sampling designs of size two. *Cal. Statist. Assoc. Bull.*, **31**, 201–205.

Subrahmanya, M. T. (1967). On the positivity of a variance estimator. *Metrika*, **12**, 43–47.

Sunter, A. (1986). Solution to the problem of unequal probability sampling without replacement. *Internat. Statist. Rev.*, **54**, 33–50.

CHAPTER SIX

Uses of Auxiliary Size Measures in Survey Sampling: Ratio and Regression Methods of Estimation

In this chapter we present two popular methods of utilization of auxiliary information primarily at the estimation stage, based on data obtained by implementation of the SRS procedure. The estimators generally suggested are not, however, unbiased. Following the available literature, we discuss some techniques for reduction of bias. We also present some modified estimators ensuring unbiasedness. In the context of one of the estimators, we then extensively deal with the case when the population mean of the auxiliary variable is unknown. We also deal with the problem of estimating the mean of the survey variable when the population is sampled on successive occasions.

6.1. USE OF AUXILIARY SIZE MEASURES AT THE ESTIMATION STAGE

As survey statisticians, we have accepted the view that we should make every effort to improve the quality of our estimators by utilizing useful auxiliary information presented to us. Such information may be furnished before the survey is conducted or may be generated during the course of data collection on the survey variable. If auxiliary information is available beforehand, it can influence our choice of the sampling strategy at both the design and estimation stages. On the other hand, if the existence of auxiliary information is discovered only in the course of data collection, then it might be possible for us to at least use this information profitably in proposing an estimator. As in Chapter Five, an auxiliary variable X will be understood to have resulted from a suitable integration of one or more distinct sources of information.

In the previous chapter we dealt extensively with the case where the values of an auxiliary variable are available beforehand on each and every unit of the population and the relation between the auxiliary variable and the survey variable can be assumed to be approximately linear and passing through the origin. There we were primarily concerned with the development of methods of sampling at the design stage which utilize the auxiliary size measures, even though their use was also implicit in the form of the estimators suggested. In the process, ΠPS, PPS, and RHC sampling schemes were presented and suitable estimators based on them were proposed.

The effective utilization of auxiliary information in the estimation stage may arise in any one of the following circumstances.

(i) Auxiliary size measures are available on all units of the population. The practitioner used them at the design stage and wants to make further use of these values at the estimation stage.

(ii) Auxiliary size measures are available on all units of the population. These values were not used at the design stage at the discretion of the survey practitioner, who, however, does want to use them at the estimation stage.

(iii) Auxiliary information was available from different sources but the practitioner was unable to integrate them for use at the design stage. He intends to explore a suitable integration in the estimation stage, using both auxiliary information and data on the survey variable.

(iv) Auxiliary information was discovered only in the course of data collection on the survey variable and the practitioner intends to use the information in the estimation stage.

Some strategies of sampling in case (i) have been proposed in Chapter Five. In the remaining cases the formation of an integrated auxiliary variable X, which we assume to be nearly proportional to the survey variable Y, is based either on the past knowledge or on a study of the data at hand. It is advisable to use the available data for verification and suitable modification of any prior knowledge with reference to the assumed relationship between the two variables.

To summarize, if we are unable to come up with a definite approximation of the explicit relation between the two variables or if information on an auxiliary variable is not available beforehand but can be gathered along with data on the survey variable, then we might follow a data-oriented approach to explore the relation between the two variables. In this way, we can perhaps, make a profitable use of auxiliary information at the estimation stage.

In this chapter we deal with the data-oriented approach for building up our estimators of the population total or the population mean. The popular estimators suggested in the literature are based on the ratio method and the regression method of estimation. We will elaborate on these estimators and study various related problems.

To start with, we assume the following setup. It is desired to estimate the population total $T(\mathbf{Y})$ or the population mean $\bar{\mathbf{Y}}$ of the survey variable Y. There is an auxiliary variable X whose values can either be made available beforehand or through the course of data collection. Specifically, a fixed-size (n) design has been implemented and a sample $s = (i_1, i_2, \ldots, i_n)$ has been drawn. Accordingly, the units in the population bearing serial numbers $i_1, i_2, \ldots, i_n$ have been surveyed and we have arrived at the data set $\{(i_j, Y_{i_j}, X_{i_j}) \mid 1 \leq j \leq n\}$. At this stage we wish to investigate if we can make a profitable use of the auxiliary information available in suggesting an unbiased estimator of $\bar{\mathbf{Y}}$ and if so, we would also like to study the properties of the suggested estimator.

Almost all auxiliary variables selected in sample survey situations have influence on (or are influenced by) the survey variable. This being the case, we may want to postulate a relation of the type $Y = f(X)$ where the functional form of $f(\cdot)$ is not completely known and now we seek an approximation to it based on a probability sample s.

One convenient, albeit adhoc, procedure for approximating $f(\cdot)$ is to look at a scatter diagram of the sample quantities $\{(X_i, Y_i \mid i \in s\}$. A close inspection of the scatter diagram might guide us in forming $\Psi(\cdot|s)$ as an approximation to $f(\cdot)$. For example, the scatter might suggest fitting a polynomial of degree m to the n data points. We discourage the temptation to fit a polynomial of degree $n-1$ passing through all the n data points. Such an approximation might behave very badly for the unknown data points not covered by the sample. Secondly, it is hard to investigate the statistical properties of an estimator based on a polynomial approximation of high degree. Moreover, since the purpose of the approximation is basically to serve as a guiding tool, it is desirable to search for a function which is comfortably usable by and comprehensible to the survey practitioner. If this is not available, we should not feel compelled to utilize the information provided by the auxiliary variable.

The simplest approximation to $f(\cdot)$ is a straight line, which may or may not pass through the origin. This has led to two popular methods of estimation, known as the ratio and the regression method of estimation. We will present the estimators based on these methods and study various properties of the estimators.

6.2. THE RATIO METHOD OF ESTIMATION OF A POPULATION MEAN

Suppose a study of the scatter diagram reveals that the n sample points are clustered around a straight line passing through the origin, that is, the ratios

Y_i/X_i, $i \in s$ are more or less the same. Then we may postulate the approximate relation

$$\sum_{i \in s} Y_i \Big/ \sum_{i \in s} X_i \doteq \sum_{i \notin s} Y_i \Big/ \sum_{i \notin s} X_i , \tag{6.1}$$

whence we can write

$$\begin{aligned} T(\mathbf{Y}) &= \sum_{i \in s} Y_i + \sum_{i \notin s} Y_i \doteq \sum_{i \in s} Y_i + \Big(\sum_{i \in s} Y_i \Big/ \sum_{i \in s} X_i\Big)\Big(\sum_{i \notin s} X_i\Big) \\ &= \Big(\sum_{i \in s} Y_i \Big/ \sum_{i \in s} X_i\Big)\Big(\sum_{i=1}^{N} X_i\Big) \end{aligned}$$

from which we can suggest an estimator of $\bar{\mathbf{Y}}$ as

$$\hat{\bar{\mathbf{Y}}} = \Big(\sum_{i \in s} Y_i \Big/ \sum_{i \in s} X_i\Big)\bar{\mathbf{X}} = (\bar{\mathbf{y}}(s)/\bar{\mathbf{x}}(s))\bar{\mathbf{X}} , \tag{6.2}$$

where $\bar{\mathbf{y}}(s)$ and $\bar{\mathbf{x}}(s)$ refer to the sample means for Y and X, respectively. In the above, we have assumed $\bar{\mathbf{X}}$ to be known beforehand. The case of $\bar{\mathbf{X}}$ unknown will be discussed later.

We now study various properties of the estimator. At this stage we introduce some notation. We denote by $\mathbf{R}$ the ratio of the two population means: $\mathbf{R} = \bar{\mathbf{Y}}/\bar{\mathbf{X}}$. This means $\bar{\mathbf{Y}} = \mathbf{R}\bar{\mathbf{X}}$ and, hence, $\bar{\mathbf{X}}$ being assumed to be known, we may write $\hat{\bar{\mathbf{Y}}} = \hat{\mathbf{R}}\bar{\mathbf{X}}$. In other words, the estimator in (6.2) may be identified as $\hat{\mathbf{R}}\bar{\mathbf{X}}$ so that in effect we are proposing an estimator of $\mathbf{R}$ of the form

$$\hat{\mathbf{R}}(s, \mathbf{Y}) = \bar{\mathbf{y}}(s)/\bar{\mathbf{x}}(s) . \tag{6.3}$$

Because $\bar{\mathbf{Y}}$ and $\mathbf{R}$ are linearly related with a multiplier $\bar{\mathbf{X}}$ (assumed to be known), the properties of $\hat{\bar{\mathbf{Y}}}$ may be studied in terms of those of $\hat{\mathbf{R}}$. The estimator in (6.2) is popularly known as the *ratio estimator*. The ratio method of estimation is usually applied when the scatter indicates a positive slope ($\rho_{X,Y} > 0$). For a negative slope, better method of estimation is available. This is discussed in Section 6.5. We denote the ratio estimator as $\hat{\bar{\mathbf{Y}}}_R(s)$ or as $\hat{\bar{\mathbf{Y}}}_R$ for brevity. Standard statistical properties of $\hat{\mathbf{R}}$ or of $\hat{\bar{\mathbf{Y}}}_R$ have been investigated with reference to an SRS(N, n) design as the underlying sampling design. We confine ourselves primarily to this sampling strategy. It is then observed that the estimator $\hat{\bar{\mathbf{Y}}}_R$ is not unbiased for $\bar{\mathbf{Y}}$ unless the X_i's are all equal. The verification is left to the reader (Exercise 6.2). At this point we may pursue our study in two different directions. The first direction is toward a common goal of achieving unbiasedness by slightly modifying the sampling design or by suitably modifying the estimator. The second direction is toward a study of the amount of bias in the proposed estimator and, subsequently, to suggest a suitable procedure for bias reduction.

6.3. UNBIASEDNESS OF THE RATIO ESTIMATOR AND THE RATIO-TYPE ESTIMATOR

As stated before, an SRS(N, n) sampling design does not ensure unbiasedness of the ratio estimator $\hat{\bar{\mathbf{Y}}}_R$ for the population mean $\bar{\mathbf{Y}}$ or, equivalently, of $\hat{\mathbf{R}}$ for $\mathbf{R}$. To overcome this drawback, we may wish to implement another sampling design which is not very different from an SRS(N, n) sampling design and yet retains the unbiasedness property of $\hat{\bar{\mathbf{Y}}}_R$ and $\hat{\mathbf{R}}$ based on such a modified design. This problem was satisfactorily resolved by Lahiri (1951). We state and prove this result below. (The reader will discover a solution to Exercise 5.29 in this theorem).

Theorem 6.1. *Suppose the first unit of the sample is drawn using a PPS sampling design given by* $P((i)) = X_i/T(\mathbf{X})$, $1 \le i \le N$. *Next, suppose the remaining units of the sample are drawn by using* SRS($N-1$, $n-1$) *sampling on the remaining units of the population. Then the estimator* $\hat{\mathbf{R}}$ *is unbiased for the population ratio* $\mathbf{R}$.

Proof. Let $s = (i_1, i_2, \ldots, i_n)$ be a typical sample drawn according to the sampling design described above. Then, the probability of the sample s in its reduced form is given by

$$P^*(s) = \sum_{i \in s} X_i / \binom{N-1}{n-1} T(\mathbf{X}) . \tag{6.4}$$

Now we verify that the estimator $\hat{\mathbf{R}}$ is unbiased under the sampling design (6.4). We have

$$\begin{aligned} E_p(\hat{\mathbf{R}}) &= \sum_s \{\bar{\mathbf{y}}(s)/\bar{\mathbf{x}}(s)\}\Big(\sum_{i \in s} X_i / \binom{N-1}{n-1} T(\mathbf{X})\Big) = \sum_s n\bar{\mathbf{y}}(s)/T(\mathbf{X})\binom{N-1}{n-1} \\ &= \sum_s \bar{\mathbf{y}}(s)/\bar{\mathbf{X}}\binom{N}{n} = \bar{\mathbf{Y}}/\bar{\mathbf{X}} = \mathbf{R} . \end{aligned}$$

Hence the result is proved for $\hat{\mathbf{R}}$ and also, consequently, for $\hat{\bar{\mathbf{Y}}}_R$. □

We recall that the scheme of sampling in (6.4) is the well-known scheme due to Midzuno (1952), discussed in Example 1.9 and also in Chapter Five. Because the reduced form of the sampling design allows the probability of a sample s to be proportional to the aggregate size $\Sigma_{i \in s} X_i$ of the units in the sample, this is also known as the method of *sampling with probability proportional to aggregate size* (*PPAS*) or as the PPAS method of sampling. The sampling design is *very close* to an SRS(N, n) sampling design. The notion of closeness has been defined and examined in Gabler (1987). We do not discuss this here. If we do not insist on a fixed-size design, then a general solution would have been a mixture of sampling designs of the type (6.4). The verification is left to the reader (Exercise 6.3).

The above procedure can also be generalized to modify some other types of sampling designs to validate the use of the ordinary ratio estimator as an unbiased estimator under the modified sampling design. See, for example, Nanjamma et al. (1959). We do not go into detail here.

Based on the above PPAS sampling design of size n, the variance of the estimator $\hat{\bar{\mathbf{Y}}}_R = \bar{\mathbf{X}}\hat{\mathbf{R}}$ is given by

$$V(\hat{\bar{\mathbf{Y}}}_R) = \sum_{i=1}^{N} a_{ii} Y_i^2 + \sum_{\substack{i \neq j \\ 1}}^{N}\sum a_{ij} Y_i Y_j \tag{6.5}$$

where

$$a_{ii} = \frac{\bar{\mathbf{X}}}{n^2\binom{N}{n}} \left(\sum_{s \ni i} \frac{1}{\bar{\mathbf{x}}(s)} \right) - \frac{1}{N^2}, \qquad 1 \leq i \leq N \tag{6.6}$$

$$a_{ij} = \frac{\bar{\mathbf{X}}}{n^2\binom{N}{n}} \left(\sum_{s \ni i,j} \frac{1}{\bar{\mathbf{x}}(s)} \right) - \frac{1}{N^2}, \qquad 1 \leq i \neq j \leq N . \tag{6.7}$$

We can now use a version of Theorem 3.4 to deduce that any nonnegative unbiased estimator of $V(\hat{\bar{\mathbf{Y}}}_R)$ must be of the form

$$\hat{V}(\hat{\bar{\mathbf{Y}}}_R) = -\sum_{\substack{i<j \\ i,j \in s}}\sum a_{ij}(s) X_i X_j \left(\frac{Y_i}{X_i} - \frac{Y_j}{X_j} \right)^2 . \tag{6.8}$$

where the coefficients satisfy the conditions of unbiasedness:

$$\sum_{s \ni i,j} a_{ij}(s) P(s) = a_{ij} , \qquad 1 \leq i < j \leq N . \tag{6.9}$$

In the particular case of $n = 2$, it follows that the only possible nnuve of $V(\hat{\bar{\mathbf{Y}}}_R)$ will assume the form (6.8) with

$$a_{ij}(s) = \frac{\bar{\mathbf{X}}}{X_i + X_j} \left\{ \frac{\bar{\mathbf{X}}}{X_i + X_j} - \left(1 - \frac{1}{N} \right) \right\} \qquad \text{for } s = (i, j) . \tag{6.10}$$

It is clear that the estimator becomes nonnegative if and only if $X_i + X_j \geq T(\mathbf{X})/(N-1)$ for $1 \leq i < j \leq N$.

In the general case J. N. K. Rao and Vijayan (1977) suggested the following choices of $a_{ij}(s)$:

(i) $a_{ij}(s) = a_{ij}/\pi_{ij}$, $\qquad i, j \in s$

(ii) $$a_{ij}(s) = \frac{\bar{\mathbf{X}}^2}{n^2 \bar{\mathbf{x}}^2(s)} - \frac{\bar{\mathbf{X}}(N-1)}{\bar{\mathbf{x}}(s) N n (n-1)} , \qquad i, j \in s . \tag{6.11}$$

The variance estimator based on (ii) would be nonnegative if and only if $\bar{\mathbf{x}}(s)(1-(1/N)) \geq \bar{\mathbf{X}}(1-(1/n))$ for all samples. For various related results of interest we refer the reader to T. J. Rao (1972), Lanke (1974), and J. N. K. Rao and Vijayan (1977).

Example 6.1. Consider the population of 15 units presented in Example 5.11. Suppose we want to select 5 units using the PPAS method of sampling and then use the ordinary ratio estimate $\hat{\mathbf{R}}(s, \mathbf{Y})$ in (6.3) to unbiasedly estimate the population ratio $\mathbf{R}$. According to Lahiri's method we first select one unit by using PPS method of sampling and, subsequently, select an additional 4 units using simple random sampling. Let $s = (3, 8, 10, 11, 15)$ denote the selected sample in its reduced form. We take the observed Y-values as given in Table 6.1. Further, from the data in Example 5.11, we have $T(\mathbf{X}) = 74.5$ so that $\bar{\mathbf{X}} = 4.97$. Computations yield

$$\bar{\mathbf{x}}(s) = 5.94\,, \qquad \bar{\mathbf{y}}(s) = 58.8\,, \qquad \hat{\mathbf{R}} = 9.90\,, \qquad \hat{\bar{\mathbf{Y}}}_R = 49.20\,.$$

We ask the reader to compute $\hat{V}(\hat{\bar{\mathbf{Y}}})$ using the formula (6.8) in combination with (6.11) (Exercise 6.8).

Table 6.1

i	3	8	10	11	15
Y_i	40	50	70	82	52
X_i	4.1	5.2	7.2	8.0	5.2

Next we try to construct an unbiased estimator of $\mathbf{R}$, different from $\bar{\mathbf{y}}(s)/\bar{\mathbf{x}}(s)$, utilizing the auxiliary size measures and assuming the sample has been drawn according to an SRS(N, n) design. The estimator given below was suggested by Hartley and Ross (1954).

Theorem 6.2. *Under an* SRS(N, n) *sampling design, the ratio-type estimator*

$$\hat{\hat{\mathbf{R}}} = \bar{\mathbf{r}}(s) + (N-1)(\bar{\mathbf{y}}(s) - \bar{\mathbf{r}}(s)\bar{\mathbf{x}}(s))/N(n-1) \tag{6.12}$$

is unbiased for $\mathbf{R}$, *where* $r_i = Y_i/X_i$, $\bar{\mathbf{r}}(s) = \Sigma_{i\in s} r_i/n$, *mean of ratios.*

Proof. First we observe that

$$\begin{aligned}
E_p(\bar{\mathbf{r}}(s)) &= \sum_{i=1}^{N} (Y_i/X_i)/N = \mathbf{R} + \sum_{i=1}^{N} (Y_i/X_i)/N - \mathbf{R} \\
&= \mathbf{R} + \sum_{i=1}^{N} (Y_i/X_i)/N - \left(\sum_{i=1}^{N} Y_i\right)\Big/ N\bar{\mathbf{X}} = \mathbf{R} - \left\{\sum_{i=1}^{N} (Y_i - r_i\bar{\mathbf{X}})\right\}\Big/ N\bar{\mathbf{X}} \\
&= \mathbf{R} - \left\{\sum_{i=1}^{N} r_i(X_i - \bar{\mathbf{X}})\right\}\Big/ N\bar{\mathbf{X}}\,.
\end{aligned}$$

Next, it is easily seen that an unbiased estimator of

$$\sum_{i=1}^{N} r_i(X_i - \bar{\mathbf{X}})/(N-1)$$

would be given by

$$\sum_{i \in s} r_i(X_i - \bar{\mathbf{x}}(s))/(n-1) .$$

Therefore,

$$E_p[\bar{\mathbf{r}}(s) + (N-1)\sum_{i \in s} r_i(X_i - \bar{\mathbf{x}}(s))/N(n-1)\bar{\mathbf{X}}] = \mathbf{R} .$$

It can be seen that the bracketted expression above coincides with $\hat{\hat{\mathbf{R}}}$ in (6.12). Hence the result is proved. □

It is algebraically complicated to give an expression for $V(\hat{\hat{\mathbf{R}}})$ and, particularly, $\hat{V}(\hat{\hat{\mathbf{R}}})$ has no simple form. This may be the reason why this estimator has not been popular even for small samples. Further, the form of the estimator in (6.12) is quite complicated compared to the estimator $\hat{\mathbf{R}} = \bar{y}(s)/\bar{\mathbf{x}}(s)$. We do not present results involving the variance and the variance estimators. The reader may consult Robson (1957), Goodman and Hartley (1958), Mickey (1959), Williams (1961), and T. J. Rao (1966) for various results in this direction.

Example 6.2. With reference to Example 6.1, we now assume that the same sample $s = (3, 8, 10, 11, 15)$ has been drawn by using an SRS(15, 5) sampling procedure. To produce an unbiased estimator of the population ratio $\mathbf{R}$, we can use the Hartley–Ross estimator $\hat{\hat{\mathbf{R}}}$ in (6.12). Computations yield the r_i-values as $r_3 = 9.76$, $r_8 = 9.61$, $r_{10} = 9.72$, $r_{11} = 10.25$, and $r_{15} = 10.00$ so that $\bar{\mathbf{r}}(s) = 9.90$ and, hence, $\hat{\hat{\mathbf{R}}} = 9.94$.

6.4. BIAS OF THE RATIO ESTIMATOR IN SRS DATA AND BIAS REDUCTION TECHNIQUES

In our study of the bias of $\hat{\mathbf{R}}$ in (6.3) based on an SRS(N, n) sample data, we shall be concerned mainly with its dominating term. Writing $\bar{\mathbf{X}}/\bar{\mathbf{x}}(s)$ as

$$\begin{aligned} \bar{\mathbf{X}}/\bar{\mathbf{x}}(s) &= \bar{\mathbf{X}}/\{\bar{\mathbf{X}} + (\bar{\mathbf{x}}(s) - \bar{\mathbf{X}})\} \\ &= [1 + (\bar{\mathbf{x}}(s) - \bar{\mathbf{X}})\bar{\mathbf{X}}^{-1}]^{-1} \qquad (6.13) \\ &\doteq 1 - (\bar{\mathbf{x}}(s) - \bar{\mathbf{X}})\bar{\mathbf{X}}^{-1} , \end{aligned}$$

we obtain

$$
\begin{aligned}
E(\hat{\mathbf{R}} - \mathbf{R}) &= E[\{\bar{\mathbf{y}}(s)/\bar{\mathbf{x}}(s)\} - \{\bar{\mathbf{Y}}/\bar{\mathbf{X}}\}] \\
&= \bar{\mathbf{X}}^{-1}E[(\bar{\mathbf{X}}/\bar{\mathbf{x}}(s))\{\bar{\mathbf{y}}(s) - \mathbf{R}\bar{\mathbf{x}}(s)\}] \doteq \bar{\mathbf{X}}^{-1}E[\{\bar{\mathbf{y}}(s) - \mathbf{R}\bar{\mathbf{x}}(s)\} \\
&\quad - \{(\bar{\mathbf{x}}(s) - \bar{\mathbf{X}})(\bar{\mathbf{y}}(s) - \mathbf{R}\bar{\mathbf{x}}(s))\bar{\mathbf{X}}^{-1}\}] \\
&= -\bar{\mathbf{X}}^{-2}E[(\bar{\mathbf{x}}(s) - \bar{\mathbf{X}})(\bar{\mathbf{y}}(s) - \mathbf{R}\bar{\mathbf{x}}(s))] \\
&= -\bar{\mathbf{X}}^{-2}(n^{-1} - N^{-1})(\mathbf{S}_{XY} - \mathbf{R}\mathbf{S}_{XX}) \\
&= -\bar{\mathbf{X}}^{-2}(1 - f)(\mathbf{S}_{XY} - \mathbf{R}\mathbf{S}_{XX})/n\,, \qquad f = n/N\,. \qquad (6.14)
\end{aligned}
$$

The next to the last expression follows from the fact that

$$E\{\bar{\mathbf{y}}(s)(\bar{\mathbf{x}}(s) - \bar{\mathbf{X}})\} = E\{(\bar{\mathbf{y}}(s) - \bar{\mathbf{Y}})(\bar{\mathbf{x}}(s) - \bar{\mathbf{X}})\} = (n^{-1} - N^{-1})\mathbf{S}_{XY}\,.$$

In the above, the notations $\mathbf{S}_{XX}$ and $\mathbf{S}_{XY}$ are used, respectively, for the population variance of X and the population covariance between X and Y. Thus

$$\mathbf{S}_{XX} = (N-1)^{-1}\sum_{i=1}^{N}(X_i - \bar{\mathbf{X}})^2\,, \qquad \mathbf{S}_{XY} = (N-1)^{-1}\sum_{i=1}^{N}(X_i - \bar{\mathbf{X}})(Y_i - \bar{\mathbf{Y}})\,. \qquad (6.15)$$

It is evident from (6.14) that the leading term in the bias of $\hat{\mathbf{R}}$ is of order n^{-1}.

We shall now discuss methods for reducing or eliminating the leading term in the bias of $\hat{\mathbf{R}}$. Three methods are discussed.

Method 1. Increasing the Sample Size

Since the leading term in the bias of $\hat{\mathbf{R}}$ is a decreasing function of n, an obvious way to reduce the bias is to increase the sample size. However, the exact amount of reduction will depend on the knowledge of

$$\bar{\mathbf{X}}^{-2}(\mathbf{S}_{XY} - \mathbf{R}\mathbf{S}_{XX})$$

which simplifies to

$$\mathbf{R}\{\rho \mathrm{C.V.}(Y) - \mathrm{C.V.}(X)\}\mathrm{C.V.}(X)\,, \qquad (6.16)$$

where C.V.(.) stands for coefficient of variation and ρ is the population correlation coefficient between Y and X.

If, in a given situation, the practitioner wants to determine the sample size (n) so that the leading term in the bias of $\hat{\mathbf{R}}$ is within $100\alpha\%$ of the true value of **R**, then the estimating equation for n is given by

$$(1-f)\text{C.V.}(X)|(\rho\text{C.V.}(Y)-\text{C.V.}(X)| \doteq n\alpha \tag{6.17}$$

Prior knowledge as to the C.V.(X), C.V.(Y), and ρ can then be utilized to solve for n for a given α. The factor $(1-f)$ can be dropped, ignoring the fpc.

Example 6.3. The purpose of this example is to familiarize the reader with the task of determining the sample size for a given margin of relative error in the estimation of the population ratio **R**. We assume C.V.(X) = C.V.$(Y) = \delta$ so that ignoring the fpc, (6.17) reduces to

$$n \doteq \delta^2(1-\rho)/\alpha \ .$$

Taking $\alpha = 0.05$, we show in Table 6.2 the values of n corresponding to various choices of the pair (δ, ρ).

Table 6.2

ρ \ δ	0.5	1.0	1.5	2.0	2.5	3.0
0	5	20	45	80	125	180
0.2	4	16	36	64	100	144
0.4	3	12	27	48	75	108
0.5	3	10	23	40	63	90
0.6	2	8	18	32	50	72
0.8	–	4	9	16	25	36

Method 2. Jackknifing the estimator

This method is fairly general and it was first discussed by Quenouille (1956) in the context of a statistical estimation problem in which an estimator (for a parameter, say θ) based on a random sample of size n has a bias of order n^{-1}. The method was further developed by Schucany et al. (1971) among others. It consists of splitting up the sample at random into a number of mutually exclusive groups and using a suitable linear combination of (1) the estimators based on the original sample and (2) the average of estimators based on the data obtained from the sample after deleting various groups. In its simplest form, each individual sample unit may be used to form a group. Durbin (1959) illustrated the use of this method for ratio estimators. Denote by $\bar{\mathbf{y}}_{-i}$ and $\bar{\mathbf{x}}_{-i}$, the sample means of Y-values and X-values based on $(n-1)$ units in the sample excluding the unit i, $i \in s$. These result in n ratios of the form $\hat{\mathbf{R}}_{-i} = \bar{\mathbf{y}}_{-i}/\bar{\mathbf{x}}_{-i}$, $i \in s$. Next compute the average $\tilde{\mathbf{R}} = \Sigma_{i\in s} \hat{\mathbf{R}}_{-i}/n$. Then a suitable linear combination $\tilde{\tilde{\mathbf{R}}}$ of $\hat{\mathbf{R}}$ and $\tilde{\mathbf{R}}$, called the jackknife estimator, can be constructed so that the term involving n^{-1} is eliminated in the expression for the bias of the estimator $\tilde{\tilde{\mathbf{R}}}$. Specifically, we have the following result.

Theorem 6.3. *In contrast to $\hat{\mathbf{R}}$, the jackknifed version of $\hat{\mathbf{R}}$ given by*

$$\tilde{\tilde{\mathbf{R}}} = \hat{\mathbf{R}} + (n-1)(1-f)(\hat{\mathbf{R}} - \tilde{\mathbf{R}}) \,, \qquad f = n/N \tag{6.18}$$

has no term of order n^{-1} in the expression for its bias.

Proof. We recall from (6.14) that the leading term in the expression for bias of $\hat{\mathbf{R}}$ is given by

$$-\bar{\mathbf{X}}^{-2}(n^{-1} - N^{-1})(\mathbf{S}_{XY} - \mathbf{R}\mathbf{S}_{XX}) \,. \tag{6.19}$$

Next note that by deleting the unit i from the sample s, we are effectively refering to an SRS($N, n-1$) sampling design. Therefore, $E(\hat{\mathbf{R}}_{-i})$ has a similar representation to $E(\hat{\mathbf{R}})$ with n replaced by $(n-1)$. Since $\tilde{\mathbf{R}}$ is the average of the estimators $\hat{\mathbf{R}}_{-i}$, $i \in s$, it is clear that we also have $E(\tilde{\mathbf{R}}) = E(\hat{\mathbf{R}}_{-i})$, $i \in s$. Hence the leading term in the expression for bias of $\tilde{\mathbf{R}}$ is given by

$$-\bar{\mathbf{X}}^{-2}((n-1)^{-1} - N^{-1})(\mathbf{S}_{XY} - \mathbf{R}\mathbf{S}_{XX}) \,. \tag{6.20}$$

This yields the following result for the term involving n^{-1} in the expression for the bias of the jackknifed estimator $\tilde{\tilde{\mathbf{R}}}$:

$$-\bar{\mathbf{X}}^{-2}(\mathbf{S}_{XY} - \mathbf{R}\mathbf{S}_{XX})[(n^{-1} - N^{-1}) + (n-1)(1-f)(n^{-1} - (n-1)^{-1})] \,.$$

This, on simplification, reduces to zero. □

Remark 6.1. It can be seen that the leading term in the expression for the bias of $\tilde{\tilde{\mathbf{R}}}$ is of order n^{-2}.

Example 6.4. We refer to the same data as in Example 6.2 and carry out computations for the jackknifing technique. The results are shown below.

i	3	8	10	11	15
$\hat{\mathbf{R}}_{-i}$	9.9219	9.9592	9.9556	9.7696	9.8775

These yield

$$\tilde{\mathbf{R}} = \sum_{i \in s} \hat{\mathbf{R}}_{-i}/5 = 9.8968 .$$

Finally, using (6.18), we obtain

$$\tilde{\tilde{\mathbf{R}}} = 9.9086 .$$

Method 3. Estimating the Leading Term in the Bias

We start with a representation of the bias in $\hat{\mathbf{R}}$ as

$$E(\hat{\mathbf{R}}) - \mathbf{R} = (n^{-1} - N^{-1})B_1(\mathbf{Y}) + (n^{-2} - N^{-2})B_2(\mathbf{Y}) + \cdots \tag{6.21}$$

where $B_1(\mathbf{Y})$ is given in (6.14). It has been suggested that an attempt can be made to estimate the parameter $B_1(\mathbf{Y})$ using sample data. Suppose that an estimator $\hat{B}_1(s, \mathbf{Y})$ is proposed for which

$$E(\hat{B}_1(s, \mathbf{Y})) = B_1(\mathbf{Y}) + (n^{-1} - N^{-1})B_{11}(\mathbf{Y}) + \cdots . \tag{6.22}$$

Then, we can modify $\hat{\mathbf{R}}$ to $\hat{\mathbf{R}}^*$ defined as

$$\hat{\mathbf{R}}^* = \hat{\mathbf{R}} - (n^{-1} - N^{-1})\hat{B}_1(s, \mathbf{Y}) . \tag{6.23}$$

It now follows directly that

$$E(\hat{\mathbf{R}}^* - R) = -(n^{-1} - N^{-1})^2 B_{11}(\mathbf{Y}) + (n^{-2} - N^{-2})B_2(\mathbf{Y}) + \cdots \tag{6.24}$$

so that the leading term in the bias of $\hat{\mathbf{R}}^*$ is of the order n^{-2}. This method, therefore, seeks to modify the original estimator $\hat{\mathbf{R}}$ by using a suitable estimator of the leading term in the bias of $\hat{\mathbf{R}}$. An estimator of $B_1(\mathbf{Y})$ that is readily applicable is given by

$$\hat{B}_1(s, \mathbf{Y}) = -\bar{\mathbf{X}}^{-2}(\mathbf{s}_{XY} - \hat{\mathbf{R}}\mathbf{s}_{XX}) \tag{6.25}$$

where $\mathbf{s}_{XX}$ and $\mathbf{s}_{XY}$ denote, respectively, the sample variance of X and the sample covariance between X and Y. Thus

$$\mathbf{s}_{XX} = (n-1)^{-1} \sum_{i \in s} (X_i - \bar{\mathbf{x}}(s))^2 , \quad \mathbf{s}_{XY} = (n-1)^{-1} \sum_{i \in s} (X_i - \bar{\mathbf{x}}(s))(Y_i - \bar{\mathbf{y}}(s)) . \tag{6.26}$$

Beale (1962) and Tin (1965) suggested other estimators of $B_1(\mathbf{Y})$.

In concluding this section, we present large sample approximations for the mean squared error (mse) of $\hat{\mathbf{R}}$ and its estimator. The results are based on the approximation $\bar{\mathbf{x}}(s) \doteq \bar{\mathbf{X}}$. We start with $\hat{\mathbf{R}} - \mathbf{R}$ and write

$$\begin{aligned}\hat{\mathbf{R}} - \mathbf{R} &= (\bar{\mathbf{y}}(s)/\bar{\mathbf{x}}(s)) - \mathbf{R} \\ &= (\bar{\mathbf{y}}(s) - \mathbf{R}\bar{\mathbf{x}}(s))/\bar{\mathbf{x}}(s) \\ &\doteq \bar{\mathbf{X}}^{-1}(\bar{\mathbf{y}}(s) - \mathbf{R}\bar{\mathbf{x}}(s))\end{aligned}$$

so that

$$\begin{aligned}\text{mse}(\hat{\mathbf{R}}) &\doteq \bar{\mathbf{X}}^{-2}V(\bar{\mathbf{y}}(s) - \mathbf{R}\bar{\mathbf{x}}(s)) \\ &\doteq \bar{\mathbf{X}}^{-2}(n^{-1} - N^{-1})\left\{\sum_{i=1}^{N}(Y_i - \mathbf{R}X_i)^2/(N-1)\right\} \\ &= \bar{\mathbf{X}}^{-2}(n^{-1} - N^{-1})[\mathbf{S}_{YY} + \mathbf{R}^2\mathbf{S}_{XX} - 2\mathbf{R}\mathbf{S}_{XY}]\,. \end{aligned} \qquad (6.27)$$

Similarly,

$$\widehat{\text{mse}}(\hat{\mathbf{R}}) = \bar{\mathbf{X}}^{-2}(n^{-1} - N^{-1})[\mathbf{s}_{YY} + \hat{\mathbf{R}}^2\mathbf{s}_{XX} - 2\hat{\mathbf{R}}\mathbf{s}_{XY}] \qquad (6.28)$$

satisfies the approximate relation

$$E(\widehat{\text{mse}}(\hat{\mathbf{R}})) \doteq \text{mse}(\hat{\mathbf{R}})\,.$$

Example 6.5. Suppose a simple random sample of size $n = 64$ has been drawn from a large population. The sample size may be assumed to be large enough for the validity of the large sample results regarding the ratio estimator and its variance estimator.

Given that $\bar{\mathbf{X}} = 5.00$, $\bar{\mathbf{x}}(s) = 5.50$, $\bar{\mathbf{y}}(s) = 60.5$, $\mathbf{s}_{XX} = 120$, $\mathbf{s}_{YY} = 14{,}500$, and $\mathbf{s}_{XY} = 270$, we obtain

$$\hat{\mathbf{R}} = 11.0 \qquad \text{and} \qquad \hat{\bar{\mathbf{Y}}}_R = 55.0$$

along with

$$\begin{aligned}\widehat{\text{mse}}(\hat{\mathbf{R}}) &= \frac{1}{25}\left(\frac{1}{64}\right)(14{,}500 + 121 \times 120 - 2 \times 11 \times 270) \\ &\doteq 14.425\end{aligned}$$

and

$$\widehat{\text{mse}}(\hat{\bar{\mathbf{Y}}}_R) \doteq 360.625\,.$$

Thus, the estimated s.e. for $\hat{\bar{\mathbf{Y}}}_R \doteq 18.99$.

6.5. FURTHER ISSUES CONCERNING THE RATIO METHOD OF ESTIMATION

Ratio Estimator of $\bar{\mathbf{Y}}$ When $\bar{\mathbf{X}}$ Is Unknown

Thus far, we have assumed that the population mean $\bar{\mathbf{X}}$ of the auxiliary size variable X is known in advance. There might be situations where no prior knowledge of $\bar{\mathbf{X}}$ is available, nor is it realistic to conduct a census just for this purpose, even though auxiliary size measures may be very cheaply available. However, if the survey practitioner has strong reasons to believe that the use of auxiliary size variable will improve the statistical properties of a suitably constructed estimator, then a sample survey may be conducted beforehand, for the purpose of deriving an estimator of $\bar{\mathbf{X}}$.

Specifically, let d' be a fixed-size (n') sampling design meant for the above purpose. If s' is the sample selected according to d', then the practitioner will base the estimate of $\bar{\mathbf{X}}$ on the units included in the sample s'. As far as data on the survey variable Y is concerned, several options are available. These include the following popular methods:

(i) Select a subsample s of size $n \leq n'$ from s'.

(ii) Select a subsample of size m from s' and also an additional sample of size $u = n - m$ from those units not included in s'. The two samples together give the final sample s of size n.

(iii) Select an independent sample s of size n from the entire population, without any regard as to the nature of the sample s'.

We denote by $\hat{\bar{\mathbf{X}}}(s')$ and $\hat{\bar{\mathbf{X}}}(s)$ the unbiased estimators of $\bar{\mathbf{X}}$ based on the samples s' and s, respectively. Also we denote by $\hat{\bar{\mathbf{Y}}}(s)$ an unbiased estimator of $\bar{\mathbf{Y}}$ based only on the Y-values in the sample s. Then, traditionally, the following ratio estimator of $\bar{\mathbf{Y}}$ is used:

$$\hat{\bar{\mathbf{Y}}}_R(s, s', \mathbf{Y}) = \frac{\hat{\bar{\mathbf{Y}}}(s)}{\hat{\bar{\mathbf{X}}}(s)} \cdot \hat{\bar{\mathbf{X}}}(s') .$$

The properties of the above estimator and its jackknifed versions have been extensively studied in the literature, particularly when the underlying sampling designs for generating the samples s' and s are both SRS designs. Some useful references are Durbin (1959), Quenouille (1956), Schucany et al. (1971), and Sengupta (1981). We will not pursue this aspect of ratio estimation in this book. Later in this chapter similar problems in the context of regression method of estimation will be studied in considerable detail.

Product Method of Estimation

So far our study of the ratio estimator has been confined to the case where the auxiliary variable and the study variable were assumed to have an approximately linear relation passing through the origin and possessing a positive slope. If, however, we are confronted with situations where the values on the survey variable Y are inversely proportional to those of the auxiliary variable, say W, then the following recommendations may be useful.

One approach would be to apply the usual ratio estimate working with $X_i = 1/W_i$, $1 \le i \le N$, with or without assuming the knowledge of the mean $(\Sigma_{i=1}^N 1/W_i)/N$. This would result in the estimator

$$\frac{n\bar{\mathbf{y}}(s)}{(\Sigma_{i\in s} 1/W_i)} \left\{\sum_{i=1}^{N} (1/W_i)/N\right\}. \tag{6.29}$$

Another approach would be to use the following modified form of the ratio estimator, as suggested by Murthy (1964) and termed by him the *product estimator*:

$$\hat{\bar{\mathbf{Y}}}_P = \bar{\mathbf{y}}(s)\bar{\mathbf{x}}(s)/\bar{\mathbf{X}}. \tag{6.30}$$

The ratio method of estimation along with its modifications discussed in the earlier sections are directly related to the estimators suggested by the first approach. We do not present results on the product estimator which is seen to be particularly useful when the slope is negative. The interested reader is referred to T. J. Rao (1981a, b, 1987) and Tripathi (1988). The product estimator along with its jackknifed version have been studied extensively by Murthy (1964), J. N. K. Rao (1967, 1969), P. S. R. S. Rao (1969, 1974, 1988), P. S. R. S. Rao and J. N. K. Rao (1971), J. N. K. Rao and Webster (1966), Skukla (1976), Tin (1965), Ray and Sahai (1980) and Sengupta (1981), among others.

6.6. THE REGRESSION METHOD OF ESTIMATION OF A POPULATION MEAN

As mentioned at the beginning of this chapter, the simplest approximation to the function $f(\cdot)$ which relates the auxiliary variable X to the survey variable Y is a straight line passing through the origin. The ratio method of estimation discussed so far in this chapter was based on the supposition that this line passes through the origin. The regression method of estimation, on the other hand, applies to situations where a scatter diagram reveals an approximately linear relation of the form

$$Y \doteq \alpha + \beta X \tag{6.31}$$

with $\alpha(\neq 0)$ and β both unknown.

Our purpose is to estimate the population total $T(\mathbf{Y})$ which can be expressed as

$$T(\mathbf{Y}) = \sum_{i \in s} Y_i + \sum_{i \notin s} Y_i . \tag{6.32}$$

The quantity $\Sigma_{i \in s} Y_i$ will become known upon collection of data. The unknown quantity $\Sigma_{i \notin s} Y_i$ will obey the approximate relation

$$\sum_{i \notin s} Y_i \doteq (N - n)\alpha + \beta \sum_{i \notin s} X_i . \tag{6.33}$$

We now use the method of least squares on the available data $\{(X_i, Y_i) \mid i \in s\}$ and estimate α and β in a way that $\hat{\alpha} + \hat{\beta}X$ is the line of best fit for the observed pairs of values of (X, Y) in the scatter diagram. Clearly, $\hat{\alpha}$ and $\hat{\beta}$ are sought to minimize

$$\sum_{i \in s} (Y_i - \hat{\alpha} - \hat{\beta}X_i)^2 . \tag{6.34}$$

As is well known, the solution is

$$\hat{\alpha}(s) = \bar{\mathbf{y}}(s) - \hat{\beta}(s)\bar{\mathbf{x}}(s) , \tag{6.35}$$

$$\hat{\beta}(s) = \sum_{i \in s} (Y_i - \bar{\mathbf{y}}(s))(X_i - \bar{\mathbf{x}}(s)) \Big/ \Big\{ \sum_{i \in s} (X_i - \bar{\mathbf{x}}(s))^2 \Big\} . \tag{6.36}$$

Hence, we can estimate $\Sigma_{i \notin s} Y_i$ by the quantity

$$(N - n)\hat{\alpha}(s) + \hat{\beta}(s) \sum_{i \notin s} X_i . \tag{6.37}$$

Therefore, an estimate of the population total is given by

$$\begin{aligned} \hat{T}(\mathbf{Y}) &= \sum_{i \in s} Y_i + (N - n)\hat{\alpha}(s) + \hat{\beta}(s) \sum_{i \notin s} X_i \\ &= n\bar{\mathbf{y}}(s) + (N - n)\{\bar{\mathbf{y}}(s) - \hat{\beta}(s)\bar{\mathbf{x}}(s)\} + \hat{\beta}(s)\Big(\sum_{i=1}^{N} X_i - \sum_{i \in s} X_i \Big) \\ &= N\bar{\mathbf{y}}(s) + \hat{\beta}(s)(N\bar{\mathbf{X}} - n\bar{\mathbf{x}}(s) - (N - n)\bar{\mathbf{x}}(s)) \\ &= N\{\bar{\mathbf{y}}(s) + \hat{\beta}(s)(\bar{\mathbf{X}} - \bar{\mathbf{x}}(s))\} . \end{aligned} \tag{6.38}$$

This leads to an estimator of the population mean as

$$\hat{\bar{\mathbf{Y}}}_{\mathrm{lr}} = \frac{1}{N} \hat{T}(\mathbf{Y}) = \bar{\mathbf{y}}(s) + b(s)(\bar{\mathbf{X}} - \bar{\mathbf{x}}(s)) . \tag{6.39}$$

In (6.39) we have replaced $\hat{\beta}(s)$ by the more familiar notation $b(s)$ which denotes the sample regression coefficient of Y on X. The estimator in (6.39) is popularly known as the *linear regression estimator* of $\bar{\mathbf{Y}}$. We have used the suffix "lr" as an abbreviation for linear regression. It must be noted that in (6.39) we are using the knowledge of $\bar{\mathbf{X}}$. In Sections 6.7 and 6.8 we will study various properties of the linear regression estimator and other related estimators. The case when $\bar{\mathbf{X}}$ is unknown will be taken up later.

6.7. UNBIASEDNESS OF THE REGRESSION ESTIMATOR AND THE REGRESSION-TYPE ESTIMATOR

It is readily seen that the estimator in (6.39) based on an SRS(N, n) design is not unbiased. The amount of bias is clearly given by

$$\text{Bias} = E_p\{b(s)(\bar{\mathbf{X}} - \bar{\mathbf{x}}(s)\} . \tag{6.40}$$

We discuss the bias later. First, we present two results on achieving unbiasedness. Singh and Srivastava (1980) suggested two sampling schemes providing unbiased regression and regression-type estimators. We state and prove their results below. It is assumed that the auxiliary size measures are available beforehand on all the units in the population.

Theorem 6.4. *Suppose the first two units of the sample are drawn using the reduced sampling design*

$$P((i, j)) \propto (X_i - X_j)^2 , \qquad 1 \le i < j \le N . \tag{6.41}$$

Next suppose the remaining $(n-2)$ units of the sample are drawn using simple random sampling from the remaining population units. Then the linear regression estimator $\hat{\bar{\mathbf{Y}}}_{\text{lr}}$ is an unbiased estimator of the population mean $\bar{\mathbf{Y}}$.

Proof. For a sample s of size n, the reduced sampling design yields, upon simplification,

$$\begin{aligned} P(s) &= \sum_{\substack{i<j \\ i,j \in s}}\sum (X_i - X_j)^2 \Big/ \binom{N-2}{n-2} \Big\{ \sum\sum_{1 \le k < l \le N} (X_k - X_l)^2 \Big\} \\ &= \mathbf{s}_X^2 \Big/ \binom{N}{n} \mathbf{S}_X^2 . \end{aligned} \tag{6.42}$$

We now proceed to compute $E(\hat{\bar{\mathbf{Y}}}_{\text{lr}})$. We have

$$\begin{aligned} E(\hat{\bar{\mathbf{Y}}}_{\text{lr}}) &= E\{\bar{\mathbf{y}}(s) + b(s)(\bar{\mathbf{X}} - \bar{\mathbf{x}}(s))\} \\ &= \sum_s P(s)\{\bar{\mathbf{y}}(s) + b(s)(\bar{\mathbf{X}} - \bar{\mathbf{x}}(s))\} \\ &= \sum_s \{\bar{\mathbf{y}}(s)\mathbf{s}_X^2 + \mathbf{s}_{XY}(\bar{\mathbf{X}} - \bar{\mathbf{x}}(s))\} / \tbinom{N}{n} \mathbf{S}_X^2 \end{aligned}$$

$$= \sum_s \{\bar{\mathbf{Y}}\mathbf{s}_X^2 + (\bar{\mathbf{y}}(s) - \bar{\mathbf{Y}})\mathbf{s}_X^2 + s_{XY}(\bar{\mathbf{X}} - \bar{\mathbf{x}}(s))\} / \binom{N}{n}\mathbf{S}_X^2$$

$$= \bar{\mathbf{Y}} + \left\{\sum_s (\bar{\mathbf{y}}(s) - \bar{\mathbf{Y}})\mathbf{s}_X^2 + \sum_s \mathbf{s}_{XY}(\bar{\mathbf{X}} - \bar{\mathbf{x}}(s))\right\} \Big/ \binom{N}{n}\mathbf{S}_X^2 \quad (6.43)$$

since $\Sigma_s \mathbf{s}_X^2/\binom{N}{n} = \mathbf{S}_X^2$, as can be easily verified. We now want to show that the bracketed quantity $\{\cdots\}$ in (6.43) is identically equal to zero. Clearly, this quantity is a linear function of $Y_1, Y_2, \ldots, Y_N$. The coefficient of Y_i is given by

$$\left(\frac{1}{n} - \frac{1}{N}\right) \sum_{s \ni i} \mathbf{s}_X^2 + \frac{1}{n-1} \sum_{s \ni i} (X_i - \bar{\mathbf{x}}(s))(\bar{\mathbf{X}} - \bar{\mathbf{x}}(s)) - \frac{1}{N} \sum_{s \not\ni i} \mathbf{s}_X^2$$

$$= \left(\frac{1}{n} - \frac{1}{N}\right) \sum_{s \ni i} \mathbf{s}_X^2 + \frac{1}{n-1} \sum_{s \ni i} (X_i - \bar{\mathbf{X}})(\bar{\mathbf{X}} - \bar{\mathbf{x}}(s))$$

$$+ \frac{1}{n-1} \sum_{s \ni i} (\bar{\mathbf{X}} - \bar{\mathbf{x}}(s))^2 - \frac{1}{N} \sum_{s \not\ni i} \mathbf{s}_X^2 . \quad (6.44)$$

In (6.44), the coefficient of X_i^2 is given by

$$\left(\frac{1}{n} - \frac{1}{N}\right)\binom{N-1}{n-1}\left(\frac{1}{n}\right) + \frac{1}{(n-1)}\left(\frac{1}{N} - \frac{1}{n}\right)^2\binom{N-1}{n-1}$$

$$+ \frac{1}{(n-1)}\left(\frac{1}{N} - \frac{1}{n}\right)\left(1 - \frac{1}{N}\right)\binom{N-1}{n-1}$$

$$= \binom{N-1}{n-1}\left(\frac{1}{n} - \frac{1}{N}\right)\left[\frac{1}{n} + \frac{1}{(n-1)}\left(\frac{1}{n} - \frac{1}{N}\right) - \frac{1}{(n-1)}\left(1 - \frac{1}{N}\right)\right]$$

$$= 0 .$$

Similarly, it can be verified that the coefficients of other square and product terms in (6.44) are zeroes. Hence the coefficient of Y_i in (6.43) is $1/N$ for every i. This settles the claim that $\hat{\bar{\mathbf{Y}}}_{\mathrm{lr}}$ is an unbiased regression estimator of $\bar{\mathbf{Y}}$ under the sampling design in (6.42). □

An exact expression for the variance for the linear regression estimator under the sampling design in (6.42) is quite complicated. Singh and Srivastava (1980) have provided an approximate variance expression and an approximate variance estimator. To order $O(n^{-2})$, these are, respectively, given by

$$V(\hat{\bar{\mathbf{Y}}}_{\mathrm{lr}}) \doteq \frac{\mathbf{S}_Y^2(1-\rho^2)}{n} + \frac{(1-\rho^2)\mathbf{S}_Y^2}{n^2} + \frac{1}{n^2\mathbf{S}_X^2}\left[\frac{2\mu_{11}\mu_{31}}{\mu_{20}} - \frac{\mu_{11}^2\mu_{40}}{\mu_{20}^2} - \mu_{22}\right] \quad (6.45)$$

and

$$\hat{V}(\hat{\bar{\mathbf{Y}}}_{\mathrm{lr}}) \doteq \hat{\bar{\mathbf{Y}}}^2_{\mathrm{lr}} - \frac{1}{Nn}\frac{\mathbf{S}_X^2}{\mathbf{s}_X^2}\left[\sum_{i\in s} Y_i^2 + \frac{N-1}{n-1}\sum_{\substack{i\neq j\\ i,j\in s}}\sum Y_iY_j\right] \tag{6.46}$$

where

$$\mu_{kl} = \frac{1}{(N-1)}\sum_{i=1}^{N}(X_i - \bar{\mathbf{X}})^k(Y_i - \bar{\mathbf{Y}})^l\,. \tag{6.47}$$

In the next theorem, we provide a regression type unbiased estimator of the population mean.

Theorem 6.5. *Assume the first unit of the sample is selected using*

$$P((i)) \propto (X_i - \bar{\mathbf{X}})^2\,, \qquad 1 \le i \le N\,, \tag{6.48}$$

and other $n-1$ units are selected using SRS from the remaining population units. Then

$$\hat{\bar{\mathbf{Y}}}^*_{\mathrm{lr}} = \frac{n(N-1)}{N(n-1)}\left\{\bar{\mathbf{y}}(s) + \frac{(\bar{\mathbf{X}} - \bar{\mathbf{x}}(s))(\Sigma_{i\in s} Y_i(X_i - \bar{\mathbf{X}}))}{\Sigma_{i\in s}(X_i - \bar{\mathbf{X}})^2}\right\} \tag{6.49}$$

is an unbiased estimator of $\bar{\mathbf{Y}}$ whenever at least $N-n+1$ of the X_i's are different from $\bar{\mathbf{X}}$.

Proof. The implication of the stated condition in the theorem is that the reduced sampling design is defined on the full support of $\binom{N}{n}$ samples of size n. We will now compute directly $E(\hat{\bar{\mathbf{Y}}}^*_{\mathrm{lr}})$ and show that it coincides with $\bar{\mathbf{Y}}$. For this, we first note that the reduced form of the sampling design under consideration is given by

$$P(s) = \sum_{i\in s}(X_i - \bar{\mathbf{X}})^2 / \binom{N-1}{n-1}\sum_{i=1}^{N}(X_i - \bar{\mathbf{X}})^2 \tag{6.50}$$

for every unordered sample s of size n. Therefore,

$$\begin{aligned}
E(\hat{\bar{\mathbf{Y}}}^*_{\mathrm{lr}}) &= \sum_s \frac{n(N-1)}{N(n-1)}\left\{\sum_{i\in s}(X_i - \bar{\mathbf{X}})^2 / \binom{N-1}{n-1}(N-1)\mathbf{S}_X^2\right\}\\
&\quad\times\left\{\bar{\mathbf{y}}(s) + \frac{(\bar{\mathbf{X}} - \bar{\mathbf{x}}(s))(\Sigma_{i\in s} Y_i(X_i - \bar{\mathbf{X}}))}{\Sigma_{i\in s}(X_i - \bar{\mathbf{X}})^2}\right\}\\
&= \frac{1}{(n-1)\binom{N}{n}\mathbf{S}_X^2}\left[\sum_s \bar{\mathbf{y}}(s)\sum_{i\in s}(X_i - \bar{\mathbf{X}})^2\right]
\end{aligned}$$

$$+\frac{1}{(n-1)\binom{N}{n}\mathbf{S}_X^2}\left[\sum_{s}(\bar{\mathbf{X}}-\bar{\mathbf{x}}(s))\sum_{i\in s}Y_i(X_i-\bar{\mathbf{X}})\right]$$

$$=\frac{1}{(n-1)\binom{N}{n}\mathbf{S}_X^2}\left[\frac{1}{n}\sum_{i=1}^{N}Y_i\sum_{s\ni i}\sum_{j\in s}(X_j-\bar{\mathbf{X}})^2\right]$$

$$+\frac{1}{(n-1)\binom{N}{n}\mathbf{S}_X^2}\left[\sum_{i=1}^{N}Y_i(X_i-\bar{\mathbf{X}})\left\{\sum_{s\ni i}(\bar{\mathbf{X}}-\bar{\mathbf{x}}(s))\right\}\right]. \tag{6.51}$$

In the above expression, the coefficient of Y_i is given by

$$\frac{1}{n(n-1)\binom{N}{n}\mathbf{S}_X^2}\left[\sum_{s\ni i}\sum_{j\in s}(X_j-\bar{\mathbf{X}})^2+(X_i-\bar{\mathbf{X}})\sum_{s\ni i}(\bar{\mathbf{X}}-\bar{\mathbf{x}}(s))\right]$$

$$=\frac{1}{(n-1)\binom{N}{n}\mathbf{S}_X^2}\left[(X_i-\bar{\mathbf{X}})^2\binom{N-1}{n-1}+\binom{N-2}{n-2}\sum_{j(\neq i)}(X_j-\bar{\mathbf{X}})^2\right.$$

$$\left.-\frac{n}{n}(X_i-\bar{\mathbf{X}})^2\binom{N-1}{n-1}-\frac{n}{n}(X_i-\bar{\mathbf{X}})\sum_{j(\neq i)}(X_j-\bar{\mathbf{X}})\binom{N-2}{n-2}\right]$$

$$=\frac{1}{n(n-1)\binom{N}{n}\mathbf{S}_X^2}\left[\binom{N-2}{n-2}\sum_{i=1}^{N}(X_i-\bar{\mathbf{X}})^2\right],\quad \text{on simplification}$$

$$=\frac{(N-1)\binom{N-2}{n-2}\mathbf{S}_X^2}{n(n-1)\binom{N}{n}\mathbf{S}_X^2}=\frac{1}{N}.$$

The same phenomenon holds for every $i, 1\le i\le N$. Thus we obtain unbiasedness of $\hat{\mathbf{Y}}_{\mathrm{lr}}^*$ for the population mean $\bar{\mathbf{Y}}$. □

The expressions for $V(\hat{\mathbf{Y}}_{\mathrm{lr}}^*)$ and its unbiased estimator, both to order n^{-2}, have been given by Singh and Srivastava (1980). The expressions are algebraically complicated. We do not present these results here.

Example 6.6. The purpose of this and the next example is to familiarize the reader with some of the computations involved in the use of the results of this section. We consider once again the population of 15 units presented in Example 5.11 and also in Example 6.1. Suppose it is felt that a linear regression estimator would be suitable for estimating the population mean. Then, according to Theorem 6.4, we can achieve unbiasedness of the linear regression estimator $\hat{\mathbf{Y}}_{\mathrm{lr}}$ in (6.39) if the sampling is done according to the sampling design displayed in (6.42). The method of implementation of the design is, of course, to select a pair of units (i, j) with selection probability proportional to $(X_i-X_j)^2$, followed by an SRS procedure to select the remaining units in the sample. We take the sample size to be $n=5$.

Table 6.3

Ordered Pairs (i, j)	$(1,2)$	$(1,3)$	$(1,4)$	$\cdots$	$(1,15)$	$\cdots$	$(14,15)$
$(X_i - X_j)^2$	0.49	3.24	0.09	$\cdots$	8.41	$\cdots$	1.96

We provide in Table 6.3 a partial list of pairs of units in the population and the corresponding values of $(X_i - X_j)^2$ in the order $(1,2)(1,3), \ldots, (1,15), (2,3), \ldots, (2,15), \ldots, (14,15)$.

Clearly,

$$\sum_{i<j}\sum (X_i - X_j)^2 = \frac{1}{2}\sum_i \sum_j (X_i - X_j)^2 = N\sum_i (X_i - \bar{\mathbf{X}})^2$$

$$= N\left(\sum X_i^2\right) - \left(\sum X_i\right)^2 = 1225.70\,.$$

To select one pair with selection probability proportional to $(X_i - X_j)^2$ for the pair (i, j), we can use the method of cumulative total. We select a random number in the interval $(0.00, 1225.70)$. Let the selected number be 157.88. By working out the cumulative totals of values in Table 6.3, we observe that $(X_1 - X_2)^2 + \cdots + (X_1 - X_{10})^2 = 140.18 < 157.38 < 172.67 = (X_1 - X_2)^2 + \cdots + (X_1 - X_{11})^2$. Therefore, the pair $(1, 11)$ is selected initially. Next, using routine procedure, we select 3 additional units from the remaining population. Let the selected units be 7, 10, and 14. Then the sample is $s = (1, 7, 10, 11, 14)$. Suppose the values on the survey variable along with those on the auxiliary variable for these units are given as

$$X_1 = 2.3\,, \quad X_7 = 2.8\,, \quad X_{10} = 7.2\,, \quad X_{11} = 8.0\,, \quad X_{14} = 3.8\,,$$
$$Y_1 = 2.5\,, \quad Y_7 = 30\,, \quad Y_{10} = 70\,, \quad Y_{11} = 82\,, \quad y_{14} = 40\,.$$

Computations yield, using $\bar{\mathbf{X}} = 4.97$,

$$\bar{\mathbf{y}}(s) = 49.4\,, \qquad \bar{\mathbf{x}}(s) = 4.82$$

$$b(s) = \sum_{i \in s} Y_i(X_i - \bar{\mathbf{x}}(s)) \Big/ \sum_{i \in s} (X_i - \bar{\mathbf{x}}(s))^2$$
$$= [1453.5 - 1190.54]/[143.41 - 116.16]$$
$$\doteq 9.74$$

and, finally,

$$\hat{\mathbf{Y}}_{\text{lr}} = \bar{\mathbf{y}}(s) + b(s)(\bar{\mathbf{X}} - \bar{\mathbf{x}}(s))$$
$$\doteq 49.4 + 9.74(4.97 - 4.82)$$
$$= 50.86\,.$$

Example 6.7. Once more we refer to the same population of 15 units as considered in the previous example. This time suppose we want to use the regression-type estimator $\hat{\mathbf{Y}}^*_{\text{lr}}$ in (6.49) based on data drawn according to the sampling design in (6.50). To implement the sampling design for a sample of size $n = 5$, we first select one unit using the probability scheme in (6.48) and then supplement it by an additional 4 units drawn according to an SRS scheme of sampling. For the first part, we prepare the values of $(X_i - \bar{\mathbf{X}})^2$ in Table 6.4.

Table 6.4

Unit i	1	2	3	$\cdots$	14	15
$(X_i - \bar{\mathbf{X}})^2$	7.1289	3.8809	0.7569	$\cdots$	1.3689	0.0529

Clearly,

$$\sum_{i=1}^{N} (X_i - \bar{\mathbf{X}})^2 = \sum_{i=1}^{N} X_i^2 - N\bar{\mathbf{X}}^2 = 81.7133 .$$

To select one unit with selection probability proportional to $(X_i - \bar{\mathbf{X}})^2$ for the ith unit, we use the method of cumulative total. We select one random number in the interval $(0.0000, 81.7133)$. Let the selected number be 17.0834. By working out the cumulative totals of values in Table 6.4, we observe that $(X_1 - \bar{\mathbf{X}})^2 + (X_2 - \bar{\mathbf{X}})^2 + (X_3 - \bar{\mathbf{X}})^2 = 11.7667 < 17.0834 < 17.3836 = (X_1 - \bar{\mathbf{X}})^2 + (X_2 - \bar{\mathbf{X}})^2 + \cdots + (X_4 - \bar{\mathbf{X}})^2$. Therefore, unit 4 is selected in the first place. Next, suppose, following the routine procedure, we have selected units 1, 7, 10, 13 by an SRS procedure. Then the final sample is $s = (1, 4, 7, 10, 13)$. Suppose the values on the survey variable along with those on the auxiliary variable for these units are given as

$$X_1 = 2.3 , \quad X_4 = 2.6 , \quad X_7 = 2.8 , \quad X_{10} = 7.2 , \quad X_{13} = 4.2$$
$$Y_1 = 25 , \quad Y_4 = 24 , \quad Y_7 = 30 , \quad Y_{10} = 70 , \quad Y_{13} = 44 .$$

Computations yield, using $\bar{\mathbf{X}} = 4.97$,

$$\bar{\mathbf{y}}(s) = 38.6 , \qquad \bar{\mathbf{x}}(s) = 3.82$$

$$\sum_{i \in s} Y_i(X_i - \bar{\mathbf{X}}) = 892.7 - 959.21 = -66.51$$

$$\sum_{i \in s} (X_i - \bar{\mathbf{X}})^2 = 23.0205$$

and, finally,

$$\hat{\mathbf{Y}}^*_{\text{lr}} = \frac{n(N-1)}{N(n-1)} \left\{ \bar{\mathbf{y}}(s) + \frac{(\bar{\mathbf{X}} - \bar{\mathbf{x}}(s))(\Sigma_{i \in s} Y_i(X_i - \bar{\mathbf{X}}))}{\Sigma_{i \in s} (X_i - \bar{\mathbf{X}})^2} \right\} \doteq 41.16 .$$

6.8. BIAS OF THE REGRESSION ESTIMATOR IN SRS DATA AND BIAS REDUCTION TECHNIQUES

Based on an SRS sample design, the linear regression estimator $\hat{\bar{\mathbf{Y}}}_{\text{lr}}$ is not unbiased for the population mean $\bar{\mathbf{Y}}$. As in (6.40), the exact expression for bias is given by

$$\text{Bias} = E_p\{b(s)(\bar{\mathbf{X}} - \bar{\mathbf{x}}(s))\} = -\text{Cov}(b(s), \bar{\mathbf{x}}(s)) , \qquad (6.52)$$

where Cov stands for the covariance. Unlike the ratio estimator, in the case of the regression estimator the leading term in the expression for bias is of order $O(n^{-2})$. This can be seen by defining $B = \mathbf{S}_{XY}/\mathbf{S}_{XX}$ and writing

$$b(s) = B + \frac{B_1(s)}{n} + \frac{B_2(s)}{n^2} + \cdots$$

and

$$\bar{\mathbf{x}}(s) = \bar{\mathbf{X}} + \frac{B_1^*(s)}{n} + \frac{B_2^*(s)}{n^2} + \cdots$$

and working out the expression for the bias in (6.52). Below we will deduce an expression for the leading term in (6.52) and then suggest an estimator of the leading term as well.

Recalling the expression for $b(s)$, we may simplify (6.52) and study the nature of the leading term in the bias of $\hat{\bar{\mathbf{Y}}}_{\text{lr}}$. We have

$$b(s) = \mathbf{s}_{XY}/\mathbf{s}_{XX} = (\mathbf{s}_{XY}/\mathbf{S}_{XX})\left(1 + \frac{\mathbf{s}_{XX} - \mathbf{S}_{XX}}{\mathbf{S}_{XX}}\right)^{-1}$$

and then the bias term can be approximated as

$$\text{Bias} \doteq -\text{Cov}(\mathbf{s}_{XY}/\mathbf{S}_{XX}, \bar{\mathbf{x}}(s)) + \text{Cov}\left(\frac{\mathbf{s}_{XY}(\mathbf{s}_{XX} - \mathbf{S}_{XX})}{\mathbf{S}_{XX}^2}, \bar{\mathbf{x}}(s)\right) . \qquad (6.53)$$

The leading term in the expression for bias is given by

$$\text{Leading term} = -\text{Cov}(\mathbf{s}_{XY}, \bar{\mathbf{x}}(s))/\mathbf{S}_{XX} . \qquad (6.54)$$

We now simplify the covariance term above. Recall that

$$B = \mathbf{S}_{XY}/\mathbf{S}_{XX} = \rho\mathbf{S}_Y/\mathbf{S}_X , \qquad \text{where} \quad \mathbf{S}_{XX} = \mathbf{S}_X^2 , \quad \mathbf{S}_{YY} = \mathbf{S}_Y^2 , \qquad (6.55)$$

that is, B is the population (linear) regression coefficient of Y on X and ρ is the population correlation coefficient between Y and X. Next, define d_i as

$$d_i = (Y_i - \bar{\mathbf{Y}}) - B(X_i - \bar{\mathbf{X}}) , \qquad 1 \le i \le N , \qquad (6.56)$$

so that

$$\sum_{i=1}^{N} d_i = 0\,, \qquad \sum_{i=1}^{N} d_i(X_i - \bar{\mathbf{X}}) = 0\,. \tag{6.57}$$

Therefore, under an SRS(N, n) design,

$$E\Big(\sum_{i\in s} d_i\Big) = 0\,, \qquad E\Big\{\sum_{i\in s} d_i(X_i - \bar{\mathbf{X}})\Big\} = 0\,. \tag{6.58}$$

We want to write the expression for $\mathbf{s}_{XY}$ in terms of $\mathbf{s}_{Xd}$. We have

$$(n-1)\mathbf{s}_{XY} = \sum_{i\in s} (X_i - \bar{\mathbf{x}}(s))(Y_i - \bar{\mathbf{y}}(s))\,. \tag{6.59}$$

Using the representation

$$d_i - \bar{\mathbf{d}}(s) = (Y_i - \bar{\mathbf{y}}(s)) - B(X_i - \bar{\mathbf{x}}(s))\,, \tag{6.60}$$

we can rewrite (6.59) as

$$\begin{aligned}(n-1)\mathbf{s}_{XY} &= \sum_{i\in s} (X_i - \bar{\mathbf{x}}(s))\{(d_i - \bar{\mathbf{d}}(s)) + B(X_i - \bar{\mathbf{x}}(s))\}\\ &= (n-1)\mathbf{s}_{Xd} + B(n-1)\mathbf{s}_{XX}\,,\end{aligned} \tag{6.61}$$

which yields

$$\mathbf{s}_{XY} = \mathbf{s}_{Xd} + B\mathbf{s}_{XX}\,. \tag{6.62}$$

Therefore,

$$\mathrm{Cov}(\mathbf{s}_{XY}, \bar{\mathbf{x}}(s)) = \mathrm{Cov}(\mathbf{s}_{Xd}, \bar{\mathbf{x}}(s)) + B\,\mathrm{Cov}(\mathbf{s}_{XX}, \bar{\mathbf{x}}(s))\,. \tag{6.63}$$

Once more we use the approximation

$$\begin{aligned}\mathbf{s}_{XX} &= \mathbf{S}_{XX} + (\mathbf{s}_{XX} - \mathbf{S}_{XX})\\ &\doteq \mathbf{S}_{XX} - \frac{\mathbf{s}_{XX} - \mathbf{S}_{XX}}{\mathbf{S}_{XX}^2}\end{aligned}$$

and approximate the second term in (6.63) as

$$B\,\mathrm{Cov}(\mathbf{s}_{XX}, \bar{\mathbf{x}}(s)) \doteq B\,\mathrm{Cov}(\mathbf{S}_{XX}, \bar{\mathbf{x}}(s)) = 0\,. \tag{6.64}$$

As regards the first term in (6.63), we have

$$\begin{aligned}(n-1)\mathbf{s}_{Xd} &= \sum_{i\in s} (X_i - \bar{\mathbf{x}}(s))(d_i - \bar{\mathbf{d}}(s))\\ &= \sum_{i\in s} d_i(X_i - \bar{\mathbf{x}}(s))\end{aligned}$$

$$= \sum_{i \in s} d_i(X_i - \bar{\mathbf{X}}) - n\bar{\mathbf{d}}(s)(\bar{\mathbf{x}}(s) - \bar{\mathbf{X}})$$

$$= n\bar{\mathbf{e}}(s) - n\bar{\mathbf{d}}(s)(\bar{\mathbf{x}}(s) - \bar{\mathbf{X}}) , \qquad (6.65)$$

where

$$e_i = d_i(X_i - \bar{\mathbf{X}}) , \qquad 1 \le i \le N . \qquad (6.66)$$

Therefore, from (6.54), (6.63), (6.64), and (6.65), we see that the leading term in the expression for the bias in the estimator $\hat{\bar{\mathbf{Y}}}_{\mathrm{lr}}$ under an SRS(N, n) sampling design is given by

$$\begin{aligned}
&-\mathrm{Cov}(\mathbf{s}_{Xd}, \bar{\mathbf{x}}(s)) / \mathbf{S}_{XX} \\
&= -n(n-1)^{-1}\mathrm{Cov}(\bar{\mathbf{e}}(s) - \bar{\mathbf{d}}(s)(\bar{\mathbf{x}}(s) - \bar{\mathbf{X}}), \bar{\mathbf{x}}(s)) / \mathbf{S}_{XX} \\
&\doteq -n(n-1)^{-1}\mathrm{Cov}(\bar{\mathbf{e}}(s), \bar{\mathbf{x}}(s)) / \mathbf{S}_{XX} \\
&= -n(n-1)^{-1}E_p(\bar{\mathbf{e}}(s)(\bar{\mathbf{x}}(s) - \bar{\mathbf{X}})) / \mathbf{S}_{XX} \\
&= -n(n-1)^{-1}\left(\frac{1}{n} - \frac{1}{N}\right)\left\{\sum_{i=1}^{N} e_i(X_i - \bar{\mathbf{X}})\right\}\Big/ (N-1)\mathbf{S}_{XX} \\
&= -(n-1)^{-1}(1-f)\left\{\sum_{i=1}^{N} d_i(X_i - \bar{\mathbf{X}})^2\right\}\Big/ (N-1)\mathbf{S}_{XX} \\
&= -(1-f)N(N-1)^{-1}n^{-1}(n-1)^{-1}\left[E\left\{\sum_{i \in s} d_i(X_i - \bar{\mathbf{X}})^2/n\right\}\Big/ \mathbf{s}_{XX}\right] .
\end{aligned} \qquad (6.67)$$

This study has justified the observation made in the beginning of this section regarding the order of the leading term in the expression for the bias. To obtain an estimator of the leading term, we now recall the expression for d_i in (6.56) and replace it by

$$d_i^* = (Y_i - \bar{\mathbf{y}}(s)) - b(s)(X_i - \bar{\mathbf{X}}) , \qquad i \in s .$$

Then (6.67) with the expression inside $[\cdots]$ replaced by

$$\left[\sum_{i \in s} \{Y_i - \bar{\mathbf{y}}(s) - b(s)(X_i - \bar{\mathbf{X}})\}(X_i - \bar{\mathbf{X}})^2/n\right] / \mathbf{s}_{XX} \qquad (6.68)$$

will provide an estimator of this leading term.

In order to reduce bias, the obvious technique is to increase the sample size n so as to validate the approximation $b(s) \doteq B$. Once this approximation is used, we have

$$\hat{\bar{\mathbf{Y}}}_{\mathrm{lr}} \doteq \bar{\mathbf{y}}(s) + B(\bar{\mathbf{X}} - \bar{\mathbf{x}}(s)) \qquad (6.69)$$

and, hence, the linear regression estimator is approximately unbiased with the large sample mse given by

$$\begin{aligned}
\text{mse}(\hat{\bar{\mathbf{Y}}}_{\text{lr}}) &\doteq V(\bar{\mathbf{y}}(s) + B(\bar{\mathbf{X}} - \bar{\mathbf{x}}(s))) \\
&= E_p((\bar{\mathbf{y}}(s) - \bar{\mathbf{Y}} - B(\bar{\mathbf{x}}(s) - \bar{\mathbf{X}}))^2 \\
&= E_p(\bar{\mathbf{u}}(s) - \bar{\mathbf{U}})^2 , \qquad (6.70)
\end{aligned}$$

where

$$u_i = (Y_i - \bar{\mathbf{Y}}) - B(X_i - \bar{\mathbf{X}}) , \qquad 1 \leq i \leq N , \qquad (6.71)$$

with $\bar{\mathbf{U}} = 0$. Clearly,

$$\begin{aligned}
E_p(\bar{\mathbf{u}}(s) - \bar{\mathbf{U}})^2 &= \left(\frac{1}{n} - \frac{1}{N}\right)\left\{\sum_{i=1}^{N} (u_i - \bar{\mathbf{U}})^2/(N-1)\right\} \\
&= \left(\frac{1}{n} - \frac{1}{N}\right)(\mathbf{S}_{YY} + B^2 S_{XX} - 2B\mathbf{S}_{XY}) \qquad (6.72)
\end{aligned}$$

and, hence, the large sample mse of $\hat{\bar{\mathbf{Y}}}_{\text{lr}}$ is given by

$$\begin{aligned}
\text{mse}(\hat{\bar{\mathbf{Y}}}_{\text{lr}}) &\doteq \left(\frac{1}{n} - \frac{1}{N}\right)(\mathbf{S}_{YY} + B^2\mathbf{S}_{XX} - 2B\mathbf{S}_{XY}) \\
&= \left(\frac{1}{n} - \frac{1}{N}\right)\mathbf{S}_{YY}(1 - \rho^2) \qquad (6.73)
\end{aligned}$$

on simplification, using the fact that $B = \rho\mathbf{S}_Y/\mathbf{S}_X$. A large sample estimator of $\text{mse}(\hat{\bar{\mathbf{Y}}}_{\text{lr}})$ would be given by

$$\widehat{\text{mse}}\,(\hat{\bar{\mathbf{Y}}}_{\text{lr}}) = \left(\frac{1}{n} - \frac{1}{N}\right)\mathbf{s}_{YY}(1 - r^2) , \qquad (6.74)$$

where r is the sample correlation coefficient between Y and X. This result is obtained by first proposing $\Sigma_{i \in s}\, (u_i - \bar{\mathbf{u}}(s))^2/(n-1)$ as an unbiased estimator of $\Sigma_{i=1}^{N}\, (u_i - \bar{\mathbf{U}})^2/(N-1)$ and, next, replacing B by $b(s)$ in the expression for $u_i - \bar{\mathbf{u}}(s)$ for every $i \in s$.

Example 6.8. The purpose of this example is to demonstrate some of the computations involved in the use of the biased estimator $\hat{\bar{\mathbf{Y}}}_{\text{lr}}$ under an SRS procedure, with the use of the estimator in (6.68) of the leading term in the expression for the bias.

Assume that the same sample $s = (1, 7, 10, 11, 14)$ as in Example 6.6 has been drawn using an SRS procedure. Then the biased estimate of $\bar{\mathbf{Y}}$ is $\hat{\bar{\mathbf{Y}}}_{\text{lr}} = 50.86$. We will correct it for bias by first computing the value of the expression in (6.68). We have $N = 15$, $n = 5$, $\mathbf{s}_{XX} = 6.8401$, $b(s) \doteq 9.74$, $\bar{\mathbf{y}}(s) = 49.4$, $\bar{\mathbf{X}} = 4.97$.

$$\sum_{i \in s} \{(Y_i - \bar{\mathbf{y}}(s) - b(s)(X_i - \bar{\mathbf{X}})\}(X_i - \bar{\mathbf{X}})^2 = 11.4947 \,.$$

The computed value of (6.67) with the expression under $[\cdots]$ replaced by (6.68) is given by -0.016. Therefore, the corrected estimate of $\bar{\mathbf{Y}} \doteq 50.86 - (-0.016) = 50.876 \doteq 50.88$.

Example 6.9. A simple random sample of 100 units was selected from a large population. Computations yield the following:

$$\bar{\mathbf{y}}(s) = 50.32\,, \qquad \bar{\mathbf{x}}(s) = 5.23, \qquad b(s) = 10.25\,, \qquad \mathbf{s}_{YY} = 14{,}280$$

$$\mathbf{s}_{XX} = 120.85\,, \qquad \mathbf{s}_{XY} = 290.35\,, \qquad r = 0.2210\,.$$

It is known that $\bar{\mathbf{X}} = 5.00$. A large sample linear regression estimate of the population mean is given by

$$\hat{\bar{\mathbf{Y}}}_{\mathrm{lr}} = \bar{\mathbf{y}}(s) + b(s)(\bar{\mathbf{X}} - \bar{\mathbf{x}}(s)) \doteq 47.96$$

and its large sample estimated s.e. is given by $\sqrt{\hat{V}(\hat{\bar{\mathbf{Y}}}_{\mathrm{lr}})} = \sqrt{\mathbf{s}_{YY}(1 - r^2)/n} \doteq 11.65$.

Before concluding this section, we may make a relative comparison of the ratio and the regression estimators of the population mean with the ordinary mean-per-unit estimator (which makes no use of the auxiliary information). In large samples, using (6.27) and (6.73), we deduce that

(i) $\mathrm{mse}(\hat{\bar{\mathbf{Y}}}_{\mathrm{lr}}) \doteq ((1/n) - (1/N))\mathbf{S}_{YY}(1 - \rho^2) < V(\bar{\mathbf{y}}) = ((1/n) - (1/N))\mathbf{S}_{YY}$ under all circumstances;

(ii) $\mathrm{mse}(\hat{\bar{\mathbf{Y}}}_R) \doteq ((1/n) - (1/N))[\mathbf{S}_{YY} + \mathbf{R}^2\mathbf{S}_{XX} - 2\mathbf{R}\mathbf{S}_{XY}] < V(\bar{\mathbf{y}})$ whenever $\rho_{XY} > \frac{1}{2} \cdot \mathrm{C.V.}(X)/\mathrm{C.V.}(Y)$ where C.V. = coefficient of variation;

(iii) $\mathrm{mse}(\hat{\bar{\mathbf{Y}}}_{\mathrm{lr}}) \leq \mathrm{mse}(\hat{\bar{\mathbf{Y}}}_R)$ under all circumstances, with equality whenever $\rho_{XY} = \mathrm{C.V.}(X)/\mathrm{C.V.}(Y)$.

Thus regression method of estimation is likely to do better in large samples. In small samples, however, the inequalities may get reversed.

6.9. DOUBLE SAMPLING PROCEDURE IN THE REGRESSION METHOD OF ESTIMATION

So far we have assumed that the population mean, $\bar{\mathbf{X}}$, of the auxiliary variable values is known beforehand. However, this may not always be the case. As we mentioned in Section 6.5, in such a situation we must replace $\bar{\mathbf{X}}$ by a suitable estimate obtained from another source. Usually, a large initial sample s' of size n' is taken and the mean $\bar{\mathbf{X}}$ is estimated using the sample mean $\bar{\mathbf{x}}(s')$. It is implicitly assumed that measurements on the auxiliary variable under consideration are more easily obtained and with less cost. Then the question of selecting the sample s arises. We discuss three approaches for this purpose. One is to take an independent sample s of size n from the entire population using SRS(N, n) sampling design. As an alternative to this, a random subsample s of size n is taken from the initial sample s' of size n' using an SRS sampling procedure. A distinct advantage with the second procedure is that the values on the auxiliary variable on the units included in the second sample s are *already* available. We will denote the resultant estimator by $\hat{\hat{\mathbf{Y}}}^*_{\mathrm{lr}}$. Recalling (6.39), we may write

$$\hat{\hat{\mathbf{Y}}}^*_{\mathrm{lr}} = \bar{\mathbf{y}}(s) - b(s)(\bar{\mathbf{x}}(s) - \bar{\mathbf{x}}(s')) \tag{6.75}$$

as a proposed estimator for $\bar{\mathbf{Y}}$ based on data collected via what is popularly known as a *double sampling* procedure. It is true that if the value of $B = \mathbf{S}_{XY}/\mathbf{S}_{XX}$ is known, then we would propose the estimator

$$\hat{\bar{\bar{\mathbf{Y}}}}_{\mathrm{lr}} = \bar{\mathbf{y}}(s) - B(\bar{\mathbf{x}}(s) - \bar{\mathbf{x}}(s')) \tag{6.76}$$

as opposed to $\hat{\hat{\mathbf{Y}}}^*_{\mathrm{lr}}$ in (6.75). Again, if $\bar{\mathbf{X}}$ is known in advance, we would replace $\bar{\mathbf{x}}(s')$ in (6.76) by $\bar{\mathbf{X}}$. We first study the properties of $\hat{\bar{\bar{\mathbf{Y}}}}_{\mathrm{lr}}$ below.

Theorem 6.6. *The estimator $\hat{\bar{\bar{\mathbf{Y}}}}_{\mathrm{lr}}$ is unbiased for $\bar{\mathbf{Y}}$ with*

$$V(\hat{\bar{\bar{\mathbf{Y}}}}_{\mathrm{lr}}) = (n^{-1} - n'^{-1})\mathbf{S}^2_Y(1 - \rho^2) + (n'^{-1} - N^{-1})\mathbf{S}^2_Y \tag{6.77}$$

Proof. Conditional on the sample s', we have

$$E_2(\bar{\mathbf{y}}(s)|s') = \bar{\mathbf{y}}(s') \qquad \text{and} \qquad E_2(\bar{\mathbf{x}}(s)|s') = \bar{\mathbf{x}}(s')\,. \tag{6.78}$$

Therefore, writing $E = E_1E_2$, we have

$$E(\hat{\bar{\bar{\mathbf{Y}}}}_{\mathrm{lr}}) = E_1E_2(\hat{\bar{\bar{\mathbf{Y}}}}_{\mathrm{lr}}) = E_1(\bar{\mathbf{y}}(s') - B(\bar{\mathbf{x}}(s') - \bar{\mathbf{x}}(s'))) = E_1(\bar{\mathbf{y}}(s')) = \bar{\mathbf{Y}} \tag{6.79}$$

since s' is obtained via an SRS(N, n') sampling procedure. Next, we evaluate the variance of $\hat{\bar{\bar{\mathbf{Y}}}}_{\mathrm{lr}}$ in the usual manner. Writing $V = V_1E_2 + E_1V_2$, we first observe that $E_2(\hat{\bar{\bar{\mathbf{Y}}}}_{\mathrm{lr}}) = \bar{\mathbf{y}}(s')$ as noted before. Therefore, by definition,

$$\begin{aligned} V_2 &= E_2\{\hat{\bar{\bar{\mathbf{Y}}}}_{\mathrm{lr}} - E_2(\hat{\bar{\bar{\mathbf{Y}}}}_{\mathrm{lr}})\}^2 = E_2\{(\bar{\mathbf{y}}(s) - \bar{\mathbf{y}}(s')) - B(\bar{\mathbf{x}}(s) - \bar{\mathbf{x}}(s'))\}^2 \\ &= (n^{-1} - n'^{-1})\{\mathbf{s}_{YY}(s') + B^2\mathbf{s}_{XX}(s') - 2B\mathbf{s}_{XY}(s')\}\,, \end{aligned} \tag{6.80}$$

where $\mathbf{s}_{..}(s')$ refers to the $\mathbf{s}_{..}$ computed, based on the initial sample s' of size n'. To compute E_1V_2 it is enough to observe that $E_1\mathbf{s}_{..}(s') = \mathbf{S}_{..}$, so that

$$E_1V_2 = (n^{-1} - n'^{-1})\{\mathbf{S}_{YY} + B^2\mathbf{S}_{XX} - 2B\mathbf{S}_{XY}\} . \tag{6.81}$$

Next we already know that

$$V_1(E_2(\hat{\hat{\bar{\mathbf{Y}}}}_{\text{lr}})) = V_1(\bar{\mathbf{y}}(s')) = \mathbf{S}_{YY}(n'^{-1} - N^{-1}) . \tag{6.82}$$

Therefore, combining the two terms, we have

$$V(\hat{\hat{\bar{\mathbf{Y}}}}_{\text{lr}}) = \mathbf{S}_{YY}(n'^{-1} - N^{-1}) + (n^{-1} - n'^{-1})[\mathbf{S}_{YY} + B^2\mathbf{S}_{XX} - 2B\mathbf{S}_{XY}] . \tag{6.83}$$

Recall now that

$$B = \mathbf{S}_{XY}/\mathbf{S}_{XX} = \rho\mathbf{S}_Y/\mathbf{S}_X . \tag{6.84}$$

Then we can write $V(\hat{\hat{\bar{\mathbf{Y}}}}_{\text{lr}})$ in a more convenient expression as follows.

$$\begin{aligned} V(\hat{\hat{\bar{\mathbf{Y}}}}_{\text{lr}}) &= \mathbf{S}_Y^2(n'^{-1} - N^{-1}) + (n^{-1} - n'^{-1})\mathbf{S}_Y^2(1 - \rho^2) \\ &= (n^{-1} - n'^{-1})\mathbf{S}_y^2(1 - \rho^2) + (n'^{-1} - N^{-1})\mathbf{S}_Y^2 . \end{aligned} \tag{6.85}$$

This coincides with (6.77) and, hence, establishes the result. □

The expression for $V(\hat{\hat{\bar{\mathbf{Y}}}}_{\text{lr}})$ has two components. The first component coincides with $(n^{-1} - N^{-1})\mathbf{S}_Y^2(1 - \rho^2)$ when $n' = N$, in which case the second component reduces to 0. Notice that the implication of $n' = N$ is that we are in the earlier setup where $\bar{\mathbf{X}}$ is known and we have only one sample s drawn from the population.

Next we comment on the properties of the estimator $\hat{\hat{\bar{\mathbf{Y}}}}_{\text{lr}}^*$ which is more frequently encountered in practice. The exact small sample properties of this estimator are hard to study. It is a biased estimator, the bias term arising out of the second term $b(s)(\bar{\mathbf{x}}(s) - \bar{\mathbf{x}}(s'))$. Assuming the initial sample size n' to be large, we may approximate $\bar{\mathbf{x}}(s')$ by $\bar{\mathbf{X}}$ and hence the problem may be reduced to that of examining the bias of $\hat{\bar{\mathbf{Y}}}_{\text{lr}}$ in (6.39) which we have already undertaken. In any case, as a large sample approximation, $E(\hat{\hat{\bar{\mathbf{Y}}}}_{\text{lr}}^*) \simeq \bar{\mathbf{Y}}$ and mse $(\hat{\hat{\bar{\mathbf{Y}}}}_{\text{lr}}^*) \simeq V(\hat{\hat{\bar{\mathbf{Y}}}}_{\text{lr}})$.

Remark 6.2. In practice, it is not uncommon to meet with situations where a given population is sampled on different occasions with respect to the same characteristic of interest or related characteristics of interest. In particular, if the same survey variable is studied on different occasions, the values on the auxiliary variable X_i might be taken as those of the survey variable Y_i at an earlier point in time. In such a situation, it is quite possible that the extent of correlation between X and Y is sufficiently high, so as to justify the use of the regression estimator $\hat{\hat{\mathbf{Y}}}_{\text{lr}}$ in (6.76) rather than the ordinary sample mean estimator. This is clear from a comparison of the variance expressions for $\bar{\mathbf{y}}(s)$ and $\hat{\hat{\mathbf{Y}}}_{\text{lr}}$. We have $V(\hat{\hat{\mathbf{Y}}}_{\text{lr}}) < V(\bar{\mathbf{y}}(s))$ provided

$$(n^{-1} - n'^{-1})\mathbf{S}_Y^2(1-\rho^2) + (n'^{-1} - N^{-1})\mathbf{S}_Y^2 < (n^{-1} - N^{-1})\mathbf{S}_Y^2$$

which also holds trivially.

There is a third possibility regarding the choice of the sample s for measuring the values of the survey variable. We could possibly split the total sample size n into two parts: $n = m + u$, where m refers to the size of a sample s_m to be drawn from the initial sample s' of size n' while u refers to the size of a sample s_u of size u to be drawn independently from the part $\mathcal{U} - s'$ of the population *not* yet sampled. We propose to study this sampling procedure and the inference aspect of it in considerable detail. In this context we refer the reader to some results deduced in Section 4.2.

To study the results in proper perspective we consider the following setup. We have a population of N units. For the first time at time t_0 the population was sampled and a sample s' of size n' was drawn. Accordingly the values of a survey variable Y were collected and the sample mean $\bar{\mathbf{y}}(s')$ was used to estimate the population mean $\bar{\mathbf{Y}}^{(0)}$ at that point in time. Now at time t_1 the population mean has to be estimated again, possibly because there has been some change in the nature of the distribution of the survey variable in the population over time. At this point in time we decide to use a sample size n but the sampling procedure is to be properly thought out. Quite naturally, we might want the two samples to have some common units and some separate units. In implementing this idea, we would, therefore, divide the total sample size n into two parts: $n = m + u$, m denoting the size of the "matched" sample. Accordingly, we suggest the following sampling procedure. From the initial sample s' of size n', we take a random subsample s_m of size m using an SRS sampling procedure. From the complementary population $\mathcal{U} - s'$ of size $N - n'$ we take a random sample s_u of size u, again using an SRS sampling procedure. Then we collect data on the survey variable from the combined sample $s_m \cup s_u$ of size n at time t_1. Next comes the question of suggesting a suitable estimator of the population mean $\bar{\mathbf{Y}}^{(1)}$ at the current occasion, that is, at time t_1. We observe the following:

(a) The sample mean $\bar{\mathbf{y}}^{(1)}(s_u)$ is an unbiased estimator of $\bar{\mathbf{Y}}^{(1)}(\mathcal{U} - s')$ conditional on s'. In other words, $E(\bar{\mathbf{y}}^{(1)}(s_u)|s') = \bar{\mathbf{Y}}^{(1)}(\mathcal{U} - s')$. Note that since s' is a random sample of size n' from the population $\mathcal{U}$, it is true that the complementary part $\mathcal{U} - s'$ can also be regarded as a random sample of size $N - n'$ from the same population. In that case, $E(\bar{\mathbf{Y}}^{(1)}(\mathcal{U} - s')) = \bar{\mathbf{Y}}^{(1)}$ by the usual result in the context of an SRS sampling design. Thus in terms of the expectation operators E_1 and E_2, we may write

$$E(\bar{\mathbf{y}}^{(1)}(s_u)) = E_1 E_2(\bar{\mathbf{y}}^{(1)}(s_u)) = E_1(\bar{\mathbf{Y}}^{(1)}(\mathcal{U} - s')) = \bar{\mathbf{Y}}^{(1)} . \tag{6.86}$$

Further,

$$V_1 E_2(\bar{\mathbf{y}}^{(1)}(s_u)) = V_1(\bar{\mathbf{Y}}^{(1)}(\mathcal{U} - s')) = ((N - n')^{-1} - N^{-1})\mathbf{S}_{YY}^{(1)} \tag{6.87}$$

and

$$V_2(\bar{\mathbf{y}}^{(1)}(s_u)) = (u^{-1} - (N - n')^{-1})\mathbf{S}_{YY}^{(1)}(\mathcal{U} - s') , \tag{6.88}$$

where $\mathbf{S}_{YY}^{(1)}(\mathcal{U} - s')$ refers to the population variance based on the complementary population $\mathcal{U} - s'$ of size $N - n'$ at time t_1. Once again we use our earlier result that the sample variance in SRS sampling unbiasedly estimates the population variance. This would mean in this case that $E(\mathbf{S}_{YY}^{(1)}(\mathcal{U} - s')) = \mathbf{S}_{YY}^{(1)}$. Hence, ultimately, we obtain,

$$\begin{aligned} &E(\bar{\mathbf{y}}^{(1)}(s_u)) = \bar{\mathbf{Y}}^{(1)} , \\ &V(\bar{\mathbf{y}}^{(1)}(s_u)) = (u^{-1} - N^{-1})\mathbf{S}_{YY}^{(1)} . \end{aligned} \tag{6.89}$$

It may be noted that the same result could be obtained by yet another direct argument. This is that the sample s_u can be regarded as a random sample of size u from the entire population.

(b) In a similar manner, the sample mean $\bar{\mathbf{y}}^{(1)}(s_m)$ is an unbiased estimator of $\bar{\mathbf{Y}}^{(1)}$ with variance $V(\bar{\mathbf{y}}^{(1)}(s_m)) = (m^{-1} - N^{-1})\mathbf{S}_{YY}^{(1)}$. However, it is also true that we have observations on the units included in the sample s' on the past occassion, that is, at the time t_0. This suggests the use of the regression estimator instead of the simple mean estimator so far as this part of the sample is concerned. We, therefore, proceed to use the regression estimator of the type (6.75) suitably modified as follows. We write s_m for s and $\bar{\mathbf{y}}^{(0)}(s')$ for $\bar{\mathbf{x}}(s')$. Here the notation $\bar{\mathbf{y}}^{(0)}$ refers to the sample mean at time t_0. Similarly, we will denote by $\bar{\mathbf{Y}}^{(0)}$ the population mean at time t_0. Thus the proposed estimator is

$$\hat{\bar{\mathbf{Y}}}_{\mathrm{lr}}^{(1)}(s_m) = \bar{\mathbf{y}}^{(1)}(s_m) - b(s_m)(\bar{\mathbf{y}}_{(s_m)}^{(0)} - \bar{\mathbf{y}}^{(0)}(s')) . \tag{6.90}$$

We now make some assumptions regarding the distribution of the values of the survey variable on the two occasions. Specifically, we assume that the population variances $\mathbf{S}_{YY}^{(0)}$ and $\mathbf{S}_{YY}^{(1)}$ are the same for both time periods. Further, we assume that the sample size m is large enough so as to enable us justify the approximation $b(s_m) \simeq B = \rho\mathbf{S}_Y^{(1)}/\mathbf{S}_Y^{(0)} = \rho$ in view of the assumption $\mathbf{S}_{YY}^{(1)} = \mathbf{S}_{YY}^{(0)}$. Thus, we are actually using the large sample results

$$E(\hat{\hat{\bar{\mathbf{Y}}}}_{\mathrm{lr}}^{(1)}(s_m)) \doteq \bar{\mathbf{Y}}^{(1)}$$

$$\mathrm{mse}(\hat{\hat{\bar{\mathbf{Y}}}}_{\mathrm{lr}}^{(1)}(s_m)) \doteq (m^{-1} - n'^{-1})\mathbf{S}_{YY}(1-\rho^2) + (n'^{-1} - N^{-1})\mathbf{S}_{YY}$$

from (6.77). Assuming N to be large, we may neglect the term $N^{-1}\mathbf{S}_{YY}$ and rewrite the above expression for the mse in a slightly more convenient form

$$\mathrm{mse}(\hat{\hat{\bar{\mathbf{Y}}}}_{\mathrm{lr}}^{(1)}(s_m)) \doteq m^{-1}\mathbf{S}_{YY}(1-\rho^2) + n'^{-1}\mathbf{S}_{YY}\rho^2 . \tag{6.91}$$

We are now in a position to summarize the initial results in our attempt to provide an estimator of $\bar{\mathbf{Y}}^{(1)}$, the population mean of the survey variable values at time t_1. We have come up with two estimators, namely $\bar{\mathbf{y}}(s_u)$ and $\hat{\hat{\bar{\mathbf{Y}}}}_{\mathrm{lr}}^{(1)}(s_m)$, respectively, based on the samples s_u and s_m of sizes u and m, both drawn independently from the two disjoint parts of the population, namely $\mathcal{U} - s'$ and s'. It is true that conditional on s', the two samples are drawn independently; and, therefore, the two estimators may be regarded as being *conditionally* independent. In that situation, we might as well try to combine the two using the known technique of weighting them with reciprocals of the variances. In this part and the following sections, we will use the term variance in place of the mse for the biased estimator. This, in effect, produces the combined estimator

$$\hat{\bar{\mathbf{Y}}}_c^{(1)} = W_1\bar{\mathbf{y}}(s_u) + W_2\hat{\hat{\bar{\mathbf{Y}}}}_{\mathrm{lr}}^{(1)}(s_m) \tag{6.92}$$

where

$$W_1 \propto 1/V(\bar{\mathbf{y}}(s_u)) \qquad \text{and} \qquad W_2 \propto 1/V(\hat{\hat{\bar{\mathbf{Y}}}}_{\mathrm{lr}}^{(1)}(s_m)) . \tag{6.93}$$

We now use the approximation (6.91) for the variance of $\hat{\hat{\bar{\mathbf{Y}}}}_{\mathrm{lr}}^{(1)}(s_m)$. Of course, for $V(\bar{\mathbf{y}}(s_u))$ we have the exact expression given in (6.89). It follows that under the approximation we are using in this context, the estimator $\hat{\bar{\mathbf{Y}}}_c^{(1)}$ is unbiased for the population mean $\bar{\mathbf{Y}}^{(1)}$ and its variance is given by

$$\begin{aligned} V(\hat{\bar{\mathbf{Y}}}_c^{(1)}) &= [1/V(\bar{\mathbf{y}}(s_u)) + 1/V(\hat{\hat{\bar{\mathbf{Y}}}}_{\mathrm{lr}}^{(1)}(s_m))]^{-1} \\ &\doteq \mathbf{S}_{YY}[u + \{(1-\rho^2)m^{-1} + \rho^2 n^{-1}\}]^{-1} \end{aligned} \tag{6.94}$$

using the approximations mentioned before.

The problem of allocation of the sample size m and u to the matched and unmatched portions of the sample may now be taken up. We can see from (6.94) how the variance of the combined estimator depends on the values of m and u. It is a straightforward computaion to minimize the expression for the variance in (6.94) for the variation of m and u subject to $m, u \geq 1$, $m + u = n$. We can apply the usual differentiation rule and conclude the following:

$$m_{\text{opt}}/n = \frac{\sqrt{1-\rho^2}}{1+\sqrt{1-\rho}}, \qquad u_{\text{opt}}/n = \frac{1}{1+\sqrt{1-\rho^2}}. \tag{6.95}$$

Accordingly, the minimum value of the resulting variance is given by

$$V_{\min} = \mathbf{S}_{YY}(1 + \sqrt{1-\rho^2})/2n. \tag{6.96}$$

The above result suggests that we have to pay attention in matching a certain number of units of the sample at time t_1 with those at time t_0 to maximize the precision of the combined estimator. The optimum matching proportions depend on the unknown parameter ρ which determines the extent of correlation between the pairs of values $\{(Y_i^{(0)}, Y_i^{(1)})\ 1 \leq i \leq N\}$ over the two periods of time.

Remark 6.3. In this context, the reader may revise the problem in Exercise 4.18 of Chapter 4. We could, likewise, take up the problem of estimating the change in the mean values $\bar{\mathbf{Y}}^{(1)} - \bar{\mathbf{Y}}^{(0)}$ from time t_1 to time t_0.

In the next section, we examine the natural extension of this problem to multiple repeated sampling of the same population on various occasions.

6.10. SAMPLING ON SUCCESSIVE OCCASIONS AND THE REGRESSION ESTIMATOR

Assume that the finite population under consideration was first studied at some time t_0 and since then it has been studied on many successive occasions at different times $t_1, t_2, \ldots$. Suppose the current time is t_h and we wish to estimate the mean $\bar{\mathbf{Y}}^{(h)}$ on the current occasion of the same survey variable Y. We assume that on occasion $h-1$, we ended up with an estimator $\hat{\bar{\mathbf{Y}}}^{(h-1)}$ of $\bar{\mathbf{Y}}^{(h-1)}$ based on data obtained on that occasion as also on the previous occasions.

If the sample size on the current occasion as well as on other previous occasions is n and we split n as usual into m_h (for matched units) and u_h (for unmatched units) then the sampling procedure is to select a sample s_{m_h} of m_h units from those selected on occasion $h-1$ and to select a sample s_{u_h} of u_h units independently from those not selected on occasion $h-1$.

As before, we will arrive at two independent estimators of $\bar{\mathbf{Y}}^{(h)}$. These are shown in Table 6.5.

We will assume for simplicity that the population variance $\mathbf{S}_Y^2$ and the correlation coefficient ρ between the values of Y on any two occasions

Table 6.5

Estimator based on unmatched units	$\hat{\mathbf{Y}}_1^{(h)} = \bar{\mathbf{y}}^{(h)}(s_{u_h})$
Estimator based on matched units	$\hat{\mathbf{Y}}_2^{(h)} = \bar{\mathbf{y}}^{(h)}(s_{m_h}) + b(s_{m_h})\{\hat{\hat{\mathbf{Y}}}^{(h-1)} - \bar{\mathbf{y}}^{(h-1)}(s_{m_h})\}$

remain the same over all occasions. It is evident that

$$V(\hat{\mathbf{Y}}_1^{(h)}) = (u_h^{-1} - N^{-1})\mathbf{S}_Y^2 . \tag{6.97}$$

Next, we obtain the large sample expression for the variance of $\hat{\mathbf{Y}}_2^{(h)}$ in the following way:

$$\begin{aligned} V(\hat{\mathbf{Y}}_2^{(h)}) &\doteq V(\bar{\mathbf{y}}^{(h)}(s_{m_h}) + B(\hat{\hat{\mathbf{Y}}}^{(h-1)} - \bar{\mathbf{y}}^{(h-1)}(s_{m_h}))) \\ &= V(\bar{\mathbf{y}}^{(h)}(s_{m_h}) - B\bar{\mathbf{y}}^{(h-1)}(s_{m_h}) + B\hat{\hat{\mathbf{Y}}}^{(h-1)}) . \end{aligned} \tag{6.98}$$

In order to use the symbolic operator $V = V_1E_2 + E_1V_2$, we first note that

$$\begin{aligned} &E_2(\bar{\mathbf{y}}^{(h)}(s_{m_h}) - B\bar{\mathbf{y}}^{(h-1)}(s_{m_h}) + B\hat{\hat{\mathbf{Y}}}^{(h-1)}) \\ &= \bar{\mathbf{y}}^{(h)}(s_n^{(h-1)}) - B\bar{\mathbf{y}}^{(h-1)}(s_n^{(h-1)}) + B\hat{\hat{\mathbf{Y}}}^{(h-1)} , \end{aligned} \tag{6.99}$$

where $s_n^{(h-1)}$ refers to the sample of size n drawn on occasion $h-1$ and held conditionally fixed for the purpose of subsampling. Therefore,

$$\begin{aligned} &V_1E_2(\bar{\mathbf{y}}^{(h)}(s_{m_h}) - B\bar{\mathbf{y}}^{(h-1)}(s_{m_h}) + B\hat{\hat{\mathbf{Y}}}^{(h-1)}) \\ &= V_1(\bar{\mathbf{y}}^{(h)}(s_n^{(h-1)}) - B\bar{\mathbf{y}}^{(h-1)}(s_n^{(h-1)})) + B^2V_1(\hat{\hat{\mathbf{Y}}}^{(h-1)}) \\ &\quad + 2\,\mathrm{Cov}(\bar{\mathbf{y}}^{(h)}(s_n^{(h-1)}) - B\bar{\mathbf{y}}^{(h-1)}(s_n^{(h-1)}), B\hat{\hat{\mathbf{Y}}}^{(h-1)}) \\ &\doteq (n^{-1} - N^{-1})\mathbf{S}_y^2(1-\rho^2) + B^2V(\hat{\hat{\mathbf{Y}}}^{(h-1)}) \end{aligned} \tag{6.100}$$

since the covariance term is seen to produce terms of smaller order. Again,

$$\begin{aligned} &V_2(\bar{\mathbf{y}}^{(h)}(s_{m_h}) - B\bar{\mathbf{y}}^{(h-1)}(s_{m_h}) + B\hat{\hat{\mathbf{Y}}}^{(h-1)}) \\ &= V_2(\bar{\mathbf{y}}^{(h)}(s_{m_h}) - B\bar{\mathbf{y}}^{(h-1)}(s_{m_h})) \end{aligned} \tag{6.101}$$

and, hence, following our earlier analysis in Theorem 6.6, we obtain

$$E_1V_2(\bar{\mathbf{y}}^{(h)}(s_{m_h}) - B\bar{\mathbf{y}}^{(h-1)}(s_{m_h}) + B\hat{\hat{\mathbf{Y}}}^{(h-1)}) = (m_h^{-1} - n^{-1})\mathbf{S}_Y^2(1-\rho^2) . \tag{6.102}$$

Thus, combining (6.100) and (6.102), we finally obtain

$$V(\hat{\mathbf{Y}}_2^{(h)}) \doteq \frac{\mathbf{S}_Y^2(1-\rho^2)}{m_h} + B^2 V(\hat{\hat{\mathbf{Y}}}^{(h-1)}) - \frac{\mathbf{S}_Y^2(1-\rho^2)}{N}$$

$$\doteq \frac{\mathbf{S}_Y^2(1-\rho^2)}{m_h} + \rho^2 V(\hat{\hat{\mathbf{Y}}}^{(h-1)}) \qquad (6.103)$$

since B coincides with ρ under the assumption that $\mathbf{S}_Y^2$ remains the same on all occasions.

At this stage, we introduce the following notation:

$$V(\hat{\hat{\mathbf{Y}}}^{(h-1)}) = \mathbf{S}_Y^2 g_{h-1}(n^{-1} - N^{-1})$$

$$\doteq \mathbf{S}_Y^2 g_{h-1}/n\,. \qquad (6.104)$$

Likewise,

$$V(\hat{\hat{\mathbf{Y}}}^{(h)}) \doteq \mathbf{S}_Y^2 g_h/n\,, \qquad (6.105)$$

where $\hat{\hat{\mathbf{Y}}}^{(h)}$ is to be obtained as a linear combination of the two estimators in Table 6.5. By using reciprocals of the variances as weights, we would obtain, using (6.97), (6.103), and (6.104),

$$\mathbf{S}_Y^2 g_h/n \doteq V(\hat{\hat{\mathbf{Y}}}^{(h)}) = \left(\frac{1}{V(\hat{\mathbf{Y}}_1^{(h)})} + \frac{1}{V(\hat{\mathbf{Y}}_2^{(h)})}\right)^{-1}$$

$$\doteq \left[\frac{u_h}{\mathbf{S}_Y^2} + \left(\frac{\mathbf{S}_Y^2(1-\rho^2)}{m_h} + \rho^2 \mathbf{S}_Y^2 g_{h-1}/n\right)^{-1}\right]^{-1} \qquad (6.106)$$

which yields

$$\frac{n}{\mathbf{S}_Y^2 g_h} = \frac{u_h}{\mathbf{S}_Y^2} + \left(\frac{(1-\rho^2)}{m_h} + \frac{\rho^2 g_{h-1}}{n}\right)^{-1} \Big/ \mathbf{S}_Y^2\,. \qquad (6.107)$$

Clearly, (6.107) yields the relation

$$\frac{n}{g_h} = u_h + \left(\frac{(1-\rho^2)}{m_h} + \frac{\rho^2 g_{h-1}}{n}\right)^{-1}. \qquad (6.108)$$

An optimum choice of m_h is based on minimization of $V(\hat{\hat{\mathbf{Y}}}^{(h)})$ in (6.105). From (6.108), it is then evident that the quantity

$$u_h + \left(\frac{(1-\rho^2)}{m_h} + \frac{\rho^2 g_{h-1}}{n}\right)^{-1} \qquad (6.109)$$

has to be maximized by a proper choice of m_h. Using the fact that $u_h = n - m_h$, usual differentiation yields

$$m_{h\text{ opt}} = \frac{n\sqrt{1-\rho^2}}{g_{h-1}(1+\sqrt{1-\rho^2})} . \tag{6.110}$$

Accordingly, the corresponding minimum value of $V(\hat{\hat{\mathbf{Y}}}^{(h)})$ can be obtained by substituting the value of $m_{h\text{ opt}}$ from (6.110) in the expresion for m_h in (6.108), after replacing u_h by $u_{h\text{ opt}} = n - m_{h\text{ opt}}$. The resulting expression simplifes to

$$\frac{1}{g_h} = 1 + \frac{1-\sqrt{1-\rho^2}}{g_{h-1}(1+\sqrt{1-\rho^2})} . \tag{6.111}$$

which provides a recurrence relation involving two consecutive g_h-values. Repeated use of (6.111) yields, in a routine manner,

$$\frac{1}{g_h} = \frac{1-\alpha^h}{1-\alpha} , \quad \alpha = (1-\sqrt{1-\rho^2})/(1+\sqrt{1-\rho^2}) . \tag{6.112}$$

It turns out that the limiting value of g_h is

$$g_\infty = 1 - \alpha = \frac{2\sqrt{1-\rho^2}}{1+\sqrt{1-\rho^2}} .$$

The corresponding limiting expression for the variance of the estimator is

$$V(\hat{\hat{\mathbf{Y}}}^{(\infty)}) = 2\,\frac{\mathbf{S}_Y^2\sqrt{1-\rho^2}}{n(1+\sqrt{1-\rho^2})} . \tag{6.113}$$

Further, the limiting value of the optimum matching proportion, as obtained from (6.110), is given by

$$m_\infty/n = 1/2 . \tag{6.114}$$

Remark 6.4. The above statistical analysis of the problem of repeated sampling from the same population illustrates that an optimum procedure to estimate the mean on the latest occasion is to match 50% of the sampling units with those studied on the immediately preceding occasion. This is seen to hold irrespective of the value of ρ in the population.

Remark 6.5. To give the reader a taste of the history of work in the area of sampling on successive occasions, we may mention that Jessen (1942) first introduced the idea of using information obtained on a previous occasion in framing estimators on a current occasion. He examined the technique of partial replacement (matching and unmatching) of units on two occasions. Later, this technique was generalized by Patterson (1950), Yates (1960), and Cochran (1977), among others. While the above authors examined the problem using data based on SRS designs applied on all occasions, Des Raj (1965) looked into the same when unequal probability selection schemes are applied but his study was confined to the case of two occasions only. Later, Tripathi and Srivastava (1979) considered the general case.

EXERCISES

6.1. The purpose is to estimate the average score in the final examination of a group of 500 schoolchildren in a local county. It is stated that information on Z_1 = number of hours weekly spent on watching television and Z_2 = number of hours weekly being coached by parents can be made available on request. Table 6.A summarizes such information along with the actual score (Y) for a random sample of 50 schoolchildren of that county. Past experience indicates that the actual score is approximately an increasing function of Z_2 and a decreasing function of Z_1. The following forms of integration of the information are suggested. Draw scatter diagrams to investigate which functional form, if any, fits the approximate relation $Y \doteq f(Z_1, Z_2)$.

Table 6A. Two-Way Classified Score Distribution of 50 Schoolchildren[a]

$Z_1 \backslash Z_2$	≤4	4–5	5–6	6–7	7–8	8–9	>9
≤10	2 65, 64	1 70	—	—	—	—	—
10–12	1 64	2 69, 70	1 75	1 77	1 80	—	—
12–14	3 62, 63, 64	2 68, 70	2 70, 71	1 73	1 75	—	—
14–16	1 64	1 69	2 70, 71	2 72, 73	1 75	1 77	—
16–18	3 63, 64, 65	1 69	2 71, 72	3 72, 73, 74	2 74, 75	1 76	1 79
18–20	—	—	3 77, 77, 75	2 75, 80	1 81	1 72	—
Over 20	—	—	—	—	—	2 77, 79	2 78, 81

[a]In each cell the top figure corresponds to the frequency and the bottom figures correspond to the actual score of the schoolchildren.

(i) $f_1(Z_1, Z_2) = 50 + 4Z_2 - 3.5Z_1$,
(ii) $f_2(Z_1, Z_2) = 60 + 3Z_1^{-1} + 2Z_2$,
(iii) $f_3(Z_1, Z_2) = 80 - 2Z_1 + 3Z_2$.

6.2. Show that the ratio estimator $\hat{\bar{\mathbf{Y}}}_R$ of $\bar{\mathbf{Y}}$ based on an SRS(N, n) sample design will be unbiased if and only if all the X_i's are equal.

6.3. (a) Show that a sampling design d of the type

$$d = \sum_{k=1}^{N} p_k d_k, \qquad 0 \le p_k \le 1, \qquad \sum_{k=1}^{N} p_k = 1,$$

where

$$d_k(\text{reduced}): \text{ PPAS of fixed-size } (k), \qquad 1 \le k \le N,$$

allows $\hat{\mathbf{R}}(s, \mathbf{Y})$ in (6.3) to be unbiased for $\mathbf{R}$.

(b) Deduce an expression for the variance of the estimator based on the design d. For what choice of the p_k's, can you estimate the variance of the estimator $\hat{\mathbf{R}}(s, \mathbf{Y})$?

6.4. By unbiasedly estimating the square terms and the product terms separately, provide an unbiased estimator of $V(\hat{\bar{\mathbf{Y}}}_R)$ in (6.5). Derive a condition for nonnegativeness of this variance estimator.

6.5. Show that $V(\hat{\bar{\mathbf{Y}}}_R)$ in (6.5) can be put into the form

$$V(\hat{\bar{\mathbf{Y}}}_R) = -\sum_{i<j}\sum a_{ij} X_i X_j \left(\frac{Y_i}{X_i} - \frac{Y_j}{X_j}\right)^2.$$

6.6. (a) Show that the choices in (i) and (ii) in (6.11) indeed satisfy the conditions of unbiasedness of $\hat{V}(\hat{\bar{\mathbf{Y}}}_R)$ in (6.8) for (6.5).

(b) For both the choices in (i) and (ii), give numerical examples of values of an auxiliary variable so that the resulting estimate turns out to be nonnegative.

6.7. Show that under the PPAS method of sampling given by (6.4), an unbiased estimator of $\mathbf{R}^2$ is

$$\widehat{\mathbf{R}^2} = \frac{1}{Nn\bar{\mathbf{x}}(s)\bar{\mathbf{X}}}\left[\sum_{i\in s} Y_i^2 + \frac{N-1}{n-1}\sum_{\substack{i \ne j\\ i,j\in s}}\sum Y_i Y_j\right].$$

Hence deduce an expression for an unbiased estimator of $V(\hat{\mathbf{R}})$. Use this to deduce the corresponding expression for an unbiased variance estimator of $\bar{\mathbf{Y}}$. Does it coincide with any of the forms given in the text?

6.8. From the data in Example 6.1, compute an unbiased variance estimate of the ratio estimate of the population mean for the population under consideration.

6.9. Writing

$$\hat{\bar{\mathbf{Y}}}_R = \bar{\mathbf{y}}(s)\bar{\mathbf{x}}(s)/\bar{\mathbf{X}} \quad \text{and} \quad \hat{\bar{\mathbf{Y}}}_R^* = \sum_{i \in s} X_i Y_i / n\bar{\mathbf{X}},$$

show that in simple random sampling,

(a)
$$\text{Bias of } \hat{\bar{\mathbf{Y}}}_R^* = \frac{n(n-1)}{(N-n)} (\text{Bias of } \hat{\bar{\mathbf{Y}}}_R),$$

(b)
$$E_p\left[k\hat{\bar{\mathbf{Y}}}_R - \frac{k(N-n)}{n(N-1)}\hat{\bar{\mathbf{Y}}}_R^* + \left(1 - \frac{kN(n-1)}{n(N-1)}\right)\bar{\mathbf{y}}(s)\right] = \bar{\mathbf{Y}}$$

for any arbitrary constant k.

6.10. Under an SRS(N, n) procedure, derive conditions on the constants ϕ_1, ϕ_2, and ϕ_3 so that

$$\phi_1\bar{\mathbf{y}}(s) + \phi_2\hat{\bar{\mathbf{Y}}}_R + \phi_3\hat{\bar{\mathbf{Y}}}_R^*$$

is exactly unbiased for the population mean $\bar{\mathbf{Y}}$. In particular, show that the coefficients in (b) in Exercise 6.9 satisfy the stated conditions. Suggest some other choices of the ϕ_i's making one of them 0 or 1.

6.11. Show that the expression in (6.44) is identically equal to zero.

6.12. Verify the results in (6.58).

6.13. The purpose is to estimate the median net annual income exclusively from farms of 110 farmers in a certain state. For a simple random sample of 9 farmers, the following data relate to X: farm size, and Y: net annual income from farm.

Suggest a suitable procedure based on regression to resolve this problem and justify your procedure. You may or may not assume availability of advance information on all 110 farm sizes.

Sampled farm (i)	07	13	34	45	56	69	71	89	103
X_i (in acres)	8.3	7.2	13.2	4.3	7.0	5.3	15.8	2.4	7.5
Y_i (in \$1000)	12.6	12.2	14.2	11.3	11.8	11.6	15.1	10.8	12.1

How would you handle this problem if 10 farmers provide data?

REFERENCES

Beale, E. M. L. (1962). Some uses of computation in operational research. *Industrielle Organisation*, **31**, 51–52.

Cochran, W. G. (1977). *Sampling Techniques*. Third edition, Wiley, New York.

Durbin, J. (1959). A note on the application of Quenouille's method of bias reduction to the estimation of ratios. *Biometrika*, **46**, 477–480.

Gabler, S. (1987). The nearest proportional to size sampling designs. *Commun. Statist.-Theory Methods*, **16**, 1117–1131.

Goodman, L. A., and Hartley, H. O. (1958). The precision of unbiased ratio-type estimators. *J. Amer. Statist. Assoc.*, **53**, 491–508.

Hartley, H. O., and Ross, A. (1954). Unbiased ratio estimators. *Nature*, **174**, 270–271.

Jessen, R. J. (1942). Statistical investigation of a sample survey for obtaining farm facts. *Iowa Agric. Exp. Sta. Res. Bull.*, 304.

Lahiri, D. B. (1951). A method of sample selection providing unbiased ratio estimates. *Bull. Internat. Statist. Inst.*, **33**, 133–140.

Lanke, J. (1974). On nonnegative variance estimators in survey sampling. *Sankhyā*, **C36**, 33–42.

Mickey, M. R. (1959). Some finite population unbiased ratio and regression estimators. *J. Amer. Statist. Assoc.*, **54**, 594–612.

Midzuno, H. (1952). On the sampling system with probability proportional to sum of sizes. *Ann. Inst. Statist. Math.*, **3**, 99–107.

Murthy, M. N. (1964). Product method of estimation. *Sankhyā*, **A26**, 69–74.

Nanjamma, N. S., Murthy, M. N., and Sethi, V. K. (1959). Some sampling systems providing unbiased ratio estimators. *Sankhyā*, **21**, 299–314.

Patterson, H. D. (1950). Sampling on successive occasions with partial replacement of units. *J. Roy. Stat. Soc.*, **B12**, 241–255.

Quenouille, M. H. (1956). Notes on bias in estimation. *Biometrika*, **43**, 353–360.

Raj, D. (1965). On sampling over two occasions with pps. *Ann. Math. Stat.*, **36**, 327–330.

Rao, J. N. K. (1967). The precision of Mickey's unbiased ratio estimator. *Biometrika*, **54**, 321–324.

Rao, J. N. K. (1969). Ratio and regression estimators. In: *New Developments in Survey Sampling* (Johnson, N. L., and Smith, H. Jr., Eds.), 213–234. Wiley, New York.

Rao, J. N. K., and Vijayan, K. (1977). On estimating the variance in sampling with probability proportional to aggregate size. *J. Amer. Statist. Assoc.*, **72**, 579–584.

Rao, J. N. K., and Webster, J. T. (1966). On two methods of bias reduction in the estimation of ratios. *Biometrika*, **53**, 315–321.

Rao, P. S. R. S. (1969). Comparison of four ratio-type estimates under a model. *J. Amer. Statist. Assoc.*, **64**, 574–580.

Rao, P. S. R. S. (1974). Jackknifing the ratio estimator. *Sankhyā*, **C36**, 84–97.

Rao, P. S. R. S. (1988). Ratio and regression estimators. In: *Handbook of Statistics, Vol. 6, Sampling* (Krishnaiah, P. R., and Rao, C. R., Eds.), 449–468. North-Holland, Amsterdam.

Rao, P. S. R. S., and Rao, J. N. K. (1971). Small sample results for ratio estimators. *Biometrika*, **58**, 625–630.

Rao, T. J. (1966). On the variance of the ratio estimator for Midzuno–Sen sampling scheme. *Metrika*, **10**, 89–91.

Rao, T. J. (1972). On the variance of the ratio estimator. *Metrika*, **18**, 209–215.

Rao, T. J. (1981a). A note on unbiasedness in ratio estimator. *J. Statist. Plann. Infer.*, **5**, 335–340.

Rao, T. J. (1981b). On a class of almost unbiased ratio estimators. *Ann. Inst. Statist. Math.*, **33**, 225–231.

Rao, T. J. (1987). On certain unbiased product estimators. *Commun. Statist. Theory Methods* **16**, 3631–3640.

Ray, S. K., and Sahai, A. (1980). Efficient families of ratio and product type estimators. *Biometrika*, **67**, 211–215.

Robson, D. S. (1957). Applications of multivariate polykeys to the theory of unbiased ratio-type estimation. *J. Amer. Statist. Assoc.*, **52**, 511–522.

Schucany, W. R., Gray, H. L., and Owen, D. B. (1971). On bias reduction in estimation. *J. Amer. Statist. Assoc.*, **66**, 524–533.

Sengupta, S. (1981). Jackknifing the ratio and the product estimators in double sampling. *Metrika*, **28**, 245–256.

Shukla, N. D. (1976). Almost unbiased product estimators. *Metrika*, **23**, 127–133.

Singh, P., and Srivastava, A. K. (1980). Sampling schemes providing unbiased regression estimators. *Biometrika*, **67**, 205–209.

Tin, M. (1965). Comparison of some ratio estimators. *J. Amer. Statist. Assoc.*, **60**, 294–307.

Tripathi, T. P. (1988). Estimation of domains in sampling on two occasions. *Sankhyā*, **B50**, 103–110.

Tripathi, T. P., and Srivastava, O. P. (1979). Estimation on successive occasions using ppswr sampling. *Sankhyā*, **C41**, 84–91.

Williams, W. H. (1961). Generating unbiased ratio and regression estimators. *Biometrics*, **17**, 267–274.

Yates, F. (1960). *Sampling Methods for Censuses and Surveys*. Third Edition. Charles Griffin and Co., London.

CHAPTER SEVEN

Cluster Sampling Designs

There are typical survey situations where the frame of the population under study is either not readily available or largely inadequate for the purpose of sampling. Instead, the population units are found to be conveniently grouped into several natural clusters. In such instances, a class of sampling designs, called cluster sampling designs, can be successfully applied. In this chapter we study various aspects of such designs.

7.1. SAMPLING DESIGNS FOR POPULATIONS WITH UNITS ARRANGED IN CLUSTERS

Some special types of sampling designs are needed in practice in situations where a complete frame of the population units may be lacking while the units are to be found in clusters. For example, in a study conducted by the state department of health relating to the adult citizens' awareness of community services rendered to the senior citizens in a given community, the households may be properly listed and identified but a complete frame of the entire adult population in the community may be lacking. Again, in a study conducted by a safety regulating agency regarding the seriousness of bodily injuries caused to butchers while at work in a metropolitan area, it is highly unlikely that a complete listing of all the butchers regularly engaged in the profession will be available. On the other hand the food stores employing them can easily be identified and listed. As a third example, in a passenger opinion survey regarding the handling of check-in baggage in an airport, it is impossible to conceive of a frame of the mobile population under study. However, in any predetermined time interval, say between eight to nine a.m., passengers will be checking in and some or all of them can possibly be interviewed but as such we have no idea regarding a listing of such passengers. These are some of the cases where the sampling units are already grouped or may be conveniently grouped into several clusters which are clearly identifiable.

The above examples demonstrate that on some occasions it may not be economical to make any effort to prepare a frame for the population under study, while on some others, the construction of a frame is not feasible at all. For such populations, some special types of sampling designs are suggested in the literature. In this and the following two chapters we will present some such designs along with a study of estimators based on them in considerable detail. We may mention that the entire study will be confined to estimation of a finite population total or mean and its variance estimator, making use of the HTE. The problem of estimation of the population variance is incidental to this study and it will not be discussed explicitly. Further, no specific results concerning a population proportion will be given since the estimation problems for such a population parameter can be understood in terms of the population mean, as has been shown previously in Chapter Four.

To start with, we assume that a given finite population $\mathcal{U}$ involving N sampling units is already partitioned or can be conveniently partitioned into M nonoverlapping clusters $\mathcal{C}_1, \mathcal{C}_2, \ldots, \mathcal{C}_M$ having sizes $N_1, N_2, \ldots, N_M$, respectively, where $N_i \geq 1$, $\Sigma_{i=1}^{M} N_i = N$. It is quite possible that the total number of sampling units and also the number of units in different clusters may *not* be known beforehand but we must know the total number of clusters (M) constituting the population and these clusters must be clearly labelled. As before, we will denote by Y the survey variable which assumes the value Y_{ij} on the jth unit in the ith cluster, $1 \leq j \leq N_i$, $1 \leq i \leq M$. Further

$$\mathbf{T}_i = \sum_{j=1}^{N_i} Y_{ij}\,, \qquad \bar{\mathbf{Y}}_i = \mathbf{T}_i / N_i\,, \qquad 1 \leq i \leq M\,, \tag{7.1}$$

will denote the cluster totals and the cluster means, respectively. Clearly the population total is $T(\mathbf{Y}) = \Sigma\, \mathbf{T}_i = \Sigma\, \Sigma\, Y_{ij}$. We will refer to

$$\bar{\mathbf{Y}} = T(\mathbf{Y})/N\,, \qquad \bar{\mathbf{Y}}_c = T(\mathbf{Y})/M \tag{7.2}$$

as the *mean per unit* and *mean per cluster*, respectively. It is *not* uncommon to meet situations in practice where one is primarily interested in the mean per cluster only. The population variance $\mathbf{S}_Y^2$ admits the usual decomposition (recall ANOVA)

$$\begin{aligned}(N-1)\mathbf{S}_Y^2 &= \sum_i \sum_j (Y_{ij} - \bar{\mathbf{Y}})^2 \\ &= \sum_i N_i(\bar{\mathbf{Y}}_i - \bar{\mathbf{Y}})^2 + \sum_i \sum_j (Y_{ij} - \bar{\mathbf{Y}}_i)^2 \\ &= (M-1)\mathbf{S}_B^2 + (N-M)\mathbf{S}_W^2\,, \end{aligned} \tag{7.3}$$

where $\mathbf{S}_B^2$ and $\mathbf{S}_W^2$ are referred to as the *between clusters* and *within clusters variances*, respectively.

In case all the clusters are of the same size N_0, we have $N = MN_0$. When $N_0 = 1$, we end up with the original set-up of a labelled population of N units.

7.2. SINGLE-STAGE CLUSTER SAMPLING DESIGNS

A single-stage cluster sampling design for a population decomposed into M clusters is the same as an ordinary sampling design where the clusters themselves are regarded as sampling units, sometimes called the *primary sampling units* (PSU) or the whole units, in contrast to the original units, called the *secondary sampling units* (SSU) or the ultimate units.

For the sake of completeness, we give the following definition.

Definition 7.1. A single-stage cluster sampling design for a population $\mathcal{U}$ decomposed into M nonoverlapping clusters $\mathcal{C}_1, \mathcal{C}_2, \ldots, \mathcal{C}_M$ is a sampling design based on $\mathcal{U}$ whose units are M clusters.

In other words, a single-stage cluster sampling design d is a pair (S_d, P_d) where S_d is a subset of all possible samples s formed of the $\mathcal{C}_i$'s themselves and P_d is a probability distribution on S_d such that

(i) $P_d(s) > 0$ for each s in S_d,
(ii) for every cluster in the population, there is at least one sample $s \in S_d$ containing the cluster.

As to the nature of the single-stage cluster sampling designs employed in practice, it may be said that SRS designs as well as ΠPS or PPS sampling designs are quite commonly found. Here size refers to the cumulative sum of the auxiliary size measures of the units in a cluster. When no auxiliary information is available, the cluster sizes themselves are often regarded as auxiliary size measures of the clusters in a population.

It is to be understood that upon implementation of the sampling design d, each of the selected clusters is completely enumerated for the purpose of data collection. In other words, the sampling operation is, in effect, carried out at *one* single stage, that is, at the stage of selection of the clusters. Thereafter, we collect data on all units in each of selected clusters.

Sometimes the sampling units of interest, namely, those sought for the purpose of listing and data collection, may remain embedded in a larger cluster and we need to do *filtering* or *screening* at the stage of data collection. The example below brings out this important feature.

Example 7.1. It is stated that there is an even proportion of left-handed males and females in the society. To verify this, a statistician confines his study to a neighboring township which happens to have eight elementary schools. Clearly, a complete frame of the population of left-handed students

Table 7.1

School No.	1	2	3	4	5	6	7	8
No. of registered students	450	370	400	280	370	430	500	200

in all the schools combined is *not* readily available. However, the school district can furnish prior information on the number of registered students in each school. The survey statistician decided to use a ΠPS sampling design using the above figures as the size measures for various schools. Suppose the school district made the data in Table 7.1 available. Then the statistician came up with a ΠPS sampling design shown in Table 7.2. If the sample $(1, 3, 7)$ happens to be selected, the statistician is supposed to visit the schools numbered $1, 3, 7$. In each school, he has to first of all filter (screen out) the left-handers among all students and then count the number of boys and girls among the left-handers. These will then constitute the relevant data for further study.

Table 7.2

s	(1, 2, 4)	(2, 3, 5)	(3, 4, 6)	(4, 5, 7)	(1, 5, 6)	(2, 6, 7)	(1, 3, 7)	(1, 2, 8)	(3, 4, 8)	(5, 6, 8)	(2, 7, 8)
$P(s)$	0.05	0.07	0.08	0.10	0.15	0.15	0.20	0.05	0.05	0.05	0.05

In case the selected clusters are to be subsampled, we get involved with the notion of two-stage cluster sampling design which will be taken up in Section 7.3. The usefulness of a cluster sampling design is clearly felt in situations where there exists hardly any frame for the population units. It is enough to get hold of the clusters, label them and sample some of them. Then, of course, unless the selected clusters are to be subsampled, we do *not* have to bother about listing or labelling the ultimate units in them. In any case, the hard task of listing the ultimate units in the clusters *not* selected has been avoided outright and that is indeed desirable from considerations of time and cost.

We have conveniently denoted a single-stage sample of clusters by the symbol s. We use the notation $i \in s$ to actually denote $\mathscr{C}_i \in s$, since the whole cluster $\mathscr{C}_i$ is to be treated as one single unit (PSU). We will denote by $\pi_i^*(d)$ and $\pi_{ij}^*(d)$ the inclusion probability of the ith cluster and the joint inclusion probabilities of the ith and the jth clusters, respectively, based on the chosen design d. Frequently, we will use π_i^* and π_{ij}^* for simplicity in our presentation of the results.

We now turn to the problems of estimation of the mean per unit $\bar{\mathbf{Y}}$ (also known as the mean per ultimate unit or per SSU) and the mean per cluster $\bar{\mathbf{Y}}_c$ (also known as the mean per PSU). Both can be easily obtained from an estimator of the entire population total $T(\mathbf{Y})$ which we provide below. However, for $\bar{\mathbf{Y}}$ we need a knowledge of N, the total size of the population in terms of the number of SSU.

Theorem 7.1. *An unbiased estimator of the population total $T(\mathbf{Y})$ based on a single-stage cluster sampling design is given by*

$$\hat{T}(\mathbf{Y}) = \sum_{i \in s} \mathbf{T}_i / \pi_i^* \tag{7.4}$$

with variance

$$V(\hat{T}(\mathbf{Y})) = \sum_{i=1}^{M} \mathbf{T}_i^2 \Big(\frac{1}{\pi_i^*} - 1 \Big) + \sum_{\substack{i \neq i' \\ 1}}^{M} \sum \mathbf{T}_i \mathbf{T}_{i'} \Big(\frac{\pi_{ii'}^*}{\pi_i^* \pi_{i'}^*} - 1 \Big) . \tag{7.5}$$

Further, the above variance can be estimated unbiasedly if and only if $\pi_{ii'}^* > 0$, $1 \leq i \neq i' \leq N$. When this holds, one unbiased estimator is given by

$$\hat{V}(\hat{T}(\mathbf{Y})) = \sum_{i \in s} \mathbf{T}_i^2 \Big(\frac{1}{\pi_i^*} - 1 \Big) \Big/ \pi_i^* + \sum_{\substack{i \neq i' \\ i, i' \in s}} \sum \frac{\mathbf{T}_i \mathbf{T}_{i'}}{\pi_{ii'}^*} \Big(\frac{\pi_{ii'}^*}{\pi_i^* \pi_{i'}^*} - 1 \Big) . \tag{7.6}$$

The reader can easily notice that we have used the HTE, $V_{\mathrm{HT}}(\mathrm{HTE})$, and $\hat{V}_{\mathrm{HT}}(\mathrm{HTE})$ in the above. We omit the proof. Further, we note that in case the single-stage cluster sampling design is a fixed-size design, we can replace (7.5) and (7.6) by the corresponding Sen–Yates–Grundy expressions given in Chapter Three and the variance estimator becomes nonnegative if $\pi_{ii'}^* \leq \pi_i^* \pi_{i'}^*$, $1 \leq i \neq i' \leq M$.

Before proceeding further, we emphasize that in case the single-stage cluster sampling design results in the selection of *only* one cluster, no unbiased variance estimator can be provided since $\pi_{ii'}^* = 0$ for all pairs (i, i'). In our subsequent discussions, we will, therefore, assume that at least two clusters are selected by the single-stage design.

Specializing on an SRS(M, m) design, we have the following result.

Corollary 7.2. *If the adopted single-stage cluster sampling design is an* SRS(M, m) *design, then an unbiased estimator of the population total $T(\mathbf{Y})$ is given by*

$$\hat{T}(\mathbf{Y}) = M \bar{\mathbf{y}}_c(s) , \tag{7.7}$$

where

$$\bar{\mathbf{y}}_c(s) = \sum_{i \in s} T_i / m \tag{7.8}$$

is the sample mean per cluster or per PSU. The estimator $\hat{T}(\mathbf{Y})$ has variance

$$V(\hat{T}(\mathbf{Y})) = M^2 \Big(\frac{1}{m} - \frac{1}{M} \Big) \Big\{ \sum_{i=1}^{M} (\mathbf{T}_i - \bar{\mathbf{Y}}_c)^2 / (M-1) \Big\} . \tag{7.9}$$

Further, if $m \geq 2$, an unbiased estimator of $V(\hat{T}(\mathbf{Y}))$ is given by

$$\hat{V}(\hat{T}(\mathbf{Y})) = M^2\left(\frac{1}{m} - \frac{1}{M}\right)\left\{\sum_{i \in s} (\mathbf{T}_i - \bar{\mathbf{y}}_c(s))^2/(m-1)\right\}. \tag{7.10}$$

In addition to the above, if the clusters are of the same size N_0, then we have the following simplified result.

Corollary 7.3. *If an SRS(M, m) single-stage cluster sampling design is adopted for a population of M clusters of size N_0 each, then an unbiased estimator of the population total $T(\mathbf{Y})$ is given by*

$$\hat{T}(\mathbf{Y}) = MN_0\bar{\mathbf{y}}(s)\,, \tag{7.11}$$

where

$$\bar{\mathbf{y}}(s) = \sum_{i \in s} \mathbf{T}_i/mN_0 \tag{7.12}$$

is the sample mean per ultimate unit (SSU). The estimator $\hat{T}(\mathbf{Y})$ has variance

$$V(\hat{T}(\mathbf{Y})) = M^2N_0^2\left(\frac{1}{m} - \frac{1}{M}\right)\left\{\sum_{i=1}^{M} (\bar{\mathbf{Y}}_i - \bar{\mathbf{Y}})^2/(M-1)\right\}. \tag{7.13}$$

Further, for $m \geq 2$, an unbiased estimator of $V(\hat{T}(\mathbf{Y}))$ is given by

$$\hat{V}(\hat{T}(\mathbf{Y})) = M^2N_0^2\left(\frac{1}{m} - \frac{1}{M}\right)\left\{\sum_{i \in s} (\bar{\mathbf{Y}}_i - \bar{\mathbf{y}}_{(s)})^2/(m-1)\right\}. \tag{7.14}$$

Remark 7.1. In the corollaries above, instead of using an SRS(M, m) design, we could as well use any equivalent form of it in the sense of constant first- and second-order inclusion probabilities. Recall the result stated in Section 4.9 in this context.

In Example 7.2 below we present an application of Corollary 7.3.

Example 7.2. A population of 500 households was conveniently divided into 20 clusters of size 25 each. Subsequently, it was decided to select a simple random sample of 3 clusters and completely enumerate each of them. The following data on household sizes were made available based on the above sampling procedure.

Sampled clusters	5	13	17
Cluster means	5.52	5.68	4.28

The purpose is to estimate the average household size in the population along with estimated s.e. In the notation of Corollary 7.3, we have $M = 20$,

$m = 3$, $N_0 = 25$. Using (7.12) and suitably modifying (7.14), we obtain $\hat{\bar{\mathbf{Y}}} = 5.1600$, estimated variance $= 0.1664$, and estimated standard error $=$ 0.4079.

Example 7.3. With reference to Example 7.1, suppose the schools bearing the serial numbers 1, 3, 7 were actually selected following the given ΠPS sampling design. Suppose the survey statistician obtained the relevant information shown in Table 7.3. To estimate the number of student left-handers in the township along with its estimated s.e., we use (7.4) and the Sen–Yates–Grundy estimator of variance in (7.5). Computations yield

$$\pi_1^* = 0.45,\ \pi_3^* = 0.40,\ \pi_7^* = 0.50,\ \pi_{1,3}^* = 0.20,\ \pi_{1,7}^* = 0.20,\ \pi_{3,7}^* = 0.20\,.$$

Therefore, using $\mathbf{T}_1 = 10$, $\mathbf{T}_3 = 4$, and $\mathbf{T}_7 = 2$, we obtain

$$\hat{\mathbf{T}} \doteq 36\,, \qquad \hat{V}(\hat{\mathbf{T}}) \doteq 26.56\,, \qquad \text{s.e. } (\hat{\mathbf{T}}) \doteq 5.15\,.$$

Further computations yield the following:

Estimated number of left-handed school-boys $\doteq 20$, s.e. $\doteq 1.67$
Estimated number of left-handed school-girls $\doteq 16$, s.e. $\doteq 3.24$.

Table 7.3

Selected Schools	No. of Students		No. of Left-handers	
	Boys	Girls	Boys	Girls
1	300	150	4	6
3	200	200	3	1
7	300	200	2	0

Since every cluster in its turn relates to a collection of ultimate units, it is evident that every single-stage cluster sampling design also exhibits a sampling design on the ultimate units in its own right. For a detailed analysis, we will specifically denote by $j(i)$ the jth unit in the ith cluster $\mathscr{C}_i$, $1 \le j \le N_i$, $1 \le i \le M$. In terms of the ultimate units $j(i)$, $j'(i)$, and $j'(i')$, we have

$$\pi_{j(i)} = \pi_i^*\,, \qquad \pi_{j(i)j'(i)} = \pi_i^* \qquad \text{and} \qquad \pi_{j(i)j'(i')} = \pi_{ii'}^*\,,$$

where $1 \le j(i) \ne j'(i) \le N_i$, $1 \le i \ne i' \le M$. In concluding this section, we mention that the above knowledge regarding the inclusion probabilities of the ultimate units should be helpful in rephrasing the results of Theorem 7.1 and the Corollaries 7.2 and 7.3 in terms of the HTE based on values of the survey variable on the ultimate units. In Exercises 7.1, 7.2, and 7.3, the reader is asked to verify this observation.

7.3. TWO-STAGE CLUSTER SAMPLING DESIGNS

We note that in the case of single-stage cluster sampling designs, we carry out census in each selected cluster. However, due to budgetary constraints and other considerations, it may not be practical to implement a census in a cluster which is substantially large in size. Besides this, it may not be wise to completely enumerate a selected cluster if the response of its members (ultimate units) are not believed to vary much. In such situations we can take recourse to subsampling the selected clusters.

To fix ideas, we introduce the following definition.

Definition 7.2. A two-stage cluster sampling design $d^{(2)}$ for a population $\mathcal{U}$ decomposed into M clusters $\mathcal{C}_1, \mathcal{C}_2, \ldots, \mathcal{C}_M$ consists of a first stage sampling design d based on the collection of clusters as sampling units and, subsequently, of second stage sampling designs $d_{i_1}, d_{i_2}, \ldots, d_{i_{n(s)}}$ based, respectively, on the clusters $\mathcal{C}_{i_1}, \mathcal{C}_{i_2}, \ldots, \mathcal{C}_{i_{n(s)}}$, where $s = (\mathcal{C}_{i_1}, \mathcal{C}_{i_2}, \ldots, \mathcal{C}_{i_{n(s)}})$ is the first stage sample selected according to d.

Clearly, a two-stage cluster sampling design generates a sample of the type

$$s^0 = ((\mathcal{C}_{i_1}, s_{i_1}), (\mathcal{C}_{i_2}, s_{i_2}), \ldots, (\mathcal{C}_{i_{n(s)}}, s_{i_{n(s)}})) \tag{7.15}$$

with probability

$$P_{d^{(2)}}(s^0) = P_d(s) \prod_{l=1}^{n(s)} P_{d_{i_l}}(s_{i_l}) , \tag{7.16}$$

$$s = (\mathcal{C}_{i_1}, \mathcal{C}_{i_2}, \ldots, \mathcal{C}_{i_{n(s)}}), s_{i_1} \subseteq \mathcal{C}_{i_1}, s_{i_2} \subseteq \mathcal{C}_{i_2}, \ldots, s_{i_{n(s)}} \subseteq \mathcal{C}_{i_{n(s)}} .$$

In the set-up of two-stage cluster sampling, it is assumed that the second stage sampling designs $d_1, d_2, \ldots, d_M$ based respectively on the clusters $\mathcal{C}_1, \mathcal{C}_2, \ldots, \mathcal{C}_M$ are fixed before hand and are possibly different. Moreover, the second stage sampling designs for the selected clusters are implemented independently of one another. To avoid triviality, it may be assumed that at least one second stage sampling design is different from a census.

Remark 7.2. The above concept of a two-stage sampling design could be generalized to meet complex survey situations. When the clusters are of unequal sizes or the cost of subsampling in different clusters is possibly different, one may, for the purpose of controlling the overall cost of the survey, adopt various types of second stage survey designs depending on the outcome of the first stage cluster survey. We shall *not* deal with such complex survey designs. Instead, in Example 7.5, we will illustrate this point.

First we present an example of a two-stage cluster sampling design.

Example 7.4. A population consists of $M = 7$ clusters having the sizes indicated in Table 7.4. Let the first stage sampling design be uniform on the following support:

$$(1, 3, 4),\ (2, 4, 5),\ (3, 5, 6),\ (4, 6, 7),\ (1, 5, 7),\ (1, 2, 6),\ (2, 3, 7)\,.$$

Table 7.4

Cluster i	1	2	3	4	5	6	7
Size N_i	10	20	30	40	50	60	70

Suppose for the second stage sampling it is decided to adopt SRS(N_i, n_i) design for the ith selected cluster where $n_i = 4N_i/10$, $1 \le i \le 7$. It is easy to see that this two-stage cluster sampling design yields samples for which the overall sample size varies between 32 and 68.

Example 7.5. The purpose of this example is to suitably modify the second stage sampling designs of the preceding example so that the overall sample size remains the same irrespective of the outcome of the first stage sampling design given above. This can be easily achieved by allowing the sample size in the SRS design for any given cluster to vary depending on the particular combination of the clusters selected through the first stage sampling. One solution based on approximately proportional allocation is provided in Table 7.5 for a total sample size of 40. Clearly, many other solutions are possible with this total.

Table 7.5

First Stage	Subsample sizes							Overall
Sample	1	2	3	4	5	6	7	Sample Size
(1, 3, 5)	5	–	13	–	22	–	–	40
(2, 4, 5)	–	7	–	15	18	–	–	40
(3, 5, 6)	–	–	9	–	14	17	–	40
(4, 6, 7)	–	–	–	9	–	14	17	40
(1, 5, 7)	3	–	–	–	15	–	22	40
(1, 2, 6)	4	9	–	–	–	27	–	40
(2, 3, 7)	–	7	10	–	–	–	23	40

We now turn to the estimation problem based on data arising out of adoption of a two-stage cluster sampling design. As before, we limit ourselves to the estimation of the population total $T(\mathbf{Y})$ of the survey variable Y. We again denote by π_i^* and $\pi_{ii'}^*$ the ith cluster inclusion probability and the ith and the i'th clusters joint inclusion probability based on the first stage sampling design. It is further to be assumed that each second stage sampling design d_i when applied to the cluster $\mathscr{C}_i$ provides an unbiased estimator $\hat{\mathbf{T}}_i$ of the ith cluster total $\mathbf{T}_i$ along with an unbiased estimator $\hat{V}(\hat{\mathbf{T}}_i)$ of the variance $V(\hat{\mathbf{T}}_i)$. Under this setup we have the following general result.

Theorem 7.4. *For a general two-stage cluster sampling design given by* (7.15) *and* (7.16), *an unbiased estimator of the population total* $T(\mathbf{Y})$ *is given by*

$$\hat{T}(\mathbf{Y}) = \sum_{i \in s} \hat{\mathbf{T}}_i / \pi_i^* \tag{7.17}$$

with variance

$$V(\hat{T}(\mathbf{Y})) = \left[\sum_{i=1}^{M} \mathbf{T}_i^2 \left(\frac{1}{\pi_i^*} - 1 \right) + \sum_{\substack{i=1 \\ i \neq i'}}^{M} \sum_{i'=1}^{M} \mathbf{T}_i \mathbf{T}_{i'} \left(\frac{\pi_{ii'}^*}{\pi_i^* \pi_{i'}^*} - 1 \right) \right] + \sum_{i=1}^{M} V(\hat{\mathbf{T}}_i) / \pi_i^* . \tag{7.18}$$

Further, assuming that $\pi_{ii'}^* > 0$, $1 \leq i \neq i' \leq m$, *an unbiased estimator of this variance is given by*

$$\hat{V}(\hat{T}(\mathbf{Y})) = \left[\sum_{i \in s} \frac{\hat{\mathbf{T}}_i^2}{\pi_i^*} \left(\frac{1}{\pi_i^*} - 1 \right) + \sum_{\substack{i \neq i' \\ i, i' \in s}} \sum \frac{\hat{\mathbf{T}}_i \hat{\mathbf{T}}_{i'}}{\pi_{ii'}^*} \left(\frac{\pi_{ii'}^*}{\pi_i^* \pi_{i'}^*} - 1 \right) \right] . \tag{7.19}$$

$$+ \sum_{i \in s} \hat{V}(\hat{\mathbf{T}}_i) \pi_i^* .$$

Proof. We first observe that in a two-stage cluster sampling design, two probability sampling designs are acting at different stages. We will denote by E_1 and V_1 the operators of expectation and variance with respect to the first stage sampling design and by E_2 and V_2 the corresponding operators for the second stage sampling design. Accordingly, $E_2(\hat{\mathbf{T}}_i) = \mathbf{T}_i$, $E_2(\hat{V}(\hat{\mathbf{T}}_i)) = V(\hat{\mathbf{T}}_i)$. Let E and V denote the operators for the mean and variance, respectively, based on the overall sampling design. We recall that symbolically, $E = E_1 E_2$ and $V = V_1 E_2 + E_1 V_2$ (see Section 2b.3 of C. R. Rao (1965), for example, for an explanation and use of these notations). We now have

$$E(\hat{T}(\mathbf{Y})) = E_1 E_2(\hat{T}(\mathbf{Y})) = E_1 \left\{ \sum_{i \in s} E_2(\hat{\mathbf{T}}_i) / \pi_i^* \right\}$$

$$= E_1 \left\{ \sum_{i \in s} \mathbf{T}_i / \pi_i^* \right\} = \sum_{i=1}^{M} \mathbf{T}_i = T(\mathbf{Y}) .$$

Next,

$$V(\hat{T}(\mathbf{Y})) = (V_1 E_2 + E_1 V_2)(\hat{T}(\mathbf{Y}))$$

$$= V_1(E_2(\hat{T}(\mathbf{Y}))) + E_1(V_2(\hat{T}(\mathbf{Y}))) .$$

Now, as seen above, $E_2(\hat{T}(\mathbf{Y})) = \Sigma_{i\in s} \mathbf{T}_i/\pi_i^*$ and this represents the HTE (based on the cluster totals as the values of the survey variable) with respect to the first stage sampling design. Therefore, $V_1(E_2(\hat{T}(\mathbf{Y})))$ can be identified as $V_{\mathrm{HT}}(\mathrm{HTE})$ with respect to the first stage sampling design and this is given under $[\cdots]$ in (7.18). Finally,

$$V_2(\hat{T}(\mathbf{Y})) = V_2\Big(\sum_{i\in s} \hat{\mathbf{T}}_i/\pi_i^*\Big) = \sum_{i\in s} V(\hat{\mathbf{T}}_i)/\pi_i^{*2}$$

$$= \sum_{i\in s} \frac{1}{\pi_i^*} \{V(\hat{\mathbf{T}}_i)/\pi_i^*\}$$

which resembles in form an HTE (with values of the survey variable taken as $V(\hat{\mathbf{T}}_i)/\pi_i^*$). Therefore, $E_1(V_2(\hat{T}(\mathbf{Y}))) = \Sigma_{i=1}^M V(\hat{\mathbf{T}}_i)/\pi_i^*$. This is the second term in the expression for $V(\hat{T}(\mathbf{Y}))$ in (7.18). As regards $\hat{V}(\hat{T}(\mathbf{Y}))$, by direct evaluation we have

$$E(\hat{V}(\hat{T}(\mathbf{Y}))) = E_1[E_2(\hat{V}(\hat{T}(\mathbf{Y})))] .$$

Now

$$E_2(\hat{V}(\hat{T}(\mathbf{Y}))) = E_2\Big[\sum_{i\in s} \hat{\mathbf{T}}_i^2/\pi_i^*\Big(\frac{1}{\pi_i^*} - 1\Big) + \sum_{\substack{i\neq i' \\ i,i'\in s}}\sum \frac{\hat{\mathbf{T}}_i\hat{\mathbf{T}}_{i'}}{\pi_{ii'}^*}\Big(\frac{\pi_{ii'}^*}{\pi_i^*\pi_{i'}^*} - 1\Big)$$

$$+ \sum_{i\in s} \hat{V}(\hat{\mathbf{T}}_i)/\pi_i^*\Big]$$

$$= \sum_{i\in s} \frac{V(\hat{\mathbf{T}}_i) + \mathbf{T}_i^2}{\pi_i^*}\Big(\frac{1}{\pi_i^*} - 1\Big) + \sum_{\substack{i\neq i' \\ i,i'\in s}}\sum \frac{\mathbf{T}_i\mathbf{T}_{i'}}{\pi_{ii'}^*}\Big(\frac{\pi_{ii'}^*}{\pi_i^*\pi_{i'}^*} - 1\Big)$$

$$+ \sum_{i\in s} V(\hat{\mathbf{T}}_i)/\pi_i^*$$

$$= \sum_{i\in s} V(\hat{\mathbf{T}}_i)/\pi_i^{*2} + \hat{V}_{\mathrm{HT}}(\mathrm{HTE}) ,$$

where HTE refers to $\Sigma_{i\in s} \mathbf{T}_i/\pi_i^*$ based on the first stage sampling design. Therefore,

$$E_1[E_2(\hat{V}(\hat{T}(\mathbf{Y})))] = E_1\Big\{\sum_{i\in s} V(\hat{\mathbf{T}}_i)/\pi_i^{*2}\Big\} + E_1(\hat{V}_{\mathrm{HT}}(\mathrm{HTE}))$$

$$= \sum_{i=1}^{M} V(\hat{\mathbf{T}}_i)/\pi_i^* + V_{\mathrm{HT}}(\mathrm{HTE}) ,$$

where $V_{\mathrm{HT}}(\mathrm{HTE})$ refers to the variance of the HTE based on the cluster totals for the first stage sampling design. Hence the theorem is proved. □

Corollary 7.5. *If the first stage design in a two-stage sampling design is a fixed-size* (m) *sampling design, then* (7.18) *can be expressed as*

$$V(\hat{T}(\mathbf{Y})) = \sum_{i<i'}^{M}\sum \left(\frac{\mathbf{T}_i}{\pi_i^*} - \frac{\mathbf{T}_{i'}}{\pi_{i'}^*}\right)^2 (\pi_i^*\pi_{i'}^* - \pi_{ii'}^*) + \sum_{i=1}^{M} V(\hat{\mathbf{T}}_i)/\pi_i^* \quad (7.20)$$

and (7.19) *can be replaced by*

$$\hat{V}(\hat{T}(\mathbf{Y})) = \sum_{\substack{i<i' \\ i,i'\in s}}\sum \left(\frac{\hat{\mathbf{T}}_i}{\pi_i^*} - \frac{\hat{\mathbf{T}}_{i'}}{\pi_{i'}^*}\right)^2 \left(\frac{\pi_i^*\pi_{i'}^* - \pi_{ii'}^*}{\pi_{ii'}^*}\right) + \sum_{i\in s} \hat{V}(\hat{\mathbf{T}}_i)/\pi_i^* . \quad (7.21)$$

In (7.20) and (7.21) above, Sen–Yates–Grundy expressions have been used. In the next result, the HTE and its variance estimator are used with every second stage sample data.

Corollary 7.6. *If all the clusters are of equal size, say* N_0 *and an* SRS(M, m) *design is applied to the* M *clusters of the population, followed by simultaneous application of* SRS(N_0, n_0) *designs in the* m *selected clusters, then an unbiased estimator of the overall population mean* $\bar{\mathbf{Y}}$ *is given by*

$$\hat{\bar{\mathbf{Y}}} = \bar{\mathbf{y}} = \textit{sample mean per ultimate unit}\,, \quad (7.22)$$

with variance

$$V(\bar{\mathbf{y}}) = \left(\frac{1}{m} - \frac{1}{M}\right)\left\{\sum_{i=1}^{M} (\bar{\mathbf{Y}}_i - \bar{\mathbf{Y}})^2/(M-1)\right\} + \frac{1}{Mm}\left(\frac{1}{n_0} - \frac{1}{N_0}\right)\left\{\sum_{i=1}^{M}\sum_{j=1}^{N_0} (Y_{ij} - \bar{\mathbf{Y}}_i)^2/(N_0 - 1)\right\}, \quad (7.23)$$

and an unbiased variance estimator (assuming $m \geq 2$, $n_0 \geq 2$) *is given by*

$$\hat{V}(\bar{\mathbf{y}}) = \left(\frac{1}{m} - \frac{1}{M}\right)\left\{\sum_{i\in s} (\hat{\bar{\mathbf{Y}}}_i - \hat{\bar{\mathbf{Y}}})^2/(m-1)\right\} + \frac{1}{Mm}\left(\frac{1}{n_0} - \frac{1}{N_0}\right)\left\{\sum_{i\in s}\sum_{j\in s_i} (Y_{ij} - \hat{\bar{\mathbf{Y}}}_i)^2/(n_0 - 1)\right\}. \quad (7.24)$$

We may further simplify (7.23) to

$$V(\bar{\mathbf{y}}) = \left(\frac{1}{m} - \frac{1}{M}\right)\mathbf{S}_B^2/N_0 + \frac{1}{m}\left(\frac{1}{n_0} - \frac{1}{N_0}\right)\mathbf{S}_W^2 \quad (7.25)$$

and (7.24) to

$$\hat{V}(\bar{\mathbf{y}}) = \left(\frac{1}{m} - \frac{1}{M}\right)\mathbf{s}_B^2/n_0 + \frac{1}{M}\left(\frac{1}{n_0} - \frac{1}{N_0}\right)\mathbf{s}_W^2\,, \quad (7.26)$$

where, referring to (7.3), $\mathbf{S}_W^2$ and $\mathbf{S}_B^2$ represent the within clusters and between clusters variances in the population and similarly $\mathbf{s}_W^2$ and $\mathbf{s}_B^2$ represent the corresponding quantities in the sample. In the special situation of equal cluster sizes and equal subsample sizes we have

$$\mathbf{S}_B^2 = N_0 \sum_{i=1}^{M} (\bar{\mathbf{Y}}_i - \bar{\mathbf{Y}})^2/(M-1)\,, \qquad \mathbf{S}_W^2 = \sum_{i=1}^{M} \sum_{j=1}^{N_0} (Y_{ij} - \bar{\mathbf{Y}}_i)^2/M(N_0 - 1)$$

$$\mathbf{s}_B^2 = n_0 \sum_{i \in s} (\hat{\bar{\mathbf{Y}}}_i - \hat{\bar{\mathbf{Y}}})^2/(m-1)\,, \qquad \mathbf{s}_W^2 = \sum_{i \in s} \sum_{j \in s_i} (Y_{ij} - \hat{\bar{\mathbf{Y}}}_i)^2/m(n_0 - 1)\,.$$

A commonly used two-stage cluster sampling design in practice is the combination of a suitable ΠPS, PPS, or PPAS sampling design on the first-stage units (clusters) together with simple random sampling at the second stage.

We present an example to illustrate the nature of computations involved.

Example 7.6. A large agricultural farm consisting of 507 plots of land was conveniently divided into 7 clusters serially numbered from 1 to 7. For the purpose of estimating the total agricultural production of the farm, 3 clusters were selected using the following first stage ΠPS sampling design with the sum total of agricultural land of various plots in a cluster as the auxiliary size measure:

$$P((1,3,7)) = P((2,6,7)) = 3/17\,,$$

$$P((2,3,5)) = P((1,5,6)) = P((1,2,4)) = P((3,4,6)) = 2/17\,,$$

$$P((3,4,7)) = P((4,5,7)) = P((4,6,7)) = 1/17\,.$$

Suppose the clusters numbered 3, 4, and 6 were actually selected and each of them was subsampled using simple random sampling, resulting in the data shown in Table 7.6. We have $M = 7$, $m = 3$, $s = (3, 4, 6)$, $\pi_3^* = 8/17$, $\pi_4^* = 7/17$, $\pi_6^* = 8/17$, $\pi_{3,4}^* = 3/17$, $\pi_{3,6}^* = 2/17$, $\pi_{4,6}^* = 3/17$. Further, $\hat{\mathbf{T}}_3 = 30{,}408$, $\hat{\mathbf{T}}_4 = 18{,}075$, $\hat{\mathbf{T}}_6 = 7.924$, $\hat{V}(\hat{\mathbf{T}}_3) = 3{,}954.9195$, $\hat{V}(\hat{\mathbf{T}}_4) = 3{,}215.5350$, $\hat{V}(\hat{\mathbf{T}}_6) = 2{,}748.3456$. We now employ the formula (7.17) and compute

$$\hat{T}(\mathbf{Y}) = 17[(30{,}408/8) + (18{,}075/7) + (7{,}924/8)]$$

$$\doteq 125.352\,.$$

Table 7.6

Sampled Clusters	Number of Plots in the Cluster	Number of Sampled Plots	Subsample Mean	Subsample S.D.
3	56	22	543	6.76
4	25	10	723	9.26
6	28	12	283	8.58

Next, we provide also an estimated s.e. of our estimate using (7.21). We have

$$\hat{V}(\hat{T}(\mathbf{Y})) = 289\left[\left(\frac{30{,}408}{8} - \frac{18{,}075}{7}\right)^2\left(\frac{56}{51} - 1\right)\right.$$
$$+ \left(\frac{30.408}{8} - \frac{7{,}924}{8}\right)^2\left(\frac{64}{34} - 1\right) + \left.\left(\frac{18{,}075}{7} - \frac{7{,}924}{8}\right)^2\left(\frac{56}{51} - 1\right)\right]$$
$$+ 17[(3{,}954.9195/8) + (3{,}215.5350/7) + (2{,}748.3456/8)]$$
$$= 289 \times 7{,}363{,}716.4$$

so that the estimated s.e. $\doteq 46{,}131$.

7.4. INCLUSION PROBABILITIES OF ULTIMATE UNITS IN TWO-STAGE CLUSTER SAMPLING DESIGNS

The present study may be regarded as a generalization of the study we made at the end of Section 7.2. Our purpose is to elucidate some properties of estimators based on two-stage cluster sampling designs by first analyzing the expressions for the inclusion probabilities of the ultimate units. This we do with reference to quite a general two-stage sampling design $d^{(2)}$ introduced in Definition 7.2. As before, we denote by π_i^* the inclusion probability of the ith cluster based on the first stage sampling design (which uses the clusters as the sampling units). Further, we will denote by $\pi_{j(i)}(d_i)$ the inclusion probability of the unit $j(i)$ and by $\pi_{j(i)j'(i)}(d_i)$ the joint inclusion probability of the units $j(i)$ and $j'(i)$ based on the second stage sampling design d_i adopted for sampling from the ith cluster $\mathscr{C}_i$. In the above, $j(i)$ stands for the jth unit in the ith cluster, $1 \le j(i) \le N_i$, $1 \le i \le M$.

We begin by stating the following result whose proof is straightforward and hence omitted.

Theorem 7.7. *If the second stage sampling designs are implemented independently of the first stage design, then*

$$\pi_{j(i)}(d^{(2)}) = \pi_i^* \pi_{j(i)}(d_i)\,, \tag{7.27}$$

$$\pi_{j(i)j'(i)}(d^{(2)}) = \pi_i^* \pi_{j(i)j'(i)}(d_i)\,, \tag{7.28}$$

$$\pi_{j(i)j'(i')}(d^{(2)}) = \pi_{ii'}^* \pi_{j(i)}(d_i)\pi_{j'(i')}(d_{i'})\,, \tag{7.29}$$

$$1 \le i \ne i' \le M,\ 1 \le j(i) \ne j'(i) \le N_i,\ 1 \le j'(i') \le N_{i'}\,.$$

It is evident from the above representation that neither the first-order inclusion probabilities nor the joint second-order inclusion probabilities for the ultimate units in an arbitrary two-stage cluster sampling design are the same. For practical reasons (and theoretical interest, too) it would be useful to spell out situations where these inclusion probabilities might remain constant. Below we will present some results in this direction. The proofs are straightforward and are omitted altogether.

Theorem 7.8. *A necessary condition for the first-order inclusion probabilities of the ultimate units in the two-stage design $d^{(2)}$ to be all equal is that for every i, $\pi_{j(i)}(d_i)$ is the same for all $j(i)$, $1 \leq j(i) \leq N_i$, $1 \leq i \leq M$.*

Theorem 7.9. *Assume the first stage sampling design d to be a fixed-size (m) design. Assume further that d_i is a fixed-size (n_i) design for each i, $1 \leq i \leq M$. Then the first-order inclusion probabilities for the ultimate units is the same if and only if*

$$\pi_{j(i)}(d_i) = n_i/N_i\ , \tag{7.30}$$

$$\pi_i^* = (mN_i/n_i)\Big/\Big\{\sum (N_i/n_i)\Big\}\ , \quad 1 \leq j(i) \leq N_i,\ 1 \leq i \leq M\ . \tag{7.31}$$

A useful corollary to Theorem 7.9 is the following.

Corollary 7.10. *If the first stage sampling design is an* SRS(M, m) *design (or its equivalent) and if d_i is an* SRS(N_i, n_i) *design (or its equivalent) for each i, $1 \leq i \leq M$, then all first-order inclusion probabilities of the ultimate units are equal if and only if n_i is proportional to N_i.*

In this context we can look back to Example 7.4. It may be verified that the first stage design is equivalent to SRS(7, 3) design. Further, each and every second stage design is an SRS design with $n_i = 4N_i/10$, $1 \leq i \leq 7$. The conditions of the above corollary are satisfied and, hence, for all ultimate units, the inclusion probability is the same and it is given by 6/35.

It is only natural to investigate the constancy of the second-order inclusion probabilities in situations where the first-order inclusion probabilities are all equal. Of course, it is highly unlikely to come up with a case for which the second-order inclusion probabilities will remain the same. We present below one special case in the setup of Corollary 7.10.

Theorem 7.11. *Assume the following in regard to the designs involved at the first and the second stage of sampling:*

$$d = \text{SRS}(M, m)\ , \qquad d_i \equiv \text{SRS}(N_i, n_i)\ , \qquad 1 \leq i \leq M\ .$$

Then all the second-order inclusion probabilities for the ultimate units will be the same if and only if

$$n_1 = n_2 = \cdots = n_M = n_0 \, , \tag{7.32}$$

$$N_1 = N_2 = \cdots = N_M = N_0 \, , \tag{7.33}$$

and, moreover, m, M, n_0, N_0 *satisfy the relation*

$$m = 1 + N_0(n_0 - 1)(M - 1) / n_0(N_0 - 1) \, . \tag{7.34}$$

Proof. Specializing to the present setup, to ensure equality of all the second-order inclusion probabilities involving pairs of ultimate units, we must have

$$mn_i(n_i - 1) / MN_i(N_i - 1) = m(m - 1)n_i n_{i'} / M(M - 1)N_i N_{i'} \, . \tag{7.35}$$

This holds if and only if

$$n_1/N_1 = n_2/N_2 = \cdots = n_M/N_M \tag{7.36}$$

and further

$$(n_i - 1)/(N_i - 1) = n_i(m - 1)/N_i(M - 1) \, , \qquad 1 \leq i \leq M \, . \tag{7.37}$$

From the above, we deduce (7.32) and (7.33). Finally, using n_0 and N_0 in (7.37), we deduce (7.34). □

A combination of values of M, m, N_0, and n_0 satisfying (7.34) is given by $M = 100$, $m = 89$, $N_0 = 10$, and $n_0 = 5$.

Remark 7.3. Once more we mention that the conclusion of Theorem 7.11 remains true if the SRS designs are replaced by their corresponding equivalent forms. This flexibility could be beneficial to a practitioner who wants to reduce the support of the survey design or manipulate the probability distribution over the support. Another useful point to be noted is that even though the ultimate units are selected by adopting two-stage simple random sampling, whenever the stated conditions in (7.32)–(7.34) hold, the survey practitioner is, in effect, dealing with simple random sample data for all practical purposes. Thus, the sample mean and the sample variance would provide unbiased estimates of the population mean and the population variance.

Exercises 7.6 and 7.7 deal with some problems related to the material presented in this section.

7.5. OPTIMUM CHOICE OF SAMPLE SIZES IN CLUSTER SAMPLING

A cluster sampling design eventually leads to a number of ultimate units on which the relevant data are collected and subsequently analyzed. However, the number of ultimate units to be surveyed depends on the available budget and the precision we desire to achieve on the estimator to be used. Due to the hierarchical nature of such designs, the available budget will be shared by components like the overhead cost, the cost of preparation of frames (with possible filtering) for the units at different stages and finally the cost of data collection and analysis on the ultimate units.

In this section, we wish to acquaint the reader with the problem of choice of sample sizes in cluster sampling designs. To simplify the presentation, we discuss the results with reference to a special type of two-stage cluster sampling design. To fix ideas, we denote by C_0 the overhead cost and by c_1 the cost of selection of a cluster and preparation of a frame of the ultimate units belonging to it. Further, we denote by c_2 the unit cost for collection of data on the ultimate units and their analyses.

We assume that the population consists of M clusters each of size N_0 and the first stage sampling design is an SRS(M, m) design while each subsequent second stage sampling design is an SRS(N_0, n_0) design with $m, n_0 \geq 2$. In that case, the number of clusters selected is m and the number of ultimate units selected is mn_0, thereby resulting in the total cost of operation of the survey to $C_0 + c_1 m + c_2 m n_0$, taking the cost components as additive. Next suppose that the total budget available is C^* so that the choice of (m, n_0) is restricted by the condition:

$$C_0 + c_1 m + c_2 m n_0 \leq C^* , \qquad 2 \leq m \leq M , \qquad 2 \leq n_0 \leq N_0 . \quad (7.38)$$

Subject to this constraint, if the choice of (m, n_0) is not unique, then we make a judicious selection of them which maximizes the precision of our estimator of the population total or the population mean per ultimate unit.

We now refer to the setup of Corollary 7.6. We will use the sample mean per ultimate unit $\bar{\mathbf{y}}$ in (7.22) as an estimator of the population mean per ultimate unit $\bar{\mathbf{Y}}$. In that case, we want to minimize $V(\bar{\mathbf{y}})$ given in (7.23) or equivalently in (7.25). First we rewrite this variance expression in the following form:

$$V(\bar{\mathbf{y}}) = \mathbf{S}_W^2[1/mn_0 + (\mathbf{S}_B^2/\mathbf{S}_W^2 - 1)/mN_0] - \mathbf{S}_B^2/MN_0 . \qquad (7.39)$$

We now develop a guiding rule to resolve the issue of optimum choice of m and n_0. It is a discrete optimization problem and no closed form solution can be given. We separate out the two cases of $\mathbf{S}_B^2 \leq \mathbf{S}_W^2$ and $\mathbf{S}_B^2 > \mathbf{S}_W^2$ for the sake of simplicity.

Case I. $\mathbf{S}_B^2 \le \mathbf{S}_W^2$. When $\mathbf{S}_B^2 = \mathbf{S}_W^2$, we make a choice of (m, n_0) so that mn_0 is a maximum subject to (7.38). When $\mathbf{S}_B^2 < \mathbf{S}_W^2$, the variance expression in (7.39) is decreasing in mn_0 and increasing in m separately in the two components inside [ ]. Therefore, starting with the smallest feasible value of m, we should find the largest possible value of n_0 for which $C_0 + c_1 m + c_2 mn_0$ is as close to C^* as possible. It is generally true that an increase in the value of m will lead to a reduction in the value of mn_0 because of the restriction in (7.38). However, an exception to this is *not* uncommon. See Example 7.7 in this context. In any case, our search should start with the least feasible value of m for which $2 \le (C^* - C_0 - c_1 m)/c_2 m \le N_0$.

Case II. $\mathbf{S}_B^2 > \mathbf{S}_W^2$. In this case, the optimum choice of (m, n_0) depends very much on the ratio $\mathbf{S}_B^2/\mathbf{S}_W^2 = \alpha$. Minimization of (7.39) subject to (7.38) using the techniques of differentiation does *not* help all the time, due to the discrete nature of the optimization problem. Below we develop an easy algorithm which helps to discard many combinations of values of m and n_0.

A combination (m', n_0') is preferred to another combination (m'', n_0''), both satisfying (7.38), if $m' \ge m''$ and $m'n_0' \ge m''n_0''$ with strict inequality in at least one case. This is clear from the expression (7.39), remembering $\alpha > 1$.

In the following example, we illustrate in detail the computations and the steps involved in the search for an optimum choice of (m, n_0).

Example 7.7. Consider a population consisting of $N = 160$ ultimate units and divided into $M = 20$ clusters, each of size $N_0 = 8$. The cost components are $C^* = 1000$, $C_0 = 200$, $c_1 = 10$, and $c_2 = 20$. It is easily seen that the budget constaint on m and n_0 is $m + 2mn_0 \le 80$, $2 \le m \le 20$, $2 \le n_0 \le 8$. Further, it can be argued that a combination (m, n_0) is to be ruled out unless $m + 2mn_0 \le 80$ along with $(m+1) + 2(m+1)n_0 > 80$ or $m + 2m(n_0 + 1) > 80$. It then turns out that the acceptable combinations of (m, n_0) are (4, 8), (5, 7), (6, 6), (7, 5), (8, 4), (11, 3), and (16, 2). We now carry out the following analysis. Recall that $\alpha = \mathbf{S}_B^2/\mathbf{S}_W^2$.

(i) $\alpha = 1$: Here (6, 6) is the optimum choice for (m, n_0).

(ii) $\alpha < 1$: The criterion is to look for small values of m and large values of mn_0. All the combinations listed above, except (4, 8), (5, 7), and (6, 6), are ruled out. A direct comparison of these three combinations then shows that the optimum choice of (m, n_0) is given by

$$(m, n_0) = \begin{cases} (4, 8)\,, & \text{if } 0 < \alpha \le 4/7 \\ (5, 7)\,, & \text{if } 4/7 < \alpha \le 17/21 \\ (6, 6)\,, & \text{if } 17/21 < \alpha \le 1\,. \end{cases}$$

(iii) $\alpha > 1$: The criterion is to look for large values of m and mn_0. It is seen that (6, 6) is better than (4, 8) and (5, 7). Further, (11, 3) is better than (8, 4). It remains, therefore, to compare (6, 6), (7, 5), (11, 3), and (16, 2). We end up with the finding that the optimum choice is

$$(m, n_0) = \begin{cases} (6, 6)\,, & \text{if } 1 < \alpha \le 19/15 \\ (16, 2)\,, & \text{if } \alpha > 19/15\,. \end{cases}$$

In Exercise 7.8, the reader is asked to work out a similar problem.

7.6. EFFICIENCY AND OTHER ASPECTS OF CLUSTER SAMPLING DESIGNS

We present some further theoretical and practical aspects of cluster sampling designs. Once more we refer to Corollary 7.6 regarding the setup for our discussion. We start by giving an alternative representation of $V(\bar{\mathbf{y}})$ in (7.25) in terms of the overall population variance ($\mathbf{S}_Y^2$) and a quantity (denoted by ρ_c) called the *intracluster correlation coefficient*. Such a representation of $V(\bar{\mathbf{y}})$ is useful in (i) comparing the precision of the estimator based on the use of cluster sampling to those of the estimators based on other methods of sampling, and (ii) examining how intervariability among the population units, beyond what is expressed by $\mathbf{S}_Y^2$, affects the precision of the estimator $\bar{\mathbf{y}}$. As we shall see later, given the opportunity to form the clusters, this latter analysis provides a general guideline to be followed by the survey practitioner.

Theorem 7.12. *The variance expression $V(\bar{\mathbf{y}})$ in (7.25) has the alternative representation*

$$V(\bar{\mathbf{y}}) = \frac{(N-1)\mathbf{S}_Y^2}{mMn_0N_0(N_0-1)}\,[(N_0 - n_0) + A\{1 + (N_0 - 1)\rho_c\}]\,, \quad (7.40)$$

where

$$A = \frac{(M-1)N_0(n_0-1) - (m-1)n_0(N_0-1)}{N_0(M-1)} \quad (7.41)$$

and

$$\rho_c = \frac{\Sigma_i \,\Sigma\Sigma_{j \ne j'}\,(Y_{ij} - \bar{\mathbf{Y}})(Y_{ij'} - \bar{\mathbf{Y}})}{(N_0 - 1)\,\Sigma_i\,\Sigma_j\,(Y_{ij} - \bar{\mathbf{Y}})^2} \quad (7.42)$$

is the intracluster correlation coefficient.

Proof. Recall the representation of $\mathbf{S}_Y^2$ in terms of $\mathbf{S}_B^2$ and $\mathbf{S}_W^2$ given in (7.3). Under the present setup, after eliminating $\mathbf{S}_W^2$ from (7.25) using (7.3), we have

$$\begin{aligned} V(\bar{\mathbf{y}}) &= \frac{1}{m}\left(\frac{1}{n_0} - \frac{1}{N_0}\right)\frac{(N-1)\mathbf{S}_Y^2}{M(N_0-1)} \\ &\quad + \frac{\mathbf{S}_B^2}{mMN_0}\left[(M-m) - \frac{(M-1)(N_0-n_0)}{n_0(N_0-1)}\right] \\ &= \frac{1}{m}\left(\frac{1}{n_0} - \frac{1}{N_0}\right)(N-1)\mathbf{S}_Y^2/M(N_0-1) \\ &\quad + \frac{\mathbf{S}_B^2}{mMN_0}\left[\frac{(M-1)N_0(n_0-1) - (m-1)n_0(N_0-1)}{n_0(N_0-1)}\right]. \end{aligned} \tag{7.43}$$

Carrying the analysis further, we next note that

$$\begin{aligned} N_0(M-1)\mathbf{S}_B^2 &= N_0^2 \sum_i (\bar{\mathbf{Y}}_i - \bar{\mathbf{Y}})^2 \\ &= \sum_i (N_0\bar{\mathbf{Y}}_i - N_0\bar{\mathbf{Y}})^2 \\ &= \sum_i \left\{\sum_j (Y_{ij} - \bar{\mathbf{Y}})\right\}^2 \\ &= \sum_i \sum_j (Y_{ij} - \bar{\mathbf{Y}})^2 + \sum_i \sum_{j \neq j'} \sum (Y_{ij} - \bar{\mathbf{Y}})(Y_{ij'} - \bar{\mathbf{Y}}) \\ &= (N-1)\mathbf{S}_Y^2[1 + (N_0-1)\rho_c] \end{aligned} \tag{7.44}$$

by using the expression for ρ_c in (7.42). Using (7.44) in (7.43) and simplifying, we obtain (7.40). □

In the particular case when $m = 1$ and $n_0 = N_0 = n$, we obtain the following result.

Corollary 7.13. *If all the M clusters are of the same size n and one cluster is selected at random using SRS and completely enumerated for the purpose of data collection, then the variance of* $\bar{\mathbf{y}}$ *reduces to*

$$V(\bar{\mathbf{y}}) = \frac{\mathbf{S}_Y^2}{n}\left(\frac{N-1}{N}\right)[1 + (n-1)\rho_c]. \tag{7.45}$$

We now study several implications of these representations from the point of view of comparing the two strategies given below.

Strategy I: Two-stage cluster sampling and sample mean.

Strategy II: Simple random sampling (no clustering) and sample mean.

Of course, for a fair comparison, it is assumed that for Strategy II, the sample size $n = mn_0$ and there is no substantial difference in cost for the two methods.

Theorem 7.14. *Whenever* $\rho_c > 0$, *in the setup of Corollary* 7.13, *Strategy* II *is better than Strategy* I.

Proof. Since $\rho_c > 0$, the variance expression in (7.45) exceeds $\mathbf{S}_Y^2(N-1)/Nn$ which, in its turn, exceeds $\mathbf{S}_Y^2(N-n)/Nn$. This last expression represents the variance of a simple random sample mean based on a sample of size n in the unclustered population. □

Turning back to the general framework of Theorem 7.12, we deduce the following result regarding a comparison between the two strategies.

Theorem 7.15. *In the general framework, Strategy* I *is better than Strategy* II *whenever*

$$mn_0 \geq \frac{n_0(N-1)-(N-N_0)}{(N_0-1)} \tag{7.46}$$

provided $\rho_c = 0$.

Proof. Under $\rho_c = 0$, the variance expression in (7.40) reduces to

$$\begin{aligned} V(\bar{\mathbf{y}}) &= \frac{(N-1)\mathbf{S}_Y^2}{mn_0N(N_0-1)}[(N_0-n_0)+A] \\ &= \frac{(N-1)\mathbf{S}_Y^2}{Nn(N-N_0)}[(N-N_0)-(n-n_0)] \end{aligned} \tag{7.47}$$

where $n = mn_0$. Comparing it with $(N-n)\mathbf{S}_Y^2/Nn$, we end up with the above inequality. □

Unfortunately, in cases where $\rho_c \neq 0$, we do *not* have such explicit results. Of course, inequalities similar to (7.46) can be developed involving the quantity ρ_c. We do *not* present those results. The interested reader is referred to Cochran (1977) and Hansen et al. (1953).

We now turn back to the question of the formation of clusters and suggest a general guideline to the survey practitioner who has decided to use a two-stage cluster sampling design. Assume m, n_0, M, and N_0 to be given. Further, the phrase *optimum* will be understood in the sense of least variance of the estimator $\bar{\mathbf{y}}$.

Theorem 7.16. *A basis for optimum formation of the clusters would be to achieve* $\mathbf{S}_B^2 = 0$ *whenever the quantities* m, n_0, M, *and* N_0 *satisfy the inequality*

$$N_0(M-1)(n_0-1) > n_0(N_0-1)(m-1)\,. \tag{7.48}$$

If the reverse inequality is satisfied, the basis lies in ensuring $\mathbf{S}_W^2 = 0$.

Proof. Our purpose is to minimize the variance expression in (7.25) by suitably forming the clusters. Notice that $\mathbf{S}_Y^2$ is a fixed quantity and it is *not* affected by the mode of formation of the clusters. It is now clear from the expression for $V(\bar{\mathbf{y}})$ in (7.40) that $1+(N_0-1)\rho_c \geq 0$ and hence for $A>0$, we need to have a clustering which ensures $1+(N_0-1)\rho_c = 0$ and, in view of (7.44), this is the same as demanding $\mathbf{S}_B^2 = 0$. On the other hand, for $A<0$, we need to maximize the value of $1+(N_0-1)\rho_c$. This is achieved when $\rho_c = 1$ so that (7.44) reads as $(M-1)\mathbf{S}_B^2 = (N-1)\mathbf{S}_Y^2$. Hence, from (7.3), it is clear that $\mathbf{S}_W^2 = 0$. It may be noted that in case $A=0$, the mode of formation of clusters does *not* affect the precision of the estimator. □

Remark 7.4. Ideally, we achieve $\mathbf{S}_W^2 = 0$ if we can form the clusters so that the values of the survey variable within each cluster remain the same. Similarly, $\mathbf{S}_B^2 = 0$ is achieved if we form clusters so that the survey variable has the same mean value in all the clusters. It seems easier to form the clusters with $\mathbf{S}_B^2$ nearing zero than with $\mathbf{S}_W^2$ nearing zero. See Exercise 7.9 in this context.

Remark 7.5. It must be noted that the above result serves only as a guideline to a practitioner. If the clusters have already been formed, we may examine the validity of closeness of $\mathbf{S}_B^2$ to zero and decide accordingly regarding the optimum choice of (m, n_0). If the clusters are yet to be formed, we may use this knowledge to suitably construct them.

To illustrate these results, let us take the following example.

Example 7.8. Let $N = 120$, $m = 6$, $N_0 = 20$. Consider the following two options regarding formation of clusters and the choice of (m, n_0):

Option I: $\mathbf{S}_W^2 = 0$, $m = 4$, $n_0 = 2$.
Option II: $\mathbf{S}_B^2 = 0$, $m = 2$, $n_0 = 4$.

Direct computations yield, for the variance expression in (7.25), under the two options, the values

$$V_I = 119\mathbf{S}_Y^2/1200 \qquad \text{and} \qquad V_{II} = 119\mathbf{S}_Y^2/1140$$

thereby suggesting that Option I stands better! In Exercise 7.9, the reader is asked to resolve this issue in a general framework. The implication of this example is that the task of formation of clusters is very dependent on the combination of the first stage and the second stage sample sizes the sampler has in mind.

At this stage, it may *not* be out of context to mention some features of inequality (7.48). It holds for $m = 1$ and fails to hold when $m = M$, assuming $n_0 < N_0$. There is a smooth transition between these two extremes for a given value of the total sample size. The case of $m = M$ corresponds to the case where simple random sampling is carried out in each and every cluster. Sampling designs of this type are commonly known as *stratified simple random sampling* designs. These will be discussed in Chapter Nine wherein the general theory of stratified sampling will be presented.

The above study suggests that if the clusters were initially formed keeping $m = M$ in mind, and, if for some reason this cannot be carried out, cluster sampling with a value of m sufficiently close to M would still be desirable and should be recommended. On the other hand, if clusters are formed keeping the objective $\mathbf{S}_B^2 \ll \mathbf{S}_W^2$ in mind, then single-stage cluster sampling ($n_0 = N_0$) with an affordable value of m should be recommended.

In the above discussion, we have tacitly assumed that the cost components are not seriously influential in the sense that we can freely move among various choices of m and n_0 so long as the total number mn_0 of ultimate units (to be included in the sample) remains the same. It must be noted, however, that there are cases where access to the individual ultimate units of a population requires as much effort on the part of the sampler as is needed for a cluster as a whole. It is the intermediate process that becomes a time-consuming and costly affair and it makes very little difference whether we want to check through the whole cluster or a part of it. This is particularly true of surveys dealing with data already stored in files/tapes and an access to such a file/tape is considered a *big success*! In such a situation, it would be nice if the survey practitioner gets a strong feeling of $\mathbf{S}_B^2 \ll \mathbf{S}_W^2$ so that the study can be confined to single-stage cluster sampling with a reasonably small and affordable number of clusters.

The problem of estimation of a proportion using cluster sampling data can be handled in a routine manner. We refer to Cornfield (1951) for relevant results in this area. The study of finite population quantiles based on cluster sampling data has been made in Chapman (1970) and Blesseos (1976). For a review of the available results, we refer to Sedransk and Smith (1988).

7.7. MULTISTAGE CLUSTER SAMPLING DESIGNS

The concept of single-stage and two-stage cluster sampling designs can be generalized to multistage cluster sampling designs in a natural way. The theory we presented above can be easily extended for such designs, albeit the notations get more complicated. As an illustrative example, let us expand on the idea of Example 7.1. Suppose the survey on the incidence of left-handedness is to be conducted statewide and includes 20 counties each having two or more townships. Then a three-stage cluster sampling design can be employed using, respectively, the counties, the townships within counties and the schools within townships as the first stage, the second stage and the third stage units. In Exercise 7.10 the reader is asked to develop a parallel theory for estimation of the population total based on a specific type of three-stage cluster sampling design. Exercise 7.11 is intended for numerical computations.

EXERCISES

7.1. Treating a single-stage cluster sampling design as a sampling design based on the ultimate units of a population, use the HTE to provide a proof of Theorem 7.1.

7.2. Same as in Exercise 7.1. Establish Corollary 7.2. Comment on the result.

7.3. Same as in Exercise 7.1. Establish Corollary 7.3. Comment on the result.

7.4. With reference to Example 7.1, suppose a PPAS sampling design of size 3 was adopted and the schools bearing the serial numbers 1, 3, and 7 are actually selected. Using the data given in Example 7.1 and Example 7.3, compute estimates of the number of left-handed boys and girls in the schools of the township, along with the s.e. of the estimates.

7.5. In Example 7.4, it is stated that the overall sample size of the two-stage cluster sampling design varies between 32 and 68. Verify this statement and display a probability distribution of the overall sample size. What is the expected overall sample size?

7.6. A population consisting of 125 units is grouped into five clusters. A two-stage cluster sampling design is recommended. At stage one, two clusters are to be selected using the sampling design indicated in Table 7.A.

Table 7.A

Sample	(1, 2)	(1, 3)	(1, 4)	(1, 5)	(2, 3)	(2, 4)	(2, 5)	(3, 4)	(3, 5)	(4, 5)
Prob.	2/27	6/27	1/27	1/27	1/27	1/27	4/27	2/27	1/27	8/27

At stage two a simple random sample is to be drawn from each selected cluster. The cluster sizes and the subsample sizes are shown in Table 7.B.

Table 7.B

Cluster i	1	2	3	4	5
Size N_i	15	20	25	30	35
Subsample size n_i	3	5	5	5	5

(a) Compute π_i^*'s and show that the first-order inclusion probabilities of all the 125 ultimate units are the same.

(b) Suppose specifically that the clusters numbered 2 and 5 are selected and further subsampled resulting in data shown in Table 7.C.

Estimate the population total and also give the estimated standard error of your estimate.

Table 7.C

	Subsample	
Sampled Cluster	Mean	S.D.
2	30.8	2.47
5	20.5	3.08

7.7. A finite population of 36 units is grouped into 6 equal sized clusters. A two-stage cluster sampling design is to be implemented with the selection of 15 (ultimate) units.

(a) If simple random sampling is implemented at both stages, devise a two-stage cluster sampling design which is equivalent to an SRS(36, 15) design on the ultimate units in terms of their first- and second-order inclusion probabilities.

(b) Can you replace the SRS design at the second stage in your solution to (a) above by an equivalent sampling design which is based on no more than 10 samples in its support?

7.8. In the setup of Example 7.7, suppose that the population was divided into $M = 8$ clusters, each of size $N_0 = 20$. Taking the same cost components, work out the optimum choice of (m, n_0) for various values of the ratio $\alpha = \mathbf{S}_B^2/\mathbf{S}_W^2$.

7.9. (a) A population consists of $N = MN_0$ units. A statistician decomposed the population into M clusters of size N_0 each in a way that

$\mathbf{S}_B^2 = 0$. Then he recommended a two-stage cluster sampling design with an SRS(M, m') design at the first stage and with an SRS(N_0, n_0') design at each second stage. On the other hand, another statistician decomposed the population in a different way again into M clusters of size N_0 each in a way that $\mathbf{S}_W^2 = 0$. He then suggested the use of a two-stage cluster sampling design with an SRS(M, m'') design at the first stage and with an SRS(N_0, n_0'') design at each second stage. Of course, the values m', m'', n_0', and n_0'' were so chosen that $m'n_0' = m''n_0''$ and the total cost is the same in either approach. Which of the designs would you choose?

(b) Taking $N = 144$, $M = 6$, and $N_0 = 24$, devise a set of 144 values for the survey variable and two clustering methods so that in one case, you end up with $\mathbf{S}_B^2 = 0$ while in the other, you achieve $\mathbf{S}_W^2 = 0$.

7.10. Consider a population decomposed into m first stage units, each first stage unit consisting of N_0 second stage units and each second stage unit, in its turn, consisting of N_{00} third stage (ultimate) units. Suppose we adopt an SRS(M, m) design at the first stage, next an SRS(N_0, n_0) design at each second stage and, finally, an SRS(N_{00}, n_{00}) design at each ultimate stage. Using the HTE, suggest an unbiased estimator of the population total and compute its variance. Assuming m, n_0, $n_{00} \geq 2$, suggest an unbiased estimator of the variance.

7.11. The management of a public library wanted to estimate, on an average, the total number of misplaced books. It was decided that *shelf-reading* would be conducted at the end of a normal working day. A member-cum-statistician volunteered to offer help in the execution of the above task. The chief librarian of the library then briefed the statistician about the basic facts and figures regarding the arrangement of the books in the library. It was stated that there were 3 wall stacks, 5 double (both sided) stacks and 12 single stacks. A stack of any kind was divided into 20 sections and each section had 10 shelves. The chief librarian admitted that she had limited manpower and time and she could only afford to obtain shelf-reading on about 100 shelves. The statistician came up with the following sample survey plan.

SAMPLING DESIGN: 3-STAGE CLUSTER SAMPLING DESIGN

Stage I: Select 6 stacks out of 25 stacks using simple random sampling

Stage II: In each selected stack, select 5 sections out of 20 of them using simple random sampling

Stage III: In each selected section of each selected stack, select 3 shelves out of 10 of them using simple random sampling.

The adoption of the above three-stage cluster sampling design would result in 90 shelves for the purpose of shelf-reading. The chief librarian was excited over the matter and gladly accepted the statistician's survey plan.

At the end of one randomly chosen working day, the survey was conducted. The stacks were serially numbered 1, 2, 3 (for the wall stacks), 4(a), 4(b), . . . , 8(a), 8(b) (for the double stacks), 9, 10, . . . , 20 (for the single stacks). The sections and the shelves were already serially numbered for proper identification. The results of the survey are shown in Table 7.D.

Table 7.D

Sampled Stacks	Sampled Sections	Sampled Shelves (No. of misplaced books/ No. of books counted)		
	4	2 (0/20)	4 (3/22)	7 (2/23)
	7	3 (0/21)	6 (1/24)	8 (1/21)
3	9	1 (0/22)	3 (1/21)	6 (2/23)
	13	2 (0/21)	4 (1/22)	7 (2/22)
	16	2 (1/22)	5 (3/23)	7 (2/22)
	2	3 (1/23)	5 (0/22)	8 (0/24)
	10	1 (2/24)	4 (2/25)	7 (1/25)
7(a)	13	2 (0/21)	6 (2/22)	9 (1/24)
	17	3 (1/23)	4 (1/25)	8 (0/24)

Table 7.D (*Continued*)

Sampled Stacks	Sampled Sections	Sampled Shelves (No. of misplaced books/ No. of books counted)		
	19	2 (1/24)	5 (0/25)	7 (0/21)
10	3	1 (0/22)	6 (1/24)	9 (1/22)
	6	1 (0/23)	4 (0/24)	7 (1/22)
	11	2 (2/25)	3 (1/24)	5 (0/23)
	14	3 (0/23)	7 (1/25)	10 (2/22)
	18	4 (2/24)	7 (1/23)	9 (3/24)
14	1	2 (0/20)	4 (1/23)	6 (2/24)
	5	2 (0/23)	5 (0/22)	8 (2/21)
	10	1 (1/22)	7 (2/24)	9 (0/22)
	13	2 (2/24)	7 (1/20)	10 (1/25)
	17	3 (0/22)	5 (1/22)	7 (2/23)
17	2	4 (0/24)	8 (1/23)	10 (2/25)
	6	1 (0/23)	7 (1/24)	10 (0/26)
	11	3 (0/22)	6 (1/25)	7 (2/24)
	14	4 (1/22)	8 (2/23)	9 (1/25)

Table 7.D (***Continued***)

Sampled Stacks	Sampled Sections	Sampled Shelves (No. of misplaced books/ No. of books counted)		
	17	6 (0/22)	8 (0/23)	10 (1/25)
18	2	3 (1/24)	5 (1/24)	7 (1/23)
	12	6 (0/25)	9 (2/23)	10 (1/24)
	14	2 (0/22)	7 (1/23)	9 (2/24)
	18	3 (1/23)	5 (2/24)	10 (3/22)
	20	2 (1/22)	8 (0/23)	10 (1/24)

(i) Utilize the data to provide an unbiased estimate of the total number of misplaced books on the day the survey was conducted.

(ii) Provide an estimate of the s.e. of your estimate.

(iii) What further information, if any, would you need if you were asked to unbiasedly estimate the proportion of misplaced books in the library on that particular day?

(iv) If you were told beforehand that the information you need in (iii) could *not* be made available, would you have recommended a different sampling design to begin with?

7.12. A wholesale distributor received a large consignment of 10,000 electronic transistors in 1000 boxes of 10 each. The distributor wants to estimate the number of defective transistors in the consignment. It is a costly and time-consuming affair to test each and every transistor. Instead, it is suggested that 10 boxes can be selected using simple random sampling and total inspection can be carried out in the selected boxes. The sampling and inspection are accordingly carried out resulting in data shown in Table 7.E.

Table 7.E

Box no.	38	152	202	383	417	639	721	739	802	900
No. of defectives	0	0	1	0	1	0	1	2	0	1

Give an unbiased estimate of the total number of defective transistors in the consignment. Also compute the s.e. of your estimate. Can you unbiasedly estimate the variance of the number of defectives in a box?

Would you have recommended a different sampling design for the same overall sample size of 100? If so, justify your recommendation against the sampling plan already proposed.

REFERENCES

Blesseos, N. (1976). Distribution-free confidence intervals for quantiles in stratified and cluster sampling. Ph.D. Dissertation, Ball State Univ., Muncie, IN.

Chapman, D. W. (1970). Cluster sampling and approximate distribution-free confidence intervals. Ph.D. Dissertation, Cornell Univ., Ithaca, NY.

Cochran, W. G. (1977). *Sampling Techniques*. Third Edition. Wiley, New York.

Cornfield, J. (1951). The determination of sample size. *Amer. J. Publ. Health*, **41**, 654–661.

Hansen, M. H., Hurwitz, W. N., and Madow, W. G. (1953). *Sample Survey Methods and Theory I,II*. Wiley, New York.

Rao, C.R. (1965). *Linear Statistical Inference and Its Applications*. Wiley, New York.

Sedransk, J., and Smith, P. J. (1988). Inference for finite population quantiles. In: *Handbook of Statistics, Vol. 6, Sampling* (Krishnaiah, P. R. and Rao, C. R., Eds.), pp. 267–289. North-Holland, Amsterdam.

CHAPTER EIGHT

Systematic Sampling Designs

Systematic sampling designs are recommended and adopted in practice due to their appealing simplicity and usefulness, particularly in situations where traditional sampling designs are not applicable. In this chapter we provide a thorough discussion of various aspects of systematic sampling designs. Starting with the traditional concept, we present various modifications and illustrate them with examples. The importance of and the need for such modifications are emphasized in the context of variance estimation.

8.1. SAMPLING DESIGNS FOR POPULATIONS WITH UNITS ARRANGED IN ARRAYS OR IN SEQUENCES

Up until now we have dealt exclusively with populations which are stable or, at least possess stable natural clusters in some reasonable sense. In either case, we had access to the entire reference population for the purpose of sampling and data collection. In addition, we could prepare frames for such populations, though in some cases, it might have been a very costly and time-consuming affair. However, *not* all survey populations of interest are stable. Some are, indeed, very *mobile* and, occasionally, there are *evolving* populations, with the units in the process of appearing and disappearing in time. For such populations, it is extremely expensive, if not impossible, to prepare a frame for the purpose of sampling. Some examples of such populations are the following: patients to be treated in an emergency room; passengers to be received by an immigration office during a specified period of time; phone calls to be received in a central switch board during a specified time interval and the like.

The peculiarities of such populations definitely preclude the use of traditional sampling designs such as simple random sampling, with or without stratification, a concept which will be introduced in Chapter Nine. Even, cluster sampling may *not* be possible to implement in the case that the recording of data on the sampling units takes longer time than the units in a cluster (of successive units) remain available. Such populations call for special sampling methods. *Systematic sampling designs*, along with some modifications to be presented below form a celebrated family of sampling designs particularly suited to meet these challenging situations. As we shall see later, for any evolving population, systematic sampling designs are applicable for the purpose of data collection, even though the underlying frame is not available to start with. Moreover, the survey statistician can take advantage of the nature of the data collection procedure to prepare a frame for the referenced population.

Systematic sampling designs can also be conveniently used for a large class of stable populations where the units are found to have been arranged in sequences, arrays or lines. Some examples of such populations are: files stored in computers or cabinets; items or books arranged in shelves; cars parked in parking lots; and a queue of the first one hundred people waiting to enter a polling booth. Levy and Lemeshow (1980) describe another interesting example which involves the incoming patients in a hospital intensive care unit. In such a case the sampling units form a natural sequence and over a period of time, say one year, we may obtain a sizeable population from which to sample. For other illustrative examples, the reader may consult Murthy and T. J. Rao (1988).

In general, if the units in a population, stable or not, are already organized (or can be conceived of as appearing) in some *orderly fashion*, then we can conveniently apply systematic sampling designs. Here no rigorous definition of *order* can be given since the concept is very subjective. However, we can generally agree that units appearing on a line or in an array of two or more dimensions may be said to be in order. Some common examples are: names listed in a telephone directory, houses along a street in a residential complex, patients entering a hospital emergency room and so on. On the other hand, the population of smokers in a community cannot be taken to be in order. In a broad sense we can say that units in a population are in order if we can identify and reach them in a *systematic* way. Of course, with modern powerful computer facilities, the scope of the notion of orderly arrangements of the population units is indeed broad.

Below we give a broad definition of systematic sampling design. Subsequently, we will specialize to some of its familiar versions available in the literature and extensively used in practice. At this stage we need the concept of a *systematic selection rule with a given starting unit*. This is a rule which allows us to systematically select in an orderly and unique fashion any predetermined number of additional *distinct* units of a population starting from the given unit.

A sample in which the first unit is a starting unit and the remaining units are obtained through a systematic selection rule is called a *systematic sample*.

Definition 8.1. A *systematic sampling design* is a sampling design whose support consists exclusively of systematic samples.

We would like to emphasize that it is not a theoretical issue to judge whether or not a given sampling design is a systematic sampling design; rather this is a query to be settled with reference to the associated sample selection rule.

Some implications of the above definition are listed below.

(1) Any systematic sampling design can be conveniently characterized by the set of its starting units, the corresponding probabilities of selection and the underlying systematic selection rule. In terms of the algorithmic scheme, it follows that $q_1(i) = P(s)$ if i is the starting unit in the sample s, while $q_2(\cdot)$ and $q_3(\cdot)$ assume only values zero and one. The reader can easily verify this fact.

(2) A systematic selection rule occasionally induces a partition of the population; that is, the systematic samples in the support form *nonoverlapping* samples. Theoretically, a sampling design based on this type of systematic selection rule can be identified as a single-stage cluster sampling design resulting in the selection of only one cluster. As a consequence, various statistical aspects of single-stage cluster sampling designs as discussed in Chapter Seven are equally applicable here. Thus, for example, no unbiased variance estimator can be constructed for an unbiased estimator of the population total. Further, the representation of the variance function involving the *intracluster correlation coefficient* as given in (7.40) in Section 7.6 can be utilized for the purpose of comparing such systematic sampling designs with other sampling designs. However, we would like to emphasize that the practical motivation for clustering and systematic selection are entirely different. We will come across some selection rules of the type mentioned above in parts of this chapter.

(3) While examples of systematic sampling designs with $\pi_{ij} > 0$, $1 \le i \ne j \le N$ are *not* uncommon, most of the commonly used designs lack the above desirable feature. This is because the necessary condition for $\pi_{ij} > 0$, $1 \le i \ne j \le N$ is $M^* \ge \binom{N}{2}/\binom{n}{2} = N(N-1)/n(n-1)$ where M^* is the support size of the reduced form of the design assuming that it is of fixed-size (n). This condition is violated most often.

We now present a series of examples for illustrative purpose.

Example 8.1. An auditor wishes to sample from a collection of 120 files for auditing purposes. The files are in a cabinet, arranged in an orderly fashion with serial numbers ranging from 1 to 120. In Table 8.1, we give five systematic sampling designs each of size $n = 10$.

We now discuss some interesting features of the designs displayed in Table 8.1. The systematic selection rule for d_1 and d_4 is the same and it

Table 8.1. Five Systematic Sampling Designs Each of Size $n = 10$ Based on a Population of $N = 120$ Units

Design	Starting Unit (i)	$q_1(i)$	Systematic Selection Rule
d_1	$i = 1, 2, \ldots, 12$	1/12	$i, i + 12, i + 24, \ldots, i + 108$
d_2	$i = 1, 2, \ldots, 120$	1/120	$i, i + 12, i + 24, \ldots, i + 108 (\text{mod } 120)$
d_3	$i = 5, 6, \ldots, 115$	1/111	$i, i \pm 1, i \pm 2, i \pm 3, i \pm 4, i + 5$
d_4	$i = 1, 2, \ldots, 12$	$i/78$	$i, i + 12, i + 24, \ldots, i + 108$
d_5	$i = 1, 2, \ldots, 12$	1/12	$i, 25 - i, i + 24, 49 - i, i + 48,$ $73 - i, i + 72, 97 - i, i + 96, 121 - i$

simply says that upon selection of the starting file (among the first 12 files), the remaining 9 files are to be selected at intervals of 12. Such a selection rule may be said to generate a *linear equidistant systematic sample*. However, we note that d_1 and d_4 represent two different systematic sampling designs. On the otherhand, d_2, in its reduced form, coincides with d_1, albeit it has a different systematic selection rule. A convenient way of interpreting this latter selection rule is to momentarily imagine that the files are placed serially on the circumference of a circle. Upon selection of the starting file, the remaining 9 files are selected at intervals of 12 proceeding along the clockwise direction. Such a selection rule may be said to generate a *circular equidistant systematic sample*. The selection rule for d_3 calls for selecting 5 immediately consecutive files to the right and 4 such files to the left of the starting file. The selection rule for d_5 suggests that the population can be regarded as composed of 5 consecutive groups of 24 files each. Next, having selected the starting file from among the first 12 files, the next (second) file is located symmetrically from the other end of the first group, and the locations of these two files are then carried through in a parallel fashion in the remaining groups. Because of this feature of grouping and symmetrical location of the units, this rule is said to generate *balanced systematic samples*.

In all the above designs we observe that $\pi_{ij} = 0$ for many combinations of values of i and j. Indeed, since the sample size is 10, *no* systematic sampling design would have provided $\pi_{ij} > 0$ for *all* pairs (i, j).

Linear and circular equidistant systematic samples can be described in an obvious manner in general terms. For a population consisting of $N = nk$ units, n and k being both integers, a *linear equidistant systematic sample* (LESS) of size n is composed of a randomly selected unit among the first k units, followed by the remaining $n - 1$ units selected at intervals of length k. A sampling design whose support exclusively consists of LESS will be said to be a LESS design. If the random selection of the starting unit in a LESS design is done using uniform probability of $1/k$, we call this design a UNILESS design. This sampling method is also known in the literature as a 1 *in k linear systematic sampling method*. An application of the above LESS method to a population of size $N = nk + m$, $0 < m < k$, would clearly generate samples of two different sizes. Those having a random start between 1 to m would result in samples of size $n+1$ each. Otherwise, the resulting samples would be of size n each.

On the other hand, a *circular equidistant systematic sample* (CESS) of size n based on a population of size N is obtained by having a random starting unit from the whole population, whose units are conveniently imagined to have been displayed in a circular fashion, and then proceeding in the clockwise direction, selecting the remaining $(n-1)$ units using the sampling or the skipping interval $[N/n]$ where $[x]$ denotes the largest integer not exceeding x. See Exercises 8.1 and 8.2 in this context. A sampling design whose support exclusively consists of CESS will be said to be a CESS design. If the starting unit in a CESS design is selected with uniform probability of $1/N$ each, then the design will be called a UNICESS design. It can be readily seen that a UNICESS design in its reduced form coincides with a UNILESS design whenever $N = nk$.

A balanced systematic sampling design can also be similarly defined and expressed in general terms. We have to assume that the sample size n is even and $N = nk$, k an integer.

Even though systematic sampling designs have been introduced with reference to UNILESS, UNICESS, and balanced systematic sampling designs, we must emphasize that there is the possibility of effective utilization of auxiliary information in the choice of the sampling design. Thus, for example, instead of a UNILESS design, we could implement a PPASLESS design meaning thereby that the probability of selection of any starting unit in the LESS design is made proportional to the aggregate size of all the units in the systematic sample covered by the starting unit.

Except in some special cases, all such designs have the serious drawback of producing $\pi_{ij} = 0$ for many combinations of values of (i, j). See Exercise 8.3 in this context. Some modifications of the above procedures featuring $\pi_{ij} > 0$ for all pairs (i, j) will be presented in the next section.

In the example below, we present 2 two-dimensional systematic selection rules.

Example 8.2. In a controlled agricultural experiment, 100 plants were raised in a 10×10 square array. A plant pathologist wishes to take a sample of plants from this array.

(i) For a sample of size 10, the following two-dimensional systematic sample selection rule can be suggested. Select any plant from the first row using simple random sampling and then select the remaining 9 plants by going along a transversal (broken diagonal) parallel to the main diagonal.

(ii) For a sample of size 9, we can suggest the following rule. Select one inner plant, excluding the boundary plants on all sides, using simple random sampling for the remaining 64 plants. Then take the neighboring 8 plants to form the desired sample.

Not all π_{ij}'s are positive for the sampling design produced by the above two-dimensional systematic selection rules. See Exercise 8.4 in this context. In Section 8.2 we present some general results to rectify such drawbacks.

Before concluding this section we add a few important remarks at this stage.

(1) In the definition of a systematic sample generated by a starting point and a well-defined systematic selection rule, the starting point may be regarded as a *reference* point *without* necessarily *including* it in the sample. We will *not* address this issue theoretically. Instead, we illustrate the idea in the following example.

Example 8.3. A population consists of 16 units on a 4×4 grid. A systematic sample of size 6 is desired. The following selection rule is suggested. Selection one unit at random as the reference point, each with probability 1/16 and let the systematic sample consist of the remaining six units along the same row and column as the reference point. This systematic sampling design has the desirable property that $0 < \pi_{ij} < \pi_i \pi_j$ for all pairs (i, j). In Exercise 8.5 the reader is asked to verify this and some other additional properties of this design.

(2) The desirable feature of positivity of all second-order inclusion probabilities may be achieved in a variety of ways. Specifically,

(i) equal spacing in LESS and CESS procedures can be relaxed,
(ii) some suitable procedures of blending with other sampling designs can be developed, and
(iii) more than one random start can be considered and suitably implemented.

These issues will be taken up in considerable detail in the next section when we discuss the inference aspects of systematic sampling designs and appreciate the need for such modifications.

8.2. INFERENCE BASED ON SYSTEMATIC SAMPLING DESIGNS

We now turn to the problem of estimation of a finite population total $T(\mathbf{Y})$ or mean $\bar{\mathbf{Y}}$ based on data obtained by implementing a systematic sampling design. The HTE is a readily applicable homogeneous linear unbiased estimator (hlue) in this case also. However, there is a serious practical problem in the use of many traditional systematic sampling designs for which the conditions $\pi_{ij} > 0$, $1 \leq i \neq j \leq N$, are violated. As we saw in Chapter Three, the implication is that the variance of any hlue, including the HTE, cannot be unbiasedly estimated. Further, for such designs the population variance also cannot be unbiasedly estimated. This greatly reduces the practical importance of such designs unless some sort of modification are brought into effect.

We will first discuss below the methods listed at the end of the previous section with reference to a population whose units are arranged in an orderly fashion in *one dimension*. For operational convenience, the N units of the population may be imagined to correspond to N numbers $1, 2, \ldots, N$ serially ordered along the clockwise direction on the circumference of a circle. For ready reference, this will be referred to as a *circular arrangement* of the units.

Secondly, a general method of construction of systematic sampling designs for populations whose units are arranged in two dimensions will be presented.

Method of Multiple Distance for a Linearly Ordered Population

To start with, we imagine a circular arrangement of the units. According to this method, after choosing a random starting unit r in the range 1 to N, we systematically move in the clockwise direction at steps of $k_1, k_2, \ldots, k_{n-1}$ where k_i's constitute a set of fixed positive integers. The sample is composed of the distinct units among $r, r+k_1, r+(k_1+k_2), \ldots, r+(k_1+k_2+\cdots+k_{n-1})$, the integers being suitably reduced mod(N), with the convention that the number 0 has to be identified as N. We refer to $N, n, k_1, k_2, \ldots, k_{n-1}$ as the parameters of the method of multiple distance.

Theorem 8.1. *The method of multiple distance with parameters* $N, n, k_1, k_2, \ldots, k_{n-1}$ *yields a fixed-size* (n) *circular systematic sampling design if and only if*

$$\sum_{i=l+1}^{t} k_i \neq 0 \bmod(N)\,, \qquad 0 \leq l < t \leq n-1\,. \tag{8.1}$$

Further, in addition to (8.1), *the joint inclusion probabilities will be positive if and only if for every integer* i, $1 \leq i \leq N-1$, *there exists a pair* (l_i, t_i), $0 \leq l_i < t_i \leq n-1$ *such that*

$$\sum_{j=l_i+1}^{t_i} k_j \equiv i \text{ or } N-i \bmod(N)\,. \tag{8.2}$$

Proof. To have a sample of n distinct units, it is necessary and sufficient that $r \neq r + \Sigma_{i=1}^{l} k_i \neq r + \Sigma_{i=1}^{t} k_i$ for $0 < l < t \leq n-1$. That is, $\Sigma_{i=l+1}^{t} k_i \not\equiv 0 \bmod(N)$ for $0 \leq l < t \leq n-1$ as stated in (8.1).

Not to obscure the essential steps of reasoning, we provide a detailed derivation of (8.2). Fix arbitrarily one unit x in the population. By the

Table 8.2. Samples Generated by the Multiple Distance Method

Generating Units	Units in the Sample				
x	x	$x+k_1$	$\cdots$	$x-\sum_{i=1}^{n-2} k_i$	$x-\sum_{i=1}^{n-1} k_i$
$x-k_1$	$x-k_1$	x	$\cdots$	$x-\sum_{i=2}^{n-2} k_i$	$x-\sum_{i=2}^{n-1} k_i$
$\vdots$	$\vdots$	$\vdots$	$\cdots$	$\vdots$	$\vdots$
$x-\sum_{i=1}^{n-1} k_i$	$x-\sum_{i=1}^{n-1} k_i$	$x-\sum_{i=2}^{n-1} k_i$	$\cdots$	$x-k_{n-1}$	x

method of generation of the samples and by condition (8.1), there are precisely n samples containing the unit x. The following n starting units produce such samples: $x, x-k_1, x-k_1-k_2, \ldots, x-\Sigma_{i=1}^{n-1} k_i$. Specifically, these samples are given in Table 8.2. From the table, it is clear that the units appearing together with x are the distinct units of the set

$$\mathbf{S}_x = \left\{x \pm \sum_{i=l+1}^{t} k_i, 0 \le l < t \le n-1\right\}.$$

In order that $\pi_{xj} > 0$ for every $j \neq x$, it is necessary and sufficient that $j \in \mathbf{S}_x$, $1 \le j(\neq x) \le N$. Now note that $\{j : 1 \le j(\neq x) \le N\} \equiv \{x+1, x+2, \ldots, x+N-1, \text{mod}(N)\}$. Further, $x + i \in \mathbf{S}_x$ if and only if there exists a pair (l_i, t_i) such that

$$i \equiv \sum_{j=l_i+1}^{t_i} k_j \quad \text{or} \quad i \equiv -\sum_{j=l_i+1}^{t_i} k_j \ \text{mod}(N).$$

This is precisely the set of conditions listed in (8.2). Hence the theorem is proved. □

We now illustrate the above technique by an example.

Example 8.4. Consider a population of $N = 16$ units linearly ordered in one dimension. To generate a systematic sample of size $n = 5$ by the method of multiple distance, we can take $k_1 = 2$, $k_2 = 3$, $k_3 = 4$, and $k_4 = 1$. It can be seen that the conditions in (8.1) and (8.2) are readily satisfied. As a direct verification for $\pi_{ij} > 0$, $j = 2, 3, \ldots, 16$, it is enough to see that the five samples containing the unit 1 are: $(1, 3, 6, 10, 11)$, $(15, 1, 4, 8, 9)$, $(12, 14, 1, 5, 6)$, $(8, 10, 13, 1, 2)$ and $(7, 9, 12, 16, 1)$. The verification for other pairs is straightforward and is left to the reader.

Two special cases of the method of multiple distance which may have potential applications in practice are presented below.

(1) Singh and Singh Approach

Singh and Singh (1977) suggested the particular choice

$$k_1 = k_2 = \cdots = k_{u-1} = 1\,, \qquad k_u = \cdots = k_{n-1} = \delta \tag{8.3}$$

and showed that the condition (8.1) is satisfied if

$$u + (n - u)\delta \le N \tag{8.4}$$

and the condition (8.2) is satisfied if

$$\delta \le u\,, \qquad u + (n - u)\delta \ge \frac{N}{2} + 1\,. \tag{8.5}$$

They have also indicated that these sufficient conditions are feasible provided $n \ge \sqrt{2N + 4} - 1$.

Here is an illustrative example.

Example 8.5. Let $N = 16$ and $n = 5$. Then a choice of $u = \delta = 3$ satisfies the conditions (8.4) and (8.5). Therefore, the use of the multiple distance method with $k_1 = k_2 = 1$, $k_3 = k_4 = 3$ will result in a systematic sampling design with $\pi_{ij} > 0$ for all pairs (i, j). As a direct verification, we see that the samples containing the unit 1 are: $(1, 2, 3, 6, 9)$, $(16, 1, 2, 5, 8)$, $(15, 16, 1, 4, 7)$, $(12, 13, 14, 1, 4)$, and $(9, 10, 11, 4, 1)$. These include all pairs involving the unit 1.

The general method of multiple distance for producing systematic sampling designs does *not* automatically ensure $\pi_{ij} \le \pi_i \pi_j$, $1 \le i \ne j \le N$. To guard against this, we will shortly present another approach. We start with the following concept of a *difference set*.

Definition 8.2. Let $D = \{a_0, a_1, \ldots, a_{n-1}\}$ be a subset of integers from the set $\{0, 1, 2, \ldots, N - 1\}$. Then D is said to form a difference set with parameters (N, n, λ) provided the $n(n - 1)$ differences $\{(a_i - a_j) \bmod(N), 0 \le i \ne j \le n - 1\}$ contain every integer in $\{1, 2, \ldots, N - 1\}$ λ times.

Clearly, in any difference set with parameters (N, n, λ), the integer λ satisfies $\lambda = n(n - 1)/(N - 1)$. For example, $\{1, 2, 4\}$ forms a difference set with parameters $(7, 3, 1)$.

We shall now present another special case of the method of multiple distance.

(2) Difference Set Approach

In this approach, the distances $k_1, k_2, \ldots, k_{n-1}$ in the multiple distance method are so chosen that

$$D = \left\{0, k_1, k_1 + k_2, \ldots, \sum_{i=1}^{n-1} k_i\right\}$$

forms a (N, n, λ) difference set.

With such a choice of the k_i's and by incorporating the properties of a difference set in the proof of Theorem 8.1, we can easily argue that under an equal probability random start $\pi_i = n/N$, $\pi_{ij} = n(n-1)/N(N-1)$ for all pairs (i, j) and, consequently, $0 < \pi_{ij} < \pi_i \pi_j$ (Exercise 8.7).

To generate a set of distances $k_1, k_2, \ldots, k_{n-1}$ for a population of size N so that $D = \{0, k_1, k_1 + k_2, \ldots, \Sigma_{i=1}^{n-1} k_i\}$ is a difference set, we can start with an arbitrary difference set $\{b_0, b_1, b_2, \ldots, b_{n-1}\}$ with parameters (N, n, λ). Next we assume without any loss of generality that $b_0 < b_1 < b_2 < \cdots < b_{n-1}$. Then $k_1 = b_1 - b_0$, $k_2 = b_2 - b_1, \ldots, k_{n-1} = b_{n-1} - b_{n-2}$ form the required distances in this procedure.

We mention in passing that there is a vast literature on difference sets (see, for example, Baumert, 1971; Hall, 1986).

The following example illustrates the computations.

Example 8.6. Let $N = 11, n = 5$. Then it can be seen that $\{1, 3, 4, 5, 9\}$ is a difference set with parameters $(11, 5, 2)$. Therefore, a multiple distance systematic sampling design with an equal probability random start and with the skipping intervals $k_1 = 3 - 1 = 2$, $k_2 = 4 - 3 = 1$, $k_3 = 5 - 4 = 1$, and $k_4 = 9 - 5 = 4$ can be employed for this population. For this design, $\pi_i = 5/11$ and $\pi_{ij} = 2/11$.

We observe that the difference set approach to the construction of systematic sampling designs with an equal probability random start leads to sampling designs which are equivalent to SRS(N, n) designs, and indeed based on a much small support.

There is an intimate relationship between systematic sampling designs and block designs as has been established by Hedayat and Pesotan (1990).

Method of Composition

In this method the total sample size n is split into two parts, n_0 and $n - n_0$ where n_0 is a relatively small integer ≥ 1. Next, a fixed-size $(n - n_0)$ systematic sampling design is implemented and a systematic sample s_{n-n_0} is selected. Then a sample s_{n_0} of size n_0 is drawn from the *remaining* $N - n + n_0$ population units by adopting any sampling design. The two component samples s_{n-n_0} and s_{n_0} together constitute the ultimate sample of size n.

It is to be understood that the choice of the sampling design producing a sample of size n_0 is completely arbitrary and it may well vary from one systematic sample to another. We show below that this method of composition yields $\pi_{ij} > 0$ for all pairs (i, j).

Theorem 8.2. *Let d_1 be a fixed-size $(n - n_0)$ systematic sampling design based on $\mathcal{U}$, a population of size N. Conditional on the sample s_{n-n_0} for which $P_{d_1}(s_{n-n_0}) > 0$, let d_2 be a fixed-size (n_0) sampling design based on $\mathcal{U} - s_{n-n_0}$. Then the composite design d for which*

$$P_d(s) = P_{d_1}(s_{n-n_0}) P_{d_2}(s_{n_0} \mid s_{n-n_0}),\ s = (s_{n-n_0}, s_{n_0})$$

satisfies $\pi_{ij}(d) > 0$ for all pairs (i, j) of units in the population.

Proof. We choose an arbitrary but fixed pair of units (i, j), $1 \le i \ne j \le N$. We now distinguish between the two cases:

Case I: $i, j \in s_{n-n_0}$ for some systematic sample s_{n-n_0} with $P_{d_1}(s_{n-n_0}) > 0$. Clearly, in this case, $\pi_{ij}(d) \ge P_{d_1}(s_{n-n_0}) > 0$.

Case II: No systematic sample contains the units i and j together. In this case, we see that

$$\begin{aligned}\pi_{ij}(d) = & \sum_{s_{n-n_0} \ni i, \not\ni j} P_{d_1}(s_{n-n_0}) \Big\{ \sum_{s_{n_0} \ni j} P_{d_2}(s_{n_0} \mid s_{n-n_0}) \Big\} \\ & + \sum_{s_{n-n_0} \ni j, \not\ni i} P_{d_1}(s_{n-n_0}) \Big\{ \sum_{s_{n_0} \ni i} P_{d_2}(s_{n_0} \mid s_{n-n_0}) \Big\} \\ & + \sum_{s_{n-n_0} \not\ni i, \not\ni j} P_{d_1}(s_{n-n_0}) \Big\{ \sum_{s_{n_0} \ni i, j} P_{d_2}(s_{n_0} \mid s_{n-n_0}) \Big\}, \qquad (8.6)\end{aligned}$$

where, in the above, $P_{d_2}(s_{n_0} \mid s_{n-n_0})$ refers to the conditional probability of the sample s_{n_0} given that the systematic sampling design d_1 resulted in the sample s_{n-n_0}.

It is now clear that in the above expression for $\pi_{ij}(d)$, the first two components are positive. Hence the result is established. □

Corollary 8.3. *In the particular case when the sampling design d_2 is an* SRS$(N - n + n_0, n_0)$ *design irrespective of the choice of the systematic sample s_{n-n_0}, for any pair (i, j),*

$$\begin{aligned}\pi_{ij}(d) = & \pi_{ij}(d_1) + \{\pi_i(d_1) - \pi_{ij}(d_1)\} n_0/(N - n + n_0) \\ & + \{\pi_j(d_1) - \pi_{ij}(d_1)\} n_0/(N - n + n_0) \\ & + \{1 - \pi_i(d_1) - \pi_j(d_1) + \pi_{ij}(d_1)\} \times \\ & n_0(n_0 - 1)/(N - n + n_0) \cdot (N - n + n_0 - 1). \qquad (8.7)\end{aligned}$$

Further, if the systematic sampling design d_1 is a UNILESS design providing samples of equal size, we can simplify (8.7) to

$$\pi_{ij}(d) = \frac{(n-n_0)}{N} I_{(i,j)} + \frac{2(n-n_0)n_0}{N(N-n+n_0)}(1 - I_{(i,j)}) + \left\{1 - \frac{2(n-n_0)}{N} + \frac{(n-n_0)}{N} I_{(i,j)}\right\} \frac{n_0(n_0-1)}{(N-n+n_0)(N-n+n_0-1)}, \tag{8.8}$$

where

$$I_{(i,j)} = \begin{cases} 1, & \text{if } i \text{ and } j \text{ are members of the same systematic sample,} \\ 0, & \text{otherwise.} \end{cases}$$

It may be noted that the value of n_0 may be kept very small, say 1 or 2, so that the composite sample may be viewed mostly as a systematic sample blended with a small sample of the other type. We now give an illustrative example.

Example 8.7. To draw a sample of size 14 from a population of size 180 arranged in a row, we may decide to take a 1 in 15 UNILESS of size 12 and then follow it up with a simple random sample of size 2 drawn from the remaining 168 units. The reader may verify that this method produces

$$\pi_{ij} = \begin{cases} \dfrac{1003}{15030}, & \text{if } i \text{ and } j \text{ are members of the same systematic sample,} \\ \dfrac{347}{210420}, & \text{otherwise.} \end{cases}$$

In the context of composition of samples, properties of simple estimators such as linear functions of the systematic sample mean and the simple random sample mean have been studied by Zinger (1963, 1964) and Wu (1984).

Method of Multiple Start

As the name suggests, in this approach the ultimate systematic sample is obtained in a number of stages, each time with a new random start, in the following fashion. At stage one, a systematic sample s_1 of size n_1 is drawn from the whole population $\mathcal{U}$ of size N. At the second stage, a systematic sample s_2 of size n_2 is drawn from the remaining population $\mathcal{U} - s_1$ of size $(N - n_1)$. This procedure is continued and at the ith stage, a systematic sample s_i of size n_i is drawn from $\mathcal{U} - s_1 - \cdots - s_{i-1}$ of size $N - n_1 - \cdots - n_{i-1}$. If we stop at the lth stage, then the ultimate systematic sample is of size $n = n_1 + n_2 + \cdots + n_l$ and the sample s so formed is composed of the units in the subsamples $s_1, s_2, \ldots, s_l$.

The above procedure is said to generate a *multiphase systamtic sample* or a *repeated systematic sample*. This procedure is quite general and flexible. It gives us a wide choice in terms of the parameters $l, n_1, n_2, \ldots, n_l$ subject to a total sample size of $n = n_1 + n_2 + \cdots + n_l$, and, moreover, an arbitrary independent selection of the component systematic sampling designs. By a judicious selection of these parameters for the above procedure, in most instances we are able to come up with a multiple start systematic sampling design having the desired feature of *nonnegative variance estimation*. Below we provide an affirmative result in this direction for a population whose units are linearly ordered.

Let $n = n_1 n_2$ and $N = N_1 n_1$. We select n_2 *distinct* random starts among the first N_1 units of the population by using a fixed-size (n_2) sampling design d^*. Based on each and every selected random start r, we produce a 1 in N_1 LESS composed of the n_1 units $r, r + N_1, \ldots, r + (n_1 - 1)N_1$. Collectively, all the $n_2 n_1$ units thus produced form the ultimate sample. The resulting sampling design may be termed a *multiple start* LESS design. If the starting design d^* is an SRS(N_1, n_2) design, the resulting sampling design may be termed a *multiple start* UNILESS design. With a broad choice of the *starting design* d^*, we can conclude the following.

Theorem 8.4. *The joint inclusion probabilities in a multiple start LESS design are positive for all pairs of units in the population if and only if the starting design d^* satisfies*

$$\pi_{ij}(d^*) > 0 \,, \qquad \textit{for all pairs of starting units} \,. \tag{8.9}$$

Proof. We will write, as before, π_{ij} for the joint inclusion probability of the units i and j in the overall sampling design. Clearly, $\pi_{ij} = \pi_{ij}(d^*)$ for $1 \le i \ne j \le N_1$ where n_1, n_2, and N_1 are the parameters of the sampling procedure. This establishes the necessity part. Further, by the nature of LESS, for a pair (x, y),

$$\pi_{xy} = \begin{cases} \pi_i(d^*) \,, & \text{if } x = i + l_1 N_1,\ y = i + l_2 N_1 \,, \\ & 0 \le l_1 < l_2 \le n_1 - 1 \,; \\ \pi_{ij}(d^*) \,, & \text{if } x = i + l_1 N_1,\ y = j + l_2 N_1 \,, \\ & 0 \le l_1, l_2 \le n_1 - 1 \,. \end{cases} \tag{8.10}$$

This establishes the sufficiency part. □

Corollary 8.5. *In a multiple start UNILESS design with n_2 random starts out of the first N_1 units,*

$$\pi_{xy} = \begin{cases} n_2 / N_1 \,, & \textit{if } x \textit{ and } y \textit{ are covered by the same} \\ & \textit{starting unit,} \\ n_2(n_2 - 1)/N_1(N_1 - 1) \,, & \textit{otherwise} \,. \end{cases}$$

Remark 8.1. To maintain the spirit of a systematic selection rule in a multiple start systematic sampling design, we should keep the number of random starts close to the minimum value of 2.

Remark 8.2. It is clear that in a multiple start systematic sampling design of the type considered above, $\pi_{xy} < \pi_x \pi_y$ whenever x and y are generated by two different random starts. However, $\pi_{xy} = \pi_x > \pi_x^2 = \pi_x \pi_x$ if x and y are generated by the same random start. This apparent undesirable feature should *not* discourage us from using the above method of sampling as we show below that nonnegative variance estimation is still possible.

Theorem 8.6. *If the starting design d^* in a multiple start LESS design yields $\pi_{ij}(d^*) < \pi_i(d^*)\pi_j(d^*)$ for all pairs of starting units, then there exists a nonnegative variance estimator of the HTE of the population total.*

Proof. Consider the collection of N_1 starting units which the starting design d^* is based on. The starting unit r produces the component

$$\mathscr{C}_r = (r, r + N_1, r + 2N_1, \ldots, r + (n_1 - 1)N_1)$$

in the overall sample. As r runs over the starting units, the components $\mathscr{C}_1, \mathscr{C}_2, \ldots, \mathscr{C}_{N_1}$ form a partition of the whole population and the multiple start LESS design picks up n_2 of these components according to d^*. Since the multiple start LESS design is completely characterized by the associated starting design d^*, we conclude that the overall design is equivalent to a single-stage fixed-size (n_2) cluster sampling design where $\mathscr{C}_1, \mathscr{C}_2, \ldots, \mathscr{C}_{N_1}$ are the clusters, each of size n_1. We can now use Theorem 7.1 and the discussion following it, to conclude that a nonnegative unbiased variance estimator (namely, the Sen–Yates–Grundy variance estimator) can be constructed. □

We shall now elucidate the above constructional procedure and allied concepts in the following example.

Example 8.8. Consider a linearly ordered population of $N = 15$ units. We want a systematic sample of size $n = 6$ by the two-start LESS method. In our notation, $n_2 = 2$, $n_1 = 3$ and $N_1 = 5$. In Table 8.3, we display all pairs of starting units, the associated component subsamples (clusters) and all the 10 ultimate samples. Specifically, the clusters are:

$$\mathscr{C}_1 = (1, 6, 11),\ \mathscr{C}_2 = (2, 7, 12),\ \mathscr{C}_3 = (3, 8, 13),\ \mathscr{C}_4 = (4, 9, 14),$$
$$\mathscr{C}_5 = (5, 10, 15)\,.$$

If the starting design is uniform on the support of the 10 pairs, it can be

Table 8.3. Two-Start Linear Equidistant Systematic Samples of Size 6

Random Starting Pair (i, j)	Subsample (Cluster) Generated by Random Start i	Subsample (Cluster) Generated by Random Start j	Overall Sample
(1, 2)	(1, 6, 11)	(2, 7, 12)	(1, 2, 6, 7, 11, 12)
(1, 3)	(1, 6, 11)	(3, 8, 13)	(1, 3, 6, 8, 11, 13)
(1, 4)	(1, 6, 11)	(4, 9, 14)	(1, 4, 6, 9, 11, 14)
(1, 5)	(1, 6, 11)	(5, 10, 15)	(1, 5, 6, 10, 11, 15)
(2, 3)	(2, 7, 12)	(3, 8, 13)	(2, 3, 7, 8, 12, 13)
(2, 4)	(2, 7, 12)	(4, 9, 14)	(2, 4, 7, 9, 12, 14)
(2, 5)	(2, 7, 12)	(5, 10, 15)	(2, 5, 7, 10, 12, 15)
(3, 4)	(3, 8, 13)	(4, 9, 14)	(3, 4, 8, 9, 13, 14)
(3, 5)	(3, 8, 13)	(5, 10, 15)	(3, 5, 8, 10, 13, 15)
(4, 5)	(4, 9, 14)	(5, 10, 15)	(4, 5, 9, 10, 14, 15)

readily verified from Table 8.3 that

$$\pi_{xy} = \begin{cases} 2/5\,, & \text{if } x, y \text{ are in the same cluster,} \\ 1/10, & \text{otherwise.} \end{cases}$$

In the literature of survey sampling, there is a family of systematic sampling designs known as *interpenetrating network of subsampling designs* for populations whose units are linearly ordered. The family of such designs includes the multiple start UNILESS designs as well as the systematic designs which are *compositions of several independent single-start* UNILESS designs. These latter designs can be thought of as multiple start LESS designs when the starting units are selected by an SRS with replacement procedure.

Below is an example of such a systematic sampling design.

Example 8.9. Consider a population consisting of 15 linearly ordered units. Suppose we want a 1 in 3 systematic sample based on 2 starting units selected at random and with replacement from $\{1, 2, 3\}$. In Table 8.4, we list all possible random starting pairs (i, j), $1 \le i, j \le 3$, the subsamples based on the starting pairs and the overall samples.

As regards inference on the population total or mean based on such multiple start UNILESS designs with replacement, once more we can refer to the notion of clusters and the cluster totals based on each random start, as has been illustrated before in the context of Theorem 8.6. Then an application of the results in Section 4.8 of Chapter Four will provide an unbiased estimator of the population mean, an expression for its variance and an unbiased estimator of the variance of the estimator. We will reproduce the final results below without giving the details, after noticing the following correspondence:

Table 8.4. Composition of Two Independent One-Start Linear Equidistant Systematic Subsamples Each of Size 5

Random Starting Pair (i, j)	Selection probability	Subsample (cluster) generated by		Overall Sample
		Random Start i	Random Start j	
(1, 1)	1/9	(1, 4, 7, 10, 13)	(1, 4, 7, 10, 13)	(1, 1, 4, 4, 7, 7, 10, 10, 13, 13)
(1, 2)	1/9	(1, 4, 7, 10, 13)	(2, 5, 8, 11, 14)	(1, 2, 4, 5, 7, 8, 10, 11, 13, 14)
(1, 3)	1/9	(1, 4, 7, 10, 13)	(3, 6, 9, 12, 15)	(1, 3, 4, 6, 7, 9, 10, 12, 13, 15)
(2, 1)	1/9	(2, 5, 8, 11, 14)	(1, 4, 7, 10, 13)	(1, 2, 4, 5, 7, 8, 10, 11, 13, 14)
(2, 2)	1/9	(2, 5, 8, 11, 14)	(2, 5, 8, 11, 14)	(2, 2, 5, 5, 8, 8, 11, 11, 14, 14)
(2, 3)	1/9	(2, 5, 8, 11, 14)	(3, 6, 9, 12, 15)	(2, 3, 5, 6, 8, 9, 11, 12, 14, 15)
(3, 1)	1/9	(3, 6, 9, 12, 15)	(1, 4, 7, 10, 13)	(1, 3, 4, 6, 7, 9, 10, 12, 13, 15)
(3, 2)	1/9	(3, 6, 9, 12, 15)	(2, 5, 8, 11, 14)	(2, 3, 5, 6, 8, 9, 11, 12, 14, 15)
(3, 3)	1/9	(3, 6, 9, 12, 15)	(3, 6, 9, 12, 15)	(3, 3, 6, 6, 9, 9, 12, 12, 15, 15)

(i) Sampling units are the subsamples or clusters denoted as $\mathscr{C}_1, \mathscr{C}_2, \ldots, \mathscr{C}_{N_1}$, each of size n_1;

(ii) Values of the survey variable are the cluster totals $\mathbf{T}_i = \Sigma_{j \in \mathscr{C}_i} Y_j$, $1 \le i \le N_1$;

(iii) Population mean per ultimate unit $= \bar{\mathbf{Y}} = \Sigma_{i=1}^{N_1} \mathbf{T}_i / N_1 n_1 = \bar{\mathbf{Y}}_c / n_1$ where $\bar{\mathbf{Y}}_c$ is the population mean per cluster.

Using (4.30)–(4.33) and denoting by $\nu(s)$ the number of distinct starting units in a collection s of n_2 starting units drawn by using an SRS with replacement scheme out of the N_1 starting units, we obtain $\hat{\bar{\mathbf{Y}}}$ as an unbiased estimator of the population mean per ultimate unit where

$$\hat{\bar{\mathbf{Y}}} = \sum_{i \in s} \mathbf{T}_i / n_1 \nu(s) = \bar{\mathbf{T}}_\nu / n_1 \tag{8.11}$$

with variance

$$V(\hat{\bar{\mathbf{Y}}}) = \left\{ E\left(\frac{1}{\nu}\right) - \frac{1}{N_1} \right\} \mathbf{S}_T^2 / n_1^2 . \tag{8.12}$$

Further, an unbiased estimator of this variance is given by

$$\hat{V}(\hat{\bar{\mathbf{Y}}}) = \left\{ E\left(\frac{1}{\nu}\right) - \frac{1}{N_1} \right\} \mathbf{s}_T^2 / n_1^2 , \tag{8.13}$$

where

$$\mathbf{s}_T^2 = \begin{cases} 0 , & \text{if } \nu = 1 \\ (1 - N_1^{-n_2+1}) \mathbf{s}_\nu^2 , & \text{if } \nu > 1 \end{cases} \tag{8.14}$$

and

$$\mathbf{s}_\nu^2 = (\nu - 1)^{-1} \sum_{i \in s} (\mathbf{T}_i - \bar{\mathbf{T}}_\nu)^2 . \tag{8.15}$$

In Exercise 8.11 the reader is asked to carry out the above computations for the sampling problem described in Example 8.9.

For further results on interpenetrating networks of subsamples, the reader is referred to Cochran (1977), Koop (1967, 1988), Sengupta (1982), and Tornqvist (1963). An earlier work by Gautschi (1957) related to the method of multiple start is also suggested for further reading.

Before concluding this section, we give a general method for the construction of a family of systematic sampling designs for populations whose units are arranged in a two-dimensional rectangular array, utilizing the systematic sampling methods discussed so far for populations whose units are ordered in one dimension.

Method of Crossing Two One-Dimensional Systematic Selection Rules

Assume that one-dimensional systematic selection rules are available for each of two populations of sizes N_1 and N_2 producing samples of sizes n_1 and n_2, respectively. Then a two-dimensional systematic sample selection rule for a sample of size $n = n_1 n_2$ based on a population of size $N = N_1 N_2$ is obtained by forming samples indexed by (i, j) where i and j refer to the sample units in the one-dimensional components.

If the component systematic sampling designs satisfy $\pi_{ii'} > 0$ and $\pi_{jj'} > 0$ for all relevant pairs of units, and the two component designs are implemented independently of one another, then the same is true of the two-dimensional systematic sampling design obtained by the method of crossing described above. Let d_1 and d_2 be the two independently implemented component designs and let d be the sampling design obtained by crossing d_1 and d_2. Then we can readily verify that

$$\pi_{xy}(d) = \begin{cases} \pi_{ii'}(d_1)\pi_j(d_2)\,, & \text{if } x = (i, j),\ y = (i', j) \\ \pi_i(d_1)\pi_{jj'}(d_2)\,, & \text{if } x = (i, j),\ y = (i, j') \\ \pi_{ii'}(d_1)\pi_{jj'}(d_2)\,, & \text{if } x = (i, j),\ y = (i', j') \end{cases} . \qquad (8.16)$$

This justifies the validity of the statement made above.

We now take up an illustrative example.

Example 8.10. Consider a population of 165 units arranged in the form of a 15×11 rectangular array. It is required to draw a sample of size $n = 30$ using a suitable two-dimensional systematic selection rule so that the sampling design will have the feature $\pi_{xy} > 0$ for all pairs of units (x, y). We will use the method of crossing described above. From Example 8.6, we adopt the component design with $N_2 = 11$ and $n_2 = 5$ obtained by the multiple distance systematic sampling method. For the other component with $N_1 = 15$ and $n_1 = 6$, we may adopt the two-start LESS design presented in Example 8.8. Taking the two typical samples, say $s_1 = (1, 2, 6, 7, 11, 12)$, $s_1' = (2, 4, 5, 6, 10)$ from the two components, respectively, we now form the crossed sample of size $n = 30$ for the two-dimensional design as

$$s = ((1, 2), (1, 5), \ldots, (1, 10), (2, 2), \ldots, (2, 10), \ldots, (12, 2), \ldots, (12, 10))$$

where any pair (i, j) represents a unit of the whole population in the two-dimensional array. When the two component designs are implemented independently, it follows from (8.16) that for the overall sampling design

$$\pi_{(1,1)(1,2)} = \pi_1(d_1)\pi_{12}(d_2) = \frac{2}{5} \cdot \frac{2}{11} = \frac{4}{55} .$$

Similarly, the other joint inclusion probabilities can be worked out and can be readily verified to be positive.

8.3. SYSTEMATIC SAMPLING DESIGNS AND STRUCTURED POPULATIONS

Thus far we have dealt with populations whose units are available in an orderly fashion and we made no assumptions whatsoever regarding the nature of values of the survey variable in relation to their relative positions in the array or the sequence. There are situations where we may have some knowledge about a functional relation, deterministic or otherwise, between the labelling index i (in one or more dimensions) and the associated value Y_i of the survey variable Y, $1 \leq i \leq N$. In such a case, the population may be said to be structured in relation to the survey variable.

In the literature, various structural relations have been stipulated, mostly assuming the population units arranged in order in one dimension. Thus, for example, a population in which the value Y_i varies monotonically with its index i in a linear fashion $Y_i = a + bi$, $1 \leq i \leq N$ is known as a *population with linear trend*. Likewise, a population in which the variation between values of the survey variable in any group of contiguous units increases steadily with an increase in the size of the group, is known as an *autocorrelated population*. Similarly, a *population with periodic variation* is one in which the units with large, medium, and small values follow one another according to a regular repetitive pattern, that is, one in which Y_i is a periodic function of period, say $2h$. On the other hand, a population in which the units are arranged completely at random, is known as a *population in random order*. This means that for a fixed value of N, *any ordered* arrangement $(u_{i_1}, u_{i_2}, \ldots, u_{i_N})$ of the N population units $u_1, u_2, \ldots, u_N$ can show up and each has the same probability of $1/N!$, as there are $N!$ such arrangements.

For a population in random order, the following interesting property holds.

Theorem 8.7. *Let $N = nk$, n and k being both integers. For a population in random order, a single-start 1 in k UNILESS design coincides with an* SRS(N, n) *design.*

Proof. An appeal to symmetry settles the result. For the sake of completeness, we provide the details below. Let $(u_{i_1}, u_{i_2}, \ldots, u_{i_N})$ be a typical randomly ordered arrangement of the population units $u_1, u_2, \ldots, u_N$ occurring with probability $1/N!$. Under a single-start 1 in k UNILESS design d based on this particular order of the units, the k systematic samples are

$$s_1 = (i_1, i_{k+1}, \ldots, i_{N-k+1}), \qquad s_2 = (i_2, i_{k+2}, \ldots, i_{N-k+2}), \ldots,$$
$$s_k = (i_k, i_{2k}, \ldots, i_N)$$

each with probability $1/k$. Let $(j_1, j_2, \ldots, j_n)$ be a typical ordered sample under an SRS(N, n) design with probability $1/N^{(n)}$ where $N^{(n)} = N(N-1)\ldots(N-n+1)$, $1 \leq j_1 \neq j_2 \neq \cdots \neq j_n \leq N$. Then conditional on $(i_1, i_2, \ldots, i_N)$,

$$P_d((j_1, j_2, \ldots, j_n)) = \begin{cases} 0\,, & \text{if } (j_1, j_2, \ldots, j_n) \text{ does } \textit{not} \text{ match with any of the } s_i\text{'s,} \\ \dfrac{1}{k}\,, & \text{if } (j_1, j_2, \ldots, j_n) \text{ matches with one (and naturally only one) of the } s_i\text{'s}. \end{cases}$$

Therefore, the unconditional probability of $(j_1, j_2, \ldots, j_n)$ arising out of the above systematic sampling procedure is given by

$$\begin{aligned} P((j_1, j_2, \ldots, j_n)) &= k^{-1}P(\text{random permutation yields } s_1 = (j_1, j_2, \ldots, j_n)) \\ &\quad + \cdots + k^{-1}P(\text{random permutation yields } s_k = (j_1, j_2, \ldots, j_n))\,. \end{aligned}$$

We have k terms each of which is given by $k^{-1}(N-n)!/N!$ and so the resulting expression is $(N-n)!/N!$ which is equal to $1/N^{(n)}$, the same as under an SRS(N, n) design. Hence the theorem is proved. □

Remark 8.3. It thus turns out that with $N = nk$, a simple 1 in k systematic sampling design based on a population in random order is equivalent to an SRS(N, n) design. This fact can be used in drawing inference based on such systematic sample data. Thus the sample mean will have all the properties of a sample mean based on a simple random sample of size n. In particular, its variance expression will coincide with V_{ran} given by $((1/n) - (1/N))\mathbf{S}_Y^2$ and an unbiased variance estimator will be provided simply by $((1/n) - (1/N))\mathbf{s}_Y^2$ where $\mathbf{s}_Y^2$ is the sample variance based on the systematic sample of size n. This is true even though we have only one systematic sample based on a simple start so long as the population units are in random order. However, we would like to emphasize that, in most cases, systematic sampling is *not* equivalent to simple random sampling and we need alternative methods to be able to provide unbiased variance estimators. Some such methods have already been discussed in the previous section.

With reference to the so-called structured populations, a suitable choice of the starting unit or units and the skipping interval has to be made to maximize the precision of the estimator in using systematic sampling designs. Some such studies have been made from time to time by researchers in this area. However, we feel that a satisfactory study involves a good understanding about the nature of the structured population. We refer the reader to Cochran (1977), Singh and Singh (1977), and the review articles by Iachan (1982), Bellhouse (1988), and Murthy and T. J. Rao (1988) for further references. Some exercises along this line have been added at the end. We will *not* pursue this topic in this book.

In concluding this chapter, we mention that the question of choice of the sample size (n) has to be settled in conjunction with both the budgetary constraints and the precision desired of the estimator used. A parallel theory can be developed along the lines as in the context of cluster sampling presented in Chapter Seven and stratified sampling to be presented in Chapter Nine. To avoid repetition, we do not present results in this direction. However, an important point to be noted is that unlike in simple random sampling and stratified simple random sampling, an increase in the sample size while using systematic sampling designs does not necessarily increase the precision of the estimator. Exercise 8.17 highlights this point. In addition to this, a linear cost structure may not be appropriate in most situations.

EXERCISES

8.1. It is often suggested that the sampling interval m in CESS should be the integer closest to N/n, to have an even spread of the sampled units over the entire population. Show that

(i) for $m = [N/n]$, a CESS always consists of n distinct units,

(ii) for $m = [N/n] + 1$, a CESS consists of n distinct units if and only if $N \neq jm$, $j = 1, 2, \ldots, n-1$.

Hence show that to draw a CESS of n distinct units with m as close to N/n as possible, we should take

$$m = \begin{cases} \left[\dfrac{N}{n} + \dfrac{1}{2}\right], & \text{if } N \neq jm,\ j = 1, 2, \ldots, n-1 \\ [N/n], & \text{otherwise}. \end{cases}$$

Consider, in particular, the following values of N and n and examine the above results: $N = 48$, $n = 19$; $N = 5$, $n = 3$; $N = 11$, $n = 5$.

8.2. Prove that a CESS of size n with sampling interval m will contain all distinct units if and only if $[N, m] \geq mn$ or, equivalently, $N \geq (N, m)n$ where $[N, m]$ and (N, m) denote respectively the l.c.m and g.c.d of N and m.

Examine the following combinations of values of N and n and in each case recommend as many values of m as possible so that a CESS of size n with sampling interval m contains all distinct units: $N = 15$, $n = 5$; $N = 20$, $n = 4$; $N = 100$, $n = 12$.

8.3. (a) Apply the CESS method with the skipping interval $[N/n]$ to the following combinations of N and n. In each case verify if $\pi_{ij} > 0$ for all pairs (i, j). $N = 5$, $n = 3$; $N = 11$, $n = 5$; $n = 11$, $n = 6$; $N = 13$, $n = 6$.

(b) Can you find a functional relation between N and n so that the CESS method with the skipping interval $[N/n]$ produces a sampling design with $\pi_{ij} > 0$ for all pairs (i, j)?

8.4. (a) Display the sampling design produced by the two-dimensional systematic selection rule in Example 8.2(i). For how many pairs of units, $\pi_{ij} = 0$?

(b) How many different values are possible for π_i's and π_{ij}'s in Example 8.2(ii)? List these values along with their frequencies.

8.5. For the systematic sampling design in Example 8.3, show that $\pi_i = 3/8$ and $\pi_{ij} = 1/8$ for $1 \le i \ne j \le 16$. Further, verify that this design is equivalent to an SRS(16, 6) design in the sense defined in Chapter 4.

8.6. In Example 8.4, can you suggest some values of the parameters k_1, k_2, k_3, and k_4 so that the resulting multiple distance systematic sampling design satisfies $0 < \pi_{ij} \le \pi_i \pi_j$ for all pairs (i, j)?

8.7. Establish the properties of π_i's and π_{ij}'s as claimed in the difference set approach for producing a systematic sampling design.

8.8. Find the integer x for which $\{0, 1, 3, x\}$ forms a $(13, 4, \lambda)$ difference set. Construct a systematic sampling design based on the difference set approach using the above difference set. Display the support of the design. If the distribution of probability over the support is uniform, what useful conclusions can be drawn?

8.9. In the method of composition let d_1 be a UNILESS design of size $n - n_0$ and d_2 be an SRS$(N - n + n_0, n_0)$ design. Given N and n, find the least value of n_0 for which the composite design d satisfies $\pi_{ij}(d) \le \pi_i(d)\pi_j(d)$ for all pairs (i, j). Verify your result for the combination $N = 180$ and $n = 14$.

8.10. With reference to Example 8.8, can you suggest a suitable choice of d^* different from an SRS(5, 2) design so that the overall sampling design d will satisfy $0 < \pi_{ij}(d) \le \pi_i(d)\pi_j(d)$ for all pairs (i, j)?

8.11. Consider the setup of Example 8.9. The values of the survey variable are shown below.

$$Y_1 = 26, \quad Y_2 = 97, \quad Y_3 = 36, \quad Y_4 = 73, \quad Y_5 = 85, \quad Y_6 = 63,$$

$$Y_7 = 80, \quad Y_8 = 94, \quad Y_9 = 49, \quad Y_{10} = 39, \quad Y_{11} = 74, \quad Y_{12} = 33,$$

$$Y_{13} = 58, \quad Y_{14} = 35, \quad Y_{15} = 40.$$

For each of the 9 samples in Table 8.4, compute the estimate of the population mean and its variance estimate. Verify directly their unbiasedness property.

8.12. In Example 8.10 compute the values of all first-order inclusion probabilities. Also compute the values of the second-order inclusion probabilities for pairs of units in each of the three types shown in (8.16).

8.13. Let $Y_i = a + bi$, $i = 1, 2, \ldots, N$; $N = nM$. Further let d_1: SRS(N, n), d_2: UNILESS, d_3: select one unit from the first set of M units with uniform selection probability. Continue this procedure for the second and the remaining sets of M units independently.

(i) Show that in all the cases, the sample mean $\bar{\mathbf{y}}$ is an unbiased estimator of the population mean $\bar{\mathbf{Y}}$.,

(ii) Show further that the sample mean has the variance

$$V_{d_1}(\bar{\mathbf{y}}) = b^2(N-n)(N+1)/12n \,,$$

$$V_{d_2}(\bar{\mathbf{y}}) = b^2(N-n)(N+n)/12n^2 \,,$$

$$V_{d_3}(\bar{\mathbf{y}}) = b^2(N-n)(N+n)/12n^3 \,,$$

so that $V_{d_3}(\bar{\mathbf{y}}) \le V_{d_2}(\bar{\mathbf{y}}) \le V_{d_1}(\bar{\mathbf{y}})$ with equality if and only if $n = 1$ or N.

8.14. Consider a population of size $N = nk + m$, $0 < m < k$. Recall that a 1 in k LESS design produces m samples of size $(n+1)$ each and the remaining $(k-m)$ samples of size n each.

(a) Suggest suitable selection probabilities for the starting units so that the sample mean $\bar{\mathbf{y}}(s)$ is an unbiased estimator of the population mean $\bar{\mathbf{Y}}$. Compute an expression for $V(\bar{\mathbf{y}}(s))$.

(b) Under the 1 in k UNILESS design, show that $\bar{\mathbf{y}}'(s) = k\, n(s)\bar{y}(s)/N$ is an unbiased estimator of $\bar{\mathbf{Y}}$. Compute an expression for $V(\bar{\mathbf{y}}'(s))$.

(c) Assuming a linear trend, compare the two strategies under (a) and (b) and interpret your result.

8.15. Using systematic sample data it is desired to estimate the population variance $\mathbf{S}_Y^2 = \Sigma_{i=1}^{N} (Y_i - \bar{\mathbf{Y}})^2/(N-1)$.

(a) Using the representation $\mathbf{S}_Y^2 = \Sigma\,\Sigma_{1 \le i < j \le N} (Y_i - Y_j)^2/N(N-1)$, find necessary and sufficient conditions on a sampling design which would provide an unbiased nonnegative estimator of $\mathbf{S}_Y^2$.

(b) For a population of 21 linearly ordered units, it is desired to estimate the population variance using a sample of size 5. Indicate in which of the following types of systematic sampling designs an unbiased nonnegative estimator of the variance is available.
   (i) Multiple Start Systematic Sampling design,
   (ii) Multiple Distance Systematic Sampling design,
   (iii) Composite Sampling design using systematic and simple random sampling designs.

(c) Assuming a linear trend, compare the available strategies in (b).

8.16. Suppose a population of size $N = 6$ is linearly ordered. Prove that for the purpose of estimating the population mean, the variance of the sample mean of a UNILESS design of size $n = 2$ can be smaller than the variance of the sample mean of a UNILESS design of size $n = 3$. Comment on the result.

8.17. It is required to estimate the mean waiting time for airline passengers to receive the baggage in an airport baggage area which operates 24 hours a day. There are 4 different exit gates for the passengers to disperse with baggage and these gates are used approximately in the proportionate frequencies of 40%, 40%, 10%, 10%. On a normal day, 4 independent starting times are selected. In the busy exit gates, 1 in 10 UNILESS of size 10 is adopted and in the other 2 exit gates, 1 in 5 UNILESS of size 5 is adopted. The data in Table 8.A relate to passengers' declarations regarding the actual waiting time to collect the baggage.

Estimate the mean waiting time for an airline passenger to collect the baggage. Provide an estimate of the variance of this estimator. Indicate all the assumptions you make.

Table 8.A. Actual Waiting Time (minutes) for Passengers to Collect Baggage From the Baggage Area

Strata	Time (min)
1	20, 25, 25, 28, 27, 32, 23, 20, 25, 28
2	20, 25, 28, 30, 29, 25, 23, 28, 24, 23
3	23, 25, 27, 30, 26
4	22, 24, 28, 25, 32

8.18. The public relations office of a leading food company has received many complaints regarding the can soups produced by the company. The complaints stress the issue that the cans contain less than 12 ounces of soup as claimed. In response to these complaints, the company decides to sample the cans produced on a randomly chosen day.

Suppose 1000 cans emerge out of the production line via a conveyor belt during a day. These cans are then packed into 100 boxes of 10 each. It is decided to inspect 50 cans only. The following sampling methods are suggested.

Method I. Adopt two-start 1 in 40 UNILESS design directly on the 1000 cans in the production line.

Method II. Choose 5 boxes using simple random sampling and completely enumerate each of them.

Method III. Choose 10 boxes using simple random sampling and then inspect 5 cans from each box, again selected independently using simple random sampling.

Method IV. Divide the day's product into two groups, 500 cans produced in each of the 2 shifts forming one group each. Next select 5 boxes from each group and then inspect 5 cans from each selected box, each time using simple random sampling.

Using the set of synthetic sample data in Tables 8.B and 8.C, estimate the average soup content per can. Also give an estimate of the standard error of your estimate. Make a relative comparison of the various methods of sampling.

8.19. The administration of a particular office building is interested in knowing the opinions (satisfactory or not) of its 2000 employees regarding a recently installed telephone system. It is decided to interview about 100 employees while they are leaving the building at

Table 8.B. Synthetic Data for Methods I and II

Method I	Starting soup can serial no.	7	11.7, 11.8, 11.7, 11.7, 11.8, 11.9, 11.8, 11.4, 11.8, 11.8 11.7, 11.6, 11.6, 11.6, 11.8, 11.9, 11.9, 11.4, 11.4, 11.2 11.3, 11.1, 11.5, 11.3, 11.2
	Starting soup can serial no.	23	12.1, 11.8, 12.1, 11.9, 11.7, 11.7, 11.8, 11.9, 11.9, 12.1 11.6, 11.9, 12.1, 11.8, 11.9, 11.8, 11.5, 11.7, 11.2, 11.6 11.3, 11.4, 11.5, 11.2, 11.4
Method II	Selected boxes serial no.	11	11.9, 11.7, 11.7, 11.9, 11.8, 12.1, 11.8, 11.6, 11.9, 11.7
		27	11.8, 11.9, 11.9, 11.8, 11.7, 12.1, 11.7, 11.9, 11.8, 11.8
		38	11.6, 11.8, 11.8, 11.7, 11.3, 11.5, 11.4, 11.6, 11.5, 11.2
		55	11.7, 11.8, 11.9, 11.5, 11.2, 11.4, 11.4, 11.6, 11.3, 11.4
		87	11.9, 11.7, 11.6, 11.8, 11.3, 11.5, 11.4, 11.2, 11.3, 11.4

Table 8.C. Synthetic Data for Methods III and IV

	Selected Boxes Serial No.									
	7	15	18	32	35	42	57	63	69	80
Method III	12.1	11.8	11.6	11.5	11.5	11.9	11.6	11.5	11.9	11.3
	12.0	11.9	11.8	11.8	11.9	11.8	11.9	11.6	11.7	11.4
	11.9	11.7	11.7	11.6	11.6	11.7	11.7	11.9	11.4	11.6
	11.8	11.8	11.9	11.9	11.8	11.8	11.7	11.4	11.3	11.4
	11.8	11.9	11.8	11.7	11.9	11.7	11.8	11.3	11.7	11.7
	Selected Boxes Serial No.									
	Morning Shift					Afternoon Shift				
	3	16	25	32	43	6	10	17	38	47
Method IV	12.1	11.7	11.6	11.6	11.5	11.8	11.8	11.7	11.5	11.2
	11.9	11.8	11.6	11.5	11.2	11.6	11.6	11.4	11.3	11.3
	11.8	11.9	11.8	11.8	11.6	11.4	11.7	11.5	11.2	11.5
	11.9	11.8	11.7	11.6	11.6	11.5	11.8	11.6	11.6	11.2
	11.8	11.7	11.7	11.8	11.5	11.7	11.9	11.5	11.8	11.4

the end of a particular working day. It is observed that there are two exit doors used equally frequently by the employees. A consulting statistician suggests that 1 in 20 UNILESS of 50 employees should be selected *independently* from each exit door and the data obtained by questioning them should be pooled to give an overall estimate of the population proportion ϕ of the employees showing satisfaction with the new telephone system.

(a) Let $\hat{\phi}_1$ and $\hat{\phi}_2$ be the two sample proportions of employees showing satisfaction over the new telephone system. The statistician suggests the following estimator of ϕ and its variance estimator.

$$\hat{\phi} = \frac{\hat{\phi}_1 + \hat{\phi}_2}{2}, \qquad \hat{V}(\hat{\phi}) = \frac{(\hat{\phi}_1 - \hat{\phi}_2)^2}{4}.$$

What assumptions does the statistician have in mind if he claims the above estimators to be unbiased?

(b) The survey resulted in the frequency data shown in Table 8.D.

Table 8.D

	Opinion about new telephone system	
Exit Doors	Satisfactory	Unsatisfactory
First Exit Door	35	15
Second Exit Door	38	12

Compute $\hat{\phi}$ and $\hat{V}(\hat{\phi})$. Where did you make explicit use of the information that $N = 2000$?

(c) Do you think that the assumptions you mention in (a) are realistic in this instance?

REFERENCES

Baumert, L. D. (1971). *Cyclic Difference Sets*. Lecture Notes in Mathematics, No. 182. Springer-Verlag, Berlin.

Bellhouse, D. R. (1988). Systematic sampling. In: *Handbook of Statistics, Vol. 6, Sampling* (Krishnaiah, P. R., and Rao, C. R., Eds.), 125–145. North-Holland, Amsterdam.

Cochran, W. G. (1977). *Sampling Techniques*. Third Edition. Wiley, New York.

Gautschi, W. (1957). Some remarks on systematic sampling. *Ann. Math. Statist.*, **28**, 385–394.

Hall Jr. M. (1986). *Combinatorial Theory*. Second Edition. Wiley, New York.

Hedayat, A., and Pesotan, H. (1990). On the coexistence of a family of survey sampling and block designs. Technical Report, Statistical Laboratory, University of Illinois, Chicago.

Iachan, R. (1982). Systematic sampling: a critical review. *Internat. Statist. Rev.*, **50**, 293–303.

Koop, J. C. (1967). Replicated (or interpenetrating) samples of unequal size. *Ann. Math. Statist.*, **38**, 1142–1147.

Koop, J. C. (1988). The technique of replicated or interpenetrating samples. In: *Handbook of Statistics, Vol. 6, Sampling* (Krishnaiah, P. R., and Rao, C. R., Eds.), 333–368. North-Holland, Amsterdam.

Levy, P. S. and Lemeshow, S. (1980). *Sampling for Health Professionals*. Lifetime Learning, Belmont, California.

Murthy, M. N., and Rao, T. J. (1988). Systematic sampling with illustrative examples. In: *Handbook of Statistics, Vol. 6, Sampling* (Krishnaiah, P. R., and Rao, C. R., Eds.), 147–185. North-Holland, Amsterdam.

Sengupta, S. (1982). On interpenetrating samples of unequal sizes. *Metrika*, **29**, 175–188.

Singh, D., and Singh, P. (1977). New systematic sampling. *J. Statist. Plann. Infer.*, **1**, 163–178.

Tornqvist, L. (1963). The theory of replicated systematic cluster sampling with random start. *Rev. Internat. Statist. Inst.*, **31**, 11–23.

Wu, C. F. (1984). Estimation in systematic sampling with supplementary observations. *Sankhyā*, **B46**, 306–315.

Zinger, A. (1963). Estimation de variances avec echantillonage systematique. *Revue de Statistique Appliquee*, **11**, 89–97.

Zinger, A. (1964). Systematic sampling in forestry. *Biometrics*, **20**, 553–565.

CHAPTER NINE

Stratified Sampling Designs

In this chapter we introduce stratified sampling designs and discuss some inference problems based on them. Next we pose the question of allocation of sample to the different strata and present some results. We also discuss other related issues.

9.1. THE CONCEPT OF STRATIFICATION IN SURVEY SAMPLING

Stratified sampling designs form an interesting and useful subclass of sampling designs. Such designs combine the advantages of purposive (non-probabilistic) and probability sampling together. To define a stratified sampling design, we divide the population $\mathcal{U}$ of N units arbitrarily or purposively into L (≥ 2) nonoverlapping subpopulations or strata $\mathcal{U}_1, \mathcal{U}_2, \ldots, \mathcal{U}_L$ with the hth subpopulation or stratum $\mathcal{U}_h$ having N_h (≥ 2) units of $\mathcal{U}$, $1 \leq h \leq L$, $\Sigma N_h = N$. Then we adopt a sampling design d_h on the stratum $\mathcal{U}_h$ for every h, $1 \leq h \leq L$. The sampling designs in the different strata are implemented independently of one another. Next we combine $d_1, d_2, \ldots, d_L$ into an overall sampling design d as follows.

$$P_d(s) = P_{d_1}(s_1)P_{d_2}(s_2)\cdots P_{d_L}(s_L) \tag{9.1}$$

for every sample s, which can be viewed as a combination of s_1 from $\mathcal{U}_1$, s_2 from $\mathcal{U}_2, \ldots, s_L$ from $\mathcal{U}_L$. A sampling design of the type (9.1) is said to define a stratified sampling design. To avoid trivialities, we have assumed $N_h \geq 2$, $1 \leq h \leq L$. The strata are generally formed purposively by utilizing available relevant auxiliary information. The choice of the sampling design in a given stratum can be decided exclusively based on the nature of information available in that stratum. Thus, for example, a suitable ΠPS, PPS, or a PPAS sampling design may be suggested if auxiliary size measures are available, or else, an SRS or a cluster sampling design or even a systematic sampling design might be adequate for a given stratum. The survey statistician is at liberty to choose different types of sampling designs, as demanded by the situations prevailing in the various strata.

For example, let $\mathscr{U} = \{1, 2, \ldots, 10\}$. We take $L = 3$ and decompose $\mathscr{U}$ into $\mathscr{U}_1 = \{1, 2, 3\}$, $\mathscr{U}_2 = \{4, 5, 6, 7, 8\}$, and $\mathscr{U}_3 = \{9, 10\}$. Let d_1, d_2 and d_3 be sampling designs on $\mathscr{U}_1$, $\mathscr{U}_2$, and $\mathscr{U}_3$, respectively, in the following forms:

$$d_1\colon\quad P_1((1,2)) = P_1((1,3)) = 0.5\ ;$$

$$d_2\colon\quad P_2((4,6,8)) = 0.3,\ P_2((4,5,7)) = 0.2,\ P_2(5,7)) = 0.5\ ;$$

$$d_3\colon\quad P_3((9)) = 0.3,\ P_3((10)) = 0.4,\ P_3((9,10)) = 0.2,\ P_3((10,9)) = 0.1\ .$$

There are two samples in d_1, three in d_2, and four in d_3. These will constitute 24 samples altogether in the overall design d. A typical sample will be (1, 2, 4, 6, 8, 9) with probability

$$P_d((1,2,4,6,8,9)) = P_1((1,2))P_2((4,6,8))P_3((9)) = 0.045\ .$$

Further, since $(1, 2, 4, 7)$ does *not* show up as a combination of samples from the component designs, we will have $P_d((1, 2, 4, 7)) = 0$.

It is clear that (9.1) defines a sampling design as we can immediately verify that $\Sigma_s P_d(s) = 1$ and every unit of the population belongs to some sample s via one of its components. We observe that the component sampling designs $d_1, d_2, \ldots, d_L$ can be completely arbitrary—this being a very useful feature of stratified sampling designs from practical considerations. We list below some of the advantages of using stratification.

(a) It allows allocation of the total sample over the different strata to be controlled and suitably restricted so that the practitioner can collect information on the survey variable in a convenient and efficient way.

(b) Different procedures of data collection in different parts (strata) of the population can be implemented.

(c) Auxiliary information on the units of the population can be profitably utilized at the stage of formation of the strata and also at subsequent stages of the survey.

(d) The individual frames of the various strata can be conveniently utilized, particularly when no combined frame is available or costly to prepare for the entire population.

9.2. INFERENCE BASED ON STRATIFIED SAMPLE DATA

The implementation of a stratified sampling design would amount to implementation of the component sampling designs in each of the L strata independently and, hence, it poses no new problem. As regards inference on the stratum totals (or means, proportions, variances) *separately*, it is also evident that we do not encounter any new problem. As before, we will denote the survey variable by Y, $T(\mathbf{Y})$, and $\bar{\mathbf{Y}}$ will represent the population total and the population mean respectively. Also $\mathbf{S}_Y^2$ will denote the population variance with divisor $(N-1)$. Let $T_h(\mathbf{Y})$, $\bar{\mathbf{Y}}_h$, and $\mathbf{S}_h^2$ denote the total, the mean, and the variance of the N_h values of the survey variable in the hth stratum, $1 \le h \le L$. The following relations may be easily verified.

$$T(\mathbf{Y}) = \sum_h T_h(\mathbf{Y}) ,$$

$$\bar{\mathbf{Y}} = \sum_h W_h \bar{\mathbf{Y}}_h , \qquad W_h = N_h/N , \qquad 1 \le h \le L ,$$

$$(N-1)\mathbf{S}_Y^2 = \sum_h (N_h - 1)\mathbf{S}_h^2 + \sum_h N_h(\bar{\mathbf{Y}}_h - \bar{\mathbf{Y}})^2 . \tag{9.2}$$

Denote the unbiased sample estimator of $\bar{\mathbf{Y}}_h$ based on data on the hth stratum by $\hat{\mathbf{Y}}_h$ with its unbiased variance estimator (assuming it exists) given by $\hat{V}(\hat{\mathbf{Y}}_h)$, $1 \le h \le L$. Then we have the following basic result regarding unbiased estimation of the overall population mean $\bar{\mathbf{Y}}$.

Theorem 9.1. *Using stratified sample data, an unbiased estimator of* $\bar{\mathbf{Y}}$ *is provided by*

$$\hat{\mathbf{Y}} = \sum_h W_h \hat{\mathbf{Y}}_h \tag{9.3}$$

and its variance is given by

$$V(\hat{\mathbf{Y}}) = \sum_h W_h^2 V(\hat{\mathbf{Y}}_h) . \tag{9.4}$$

Further, an unbiased estimate of $V(\hat{\mathbf{Y}})$ *is given by*

$$\hat{V}(\hat{\mathbf{Y}}) = \sum_h W_h^2 \hat{V}(\hat{\mathbf{Y}}_h) . \tag{9.5}$$

Proof. We have

$$E_p(\hat{\mathbf{Y}}) = E_p\left(\sum_h W_h \hat{\mathbf{Y}}_h\right) = \sum_h W_h E_{ph}(\hat{\mathbf{Y}}_h)$$
$$= \sum_h W_h \bar{\mathbf{Y}}_h = \bar{\mathbf{Y}} .$$

Further,

$$V_p(\hat{\mathbf{Y}}) = V_p\left(\sum_h W_h \hat{\mathbf{Y}}_h\right) = \sum_h W_h^2 V_{ph}(\hat{\mathbf{Y}}_h)$$

since the component sampling designs are independent as is evident from the product form of $P_d(s)$ in (9.1). It is obvious that $E_p(\hat{V}(\hat{\mathbf{Y}})) = V(\hat{\mathbf{Y}})$. □

In the above, we have presented a general result on the estimation of a finite population mean based on stratified sample data. We emphasize that any combination of sampling strategies discussed so far in Chapters Four through Eight can be used for the purpose of inference. We do not elaborate on individual sampling strategies.

We now specialize on the HTE and present specific results for the estimation of the overall population mean $\hat{\mathbf{Y}}$. Denote, as usual, by π_i and π_{ij} the first- and second-order inclusion probabilities for the unit i and the units i and j, respectively. Note that for any pair (i, j) with $i \in \mathcal{U}_h$, $j \in \mathcal{U}_{h'}$, for $1 \leq h \neq h' \leq L$, $\pi_{ij} = \pi_i \pi_j$. Assume further that $\pi_{ij} > 0$ for any pair (i, j) where $i, j \in \mathcal{U}_h$, $1 \leq h \leq L$. Using the results for the HTE from Chapter Three, we now deduce the following.

Corollary 9.2. *Based on the HTE,*

$$\hat{\mathbf{Y}} = N^{-1} \sum_{h=1}^{L} \left(\sum_{i \in s_h} Y_i / \pi_i \right) \tag{9.6}$$

along with

$$V(\hat{\mathbf{Y}}) = N^{-2} \left[\sum_{h=1}^{L} \left\{ \sum_{i \in \mathcal{U}_h} Y_i^2 \left(\frac{1}{\pi_i} - 1 \right) \right\} + \sum_{h=1}^{L} \left\{ \sum_{\substack{i \neq j \\ i,j \in \mathcal{U}_h}} \sum Y_i Y_j \left(\frac{\pi_{ij}}{\pi_i \pi_j} - 1 \right) \right\} \right] \tag{9.7}$$

and

$$\hat{V}(\hat{\mathbf{Y}}) = N^{-2} \left[\sum_{h=1}^{L} \left\{ \sum_{i \in s_h} \frac{Y_i^2}{\pi_i} \left(\frac{1}{\pi_i} - 1 \right) \right\} + \sum_{h=1}^{L} \left\{ \sum_{\substack{i \neq j \\ i,j \in s_h}} \sum \frac{Y_i Y_j}{\pi_{ij}} \left(\frac{\pi_{ij}}{\pi_i \pi_j} - 1 \right) \right\} \right]. \tag{9.8}$$

In particular, when every component sampling design is of fixed size, we may rewrite (9.7) as

$$V(\hat{\mathbf{Y}}) = N^{-2} \left[\sum_{h=1}^{L} \left\{ \sum_{\substack{i < j \\ i,j \in \mathcal{U}_h}} \sum \left(\frac{Y_i}{\pi_i} - \frac{Y_j}{\pi_j} \right)^2 (\pi_i \pi_j - \pi_{ij}) \right\} \right] \tag{9.9}$$

and use (9.10) instead of (9.8) where (9.10) is given by

$$\hat{V}(\hat{\mathbf{Y}}) = N^{-2} \left[\sum_{h=1}^{L} \left\{ \sum_{\substack{i < j \\ i,j \in s_h}} \sum \left(\frac{Y_i}{\pi_i} - \frac{Y_j}{\pi_j} \right)^2 \left(\frac{\pi_i \pi_j}{\pi_{ij}} - 1 \right) \right\} \right]. \tag{9.10}$$

In Sections 9.3–9.7, we confine our attention to an important subclass of stratified sampling designs, namely stratified simple random sampling designs.

9.3. STRATIFIED SIMPLE RANDOM SAMPLING (STSRS) DESIGNS

When each component design in a stratified design is a simple random sampling design, we call the overall sampling design a stratified simple random sampling (STSRS) design. Generally, in practice, stratified sampling designs are composed of simple random sampling designs. A precise definition of an STSRS design may then be given as follows.

Definition 9.1. A stratified simple random sampling design with parameters N, L, $N_1, N_2, \ldots, N_L$, n, $n_1, n_2, \ldots, n_L$ satisfying

$$L \geq 2\,, \quad N_h \geq 2\,, \quad \sum N_h = N\,, \quad n \geq L\,, \quad 1 \leq n_h \leq N_h\,, \quad \sum n_h = n$$

is (in its reduced form) given by

$$P_d(s) = \prod \binom{N_h}{n_h}^{-1}$$

for *every* s which consists of n_1 units out of the first set of N_1 units, n_2 units out of the next set of N_2 units, $\ldots$, n_L units out of the last set of N_L units of the population.

In the above, we have assumed that the totality of N population units is divided into L strata $\{1, 2, \ldots, N_1\}$, $\{N_1 + 1, \ N_1 + 2, \ldots, N_1 + N_2\}, \ldots, \{N - N_L + 1, \ N - N_L + 2, \ldots, N\}$. We will denote an STSRS design with the above parameters as STSRS(N, n, L, $\{N_h, n_h\}$, $1 \leq h \leq L$) design. We note that such a design is naturally a fixed-size (n) sampling design.

The implementation of an STSRS design would amount to implementation of simple random sampling designs independently on each of the L strata and, hence, it poses no new problem. The determination of the first- and second-order inclusion probabilities is also quite simple. We obtain

(i) $\pi_i = (n_h/N_h)$ if unit i is in the hth stratum;

(ii) $\pi_{ij} = \{n_h(n_h - 1)/N_h(N_h - 1)\}$ if units i and j are in the hth stratum
$= (n_h n_s/N_h N_s)$ if one of the units i and j is in the hth stratum and the other is in the sth stratum.

Clearly, all π_{ij}'s are positive whenever every $n_h \geq 2$. Moreover, $\pi_i \pi_j \geq \pi_{ij}$ for *all pairs* (i, j). As the sampling design is a fixed-size (n) design, this would facilitate nonnegative variance estimation if we would decide to use the HTE for estimation of the overall population total or mean. This will be taken up in detail next.

Denote the sample mean and the sample variance by $\bar{\mathbf{y}}_h$ and $\mathbf{s}_h^2$, respectively. These are based on n_h values of survey variable available in a random sample on the hth stratum, $1 \le h \le L$. Using the HTE, we obtain the following results.

Theorem 9.3. *Using STSRS data and the HTE,*

$$\hat{\bar{\mathbf{Y}}} = \sum W_h \bar{\mathbf{y}}_h \tag{9.11}$$

$$V(\hat{\bar{\mathbf{Y}}}) = \sum W_h^2 (n_h^{-1} - N_h^{-1}) \mathbf{S}_h^2 \tag{9.12}$$

$$\hat{V}(\hat{\bar{\mathbf{Y}}}) = \sum W_h^2 (n_h^{-1} - N_h^{-1}) \mathbf{s}_h^2 , \qquad \textit{assuming } n_h \ge 2, \textit{ for every } h . \tag{9.13}$$

Proof. We have, using (9.6),

$$\hat{\bar{\mathbf{Y}}} = N^{-1} \sum_{h=1}^{L} \left(\sum_{i \in s_h} Y_i/\pi_i \right) = N^{-1} \sum_{h=1}^{L} N_h \bar{\mathbf{y}}_h = \sum_{h=1}^{L} W_h \bar{\mathbf{y}}_h .$$

Next, using (9.9),

$$N^2 V(\hat{\bar{\mathbf{Y}}}) = \sum_h \sum_{\substack{i<j \\ i,j \in \mathcal{U}_h}} \sum (Y_i/\pi_i - Y_j/\pi_j)^2 (\pi_i \pi_j - \pi_{ij})$$

which simplifies to

$$\sum_h N_h^2 n_h^{-2} \{ n_h^2 N_h^{-2} - n_h(n_h - 1) N_h^{-1} (N_h - 1)^{-1} \} \sum_{\substack{i<j \\ i,j \in \mathcal{U}_h}} \sum (Y_i - Y_j)^2$$

$$= \sum_h \{ 1 - (n_h - 1) N_h n_h^{-1} (N_h - 1)^{-1} \} \sum_{\substack{i<j \\ i,j \in \mathcal{U}_h}} \sum (Y_i - Y_j)^2 .$$

Now we use the identity $\sum \sum_{i<j, i,j \in \mathcal{U}_h} (Y_i - Y_j)^2 = N_h (N_h - 1) \mathbf{S}_h^2$ and simplify the above expression to

$$\sum N_h (N_h - n_h) n_h^{-1} \mathbf{S}_h^2 = \sum N_h^2 (n_h^{-1} - N_h^{-1}) \mathbf{S}_h^2 .$$

Thus, finally,

$$V(\hat{\bar{\mathbf{Y}}}) = \sum W_h^2 (n_h^{-1} - N_h^{-1}) \mathbf{S}_h^2 .$$

Similarly, using (9.10), we can show that

$$N^2\hat{V}(\hat{\bar{\mathbf{Y}}}) = \sum N_h^2(n_h^{-1} - N_h^{-1})\mathbf{s}_h^2$$

from which the expression for $\hat{V}(\hat{\bar{\mathbf{Y}}})$ in (9.13) follows. □

The standard error (s.e.) of the estimate of $\bar{\mathbf{Y}}$ is estimated by the square root of $\hat{V}(\hat{\bar{\mathbf{Y}}})$.

Remark 9.1. Since the STSRS design is a fixed-size (n) sampling design, we may also derive the preceding results by application of (3.1), (3.4), and (3.16).

Example 9.1. A certain county is served by 120 food stores. Suppose we are interested in estimating the average annual sales in these stores by collecting data on 20 of them. It is known that several food stores have chains in the region and the details of such information are provided in Table 9.1.

Table 9.1

Brand Food Store	No. of Centers
A	50
B	30
C	20
D	10
E	6
Others	4
Total	120

It is generally felt that the sales in stores under the same chain of a given brand are more or less comparable while between stores of different brands the sales would usually vary substantially. In such a situation, as will be shown later, stratified sampling with the brands as strata seems appropriate. Before proceeding further we note that the last two stratum sizes are small compared to the rest. We may, therefore, combine them to form a single stratum of size $6 + 4 = 10$. In effect, we then have five strata. Suppose we decide to use an allocation of the sample sizes in the five strata as 8, 5, 3, 2, 2, respectively. We would, therefore, select 8 stores out of 50 of brand A using simple random sampling and collect information in respect of the survey variable Y: annual sales (unit \$1000, say). A similar sampling technique would apply to other strata as well. In Table 9.2, we show the raw data on annual sales collected from the sampled stores.

Table 9.2

Stratum No.	Annual Sales (in $1000) in Sampled Stores
1	250, 330, 210, 280, 380, 190, 220, 305
2	420, 380, 470, 520, 440
3	200, 190, 160
4	100, 78
5	70, 58

We will estimate the population mean along with its s.e. using formulae (9.11) and (9.13). From the raw data, we compute the values of the relevant statistics, those being the sample means and the sample variances.

Means ($\bar{y}_h$)	270.625	446.000	183.333	89.000	64.000
Variances (s_h^2)	4274.5536	2780.0000	433.3333	242.0000	72.0000

Further $W_1 = 0.4167$, $W_2 = 0.2500$, $W_3 = 0.1667$, $W_4 = 0.0833$, $W_5 = 0.0833$. These yield

(i) $\hat{\bar{Y}} = 267.5759$
(ii) $\hat{V}(\hat{\bar{Y}}) = 111.1789$
(iii) estimated s.e. = 10.5441.

Next, suppose we have to estimate a population proportion ϕ with respect to a qualitative characteristic of interest using stratified simple random sampling data. We start with the relation $\phi = \Sigma\, W_h \phi_h$ where ϕ_h refers to the proportion in the hth stratum. Then we have the following results involving the use of the HTE.

(i) $\hat{\phi} = \Sigma\, W_h \hat{\phi}_h$
(ii) $V(\hat{\phi}) = \Sigma\, W_h^2 \{N_h \phi_h (1 - \phi_h)/(N_h - 1)\}(n_h^{-1} - N_h^{-1})$
(iii) $\hat{V}(\hat{\phi}) = \Sigma\, W_h^2 \{n_h \hat{\phi}_h (1 - \hat{\phi}_h)/(n_h - 1)\}(n_h^{-1} - N_h^{-1})$
where $\hat{\phi}_h$ = sample proportion in the hth stratum, $1 \le h \le L$.

Example 9.2. It is desired to estimate the proportion of foreign students in a certain university. The university offers degrees in six major areas such as (1) science, (2) engineering, (3) medicine, (4) business administration, (5) arts and literature, and (6) architecture and urban planning. In each area, courses are offered leading to B.S., M.S., and Ph.D. degrees. Table 9.3 summarizes information regarding some aspects of the student population as a whole as well as of simple random samples drawn from various cross-classified strata. The reference year is 1988.

Table 9.3

Degrees	Areas of Study						Total
	(1)	(2)	(3)	(4)	(5)	(6)	
B.S.	3000 (100, 70)	7000 (200, 120)	4000 (100, 40)	3000 (100, 25)	1000 (50, 10)	2000 (70, 20)	20,000 (620, 285)
M.S.	1000 (50, 30)	500 (50, 20)	700 (50, 10)	300 (30, 6)	200 (40, 6)	300 (30, 8)	3,000 (250, 80)
Ph.D.	300 (30, 16)	60 (10, 4)	40 (10, 3)	40 (10, 3)	20 (5, 1)	40 (10, 3)	500 (75, 30)
Total	4300 (180, 116)	7560 (260, 144)	4740 (160, 53)	3340 (140, 34)	1220 (95, 17)	2340 (110, 31)	23500 (945, 395)

In Table 9.3 top figures correspond to the number of students and the bottom figures correspond to the pair: (no. of students randomly selected, no. of foreign students in the random sample).

In the area of science, for example, an estimate of the proportion of foreign students (in 1988) and its estimated s.e. are obtained from formulae (i) and (iii) above by utilizing the first column of Table 9.3. Computations yield $W_1 = 0.70$, $W_2 = 0.23$, $W_3 = 0.07$, $\hat{\phi}_1 = 0.70$, $\hat{\phi}_2 = 0.60$, $\hat{\phi}_3 = 0.53$, estimated proportion $\hat{\phi} = 0.6651$, estimated s.e. $= 0.0366$. In Exercise 9.6, the reader is asked to carry out further computations to estimate other parameters of interest.

Remark 9.2. All the above formulae regarding variance estimation are based on the assumption that the sample size n is at least $2L$ and, moreover, every n_h is at least 2. With a moderately large value of L, this may be prohibitive from cost considerations. In that case, the number of strata should be reduced by combining the *nearby* ones.

9.4. COMPARISON WITH SIMPLE RANDOM SAMPLING

We might wonder when we should advocate stratified simple random sampling and with what benefit over unstratified simple random sampling. In practice, stratified sampling is often suggested from practical considerations which include geographical locations, administrative convenience and other factors. It is often said that stratified random sampling combines the advantages of purposive (leading to stratification) and random sampling together. We refer to Cochran (1977) for a discussion on this matter. Indeed, stratification leads to gain in precision over unstratified random sampling in *most* situations. This aspect will be studied in detail below.

Considering an SRS(N, n) design, the unstratified simple random sample mean $\bar{\mathbf{y}}(s)$ is an unbiased estimator of the population mean with variance, denoted as V_{ran}, given by

$$V_{\text{ran}} = (n^{-1} - N^{-1})\mathbf{S}_Y^2 . \tag{9.14}$$

Again a stratified simple random sample of the same overall size n would provide an estimator of the form (9.11) with variance, denoted as V_{stsrs}, given by

$$V_{\text{stsrs}} = \sum W_h^2(n_h^{-1} - N_h^{-1})\mathbf{S}_h^2 = \sum (W_h^2\mathbf{S}_h^2/n_h) - N^{-1}\sum W_h\mathbf{S}_h^2 . \tag{9.15}$$

We now recall the identity

$$(N-1)\mathbf{S}_Y^2 = \sum (N_h - 1)\mathbf{S}_h^2 + \sum N_h(\bar{\mathbf{Y}}_h - \bar{\mathbf{Y}})^2$$

so that using the approximations

$$(N_h - 1)/(N-1) \doteq W_h \qquad \text{for every } h$$

we may write

$$\mathbf{S}_Y^2 \doteq \sum W_h\mathbf{S}_h^2 + \sum W_h(\bar{Y}_h - \bar{Y})^2 . \tag{9.16}$$

Thus,

$$\begin{aligned} V_{\text{ran}} - V_{\text{stsrs}} &\doteq \left(n^{-1}\sum W_h\mathbf{S}_h^2 - \sum n_h^{-1}W_h^2\mathbf{S}_h^2\right) + n^{-1}\sum W_h(\bar{\mathbf{Y}}_h - \bar{\mathbf{Y}})^2 \\ &\quad + N^{-1}\left(\sum W_h\mathbf{S}_h^2 - \mathbf{S}_Y^2\right) \\ &= (n^{-1} - N^{-1})\sum W_h(\bar{\mathbf{Y}}_h - \bar{\mathbf{Y}})^2 \\ &\quad + \left\{\left(n^{-1}\sum W_h\mathbf{S}_h^2 - \left(\sum W_h^2\mathbf{S}_h^2/n_h\right)\right\}. \end{aligned}$$

The first term above is nonnegative. Also the second term may be made nonnegative for some choice of n_h's, for example, $n_h = nW_h$, $1 \le h \le L$. Thus, it is often possible to suggest a *suitable* stratified simple random sampling design providing a better (in the sense of less variance) estimator of the population mean than unstratified simple random sampling. In the above, we chose to use $n_h = nW_h$, $1 \le h \le L$. This is known as *proportional allocation* first suggested by Bowley (1926). Under proportional allocation, the variance of the estimate of the population mean, denoted by V_{prop}, is given by

$$V_{\text{prop}} = (n^{-1} - N^{-1})\sum W_h\mathbf{S}_h^2 . \tag{9.17}$$

Once more, we observe that under the approximations

$$(N_h - 1) \doteq W_h(N-1),\ N_h \doteq W_h(N-1)\,, \qquad 1 \le h \le L\,,$$

$$V_{\text{ran}} - V_{\text{prop}} \doteq (n^{-1} - N^{-1}) \sum W_h(\bar{\mathbf{Y}}_h - \bar{\mathbf{Y}})^2\,. \tag{9.18}$$

This is approximately the gain in using stratified simple random sampling design with proportional allocation compared to simple random sampling design of the same size. It is effectively a factor of $\Sigma\, W_h(\bar{\mathbf{Y}}_h - \bar{\mathbf{Y}})^2$ which is a measure of the extent of stratum means variations. We can interpret this result in two ways. First, if the stratification has already been done, then this measures the extent to which proportional allocation is better than overall simple random sampling. Secondly, if stratification is yet to be implemented, this suggests a basis for stratification in that we might try to put units of seemingly homogeneous nature (with respect to the survey variable or with respect to an auxiliary variable which might be highly correlated with the survey variable) in one stratum and those of different nature in a different stratum with the idea of maximizing our gain in using proportional allocation for stratified sampling.

We remark that under proportional allocation, (a) $\pi_i = n/N$, $1 \le i \le N$, (b) $\hat{\bar{\mathbf{Y}}}_{\text{prop}}$ reduces to $\bar{\mathbf{y}}$ (sample mean), and (c) the computation of $\hat{V}_{\text{prop}}$ does not need direct knowledge of N_h's. These facts may be useful in advocating the use of proportional allocation.

The above comparison of V_{ran} and V_{prop} was based on the approximations $(N_h - 1)/(N-1) \doteq W_h$, $N_h/(N-1) \doteq W_h$, $1 \le h \le L$. For small populations this may *not* hold and, accordingly, the comparison may lead to superiority of unstratified simple random sampling over stratified simple random sampling with proportional allocation. See Exercise 9.9 in this context.

9.5. ESTIMATION OF THE GAIN DUE TO STRATIFICATION

Another related practical problem is to estimate the gain, if any, from the sample data arising out of implementation of stratified simple random sampling. We will formulate the general problem as follows. Suppose we have available stratified sample data. We know that stratification has possibly led to a gain in precision in the sense of a reduction in variance compared to unstratified simple random sampling. The amount of reduction, under proportional allocation, is given by the approximation in (9.18). The exact expression is, of course, to be obtained by taking the difference of the expressions in (9.14) and (9.15) for an arbitrary given stratified simple random sampling design. We want to estimate the gain from the sample data. Towards this end, we observe that the difference of two quadratic forms is again a quadratic form and a quadratic form in $Y_1, Y_2, \ldots, Y_N$ is estimable under stratified simple random sampling design whenever every $n_h \ge 2$ (as this ensures $\pi_{ij} > 0$ for *all* pairs (i, j) for such a design). In actual estimation, however, we note that the additional task is to estimate the population variance $\mathbf{S}_Y^2$ using stratified sample data so that (9.14) is taken care of. This is because the other part (9.15), namely

$$V_{\text{stsrs}} = \sum W_h^2(n_h^{-1} - N_h^{-1})\mathbf{S}_h^2$$

is already estimated by

$$\hat{V}_{\text{stsrs}} = \sum W_h^2(n_h^{-1} - N_h^{-1})\mathbf{s}_h^2 .$$

All throughout, we assume $n_h \geq 2$ for every h, $1 \leq h \leq L$. Below we suggest an estimator of $\mathbf{S}_Y^2$. Once more we recall the identity

$$(N-1)\mathbf{S}_Y^2 = \sum (N_h - 1)\mathbf{S}_h^2 + \sum N_h(\bar{\mathbf{Y}}_h - \bar{\mathbf{Y}})^2 \tag{9.19}$$

and conclude that an estimator of $\mathbf{S}_Y^2$ will be obtained as soon as we can derive estimators for $(\bar{\mathbf{Y}}_h - \bar{\mathbf{Y}})^2$ for every h, $1 \leq h \leq L$. This is because $\hat{\mathbf{S}}_h^2 = \mathbf{s}_h^2$, $1 \leq h \leq L$. As regards $(\bar{\mathbf{Y}}_h - \bar{\mathbf{Y}})^2$, we start with

$$(\bar{\mathbf{y}}_h - \hat{\bar{\mathbf{Y}}})^2 = \{(\bar{\mathbf{y}}_h - \bar{\mathbf{Y}}_h) + (\bar{\mathbf{Y}}_h - \bar{\mathbf{Y}}) + (\bar{\mathbf{Y}} - \hat{\bar{\mathbf{Y}}})\}^2$$

and observe that

$$\begin{aligned}
E(\bar{\mathbf{y}}_h - \hat{\bar{\mathbf{Y}}})^2 &= E(\bar{\mathbf{y}}_h - \bar{\mathbf{Y}}_h)^2 + (\bar{\mathbf{Y}}_h - \bar{\mathbf{Y}})^2 + E(\hat{\bar{\mathbf{Y}}} - \bar{\mathbf{Y}})^2 \\
&\quad - 2E\{(\bar{\mathbf{y}}_h - \bar{\mathbf{Y}}_h)(\hat{\bar{\mathbf{Y}}} - \bar{\mathbf{Y}})\} \\
&= (n_h^{-1} - N_h^{-1})\mathbf{S}_h^2 + (\bar{\mathbf{Y}}_h - \bar{\mathbf{Y}})^2 + V_{\text{stsrs}} \\
&\quad - 2E\left\{(\bar{\mathbf{y}}_h - \bar{\mathbf{Y}}_h)\left(\sum W_k(\bar{\mathbf{y}}_k - \bar{\mathbf{Y}}_k)\right)\right\} \\
&= (n_h^{-1} - N_h^{-1})\mathbf{S}_h^2(1 - 2W_h) + (\bar{\mathbf{Y}}_h - \bar{\mathbf{Y}})^2 + V_{\text{stsrs}}
\end{aligned}$$

since $E(\bar{\mathbf{y}}_h - \bar{\mathbf{Y}}_h)(\bar{\mathbf{y}}_{h'} - \bar{\mathbf{Y}}_{h'}) = 0$ for all $h \neq h'$. Thus, $(\bar{\mathbf{Y}}_h - \bar{\mathbf{Y}})^2$ can be estimated by

$$(\bar{\mathbf{y}}_h - \hat{\bar{\mathbf{Y}}})^2 + (2W_h - 1)(n_h^{-1} - N_h^{-1})\mathbf{s}_h^2 - \hat{V}_{\text{stsrs}} . \tag{9.20}$$

Now we can compute an estimator of $\mathbf{S}_Y^2$ using (9.20) in (9.19). The final expression is obtained from

$$\begin{aligned}
(N-1)\hat{\mathbf{S}}_Y^2 &= \sum (N_h - 1)\mathbf{s}_h^2 \\
&\quad + \sum N_h\{(\bar{\mathbf{y}}_h - \hat{\bar{\mathbf{Y}}})^2 + (2W_h - 1)(n_h^{-1} - N_h^{-1})\mathbf{s}_h^2 - \hat{V}_{\text{stsrs}}\}
\end{aligned}$$

as

$$\begin{aligned}
\hat{\mathbf{S}}_Y^2 = (N-1)^{-1}\Big[&\sum (N_h - 1)\mathbf{s}_h^2 + \sum N_h(2W_h - 1)(n_h^{-1} - N_h^{-1})\mathbf{s}_h^2 \\
&+ \sum N_h(\bar{\mathbf{y}}_h - \hat{\bar{\mathbf{Y}}})^2 - N\hat{V}_{\text{stsrs}}\Big],
\end{aligned} \tag{9.21}$$

where $\hat{V}_{\text{stsrs}} = \Sigma\, W_h^2(n_h^{-1} - N_h^{-1})\mathbf{s}_h^2$ and $\hat{\bar{\mathbf{Y}}} = \Sigma\, W_h\bar{\mathbf{y}}_h$.

Example 9.3. Consider the annual sales estimation problem of Example 9.1. Suppose we had decided to use unstratified simple random sampling of 20 stores. How much would have been the loss of precision in that case? Equivalently, we want to estimate the gain in precision due to use of stratified simple random sampling and that too, using stratified sample data. For this purpose, we use the relevant data from Example 9.1 and compute, using (9.21),

$$\hat{\mathbf{S}}_Y^2 = 17690.789 .$$

This yields $\hat{V}_{\text{ran}} = 737.1162$. Further, we already computed $\hat{V}_{\text{stsrs}} = 111.1789$. Therefore, estimated gain in precision $= 625.9373$ and estimated relative gain $= \{(\hat{V}_{\text{ran}} - \hat{V}_{\text{stsrs}})/\hat{V}_{\text{stsrs}}\} \times 100\% = 563\%$.

9.6. ALLOCATION OF SAMPLES IN DIFFERENT STRATA

So far, we have tacitly assumed that the sample sizes $n_1, n_2, \ldots, n_L$ are arbitrarily chosen and allotted to the different strata. However, the expression for V_{stsrs} in (9.15) clearly shows that the sample sizes have a direct role in determining the accuracy of an estimator based on stratified sample data. A *good* choice of the n_h's depends on a suitably defined objective function. These have been discussed in a study by Neyman (1934) and it combines the precision and cost aspects together.

In simple setups, the total cost of a survey may be assumed to comprise of two components: (i) an overhead cost C_0, and (ii) the cost of data collection and analysis. We may assume that the cost per unit of the second component is c_h for a unit of the hth stratum, $1 \le h \le L$. Thus, we take a linear cost structure: $C = C_0 + \Sigma\, c_h n_h$. Subject to a budgetary constraint C^*, we are supposed to make a choice of the n_h's say $n_1', n_2', \ldots, n_L'$, so that $\Sigma_{h=1}^{L} c_h n_h' \le C^* - C_0$ (fixed) and V_{stsrs} as given in (9.15) is a minimum for the choice $n_1', n_2', \ldots, n_L'$ compared to any other choice, say $n_1'', n_2'', \ldots, n_L''$ which also satisfies $\Sigma\, c_h n_h'' \le C^* - C_0$ (fixed). However, it is readily verified that V_{stsrs} is strictly decreasing in each n_h, $1 \le h \le L$. Hence, the choice of n_h''s must be such that $\Sigma_{h=1}^{L} c_h n_h' = C^* - C_0$. Thus we will have, for $\Sigma\, c_h n_h'' \le C^* - C_0 = \Sigma\, c_h n_h'$,

$$\sum n_h'^{-1} W_h^2 \mathbf{S}_h^2 \le \sum n_h''^{-1} W_h^2 \mathbf{S}_h^2 .$$

Next observe that by Cauchy–Schwartz inequality,

$$\left(\sum c_h n_h''\right)\left(\sum n_h''^{-1} W_h^2 \mathbf{S}_h^2\right) \ge \left(\sum W_h \mathbf{S}_h \sqrt{c_h}\right)^2$$

so that

$$\left(\sum n_h''^{-1} W_h^2 \mathbf{S}_h^2\right) \ge \frac{(\Sigma\, W_h \mathbf{S}_h \sqrt{c_h})^2}{(\Sigma\, c_h n_h'')} \ge \frac{(\Sigma\, W_h \mathbf{S}_h \sqrt{c_h})^2}{(C^* - C_0)} .$$

However, the choice

$$n'_h = AW_h\mathbf{S}_h/\sqrt{c_h}\,, \qquad 1 \le h \le L \tag{9.22}$$

where A is to be determined from the condition $\Sigma\, c_h n'_h = C^* - C_0$, leads to

$$\sum n_h'^{-1} W_h^2 \mathbf{S}_h^2 = \frac{(\Sigma\, W_h\mathbf{S}_h\sqrt{c_h})^2}{(C^* - C_0)}\,.$$

The optimum sample sizes, denoted as $n_{h\,\text{opt}}$ for the hth stratum, are thus determined as

$$n_{h\,\text{opt}} = (C^* - C_0)(W_h\mathbf{S}_h/\sqrt{c_h})\left(\sum W_k\mathbf{S}_k\sqrt{c_k}\right)^{-1}, \qquad 1 \le h \le L \tag{9.23}$$

for a given budget C^*. Similarly, for a given limit V_0 to the value of V_{stsrs} an optimum choice of the sample sizes minimizing the total cost can be determined along the same line and is given by

$$n_{h\,\text{opt}} = (W_h\mathbf{S}_h/\sqrt{c_h})\left(\sum W_k\mathbf{S}_k\sqrt{c_k}\right)\left(V_0 + \sum W_k^2\mathbf{S}_k^2 N_k^{-1}\right)^{-1}, \qquad 1 \le h \le L\,. \tag{9.24}$$

The total sample size n_{opt} in the two cases is given by

$$n_{\text{opt}} = (C^* - C_0)\left(\sum W_h\mathbf{S}_h/\sqrt{c_h}\right)\left(\sum W_h\mathbf{S}_h\sqrt{c_h}\right)^{-1}$$

and

$$n_{\text{opt}} = \left(\sum W_h\mathbf{S}_h/\sqrt{c_h}\right)\left(\sum W_h\mathbf{S}_h\sqrt{c_h}\right)\left(V_0 + \sum W_h^2\mathbf{S}_h^2 N_h^{-1}\right)^{-1},$$

respectively.

Example 9.4. In a socioeconomic survey of the households of a certain region, it is proposed to undertake a stratified simple random sampling design using the counties in the region as strata. The relevant data are given in Table 9.4.

It is known that the s.d. of annual income per household in the region as a whole is 3.5 (in \$1000). The problem is to determine optimum sample sizes for stratified simple random sampling that would minimize the sampling variance of the estimate of the overall population mean for a given total budget of \$50,000. We assume that the overhead cost is \$5000, and the cost of data collection per household is \$30 for the first county and \$40, \$50, \$60, and \$70 for the subsequent counties in that order.

st observe that V_{stsrs} for two strata can be expressed as a only, namely $V_{\text{stsrs}} = (A/n_1) + \{B/(n-n_1)\} - C$ where A, B, ependent of n_1. This is a convex function of n_1 over the set of umbers. We readily verify that the above expression attains its ue when $n_1' = n/(1+\sqrt{(B/A)})$. If $n_1' > N_1$, V_{stsrs} will be a $n_1 = N_1$ over all positive real numbers not exceeding N_1. This proof. □

e problem of over-sampling will not occur if $N_h > n_h$ for all h, hen it does happen, however, it is generally solved by adhering 100% sampling as described above. It is not hard to defend r all circumstances. Below we provide a proof for the case of specifically, we prove the following result.

.5. *Suppose, for a given* n, $nW_1\mathbf{S}_1 > N_1(\Sigma_{h=1}^{3} W_h\mathbf{S}_h)$ *while*

$$_2 < N_2(W_2\mathbf{S}_2 + W_3\mathbf{S}_3)\,, \qquad (n-N_1)W_3\mathbf{S}_3 < N_3(W_2\mathbf{S}_2 + W_3\mathbf{S}_3)\,.$$

cation

$$_1\,, \qquad n_h^* = (n-N_1)W_h\mathbf{S}_h(W_2\mathbf{S}_2 + W_3\mathbf{S}_3)^{-1}\,, \qquad h = 2, 3$$

nong all possible allocations.

Ve show that

$$\sum W_h^2\mathbf{S}_h^2 n_h^{-1} \geq \sum W_h^2\mathbf{S}_h^2 n_h^{*-1}$$

of integers n_1, n_2, n_3 satisfying $n_1 + n_2 + n_3 = n$ and $n_h \leq N_h$, The left hand side of the above inequality exceeds

$$W_1^2\mathbf{S}_1^2 n_1^{-1} + (W_2\mathbf{S}_2 + W_3\mathbf{S}_3)^2(n-n_1)^{-1}$$

t hand side simplifies to

$$W_1^2\mathbf{S}_1^2 N_1^{-1} + (W_2\mathbf{S}_2 + W_3\mathbf{S}_3)^2(n-N_1)^{-1}\,.$$

it is enough to show that

$$_1^2(n_1^{-1} - N_1^{-1}) \geq (W_2\mathbf{S}_2 + W_3\mathbf{S}_3)^2\{(n-N_1)^{-1} - (n-n_1)^{-1}\}$$

uivalent to

$$W_1^2\mathbf{S}_1^2 N_1^{-1} n_1^{-1} \geq (W_2\mathbf{S}_2 + W_3\mathbf{S}_3)^2(n-N_1)^{-1}(n-n_1)^{-1}\,. \qquad (9.30)$$

Table 9.4

County No.	Total No. of Zones	Average No. of Households per Zone	S.D. of Annual Income per Household per Zone (in $1000)
1	19	487	2.3
2	16	829	3.7
3	13	822	4.2
4	11	1083	2.7
5	5	1956	0.8

In our notation, we have $C^* = \$50{,}000$ and $C_0 = \$5000$, $c_1 = \$30$, $c_2 = \$40$, $c_3 = \$50$, $c_4 = \$60$, and $c_5 = \$70$. Further, from Table 9.4, $N_1 = 9253$, $N_2 = 13264$, $N_3 = 10686$, $N_4 = 11913$, and $N_5 = 9780$ with a total of $N = 54896$. These yield $W_1 = 0.1686$, $W_2 = 0.2416$, $W_3 = 0.1947$, $W_4 = 0.2170$, and $W_5 = 0.1781$. Applying formula (9.23), we obtain the following values for the optimum sample sizes in the different strata: 165, 330, 270, 176, 40. On the other hand, proportional allocation would yield the sample sizes as 165, 237, 191, 213, and 175 in that order. Furthermore, $V_{\text{prop}} \doteq 0.0095$, $V_{\text{opt}} \doteq 0.0083$, and $V_{\text{ran}} \doteq \mathbf{S}_Y^2/n = (3.5)^2/981 = 0.0125$. The relative gain in precision due to use of optimum allocation compared to proportional allocation is 15% and that of proportional allocation compared to unstratified simple random sampling is 31%.

In the special case when the cost components $c_1, c_2, \ldots, c_L$ are all equal, (9.22) simplifies to

$$n_h \propto W_h\mathbf{S}_h\,. \qquad (9.25)$$

This is popularly known as *Neyman's optimal allocation*. In situations where the stratum variances are equal, (9.25) further simplifies to

$$n_h \propto W_h\,, \qquad \text{i.e.,} \qquad n_h \propto N_h \qquad (9.26)$$

which is clearly the *proportional allocation* discussed earlier. In any case, the overall sample size $n = \Sigma\, n_h$ is determined through the budgetary constraint or a lower bound to the precision in terms of an upper bound to the variance. Then, we obtain, using (9.25) and (9.26),

$$n_{h\,\text{opt}} = nW_h\mathbf{S}_h\Big/\Big(\sum W_k\mathbf{S}_k\Big)\,, \qquad 1 \leq h \leq L\,. \qquad (9.27)$$

$$n_{h\,\text{prop}} = nW_h\,, \qquad 1 \leq h \leq L\,. \qquad (9.28)$$

Sometimes the overall sample size n is specified irrespective of the relative values of the cost components $c_1, c_2, \ldots, c_L$. Then it is easy to verify that the optimum allocation of n into L strata for the purpose of minimizing the variance is also given by (9.27). Under optimum allocation, the variance of $\hat{\mathbf{Y}}_{\text{stsrs}}$ given in (9.15), denoted now as V_{opt}, reduces to

$$V_{\text{opt}} = n^{-1}\left(\sum W_h \mathbf{S}_h\right)^2 - N^{-1}\left(\sum W_h \mathbf{S}_h^2\right). \tag{9.29}$$

Clearly, in both (9.28) and (9.29), we have utilized the assumption of equal cost per unit for every stratum. We readily verify that

$$\begin{aligned} V_{\text{prop}} - V_{\text{opt}} &= n^{-1}\left\{\sum W_h \mathbf{S}_h^2 - \left(\sum W_h \mathbf{S}_h\right)^2\right\} \\ &= n^{-1} \sum W_h (\mathbf{S}_h - \bar{\mathbf{S}}_h)^2 \geq 0\,, \end{aligned}$$

where $\bar{\mathbf{S}}_h = \Sigma\, W_h \mathbf{S}_h$. Further, $V_{\text{prop}} = V_{\text{opt}}$ only when $\mathbf{S}_h$'s are all equal in which case $n_{h\ \text{opt}} = n_{h\ \text{prop}}$ as is evident from (9.27) and (9.28).

We observe that there are two *major* issues regarding implementation of such types of allocations. Though Neyman's allocation does the *best*, it is severely dependent on our prior knowledge about the true individual stratum variances or their relative values. On the other hand, proportional allocation is free from this limitation and it is also easy to understand. However, both these allocations raise another practical difficulty. This is that the allocations so determined by (9.27) and (9.28) will seldom provide integer-values to the n_h's. A *proper* rounding off of the n_h's with the intention of achieving the least variance may need some trial and error. We take up an example to illustrate this point.

Example 9.5. Assume that a population of 200 units is divided into 3 strata with $N_1 = 60$, $N_2 = 90$, $N_3 = 50$, and $\mathbf{S}_1 = 2\mathbf{S}_2 = 4\mathbf{S}_3$. For a sample of size $n = 30$, (9.27) and (9.28) would result in

$$n_{1\ \text{prop}} = 9\,, \qquad n_{2\ \text{prop}} = 13.5\,, \qquad n_{3\ \text{prop}} = 7.5\,,$$
$$n_{1\ \text{opt}} = 15.3\,, \qquad n_{2\ \text{opt}} = 11.5\,, \qquad n_{3\ \text{opt}} = 3.2\,.$$

For proportional allocation, we would compare the allocation vectors (9, 13, 8) and (9, 14, 7) and come up with the recommendation $n_1 = 9$, $n_2 = 14$, and $n_3 = 7$ as this yields the least variance. For optimum allocation, we would compare the allocation vectors (16, 11, 3), (15, 12, 3), and (15, 11, 4). Our recommendation would be to use $n_1 = 16$, $n_2 = 11$, and $n_3 = 3$ even though a rounding off would suggest $n_1 = 15$, $n_2 = 12$, and $n_3 = 3$.

Exercises 9.10 and 9.11 illustrate some further issues related to this problem.

9.7. THE PROBLEM OF MORE TH

We found that the optimum way of
strata is

$$n_{h\ \text{opt}} = \frac{(C^* - C_0) W_h \mathbf{S}_h / \sqrt{c_h}}{\Sigma\, W_k \mathbf{S}_k \sqrt{c_k}}\,,$$

or

$$n_{h\ \text{opt}} = \frac{(\Sigma\, W_k \mathbf{S}_k \sqrt{c_k}) W_h \mathbf{S}_h / \sqrt{c_h}}{V_0 + \Sigma\, W_k^2 \mathbf{S}_k^2 N_k^{-1}}\,,$$

In deducing the above expressions,
should not exceed N_h for *any h*. A qu
happen that $n_{h\ \text{opt}}$ will be larger than
nately, there are situations where this
phenomenon with the following examp

Example 9.6. Consider a populatio
strata with $N_1 = 5$, $N_2 = 10$, $\mathbf{S}_1 = 16$,
together 10 observations are to be mad
survey is the same for all the 15 units i
way of allocating $n = 10$ units into th
values of N_1 and N_2 is to set

$$n_{1\ \text{opt}} = nW_1\mathbf{S}_1/(W_1\mathbf{S}_1 +$$
$$n_{2\ \text{opt}} = nW_2\mathbf{S}_2/(W_1\mathbf{S}$$

This unrestricted optimization has produ
are faced with the problem of our final
sizes in the two strata. Clearly, in a gene
procedure and yet expect to have an
recommendation which is somewhat appe
strata whose $n_{h\ \text{opt}} > N_h$ and distribute t
the remaining strata, regarding it as a fr
Observe that for $L = 2$, this recommenda
of the strata and the allocation of the re
one. In our example, we would recomme

We now prove a theorem in support o
of two strata.

Theorem 9.4. *The optimum allocatio*
for the purpose of minimizing V_{stsrs} under
observation is 100% sampling of the stratu
sample size exceeds its total size and alloc
the other stratum.

Proof. Fi
function of n
and C are in
positive real
minimum va
minimum fo
completes th

Clearly, t
$1 \leq h \leq L$. W
to the rule
this rule und
$L = 3$. More

Theorem

$(n - N_1)W_2$

Then the al

$n_1^* =$

is optimal a

Proof.

for any set
$h = 1, 2, 3.$

and the rig

Therefore,

W_1^2

which is e

We now use the fact that $nW_1\mathbf{S}_1 > N_1(\Sigma_{h=1}^{3} W_h\mathbf{S}_h)$ which means

$$W_1\mathbf{S}_1 > N_1(W_2\mathbf{S}_2 + W_3\mathbf{S}_3)(n - N_1)^{-1}\,.$$

Squaring both sides, we obtain

$$W_1^2\mathbf{S}_1^2 > N_1^2(W_2\mathbf{S}_2 + W_3\mathbf{S}_3)^2(n - N_1)^{-2}$$

so that

$$W_1^2\mathbf{S}_1^2 N_1^{-1} n_1^{-1} > N_1 n_1^{-1}(n - N_1)^{-2}(W_2\mathbf{S}_2 + W_3\mathbf{S}_3)^2\,. \tag{9.31}$$

We now show that

$$N_1 n_1^{-1}(n - N_1)^{-2} \geq (n - N_1)^{-1}(n - n_1)^{-1}$$

which is the same as

$$N_1(n - n_1) \geq n_1(n - N_1)\,.$$

This last inequality is trivially true since $n_1 \leq N_1$. Thus the inequality is established. □

9.8. OTHER TYPES OF STRATIFIED SAMPLING STRATEGIES

Although in Sections 9.3–9.7, we provided a good deal of discussion on some aspects of inference from stratified simple random sample data, we would not like to leave the impression that other types of sampling designs are less important in the context of stratification. Quite to the contrary, the nature of survey sampling by stratification allows any combination of available sampling strategies to be used in a given context, depending on the nature of available information in various strata. This generality and freedom of choice should be properly exploited in case the survey setup calls for such actions. For example, we could blend cluster sampling with stratification in the following manner. First, we can conveniently adopt stratification and then, in each stratum, we can adopt cluster sampling. On the other hand, it is also possible that we first adopt clustering and then in the selected clusters, we adopt stratification for the purpose of subsampling. In a given situation, we could also blend systematic sampling with stratification in the sense of adopting systematic sampling for collecting observations in each stratum. If auxiliary size measures are available in some or all of the strata and their use can be justified, then we may suggest stratified IIPS, PPS, or PPAS methods of sampling as discussed in Chapter Five. Following the developments in Chapter Six we may also recommend SRS designs coupled with ratio or regression method of estimation.

The reader should not have difficulty in developing relevant theoretical results based on the implementation of various sampling strategies in the context of stratification. Results presented in the preceding chapters can be suitably blended, albeit notations and expressions would get complicated. Problems 9.14–9.16 are designed to acquaint the reader with some computations related to some such stratified sampling strategies.

9.9. DISCUSSION

There are many practical issues which we have not addressed so far. There is a lack of sound theoretical results dealing with these topics. However, we believe these topics should be listed and briefly discussed so that the reader has some familiarity with these issues. Some references are provided for further study.

Poststratification

The discussion at the end of Section 9.4 was meant to suggest that it may be possible to improve on the precision of the estimator of the population mean or the population total using stratification provided we can effectively stratify the population. The basis for a good stratification is, of course, apparent homogeneity among units within each stratum and heterogeneity between units of different strata. It thus appears that to carry out stratification satisfactorily, we require prior knowledge about the nature of distribution of the survey variable over the entire population. Moreover, the stratum sizes and the precise labelling of units (namely, membership listings) within each stratum are also needed.

In many practical situations such information may be largely incomplete and unreliable. Also there are situations where we may only know the stratum sizes but not the membership listings up to our satisfaction. Personal characteristics such as race, education, and union memberships are some such examples. In such situations, we cannot use this sort of incomplete information and stratify the population in the sampling stage so as to draw a stratified random sample. Instead, it is possible to have a simple random sample of relatively larger size from the entire population so that hopefully we end up with n_h' (>1) observations from stratum h, $1 \leq h \leq L$. This identification is possible upon collecting the data. Clearly, n_h' is a random variable. Let us now pretend that our data came from a stratified random sample with n_h' observations from the hth stratum, $1 \leq h \leq L$. Thus, in effect, we are now talking about stratification in the estimation stage, assuming that the data forces the stratification and leads to stratified sample data. This is popularly known as *poststratification* of the population. Carrying this notion further, we may estimate the population mean by

$$\hat{\bar{\mathbf{Y}}}_{\text{post}} = \sum W_h \bar{\mathbf{y}}_h , \qquad W_h = N_h/N . \tag{9.32}$$

Here the subscript post stands for poststratification. Whether we should use $\bar{\mathbf{y}}$, the sample mean of the entire random sample data or $\hat{\bar{\mathbf{Y}}}_{\text{post}}$ is not clear in general. There are, however, cases where $\hat{\bar{\mathbf{Y}}}_{\text{post}}$ is as good as an estimate of the population mean based on a proportional stratified sample of the same size. There are also cases where $\hat{\bar{\mathbf{Y}}}_{\text{post}}$ has poor performance. We have no definite recommendation to make here even though we recognize that the temptation to conduct a poststratified analysis of data is prevalent among some survey sampling practitioners. We refer the reader to Cochran (1977), Kish (1965), Konijn (1973), and Murthy (1967) for further discussion on this issue.

Number of Strata and Their Formation

As we have observed before, the amount of gain due to stratification largely depends on the degree of homogeneity within each stratum and heterogeneity among the strata means with respect to the survey variable. At the same time, there are no definite and unquestionable procedures for the formation of strata in general. This is because the survey variable usually has an unknown distribution over the units in the population. Stratification is usually based on a study of the distribution of an auxiliary variable, sometimes called the *stratification* variable. If we can assume that the stratification variable has a known density and if we know the mathematical relation between the stratification variable and the study variable, then it is possible to come up with a clear answer to the questions raised here. Some such studies have been carried out by several statisticians (see e.g., Cochran, 1977).

Stratification and Allocation for More Than One Survey Variables

If there is more than one survey variable, then it is very likely that there will be more than one characteristic available for the purpose of stratification. How then should we stratify the population and allocate the observations within the strata? These are very difficult and extremely important issues in real practical situations. A generally suggested rule is to form a new stratification variable based on the existing stratification variables. Sometimes a linear combination of the existing variables is employed. However, it may be difficult to implement this suggestion in practice. Clearly, there cannot be any hard and fast rules for a problem of this type where a possible satisfactory solution depends on many other factors concerning the real population. There are some ad hoc procedures recommended by experts in the area. For example, one such rule suggests stratifying the population with respect to that characteristic which is associated with the most important survey variable. The argument in favor of this procedure is the following. In practice, the study variables are usually related to each other and such a rule has been seen to perform relatively well in many cases. Another procedure calls for what is called *deep* stratification or *multiple* stratification. We stratify the population separately with respect to each survey variable and then consider the resulting cross-classification which is said to form *deep* stratification of the population. Next we obtain the sample sizes in each stratum on the basis of each stratification variable and suggest a compromise solution. For further details, we refer the reader to Murthy (1967).

EXERCISES

9.1. Consider the setup of Example 9.4. Suppose the cost of data collection per household from any county is $50.

(a) Determine the total sample size and the optimum values of n_h's.

(b) Compare the relative precisions of estimators based on proportional and optimal allocations.

9.2. Consider the setup of Example 9.4. Suppose this time that the problem is to determine optimum sample sizes for stratified random sampling that would minimize the overall cost of data collection and analysis for a given margin of error V_0 satisfying $V_0/V_{\text{ran}} = 65\%$. Here $V_{\text{ran}} \doteq \mathbf{S}_Y^2/n^*$ where n^* denotes the same overall sample size.

(a) Determine the total sample size n^* as well as the optimum sample sizes for each stratum. What is the total cost under optimum allocation? under proportional allocation?

(b) Compare the relative precisions of estimators based on proportional and optimum allocations.

9.3. Under what conditions would optimum allocation suggest n_h being proportional to $T_h(\mathbf{Y})$, the hth stratum total?

9.4. A certain population consists of two strata with stratum weights $W_1 = 0.25$ and $W_2 = 0.75$. We wish to estimate the proportion of population units possessing a characteristic A and we have strong reasons to believe that stratum 1 has about 75% units and stratum 2 has about 30% units possessing the characteristic A. The population is large enough and we want to take a sample of size $n = 100$. How much relative gain would stratified random sampling offer over unstratified simple random sampling? Consider both proportional and optimal allocations and assume equal cost of data collection from the two strata.

9.5. A certain population consists of three strata with the respective weights in the ratio of 1:2:3 and, further, the stratum standard deviations are in the ratio of 3:2:1.

(a) If the cost components are in the ratio of 1:2:3, show that the relative gain in precision due to use of optimum allocation compared to proportional allocation is given by 15%.

(b) Determine a relative cost structure so that $V_{\text{prop}} = V_{\text{opt}}$.

9.6. Consider the data in Example 9.2.

(a) What proportion of engineering students are foreigners? What is the s.e. of your estimate?

(b) What proportion of Ph.D. students are not foreigners? What would be the s.e. of your estimate?

(c) What proportion of students registered with the university are foreign students? What would be the s.e. of your estimate?

(d) Which area of study attracts the largest proportion of foreign students?

(e) What would be the change in precision of the estimate if instead a simple random sample of 945 students were interviewed?

9.7. Suppose a certain population is composed of three strata with sizes $N_1 = 8$, $N_2 = 22$, and $N_3 = 32$. Suppose further that $W_1\mathbf{S}_1 = 0.7$, $W_2\mathbf{S}_2 = 0.5$, and $W_3\mathbf{S}_3 = 0.3$. For a stratified random sample of size $n = 30$, show by actual computation that 100% allocation in stratum 1 and optimum reallocation in the other two strata would be the best solution among all possible allocations.

9.8. Generalize Theorem 9.5 to the case of L strata and show that the rule of 100% sampling, indeed, produces the optimum solution in case of over-sampling.

9.9. A population consists of 64 units and it is divided into 5 strata. Given the following information, show that $V_{\text{ran}} \leq V_{\text{stsrs}}$. Comment on the result. $N_1 = 19$, $N_2 = 16$, $N_3 = 13$, $N_4 = 11$, $N_5 = 5$; $n_1 = 6$, $n_2 = 5$, $n_3 = 4$, $n_4 = 3$, $n_5 = 2$, $\mathbf{S}_1 = 2.3$, $\mathbf{S}_2 = 3.7$, $\mathbf{S}_3 = 4.2$, $\mathbf{S}_4 = 2.7$, $\mathbf{S}_5 = 0.8$, and $\mathbf{S}_Y^2 = 10.00$.

9.10. It is generally argued that the theoretical optimum allocation numbers for the sample sizes are seldom integer-valued and, therefore, as soon as these are approximated by nearest integers, the minimum variance property of optimum allocation may be destroyed. In each of the following two examples, examine how far this argument is valid.

(a) $W_1 = 0.1$, $W_2 = 0.2$, $W_3 = 0.7$, $\mathbf{S}_1 = 3$, $\mathbf{S}_2 = 2$, $\mathbf{S}_3 = 1$, $n = 30$;

(b) $W_1 = 0.1$, $W_2 = 0.2$, $W_3 = 0.7$, $\mathbf{S}_1 = 1.5$, $\mathbf{S}_2 = 1.7$, $\mathbf{S}_3 = 1.8$, $n = 30$.

9.11. One common technique to get rid of the difficulty of implementing a theoretical optimum allocation solution in practice is to do randomization. This can be explained as follows. Suppose for the hth stratum the solution yields $n_h = 12.38$. Then we perform a Bernoulli trial with $P(n_h = 12) = 0.62$ and $P(n_h = 13) = 0.38$. Then depending on the outcome of the trial, we decide on the sample size for the hth stratum. Clearly the expected size of the sample in the hth stratum is 12.38. This randomization can be done in any stratum, whenever necessary, in a similar manner.

Denote by $n_h(r)$ the random integral sample size in the hth stratum and by $\bar{\mathbf{y}}_h(r)$ the mean of the $n_h(r)$ observations in the hth stratum, $1 \leq h \leq L$. Further, let $\hat{\bar{\mathbf{Y}}}(r) = \Sigma\, W_h \bar{\mathbf{y}}_h(r)$.

(a) Show that $\hat{\bar{\mathbf{Y}}}(r)$ is still an unbiased estimator for $\bar{\mathbf{Y}}$.

(b) Compute $V(\hat{\bar{\mathbf{Y}}}(r))$ and show directly that this is no less than $V(\hat{\bar{\mathbf{Y}}})$ based on theoretical solution n_h, $1 \leq h \leq L$.

(c) Starting with proportional ($n_{h\,\text{prop}}$) and optimum ($n_{h\,\text{opt}}$) allocations and suggesting the above modifications, examine the possibilities of $V_{\text{prop}}(r) < V_{\text{opt}}(r)$ where $V(r)$ refers to the randomized allocation.

(d) For the examples of Exercise 9.10, check if $V_{\text{prop}}(r) < V_{\text{opt}}(r)$. Comment on your findings.

(e) Comment on the practical limitations of this technique in the context of a fixed budget or a time constraint.

9.12. In applying Neyman's optimal allocation, a practitioner made a mistake and collected data based on the allocation $n_h = nW_h\mathbf{S}_h^2/\Sigma\, W_h\mathbf{S}_h^2$ rather than the one given in (9.27).

(a) Find the relative loss of precision due to deviation from optimal allocation.

(b) Show that the relative loss of precision is only 10% (approx), for a large population consisting of three strata with stratum weights as also the stratum standard deviations in the ratio 1:2:3.

9.13. Consider the problem of estimation of the population variance $\mathbf{S}_Y^2$ using stratified simple random sample data. Start with the identity

$$(N-1)\mathbf{S}_Y^2 = N\left[\sum W_h \bar{\mathbf{Z}}_h - \bar{\mathbf{Y}}^2\right]$$

where

$$\bar{\mathbf{Z}}_h = N_h^{-1} \sum_{i \in \mathcal{U}_h} Y_{hi}^2\,, \qquad 1 \leq h \leq L\,.$$

(a) Show that an unbiased estimator of $\mathbf{S}_Y^2$ is given by

$$\hat{\hat{\mathbf{S}}}_Y^2 = \frac{N}{N-1}\left[\sum W_h \bar{\mathbf{z}}_h - \hat{\mathbf{Y}}^2 + \hat{V}_{\text{stsrs}}\right]$$

where

$$\bar{\mathbf{z}}_h = n_h^{-1} \sum_{i \in s_h} Y_{hi}^2 , \qquad 1 \le h \le L ,$$

$\hat{\mathbf{Y}}$ and $\hat{V}_{\text{stsrs}}$ denoting, as usual, the stratified sample estimator of the population mean and its variance estimator as given in Section 9.5.

(b) Verify that the seemingly different estimator $\hat{\hat{\mathbf{S}}}_Y^2$ is indeed the same as $\hat{\mathbf{S}}_Y^2$ given in (9.21).

9.14. It was desired to estimate the total production of wheat in a large agricultural community. For administrative convenience, the area was divided into 4 strata and subsequently each stratum was divided into 6 zones, numbered 1 to 6 in increasing order of area under wheat in each zone. The single-stage fixed-size (2) cluster sampling designs as shown in Table 9.A were adopted in the various strata and altogether 8 zones were selected. The data in Table 9.B relate to the production of wheat in the selected zones of different stata. Estimate the total production of wheat in the community and also provide the s.e. of your estimate. For the use of HTE, recall Corollary 3.5.

9.15. With reference to Exercise 7.12, suppose the transistors were manufactured by two different companies, each handling one half of the order and shipping the same to the distributor in 500 boxes of 10 each. As an alternative to the single-stage cluster sampling design suggested there, it is proposed that stratified two-stage sampling can be adopted with a total of 50 transistors selected from each company's supply.

Table 9.A

	Area (in acres) under wheat						
	Zones						
Strata	1	2	3	4	5	6	Suggested Sampling Strategies
1	2.4	3.2	3.8	4.1	4.7	5.2	(ΠPS, HTE)
2	3.7	3.9	4.7	5.2	5.8	6.3	(PPS, Murthy's estimator)
3	1.9	2.2	2.5	2.8	2.9	3.2	(ΠPS, HTE)
4	0.3	0.6	1.1	1.5	1.7	2.3	(RHC scheme, Estimator in (5.71))

Table 9.B

Strata	Selected Zones	Production of Wheat (in 100,000 tons)
1	3	52.23
	5	68.81
2	1	52.09
	4	73.25
3	2	22.91
	4	32.03
4	3	18.21
	6	23.55.

Treat the data in Exercise 7.12 as arising out of such a sampling design with the first five boxes representing one stratum and the remaining ones representing the other. Give a revised estimate of the total number of defective transistors in the entire consignment and also an estimate of its s.e. Comment on your findings.

9.16. A population of N units consists of L strata with stratum sizes $N_1, N_2, \ldots, N_L$, $\Sigma N_h = N$. A stratified sampling design of size n is said to be a *stratified systematic sampling design* of size n if the sampling design adopted in each and every stratum is a systematic sampling design, with the total sample size equal to n. The component systematic sampling designs can, of course, be quite arbitrary.

Under such a sampling design, suggest an unbiased estimator of the population mean and give an expression for its variance. When can you unbiasedly estimate this variance?

REFERENCES

Bowley, A.L. (1926). Measurement of the precision attained in sampling. *Bull. Internat. Statist. Inst.*, **22**, 1–62.

Cochran, W.G. (1977). *Sampling Techniques*. Third Edition. Wiley, New York.

Kish, L. (1965). *Survey Sampling*. Wiley, New York.

Konijn, H.S. (1973). *Statistical Theory of Sample Survey Design and Analysis*. North-Holland, Amsterdam.

Murthy, M.N. (1967). *Sampling Theory and Methods*. Statist. Pub. Soc., Calcutta, India.

Neyman, J. (1934). On the two different aspects of the representative method: the method of stratified sampling and the method of purposive selection. *J. Roy. Statist. Soc.*, **97**, 558–625.

CHAPTER TEN

Superpopulation Approach to Inference in Finite Population Sampling

There are survey situations where the reference population can be viewed as emerging out of a superpopulation and the values of the survey variable may be regarded as realizations of variables associated with units in the underlying superpopulation. Consideration of such models may help us to judge the performance of various sampling strategies. In this chapter we explore this aspect of data collection and analysis.

10.1. CONCEPT OF SUPERPOPULATION MODELS

So far in the previous chapters we have assumed that the finite population under investigation is a fixed population. More explicitly, we have worked under the framework of N well-identified labelled units and a survey variable Y with fixed values $Y_1, Y_2, \ldots, Y_N$, a priori unknown, attached, respectively, to the units $1, 2, \ldots, N$. We have tacitly assumed that the population units are already labelled and the survey statistician has access to the true fixed value Y_i on the ith unit through an inquiry (unless, of course, there is nonresponse or we are addressing a sensitive characteristic, see Chapters Eleven and Twelve for such problems). When working with actual populations, some of these basic setups may not be totally justified and we may have to revise the inference problems and procedures. Hence other approaches to finite population inference have been developed from time to time.

From a theoretical point of view we have seen in earlier chapters that the usual fixed population approach most often leads to *no* definite optimal strategies. Even a comparison between only two given sampling strategies generally does not lead to emergence of any definite result. A possible *model* on structuring the values of the survey variable might lead to positive findings. This is one of the reasons to propose *superpopulation models* in the study of relative comparison of sampling strategies for finite population inference. We may thus regard such models as a means of reaching conclusive results when comparing strategies and of eventually producing *most efficient* (that is, *optimal*) strategies on various occasions. An early application of this approach in efficiency comparison between sampling methods may be found in Cochran (1946).

We now introduce the notion of a superpopulation model, which is assumed to be inherent in a given finite population. To fix ideas, suppose a certain university campus has N buildings, labelled 1 to N, and we are interested in knowing the total amount of annual telephone bills paid by the university. The very nature of the survey variable is such that it would be highly unrealistic to concentrate only on one particular reference year and estimate the relevant parameter. On the other hand, we can bring in the stochastic nature of the survey variable and try to predict the total amount at a particular time or estimate the mean amount averaged over the time domain. We, therefore, regard the values of the survey variable, denoted this time by $\mathcal{Y}_1, \mathcal{Y}_2, \ldots, \mathcal{Y}_N$, as random variables having a joint distribution $\xi(y_1, y_2, \ldots, y_N)$ and we may say that at a given time, the actual (but unknown, unless surveyed) quantities $Y_1, Y_2, \ldots, Y_N$ form a realization of the random variables $\mathcal{Y}_1, \mathcal{Y}_2, \ldots, \mathcal{Y}_N$, respectively. We say that $\xi(\cdot)$ refers to a superpopulation model. In a sense, the randomness in the Y-values is described by the model $\xi(\cdot)$. However, its specification may be grossly incomplete, featuring only the first two moments; or else, it may possess a detailed specification (such as normal with known or unknown means, variances and covariances).

A superpopulation model aims at providing a summary of our prior knowledge regarding the nature of the survey variable in the reference population. Such knowledge could be based on long term or short term experience. Various superpopulation models have been discussed in the literature. We list in Table 10.1 the models frequently studied. Another useful model, called the *Random Permutation* (RP) model will be introduced later in Section 10.5. Throughout we shall denote by $\mathcal{E}$, $\mathcal{V}$, $\mathcal{C}ov$, respectively, the mean, variance, and covariance of the random variables generated by the model $\xi(\cdot)$ while E, V, Cov (or E_p, V_p, $\mathcal{C}ov_p$, respectively) will still have their uses as design-based operators in the usual sense.

In the specification of the superpopulation models listed in Table 10.1, a_i's and b_i's are assumed to be known quantities, $a_i > 0$, $-\infty < b_i < \infty$, $1 \le i \le N$; μ, σ^2, and ρ are the unknown parameters. Under the transformation model or the exchangeable model, $\mathcal{E}(\mathcal{Y}_i) = b_i + a_i\mu$, $\mathcal{V}(\mathcal{Y}_i) = a_i^2\sigma^2$, and $\mathcal{C}ov(\mathcal{Y}_i, \mathcal{Y}_j) = a_i a_j \sigma^2 \rho$ where μ, σ^2, and ρ refer to the parameters of the joint distribution.

According to the transformation model the transformed variables $\mathcal{X}_1, \mathcal{X}_2, \ldots, \mathcal{X}_N$ possess the same means, variances, and covariances. However, their joint distribution is, otherwise, left unspecified. According to the

Table 10.1. Superpopulation Models Commonly Used in Survey Sampling

Model	Description	Model Parameters
Transformation model	$\mathscr{Z}_i = (\mathscr{Y}_i - b_i)/a_i, 1 \le i \le N$ have the same means, variances and covariances	μ, σ^2, ρ
Exchangeable model	Joint distribution of $\mathscr{Z}_1, \mathscr{Z}_2, \ldots, \mathscr{Z}_N$ is the same for all $N!$ permutations of them	μ, σ^2, ρ
Regression model (based on information on a positive-valued auxiliary variable X)	$\mathscr{E}(\mathscr{Y}_i) = \mu(X_i)$, $\mathscr{V}(\mathscr{Y}_i) = \sigma^2 v(X_i)$ $\mathscr{C}ov(\mathscr{Y}_i, \mathscr{Y}_j) = 0, 1 \le i \ne j \le N$	σ^2 and those in $\mu(X_i)$ and $v(X_i)$, $1 \le i \le N$

exchangeable model the $\mathscr{Z}_i$'s have an exchangeable distribution, that is, the joint distribution, though unspecified, is the same for all the $N!$ permutations of $(\mathscr{Z}_1, \mathscr{Z}_2, \ldots, \mathscr{Z}_N)$. This automatically forces the $\mathscr{Z}_i$'s to have the same means, variances, and covariances. Clearly the superpopulation models covered by the exchangeable nature of the $\mathscr{Z}_i$'s form a subclass of those under the transformation model. For the regression model we assume that we possess information on a positive-valued auxiliary variable X whose values $X_1, X_2, \ldots, X_N$ on the population units are fixed and known in advance. Then we stipulate a form of regression of $\mathscr{Y}$ on X given by the representation

$$\mathscr{E}(\mathscr{Y}_i \mid X_i = x_i) = \mu(x_i)\,, \qquad \mathscr{V}(\mathscr{Y}_i \mid X_i = x_i) = \sigma^2 v(x_i)\,,$$
$$\mathscr{C}ov(\mathscr{Y}_i, \mathscr{Y}_j \mid X_i = x_i, X_j = x_j) = 0\,, \qquad 1 \le i \ne j \le N\,.$$

In the particular case $\mu(x_i) = \alpha + \beta x_i$, $1 \le i \le N$, we say that the regression is *linear*. The form of $v(x_i)$ usually studied is $v(x_i) = x_i^g$, $g \ge 0$ being known. The value $g = 0$ corresponds to *homoscedastic* errors.

Throughout this chapter we will deal with two sets of notations. We clarify them below to avoid any confusion for the reader. We use $\mathscr{Y}_i$ to denote the random variable for the response associated with the unit i. The corresponding transformed variable $\mathscr{Z}_i$ is defined as $\mathscr{Z}_i = (\mathscr{Y}_i - b_i)/a_i$. Further we use Y_i to denote the observed response (realization of $\mathscr{Y}_i$) on the unit i and its transformed value is denoted by $Z_i = (Y_i - b_i)/a_i$. Finally the finite population mean is denoted as usual by $\bar{\mathbf{Y}}$ while its random counterpart under the superpopulation model is denoted by $\bar{\mathscr{Y}}$. Clearly, $\bar{\mathscr{Y}} = \Sigma_{i=1}^N \mathscr{Y}_i/N$.

Our purpose in this chapter is to discuss various optimality results available in the literature regarding superpopulation model based inference related to the prediction of $\bar{\mathscr{Y}}$ and estimation of $\bar{\mu} = \mathscr{E}(\bar{\mathscr{Y}})$. Among other things, it will be seen that the commonly used sampling strategies are in a sense best. This will provide a justification for their uses in actual surveys.

10.2. CRITERION OF OPTIMALITY

Recall that in the usual setup of a fixed population, the criterion used for comparing sampling strategies is minimizing $E_p(e - \theta(\mathbf{Y}))^2$. In the superpopulation setup, Godambe (1955) advocated minimization of (model-) expected variance or (model-) expected mean square error as the criterion of bestness. Thus for the purpose of prediction of $\bar{\mathcal{Y}}$, the *best predictor* $e^*(s, \mathbf{Y})$ would minimize model-expectation of $E_p(e^* - \bar{\mathcal{Y}})^2$, that is $\mathscr{E}E_p(e^* - \bar{\mathcal{Y}})^2$ among all predictors in a given class. For estimation of $\bar{\mu}$, the *best* estimator $e^*(s, \mathbf{Y})$ would minimize $\mathscr{E}E_p(e^* - \bar{\mu})^2$ among all its competitors. Godambe and Thompson (1973) have given the following justification for using the above criterion of bestness. In the classical setup, if t is an estimator of the parametric function $g(\theta)$, then by Chebycheff's inequality

$$P_\theta[|t - g(\theta)| \le \epsilon] \ge 1 - \frac{E_\theta(t - g(\theta))^2}{\epsilon^2} .$$

If we now assume a prior distribution $\xi(\cdot)$ for θ and integrate both sides of the above inequality with respect to $\xi(\cdot)$, then

$$P_\xi[|t - g(\theta)| \le \epsilon] \ge 1 - \frac{\mathscr{E}_\xi E_\theta(t - g(\theta))^2}{\epsilon^2} .$$

In view of the above inequality an estimator which maximizes the coverage probability (given by the left hand side above) uniformly in ϵ, may be considered a good estimator. Therefore, it is natural to consider an estimator for which $\mathscr{E}_\xi E_\theta(t - g(\theta))^2$ is minimal, as the *most efficient* or the *best* estimator. The above argument justifies the use of $\mathscr{E}E_p(e - \bar{\mathcal{Y}})^2$ as a measure of variability in the behavior of the predictor e for $\bar{\mathcal{Y}}$ or of $\mathscr{E}E_p(e - \bar{\mu})^2$ as that of the estimator e for $\bar{\mu}$. Thus the earlier criterion of minimization of $E_p(e - \bar{\mathbf{Y}})^2$ is now changed to minimization of $\mathscr{E}E_p(e - \bar{\mu})^2$ or of $\mathscr{E}E_p(e - \bar{\mathcal{Y}})^2$. Note that usually our choice of the sampling design is *not* governed by the values Y_i's so that it is in a sense *noninformative* regarding the true values of the survey variable of interest. As a result, the operators $\mathscr{E}$ and E_p can be interchanged where $\mathscr{E}$ operates on $\mathcal{Y}_i$'s following the model $\xi(\cdot)$ and E_p operates on the random samples drawn according to the noninformative sampling design resulting in $P(.)$. Moreover, a function $e(s, \mathbf{Y})$ regarded, as before, as mapping from $R^{n(s)}$ to R^1 will be referred to as an estimator or a predictor depending on whether it aims at estimating the parameter $\bar{\mu}$ or predicting the random quantity $\bar{\mathcal{Y}}$.

At this stage, we introduce the following definitions.

Definition 10.1. (a) $e(s, \mathbf{Y})$ is said to be a model-unbiased (written ξ-unbiased) predictor of $\bar{\mathscr{Y}}$ if $\mathscr{E}(e(s, \mathbf{Y}) - \bar{\mathscr{Y}}) = 0$ for every sample s with $P(s) > 0$. (b) $e(s, \mathbf{Y})$ is said to be a design-cum-model-unbiased (written $p\xi$-unbiased) predictor of $\bar{\mathscr{Y}}$ if $\mathscr{E}E_p(e(s, \mathbf{Y}) - \bar{\mathscr{Y}}) = 0$ identically in the parameters of the model.

Definition 10.2. $e(s, \mathbf{Y})$ is said to be a design-cum-model-unbiased (written $p\xi$-unbiased) estimator of $\bar{\mu}$ if $\mathscr{E}E_p(e(s, \mathbf{Y})) = \bar{\mu}$ identically in the parameters of the model.

Suppose $e(s, \mathbf{Y})$ is a p-unbiased (that is, design-unbiased) estimator of $\bar{\mathbf{Y}}$ with respect to a fixed population setup. Then clearly it is a $p\xi$-unbiased estimator of $\bar{\mu}$ and also a $p\xi$-unbiased predictor of $\bar{\mathscr{Y}}$ with respect to *every* superpopulation model $\xi(\cdot)$ imposed on it. This fact will be utilized in our search for optimal strategies.

Below we derive an expression for $\mathscr{E}E_p(e - \bar{\mu})^2$ in general terms. Interchanging the operators $\mathscr{E}$ and E_p, we may write

$$\begin{aligned}\mathscr{E}E_p(e - \bar{\mu})^2 &= E_p\mathscr{E}(e - \bar{\mu})^2 = E_p\mathscr{E}\{e - \mathscr{E}(e) + \mathscr{E}(e) - \bar{\mu}\}^2 \\ &= E_p\{\mathscr{E}(e - \mathscr{E}(e))^2 + (\mathscr{E}(e) - \bar{\mu})^2\} .\end{aligned}$$

Note, however, that yield,

$$\mathscr{E}(e - \mathscr{E}(e))^2 = \mathscr{V}(e),\ \mathscr{E}(e) - \bar{\mu} = \mathscr{B}(e) = \text{model-based bias of } e(s, \mathbf{Y}) . \tag{10.1}$$

These yield,

$$\mathscr{E}E_p(e - \bar{\mu})^2 = E_p\{\mathscr{V}(e)\} + E_p\{\mathscr{B}^2(e)\} . \tag{10.2}$$

The above expression for $\mathscr{E}E_p(e - \bar{\mu})^2$ will be utilized for the purpose of minimization in our search for the best strategies.

10.3. OPTIMAL SAMPLING STRATEGIES BASED ON THE TRANSFORMATION MODEL

Consider the transformation model with μ as the common superpopulation mean of the transformed variable $\mathscr{X}_i = (\mathscr{Y}_i - b_i)a_i^{-1}$, $1 \le i \le N$. This yields $\mathscr{E}(\bar{\mathscr{Y}}) = \bar{\mathbf{b}} + \mu\bar{\mathbf{a}} = \bar{\mu}$ as the superpopulation mean of $\bar{\mathscr{Y}}$ where $\bar{\mathbf{b}} = \Sigma_{i=1}^N b_i/N$ and $\bar{\mathbf{a}} = \Sigma_{i=1}^N a_i/N$. We will be concerned with two inference problems. The first one relates to estimation of $\bar{\mu}$ and the second relates to prediction of $\bar{\mathscr{Y}}$.

Suppose we want to estimate $\bar{\mu}$ using a suitable sampling strategy of the type (d, e), also written as (p, e). Since a_i's and b_i's are assumed to be known quantities, we may work with the observed Y_i's for $i \in s$, s being a sample selected according to the sampling design d and surveyed, or, conveniently, we may work with the transformed quantities Z_i's given by $(Y_i - b_i)/a_i$, $i \in s$. Among other results we will establish below the optimality of the strategy (p_0, e_0) where

$$p_0: \quad \text{fixed-size } (n) \text{ design with } \pi_i \propto a_i\,, \qquad 1 \le i \le N\,; \tag{10.3}$$

$$e_0(s, \mathbf{Y}) = \bar{\mathbf{b}} + (\bar{\mathbf{a}}/n) \sum_{i \in s} (Y_i - b_i)a_i^{-1}\,. \tag{10.4}$$

The first result provides an optimal model-based strategy in an appropriate subclass of linear $p\xi$-unbiased estimators of $\bar{\mu}$ based on fixed-size (n) designs.

Theorem 10.1. *For estimating the superpopulation mean $\bar{\mu}$ we restrict to the subclass of sampling strategies comprised of arbitrary sampling designs of fixed-size (n) and linear $p\xi$-unbiased estimators. Then among all such strategies (p, e) in this class, those which minimize $\mathscr{E}E_p(e - \bar{\mu})^2$ are given by (p, e_0) where p is an arbitrary fixed-size (n) design and e_0 is given in* (10.4) *above.*

Proof. For an arbitrary sampling strategy (p, e), we will write

$$e(s, \mathbf{Y}) = \beta_{s0} + \sum_{i \in s} \beta_{si} Y_i\,, \qquad s \in S_d = \{s \mid P_d(s) > 0\}\,. \tag{10.5}$$

Since e is a $p\xi$-unbiased estimator of $\bar{\mu}$, we must have

$$\mathscr{E}E_p(e(s, \mathbf{Y}) - \bar{\mu}) = 0 \text{ identically in } \mu\,. \tag{10.6}$$

But $\mathscr{E}E_p(e(s, \mathbf{Y})) = E_p\mathscr{E}(e(s, \mathbf{Y})) = E_p\{\beta_{s0} + \Sigma_{i\in s}\, \beta_{si}(b_i + \mu a_i)\}$ and equating this to $\bar{\mu} = \bar{\mathbf{b}} + \mu\bar{\mathbf{a}}$ identically in μ, we derive

$$\sum_s P(s)\left\{\beta_{s0} + \sum_{i\in s} \beta_{si} b_i\right\} = \bar{\mathbf{b}}\,, \tag{10.7}$$

$$\sum_s P(s)\left\{\sum_{i\in s} \beta_{si} a_i\right\} = \bar{\mathbf{a}}\,. \tag{10.8}$$

We now proceed to minimize $\mathscr{E}E_p(e - \bar{\mu})^2$. By (10.2), it has two nonnegative components namely, $E_p(\mathscr{V}(e))$ and $E_p(\mathscr{B}^2(e))$. We demonstrate that the strategy (p, e_0) minimizes both these components simultaneously. It is a straightforward task to verify, using (10.4), that $\mathscr{B}(e_0) = \mathscr{E}(e_0) - \bar{\mu} = \bar{\mathbf{b}} + (\bar{\mathbf{a}}/n)\mathscr{E}\{\Sigma_{i\in s}\,(\mathscr{Y}_i - b_i)a_i^{-1}\} - \bar{\mu} = \bar{\mathbf{b}} + (\bar{\mathbf{a}}/n)(\Sigma_{i\in s}\,\mu) - \bar{\mu} = \bar{\mathbf{b}} + \bar{\mathbf{a}}\mu - \bar{\mu} = 0$ for every $s \in S_d$. Next we take up $E_p\mathscr{V}(e)$ and below we show that $E_p(\mathscr{V}(e_0)) \le$

$E_p(\mathscr{V}(e))$. Using (10.5), we have

$$\mathscr{V}(e) = \sigma^2 \sum_{i \in s} \beta_{si}^2 a_i^2 + \rho\sigma^2 \sum_{i \neq j, i} \sum_{j \in s} \beta_{si}\beta_{sj}a_i a_j$$

$$= \sigma^2\Big[\Big(\sum_{i \in s} \beta_{si}a_i\Big)^2 - (1-\rho)\Big\{\Big(\sum_{i \in s} \beta_{si}a_i\Big)^2 - \Big(\sum_{i \in s} \beta_{si}^2 a_i^2\Big)\Big\}\Big].$$

Therefore,

$$E_p\mathscr{V}(e) = \sigma^2 \sum_s P(s)\Big(\sum_{i \in s} \beta_{si}a_i\Big)^2 - \sigma^2(1-\rho)\sum_s P(s)$$

$$\times \Big\{\Big(\sum_{i \in s} \beta_{si}a_i\Big)^2 - \Big(\sum_{i \in s} \beta_{si}^2 a_i^2\Big)\Big\}.$$

To minimize $E_p\mathscr{V}(e)$ subject to (10.8), let us write $c_{si} = \beta_{si}a_i$. Then we have to minimize $\Sigma_s P(s)(\Sigma_{i \in s} c_{si})^2 - (1-\rho)\,\Sigma_s P(s)\{(\Sigma_{i \in s} c_{si})^2 - \Sigma_{i \in s} c_{si}^2\}$ where the coefficients c_{si} satisfy $\Sigma_s P(s)\,\Sigma_{i \in s} c_{si} = \bar{\mathbf{a}}$ (from (10.8)). Consider the function $F = \Sigma_s P(s)(\Sigma_{i \in s} c_{si})^2 - (1-\rho)\,\Sigma_s P(s)\{(\Sigma_{i \in s} c_{si})^2 - \Sigma_{i \in s} c_{si}^2\} - 2\lambda\{\Sigma_s P(s)\,\Sigma_{i \in s} c_{si} - \bar{\mathbf{a}}\}$ where λ is the Lagrange multiplier. Differentiating F with respect to c_{si} and equating it to zero, we obtain $c_{si} =$ constant, independent of s and $i \in s$. The condition $\Sigma_s P(s)\,\Sigma_{i \in s} c_{si} = \bar{\mathbf{a}}$ then yields $c_{si} = \bar{\mathbf{a}}/n$. To see that this yields the minimum, we now differentiate F once more and obtain

$$\frac{\partial^2 F}{\partial c_{si}^2} = 2P(s)\,, \qquad \frac{\partial^2 F}{\partial c_{si}\,\partial c_{sj}} = 2\rho P(s)\,, \qquad i \neq j \in s\,,$$

$$\frac{\partial^2 F}{\partial c_{si}\,\partial c_{s'j'}} = 0\,, \qquad s \neq s'\,.$$

Thus the matrix of second derivatives is a block diagonal matrix with the components of the form $\{(1-\rho)\mathbf{I}_n + \rho\mathbf{J}_n\}2P(s)$ which are easily seen to be positive definite. In the above, $\mathbf{I}$ is an identity matrix and $\mathbf{J}$ is a matrix of all 1's. Therefore, the estimator which minimizes $E_p\mathscr{V}(e)$ for every design of fixed-size (n) is one for which $c_{si} = \bar{\mathbf{a}}/n$ which yields $\beta_{si} = \bar{\mathbf{a}}/na_i$. Finally, from (10.7), we obtain $\beta_{s0} = \bar{\mathbf{b}} - (\bar{\mathbf{a}}/n)\,\Sigma_{i \in s}(b_i/a_i)$. These lead to the form of $e(s, \mathbf{Y})$ as $e_0(s, \mathbf{Y})$ given in (10.4). Moreover, we have already verified that $\mathscr{B}(e_0) = 0$ for every $s \in S_d$. Hence the result. □

It is straightforward to verify that irrespective of the nature of the fixed-size (n) sampling design, we have

$$\mathscr{E}E_p(e_0 - \bar{\mu})^2 = E_p\mathscr{V}(e_0) = \sigma^2\bar{\mathbf{a}}^2\{1 + (n-1)\rho\}/n\,. \tag{10.9}$$

Remark 10.1. The estimator $e_0(s, \mathbf{Y})$ is a member of the family of generalized difference estimators suggested by Basu (1971).

Next we turn to the question of optimal prediction of $\bar{\mathcal{Y}}$. We have two primary options regarding a reasonable choice of the predictors namely those which are p-unbiased for $\bar{\mathbf{Y}}$ in the sense of a fixed population and those which are ξ-unbiased for $\bar{\mathcal{Y}}$ in the sense of a superpopulation. Of course, in either case it would trivially turn out to be $p\xi$-unbiased. The optimality criterion would be taken as $\mathscr{E}E_p(e - \bar{\mathcal{Y}})^2$. This time we will show that among all strategies (p, e) which consist of fixed-size (n) designs and linear p-unbiased estimators of $\bar{\mathbf{Y}}$, the optimal one is (p_0, e_0) where p_0 is given in (10.3) and e_0 is given in (10.4). This result is due to Cassel et al. (1976). (For slightly more general result, the interested reader is referred to Arnab (1986) and Mukerjee and Sengupta (1989)).

Theorem 10.2. *For predicting the population mean $\bar{\mathcal{Y}}$ suppose we restrict to the subclass of all strategies (p, e) where p is a noninformative design with the fixed sample size n, and e is a linear p-unbiased estimator for $\bar{\mathbf{Y}}$. Then among all competing strategies in this subclass the one which minimizes $\mathscr{E}E_p(e - \bar{\mathcal{Y}})^2$ is given by (p_0, e_0) where p_0 and e_0 are as in* (10.3) *and* (10.4), *respectively.*

Proof. First we shall derive a convenient expression for $\mathscr{E}E_p(e - \bar{\mathcal{Y}})^2$ with reference to an arbitrary competing strategy (p, e). We have

$$\begin{aligned}
\mathscr{E}E_p(e - \bar{\mathcal{Y}})^2 &= E_p \mathscr{E}\{e - \bar{\mu} + \bar{\mu} - \bar{\mathcal{Y}}\}^2 \\
&= E_p[\mathscr{E}(e - \bar{\mu})^2 + \mathscr{E}(\bar{\mathcal{Y}} - \bar{\mu})^2 - 2\mathscr{E}(e - \bar{\mu})\cdot(\bar{\mathcal{Y}} - \bar{\mu})] \\
&= E_p[\mathscr{E}(e - \bar{\mu})^2 + \mathscr{V}(\bar{\mathcal{Y}}) - 2\mathscr{C}ov(e, (\bar{\mathcal{Y}} - \bar{\mu}))] \\
&= \mathscr{E}E_p(e - \bar{\mu})^2 + \mathscr{V}(\bar{\mathcal{Y}}) - 2\mathscr{C}ov(E_p(e), \bar{\mathcal{Y}} - \bar{\mu})) .
\end{aligned}$$

Since e is p-unbiased for $\bar{\mathbf{Y}}$, $\mathscr{C}ov(E_p(e), \bar{\mathcal{Y}} - \bar{\mu})) = \mathscr{V}(\bar{\mathcal{Y}})$ and, therefore,

$$\mathscr{E}E_p(e - \bar{\mathcal{Y}})^2 = \mathscr{E}E_p(e - \bar{\mu})^2 - \mathscr{V}(\bar{\mathcal{Y}}) . \tag{10.10}$$

This shows that the strategy which minimizes $\mathscr{E}E_p(e - \bar{\mu})^2$ also minimizes $\mathscr{E}E_p(e - \bar{\mathcal{Y}})^2$ among all relevant strategies in the subclass under consideration. By Theorem 10.1, we know that $\mathscr{E}E_p(e - \bar{\mu})^2 \geq \mathscr{E}E_p(e_0 - \bar{\mu})^2 = \mathscr{E}E_{p_0}(e_0 - \bar{\mu})^2$ among all $p\xi$-unbiased estimators of $\bar{\mu}$. From (10.3) and (10.4) it is not difficult to verify that the estimate e_0 is p_0-unbiased for $\bar{\mathbf{Y}}$ and hence, e_0 is trivially $p_0\xi$-unbiased for $\bar{\mu}$. This establishes the result. □

In the particular case $a_i = X_i$, $b_i = 0$, $1 \le i \le N$, the model assumes the simplified form

$$\mathscr{E}(\mathscr{Y}_i) = \mu \,, \qquad \mathscr{V}(\mathscr{Y}_i) = \sigma^2 X_i^2 \,, \qquad \mathscr{C}ov(\mathscr{Y}_i, \mathscr{Y}_j) = \rho\sigma^2 X_i X_j \,, \quad 1 \le i \ne j \le N \,. \tag{10.11}$$

The optimal strategy then assumes the form

$$p_0\text{: fixed-size } (n) \text{ design with } \pi_i = \frac{nX_i}{T(\mathbf{X})} \,, \qquad 1 \le i \le N \,;$$

$$e_0(s, \mathbf{Y}) = \bar{\mathbf{X}}\Big(\sum_{i \in s} Y_i / X_i\Big)\Big/ n \,. \tag{10.12}$$

Thus, the well-known (ΠPS, HTE) strategy would provide optimal prediction about the population mean under the simplified form (10.11) of the superpopulation model. It may be noted that under a fixed population setup, earlier in Section 3.3 we discussed the availability of admissible strategies since optimal ones are not available. The strategy (10.12) which was earlier suggested as *admissible* now turns out to be *optimal*. The reader will find a stronger version of the above result in Corollary 10.12. Interestingly enough, this also settles optimality of the strategy (SRS(N, n), sample mean) which follows from the above when in the model we stipulate that $a_i = 1$, $b_i = 0$, $1 \le i \le N$.

Finally we turn to the question of optimal prediction of $\bar{\mathscr{Y}}$ using a ξ-unbiased predictor. This time we propose an optimal predictor for $\bar{\mathscr{Y}}$ by choosing the sampling design in advance. The reason for doing so will be made clear right after Theorem 10.3. Here optimality will again be interpreted in the sense of minimizing $\mathscr{E}E_p(e - \bar{\mathscr{Y}})^2$ for a *fixed* p. Moreover, we will restrict ourselves to the class of linear ξ-unbiased predictors but the sampling design, though fixed in advance, need not be taken to be of fixed size. We will present the following result due to Royall (1970).

Theorem 10.3. *Among all linear ξ-unbiased predictors of the population mean $\bar{\mathscr{Y}}$ based on a given sampling design, the one, say $e^*(s, \mathbf{Y})$, which minimizes $\mathscr{E}E_p(e - \bar{\mathscr{Y}})^2$ is given by*

$$e^*(s, \mathbf{Y}) = f_s \bar{\mathscr{Y}}_s + (1 - f_s)\{\bar{\mathscr{X}}_s \bar{\mathbf{a}}_{\bar{s}} + \bar{\mathbf{b}}_{\bar{s}}\} \,, \qquad s \in S_d \tag{10.13}$$

where

$$s \cup \bar{s} = \mathscr{U} \,, \qquad \bar{\mathscr{Y}}_s = \sum_{i \in s} \mathscr{Y}_i / n(s) \,,$$

$$f_s = n(s)/N \,, \qquad \bar{\mathbf{a}}_{\bar{s}} = \sum_{i \notin s} a_i / (N - n(s)) \,,$$

$$\bar{\mathbf{b}}_{\bar{s}} = \sum_{i \notin s} b_i/(N - n(s)) \qquad \textit{and} \qquad \bar{\mathscr{X}}_s = \left\{\sum_{i \in s} (\mathscr{Y}_i - b_i)a_i^{-1}\right\} \Big/ (n(s)) .$$

Proof. Observe that $\bar{\mathscr{Y}} = f_s \bar{\mathscr{Y}}_s + (1 - f_s)\bar{\mathscr{Y}}_{\bar{s}}$ for any $s \in S_d$. Without any loss of generality any linear predictor $e(s, \mathbf{Y})$ for $\bar{\mathscr{Y}}$ can be expressed as

$$e(s, \mathbf{Y}) = f_s \bar{\mathscr{Y}}_s + (1 - f_s)U(s, \mathbf{Y}) \tag{10.14}$$

with

$$U(s, \mathbf{Y}) = u_{s0} + \sum_{i \in s} u_{si} \mathscr{Y}_i . \tag{10.15}$$

Thus, in a sense, the purpose is to predict $\bar{\mathscr{Y}}_{\bar{s}} = \Sigma_{i \notin s} \mathscr{Y}_i/(N - n(s))$ from the given data using the predictor $U(s, \mathbf{Y})$. Since $e(s, \mathbf{Y})$ is assumed to be a ξ-unbiased predictor of $\bar{\mathscr{Y}}$, we have

$$\mathscr{E}(U(s, \mathbf{Y}) - \bar{\mathscr{Y}}_{\bar{s}}) = 0 , \qquad s \in S_d \tag{10.16}$$

which yields

$$u_{s0} + \sum_{i \in s} u_{si} b_i = \bar{\mathbf{b}}_{\bar{s}} , \qquad \sum_{i \in s} u_{si} a_i = \bar{\mathbf{a}}_{\bar{s}} , \qquad s \in S_d . \tag{10.17}$$

Now consider the optimality functional $\mathscr{E} E_p(e - \bar{\mathscr{Y}})^2$ which readily simplifies to

$$\begin{aligned} \mathscr{E} E_p[(1 - f_s)^2\{U(s, \mathbf{Y}) - \bar{\mathscr{Y}}_{\bar{s}}\}^2] &= E_p[(1 - f_s)^2 \mathscr{E}(U(s, \mathbf{Y}) - \bar{\mathscr{Y}}_{\bar{s}})^2] \\ &= E_p[(1 - f_s)^2 \mathscr{V}(U(s, \mathbf{Y}) - \bar{\mathscr{Y}}_{\bar{s}})] \\ &= E_p[(1 - f_s)^2\{\mathscr{V}(U) + \mathscr{V}(\bar{\mathscr{Y}}_{\bar{s}}) \\ &\quad - 2\,\mathscr{C}ov(U, \bar{\mathscr{Y}}_{\bar{s}})\}] . \end{aligned} \tag{10.18}$$

In the above we have replaced $\mathscr{E}(U(s, \mathbf{Y}) - \bar{\mathscr{Y}}_{\bar{s}})^2$ by $\mathscr{V}(U(s, \mathbf{Y}) - \bar{\mathscr{Y}}_{\bar{s}})$ since $\mathscr{E}(U(s, \mathbf{Y}) - \bar{\mathscr{Y}}_{\bar{s}}) = 0$, $s \in S_d$. It is now straightforward to verify that $\mathscr{C}ov(U, \bar{\mathscr{Y}}_{\bar{s}}) = \rho\sigma^2 \bar{\mathbf{a}}_{\bar{s}}^2$ for any $s \in S_d$ (Exercise 10.1). Accordingly, the first task is to minimize $\mathscr{V}(U(s, \mathbf{Y}))$ for a given $s \in S_d$. We have $\mathscr{V}(U(s, \mathbf{Y})) = \sigma^2 \Sigma_{i \in s} u_{si}^2 a_i^2 + \rho\sigma^2 \Sigma\Sigma_{i \neq j,\, i,j \in s} u_{si} u_{sj} a_i a_j = \sigma^2[(\Sigma_{i \in s} u_{si} a_i)^2 - (1 - \rho)\{(\Sigma_{i \in s} u_{si} a_i)^2 - \Sigma_{i \in s} u_{si}^2 a_i^2)\}]$ and, using (10.17), this can be rewritten as

$$\mathscr{V}(U(s, \mathbf{Y})) = \sigma^2\left[\bar{\mathbf{a}}_{\bar{s}}^2 - (1 - \rho)\left\{\bar{\mathbf{a}}_{\bar{s}}^2 - \sum_{i \in s} u_{si}^2 a_i^2\right\}\right] \geq \sigma^2\{\bar{\mathbf{a}}_{\bar{s}}^2 \rho + (1 - \rho)\bar{\mathbf{a}}_{\bar{s}}^2/n(s)\}$$

by the Cauchy–Schwartz inequality. Thus, finally,

$$\mathscr{V}(U(s, \mathbf{Y})) \geq \sigma^2 \bar{\mathbf{a}}_{\bar{s}}^2[\rho + \{(1 - \rho)/n(s)\}] , \qquad s \in S_d . \tag{10.19}$$

Here equality arises if and only if for every $s \in S_d$, $u_{si}a_i = \bar{\mathbf{a}}_{\bar{s}}/n(s)$, $i \in s$. This gives $u_{si} = \bar{\mathbf{a}}_{\bar{s}}/a_i n(s)$ and, hence, from (10.17), $u_{s0} = \bar{\mathbf{b}}_{\bar{s}} - (\bar{\mathbf{a}}_{\bar{s}}/n(s))(\Sigma_{i\in s} b_i a_i^{-1})$. These result in the predictor

$$\begin{aligned} e^*(s, \mathbf{Y}) &= f_s \bar{\mathcal{Y}}_s + (1 - f_s)[\bar{\mathbf{b}}_{\bar{s}} + (\bar{\mathbf{a}}_{\bar{s}}/n(s))\{\Sigma_{i\in s} (\mathcal{Y}_i - b_i)a_i^{-1}\}] \\ &= f_s \bar{\mathcal{Y}}_s + (1 - f_s)[\bar{\mathbf{b}}_{\bar{s}} + \bar{\mathbf{a}}_{\bar{s}} \bar{\mathcal{X}}_s] , \qquad s \in S_d \end{aligned}$$

as stated in (10.13). Clearly, e^* minimizes $\mathscr{E}(e - \bar{\mathcal{Y}})^2$ for every $s \in S_d$ and, hence, also $\mathscr{E}E_p(e - \bar{\mathcal{Y}})^2$. Thus the theorem is established. □

For the optimal predictor e^* the value of $\mathscr{E}E_p(e^* - \bar{\mathcal{Y}})^2$ can now be written down using (10.18) and (10.19). An explicit expression for $\mathscr{C}ov(U(s, \mathbf{Y}), \bar{\mathcal{Y}}_{\bar{s}})$ has already been evaluated. We state the final result below where $A = \Sigma_{i=1}^N a_i^2/N$.

$$\begin{aligned} \mathscr{E}E_p(e^* - \bar{\mathcal{Y}})^2 = N^{-2}\sigma^2(1 - \rho)E_p\Big[A - \sum_{i\in s} (a_i - \bar{\mathbf{a}}_s)^2 \\ + \bar{\mathbf{a}}(2\bar{\mathbf{a}}_{\bar{s}} - \bar{\mathbf{a}})N^2 n^{-1}(s) - 2N\bar{\mathbf{a}}_{\bar{s}}\bar{\mathbf{a}} \Big] . \end{aligned} \tag{10.20}$$

We can now appreciate the difficulty in relaxing the condition that the sampling design has been chosen in advance. It is not easy to minimize the above expression by making a suitable choice of the sampling design even upon restricting to a suitable subclass, for example, of fixed-size designs.

At this stage, we may observe that the optimal ξ-unbiased linear predictor of $\bar{\mathcal{Y}}$ based on a given sampling design is given by

$$e^*(s, \mathbf{Y}) = f_s \bar{\mathcal{Y}}_s + (1 - f_s)[\bar{\mathbf{b}}_{\bar{s}} + \bar{\mathbf{a}}_{\bar{s}} \bar{\mathcal{X}}_s] , \qquad s \in S_d .$$

On the other hand, the optimal p-unbiased linear estimator of $\bar{\mathbf{Y}}$ is given by

$$e_0(s, \mathbf{Y}) = \bar{\mathbf{b}} + (\bar{\mathbf{a}}/n) \sum_{i\in s} (Y_i - b_i)a_i^{-1} = \bar{\mathbf{b}} + \bar{\mathbf{a}}\bar{\mathbf{z}}(s) , \qquad s \in S_d .$$

It is interesting to examine the status of e_0 for the purpose of prediction of $\bar{\mathcal{Y}}$. This is done below by replacing Y_i by $\mathcal{Y}_i$, $1 \le i \le N$, in the expression for e_0.

Note that $\mathcal{Y}_i = b_i + a_i \mathcal{X}_i$, $1 \le i \le N$ so that

$$\bar{\mathcal{Y}}_s = \bar{\mathbf{b}}_s + \sum_{i\in s} a_i \mathcal{X}_i/n , \qquad \bar{\mathcal{Y}}_{\bar{s}} = \bar{\mathbf{b}}_{\bar{s}} + \sum_{i\notin s} a_i \mathcal{X}_i/(N - n) . \tag{10.21}$$

Therefore, we may write

$$\begin{aligned} e^*(s, \mathbf{Y}) &= f_s \bar{\mathbf{b}}_s + \sum_{i\in s} a_i \mathcal{X}_i/N + (1 - f_s)(\bar{\mathbf{b}}_{\bar{s}} + \bar{\mathcal{X}}_s \bar{\mathbf{a}}_{\bar{s}}) \\ &= \bar{\mathbf{b}} + N^{-1}\Big[\sum_{i\in s} a_i \mathcal{X}_i + \bar{\mathcal{X}}_s \sum_{i\notin s} a_i \Big] \end{aligned}$$

$$= \bar{\mathbf{b}} + N^{-1}\Big[\sum_{i\in s} a_i\mathscr{X}_i + \bar{\mathscr{X}}_s\Big(N\bar{\mathbf{a}} - \sum_{i\in s} a_i\Big)\Big]$$

$$= \bar{\mathbf{b}} + \bar{\mathbf{a}}\bar{\mathscr{X}}_s + N^{-1}\Big[\sum_{i\in s} a_i\Big(\mathscr{X}_i - \bar{\mathscr{X}}_s\Big)\Big].$$

Thus we may express the difference between e_0 and e^* as

$$e_0(s, \mathbf{Y}) - e^*(s, \mathbf{Y}) = -N^{-1}\sum_{i\in s} a_i(\mathscr{X}_i - \bar{\mathscr{X}}_s)\,, \qquad s \in S_d \tag{10.22}$$

so that $\mathscr{E}(e_0 - e^*) = 0$ and, hence,

$$\mathscr{E}(e_0 - e^*)^2 = \mathscr{V}(e_0 - e^*) = \sigma^2(1-\rho)\Big[\sum_{i\in s}(a_i - \bar{\mathbf{a}}_s)^2\Big]\Big/N^2\,. \tag{10.23}$$

It is now possible for us to resolve the following interesting issues. How does $e_0(s, \mathbf{Y})$ compare with $e^*(s, \mathbf{Y})$ for prediction purposes? In other words, by how much does $\mathscr{E}E_p(e_0(s, \mathbf{Y}) - \bar{\mathscr{Y}})^2$ exceed $\mathscr{E}E_p(e^*(s, \mathbf{Y}) - \bar{\mathscr{Y}})^2$? Below we give the precise result.

Theorem 10.4. *For any noninformative design p, using $e_0(s, \mathscr{Y})$,*

$$\mathscr{E}E_p(e_0 - \bar{\mathscr{Y}})^2 = \mathscr{E}E_p(e^* - \bar{\mathscr{Y}})^2 + N^{-2}\sigma^2(1-\rho)E_p\Big[\sum_{i\in s}(a_i - \bar{a}_s)^2\Big]. \tag{10.24}$$

Proof. We have $\mathscr{E}E_p(e_0 - \bar{\mathscr{Y}})^2 = E_p\mathscr{E}(e_0 - \bar{\mathscr{Y}})^2 = E_p\mathscr{E}(e_0 - e^* + e^* - \bar{\mathscr{Y}})^2 = E_p\mathscr{E}(e^* - \bar{\mathscr{Y}})^2 + E_p\mathscr{E}(e_0 - e^*)^2 + 2E_p\mathscr{E}\{(e_0 - e^*)(e^* - \bar{\mathscr{Y}})\}$. Therefore, the result will follow from (10.23) if we can show that $E_p\mathscr{E}\{(e_0 - e^*)(e^* - \bar{\mathscr{Y}})\} = 0$. This is, indeed, true in a more general form and we leave it as an exercise for the reader (Exercise 10.2).

Combining (10.20) and (10.24), we may write

$$\mathscr{E}E_p(e_0 - \bar{\mathscr{Y}})^2 = \sigma^2(1-\rho)E_p[AN^{-2} + \bar{\mathbf{a}}(2\bar{\mathbf{a}}_{\bar{s}} - \bar{\mathbf{a}})n^{-1}(s) - 2\bar{\mathbf{a}}_{\bar{s}}\bar{\mathbf{a}}N^{-1}] \tag{10.25}$$

where $A = \Sigma_{i=1}^N a_i^2/N$ as before.

10.4. OPTIMAL SAMPLING STRATEGIES BASED ON THE EXCHANGEABLE MODEL

We now turn our attention to the exchangeable model shown in Table 10.1. Consider again the problems of the estimation of $\bar{\mu}$ and the prediction of $\bar{\mathscr{Y}}$. It will be noted that the optimal strategies are the same as those obtained under the transformation model However, the optimality results are in a sense stronger namely, the competing estimators need *not* be linear any more. Precise statements are given below.

Theorem 10.5. *Under the exchangeable model, in the class of strategies comprised of fixed-size* (n) *sampling designs and all* $p\xi$*-unbiased estimators of* $\bar{\mu}$, *the estimator* e_0 *given in* (10.4) *based on an arbitrary fixed-size* (n) *design minimizes* $\mathscr{E}E_p(e-\bar{\mu})^2$. *Further, the best strategy in the above class for p-unbiased prediction of* $\bar{\mathscr{Y}}$ *is given by* (p_0, e_0) *where* p_0 *is as in* (10.3).

Proof. We prove this result in two steps elaborating on all the technical detail. First we show that for an arbitrary fixed-size (n) design the estimator $e_0(s.\mathbf{Y})$ is the best estimator of $\bar{\mu}$ among all linear and nonlinear $p\xi$-unbiased estimators of $\bar{\mu}$. Next we assert that we can improve on the strategy (p, e_0) by replacing the sampling design p by p_0 in case of p-unbiased prediction of $\bar{\mathscr{Y}}$. Our first step consists of arguments elaborated in (a)–(c).

(a) We claim that in our search for the best $p\xi$-unbiased estimator of $\bar{\mu}$ we can restrict to the subclass of estimators which are symmetric in $\{\mathscr{X}_i \mid i \in s\}$ for every s with $P(s) > 0$. More specifically, we claim that a $p\xi$-unbiased estimator $e(s, \mathscr{X})$ which is not symmetric in $\{\mathscr{X}_i \mid i \in s\}$ can be uniformly improved by its *symmetrized* version $e^*(s, \mathscr{X})$ defined as $e^*(s, \mathscr{X}) = \Sigma_{g\in\mathscr{G}}\, e(s, g\mathscr{X})/n!$ where $\mathscr{G}$ refers to the symmetric group of permutations of $\{\mathscr{X}_i \mid i \in s\}$ and $g\mathscr{X}$ defines a permutation of $\{\mathscr{X}_i \mid i \in s\}$ induced by g. This can be rigorously proved as follows.

Since $\{\mathscr{X}_1, \mathscr{X}_2, \ldots, \mathscr{X}_N\}$ are exchangeable, so are $\{\mathscr{X}_i \mid i \in s\}$ for every s with $P(s) > 0$ and, moreover, all such n-dimensional marginals are the same. Therefore, we may denote by F_n the n-dimensional marginal distribution of *any* subset of n of the random variables. We may also conveniently use the random variables $\{\mathscr{X}_1, \mathscr{X}_2, \ldots, \mathscr{X}_n\}$ for our purpose of specifying the form of F_n. Suppose now that $e(s, \mathscr{X})$ is a $p\xi$-unbiased estimator of $\bar{\mu}$. Then using only the model $\xi(\cdot)$, we deduce by appealing to exchangeability that

$$\begin{aligned}\mathscr{E}(e(s, \mathscr{X})) &= \mathscr{E}(e(s, g\mathscr{X})) \text{ for every } g \in \mathscr{G} \\ &= \sum_{g\in\mathscr{G}} \mathscr{E}(e(s, g\mathscr{X}))/n! = \mathscr{E}\Big(\sum_{g\in\mathscr{G}} e(s, g\mathscr{X})/n!\Big) \\ &= \mathscr{E}(e^*(s, \mathscr{X}))\,.\end{aligned}$$

Moreover, by a direct application of the Cauchy–Schwartz inequality, we may further deduce that $\mathscr{E}(e^2) \ge \mathscr{E}(e^{*2})$. This yields $\mathscr{E}(e-\bar{\mu})^2 \ge \mathscr{E}(e^*-\bar{\mu})^2$ and hence, trivially,

(i) $E_p\mathscr{E}(e) = E_p\mathscr{E}(e^*)$,
(ii) $E_p\mathscr{E}(e-\bar{\mu})^2 \ge E_p\mathscr{E}(e^*-\bar{\mu})^2$.

(b) We now start with a symmetric $p\xi$-unbiased estimator $e^*(s, \mathscr{X})$ of $\bar{\mu}$ (in the sense explained above) and write

$$\begin{aligned}\bar{\mu} &= E_p \mathscr{E}(e^*(s, \mathscr{X})) = \sum_s P(s)\mathscr{E}(e^*(s, \mathscr{X})) \\ &= \sum_s P(s)\mathscr{E}(e^*(s, g_s\mathscr{X})) \\ &= \mathscr{E}\left\{\sum_s P(s)e^*(s, g_s\mathscr{X})\right\}\end{aligned}$$

where in the last two summations above g_s represents the permutation which transforms $\{Z_i \mid i \in s\}$ to $\{\mathscr{X}_i \mid 1 \leq i \leq n\}$. Since for every sample s, with $P(s) > 0$, $e^*(s, g_s\mathscr{X})$ is symmetric in $\mathscr{X}_1, \mathscr{X}_2, \ldots, \mathscr{X}_n$, it turns out that the random variable $\Sigma_s P(s)e^*(s, g_s\mathscr{X})$ is also symmetric in $\mathscr{X}_1, \mathscr{X}_2, \ldots, \mathscr{X}_n$ irrespective of the nature of the fixed-size (n) design. Again referring to the model we readily observe that

$$\mathscr{E}\left\{\bar{\mathbf{b}} + \bar{\mathbf{a}}\left(\sum_{i=1}^{n} \mathscr{X}_i/n\right)\right\} = \bar{\mathbf{b}} + \bar{\mathbf{a}}\mu = \bar{\mu} \, .$$

Thus there are two symmetric functions of the random variables $\mathscr{X}_1, \ldots, \mathscr{X}_n$ which have the same expectation. But the joint distribution of $\mathscr{X}_1, \ldots, \mathscr{X}_n$ is exchangeable and hence we can apply a result of Fraser (1957) which states that for the family of exchangeable distributions the symmetric functions form a set of *complete sufficient statistics*. Applied to the present problem, this statement implies that the two estimators must be identical almost everywhere (a.e.). We have thus deduced that

$$\sum_s P(s)e^*(s, g_s\mathscr{X}) = \bar{\mathbf{b}} + \bar{\mathbf{a}}\left(\sum_{i=1}^{n} \mathscr{X}_i/n\right) \quad \text{a.e.}$$

(c) We now proceed to compute $E_p\mathscr{E}(e^{*2})$. Applying the Cauchy–Schwartz inequality,

$$\begin{aligned}\mathscr{E}E_p(e^{*2}) &= \mathscr{E}\left\{\sum_s P(s)e^{*2}(s, g_s\mathscr{X})\right\} \\ &\geq \mathscr{E}\left\{\sum_s P(s)e^*(s, g_s\mathscr{X})\right\}^2 = \mathscr{E}\left\{\bar{\mathbf{b}} + \bar{\mathbf{a}}\left(\sum_{i=1}^{n} \mathscr{X}_i/n\right)\right\}^2 .\end{aligned}$$

This yields

$$\begin{aligned}\mathscr{E}E_p(e^* - \bar{\mu})^2 &\geq \mathscr{E}\left\{\bar{\mathbf{b}} + \bar{\mathbf{a}}\left(\sum_{i=1}^{n} \mathscr{X}_i/n\right) - \bar{\mu}\right\}^2 \\ &= \mathscr{E}\left\{\bar{\mathbf{b}} + \bar{\mathbf{a}}\left(\sum_{i\in s} \mathscr{X}_i/n\right) - \bar{\mu}\right\}^2 \qquad \text{for every } s \text{ with } P(s) > 0\end{aligned}$$

$$= \mathscr{E}\left\{\sum_s P(s)(\bar{\mathbf{b}} + \bar{\mathbf{a}}\bar{\mathscr{X}}_s - \bar{\mu})^2\right\} \qquad \text{where } \bar{\mathscr{X}}_s = \sum_{i \in s} \mathscr{X}_i / n$$

$$= \mathscr{E} E_p(e_0 - \bar{\mu})^2 .$$

We have thus completed the first part of the proof.

In the second part our object is to verify that $E_p \mathscr{E}(e - \bar{\mathscr{Y}})^2 \geq E_{p_0} \mathscr{E}(e_0 - \bar{\mathscr{Y}})^2$ in the class of all p-unbiased strategies. This follows essentially along the same line of arguments we developed in the proof of Theorem 10.2. We leave it as an exercise (Exercise 10.3). □

Remark 10.2. The reader may observe that the argument given in (a) above in favor of using only symmetric functions of $\{\mathscr{X}_i \mid i \in s\}$ holds true even if the sampling design adopted is not a fixed-size (n) design. However, for such a sampling design, we will not be able to develop the arguments as in (b). In Exercise 10.16 the reader will find a problem involving such a case. This was pointed out by Godambe and Thompson (1973).

Finally, we consider the problem of ξ-unbiased prediction of $\bar{\mathscr{Y}}$. This time, we wish to minimize $\mathscr{E} E_p(e - \bar{\mathscr{Y}})^2$ by properly choosing an optimal strategy (p, e) where $e(s, \mathbf{Y})$ is ξ-unbiased for $\bar{\mathscr{Y}}$ in the sense of $\mathscr{E}(e - \bar{\mathscr{Y}}) = 0$ for every sample s with $P(s) > 0$. By following the approach in the proof of Theorem 10.2, we can write the expression for $\mathscr{E} E_p(e - \bar{\mathscr{Y}})^2$ as $\mathscr{E} E_p(e - \bar{\mathscr{Y}})^2 = \mathscr{E} E_p(e - \bar{\mu})^2 + \mathscr{V}(\bar{\mathscr{Y}}) - 2\mathscr{C}ov(E_p(e), \bar{\mathscr{Y}} - \bar{\mu})$. Unfortunately, however, this time $\mathscr{C}ov(E_p(e), \bar{\mathscr{Y}} - \bar{\mu})$ does not vanish nor does it reduce to a design-independent quantity such as $\mathscr{V}(\bar{\mathscr{Y}})$ unless, of course, $e(s, \mathbf{Y})$ is linear. Therefore the minimization problem does not possess any easy solution. This problem has not been tackled satisfactorily in the published literature.

10.5. RANDOM PERMUTATION MODEL AND RELATED OPTIMAL SAMPLING STRATEGIES

Consider the situation where a given population $\mathscr{U}$ is known to be of size N but the units have *not* yet been labelled for the purpose of identification. Further, suppose the values of the survey variable which are fixed unknown quantities comprise the set $(a_1 \leq a_2 \leq \cdots \leq a_N)$ of values written in ascending order of magnitude. What happens if the statistician labels the units as $1, 2, \ldots, N$ at random? In effect, for any permutation $(j_1, j_2, \ldots, j_N)$ of the labels, according to the random permutation law of probability, we then stipulate that

$$P(\mathscr{Y}_1 = a_{j_1}, \mathscr{Y}_2 = a_{j_2}, \ldots, \mathscr{Y}_N = a_{j_N}) = 1/N! \qquad (10.26)$$

This shows that even if the quantities $a_1, a_2, \ldots, a_N$ are fixed, random labelling of the units generates randomness in respect of the values assumed by $(\mathcal{Y}_1, \mathcal{Y}_2, \ldots, \mathcal{Y}_N)$. Therefore, under such a framework, it is no longer true that the reference population admits of a set of fixed unknown values $Y_1, Y_2, \ldots, Y_N$. Instead, the values assumed by the study variable behave in a random fashion with the joint distribution given by (10.26). We may say that the above considerations have led to a Random Permutation (RP) model specified by (10.26). We shall continue to denote by $\mathscr{E}, \mathscr{V}, \mathscr{C}ov$, respectively, the mean, variance, and covariance of the random variables generated by the RP model (10.26). It is evident that

$$\mathscr{E}(\mathcal{Y}_i) = \bar{\mathbf{a}}\,, \qquad \mathscr{V}(\mathcal{Y}_i) = \sigma_a^2\,, \qquad \mathscr{C}ov(\mathcal{Y}_i, \mathcal{Y}_j) = -\sigma_a^2/(N-1)\,, \qquad i \neq j \tag{10.27}$$

where

$$\bar{\mathbf{a}} = \sum_{j=1}^{N} a_j/N\,, \qquad \sigma_a^2 = \sum_{j=1}^{N} (a_j - \bar{\mathbf{a}})^2/N\,. \tag{10.28}$$

We may illustrate the concept of a RP model in the following way. Suppose a certain university campus has 30 buildings and we want to estimate the total number of telephones installed in the campus buildings. The values $a_1 \leq a_2 \leq \cdots \leq a_{30}$, written in ascending order, are fixed but unknown to the sampler beforehand. The statistician may decide to randomly label the buildings as $1, 2, \ldots, 30$ before adopting a particular sampling strategy. Therefore, in the process, the statistician is introducing an RP model as in (10.26) with $N = 30$, on the population of buildings under study. However, if the statistician decides to use a particular labelling of the units, determined in a nonprobabilistic fashion (such as labelling from the largest to the smallest), then we are again back to the usual fixed population setup. In either case, the randomness introduced through the sampling design is, of course, still present.

It is evident from (10.26) that the RP model generates a discrete exchangeable distribution for $\mathcal{Y}_1, \mathcal{Y}_2, \ldots, \mathcal{Y}_N$ with the parameters given in (10.27). Here the so-called superpopulation mean coincides with the actual finite population mean $\bar{\mathbf{Y}}$ and hence only the estimation problem becomes relevant. The optimality result contained in Theorem 10.5 extends itself to cover the RP model as well. Since for this model $\rho = -1/(N-1)$, then by (10.9) for the optimal strategy (p_0, e_0), $\mathscr{E}E_{p_0}(e_0 - \bar{\mu})^2 = \sigma^2\bar{\mathbf{a}}^2\{1 + (n-1)\rho\}/n$ simplifies to $\mathscr{E}E_{p_0}(\bar{\mathbf{y}} - \bar{\mathbf{Y}})^2 = (N-n)\sigma^2/n(N-1)$. At this stage, it is interesting to note that under the RP model, $\mathscr{E}(\bar{\mathbf{y}} - \bar{\mathbf{Y}})^2 = (N-n)\sigma^2/n(N-1)$ irrespective of the specific form of the fixed-size (n) design. Further, for any fixed-size (n) design optimality of the strategy (p, e_0) for estimation of $\bar{\mathbf{Y}}$ can be directly established for the RP model in appropriate subclasses of $p\xi$-unbiased strategies without appealing to Theorem 10.5. Kempthorne (1969) provided the optimality result restricting to linear p-unbiased and translation invariant estimators of $\bar{\mathbf{Y}}$ while C. R. Rao (1971) established the same, doing away with translation invariance. Ramakrishnan (1970), on the other hand, dispensed with p-unbiasedness but restricted to linear translation invariance to come up with the same result. Below we present the most general optimality result, due to J. N. K. Rao (1975), in an RP model involving linear estimators of $\bar{\mathbf{Y}}$.

Theorem 10.6. *For an RP model, in the class of homogeneous linear $p\xi$-unbiased estimators of $\bar{\mathbf{Y}}$ based on any given fixed-size (n) design, the one which minimizes $\mathscr{E}E_p(e-\bar{\mathbf{Y}})^2$ is given by $\bar{\mathbf{y}}$.*

Proof. A typical estimator in the given class can be written as $e(s, \mathbf{Y}) = \Sigma_{i\in s} b_{si}Y_i$ where $p\xi$-unbiasedness amounts to the restriction $\Sigma_s \Sigma_{i\in s} b_{si}P(s) = 1$ on the coefficients b_{si}, $i \in s$, $s \in S_d$. We proceed with the computation of $\mathscr{E}E_p(e-\bar{\mathbf{Y}})^2$ under the RP model (10.26) in combination with (10.27).

We will conveniently use (10.27).

$$
\begin{aligned}
\mathscr{E}E_p(e-\bar{\mathbf{Y}})^2 &= \mathscr{E}E_p\left[\sum_{i\in s}\left(b_{si}-\frac{1}{N}\right)Y_i - \sum_{j\notin s} Y_j/N\right]^2 \\
&= E_p\mathscr{E}\left[\sum_{i\in s}\left(b_{si}-\frac{1}{N}\right)Y_i - \sum_{j\notin s} Y_j/N\right]^2 \\
&= E_p\left[\mathscr{V}\left\{\sum_{i\in s}\left(b_{si}-\frac{1}{N}\right)Y_i - \sum_{j\notin s} Y_j/N\right\}\right. \\
&\quad \left. + \left\{\mathscr{E}\sum_{i\in s}\left(b_{si}-\frac{1}{N}\right)Y_i - \mathscr{E}\sum_{j\notin s} Y_j/N\right\}^2\right] \\
&= E_p\left[\frac{\sigma_a^2}{N-1}\left\{N\sum_{i\in s} b_{si}^2 - \left(\sum_{i\in s} b_{si}\right)^2\right\} + \bar{\mathbf{Y}}^2\left(\sum_{i\in s} b_{si}-1\right)^2\right] \\
&= \bar{\mathbf{Y}}^2\left[\sum_s P(s)\left(\sum_{i\in s} b_{si}-1\right)^2\right] + \frac{\sigma_a^2}{N-1}\sum_s P(s) \\
&\quad \times\left\{N\sum_{i\in s} b_{si}^2 - \left(\sum_{i\in s} b_{si}\right)^2\right\}.
\end{aligned}
$$

Next we minimize the second expression namely, $\Sigma_s P(s)\{N\Sigma_{i\in s} b_{si}^2 - (\Sigma_{i\in s} b_{si})^2\}$ subject to the condition $\Sigma_s \Sigma_{i\in s} b_{si}P(s) = 1$ as indicated above. It is not difficult to verify that the best choice of b_{si} is $b_{si} = 1/n$, $i \in s$, $s \in S_d$. Further, it is trivially true that this choice also minimizes the first part in the above. Hence, $\bar{\mathbf{y}}$ is the best linear estimator of $\bar{\mathbf{Y}}$ in the sense of minimizing $\mathscr{E}E_p(e-\bar{\mathbf{Y}})^2$ for any given fixed-size (n) design. □

Some further aspects of RP models have been studied in T. J. Rao (1984) and J. N. K. Rao and Bellhouse (1978).

10.6. OPTIMAL SAMPLING STRATEGIES BASED ON THE REGRESSION MODEL

The regression model has been extensively studied in the literature. We will discuss some results with reference to the following simplified form of the model shown in Table 10.1:

$$\mathscr{E}(\mathscr{Y}_i) = \beta X_i\,, \qquad \mathscr{V}(\mathscr{Y}_i) = \sigma^2 v(X_i)\,, \qquad \mathscr{C}ov(\mathscr{Y}_i, \mathscr{Y}_j) = 0\,, \qquad i \neq j\,. \tag{10.29}$$

The model (10.29) is known as the *linear regression* model (through the origin). A special form of $v(X)$ is

$$v(X) = X^g\,, \qquad g \in [0, 2]\,; \tag{10.30}$$

$g = 0$ corresponds to homoscedastic errors. Under model (10.29),

$$\bar{\mu} = \mathscr{E}(\bar{\mathscr{Y}}) = \beta \bar{\mathbf{X}}\,, \qquad \mathscr{E}(\mathscr{Y}_i/X_i) = \beta\,, \qquad \mathscr{V}(\mathscr{Y}_i/X_i) = \sigma^2 v(X_i) X_i^{-2}\,,$$
$$\mathscr{C}ov(\mathscr{Y}_i/X_i, \mathscr{Y}_j/X_j) = 0\,, \qquad 1 \le i \neq j \le N\,. \tag{10.31}$$

First we consider the prediction problem regarding $\bar{\mathscr{Y}}$. We restrict ourselves to the class of linear ξ-unbiased predictors. The following result was discussed by Royall (1970).

Theorem 10.7. *Under the regression model* (10.29), *the best linear ξ-unbiased predictor of $\bar{\mathscr{Y}}$ for a given sampling design d is given by $e_0(s, \mathbf{Y})$ where for every s with $P(s) > 0$*

$$e_0(s, \mathbf{Y}) = f_s \bar{\mathbf{y}}_s + (1 - f_s)\Big\{\sum_{i \in s} Y_i X_i / v(X_i)\Big\}\Big\{\sum_{i \in s} X_i^2 / v(X_i)\Big\}^{-1} \bar{\mathbf{x}}_{\tilde{s}}\,. \tag{10.32}$$

and $s \cup \tilde{s} = \mathscr{U}$, $\bar{\mathbf{y}}_s = \Sigma_{i \in s} Y_i / n(s)$, $f_s = n(s)/N$, $\bar{\mathbf{x}}_{\tilde{s}} = \Sigma_{i \notin s} X_i / (N - n(s))$.

Proof. We proceed as in the proof of Theorem 10.3. Writing $e(s, \mathbf{Y}) = f_s \bar{\mathbf{y}}_s + (1 - f_s) U(s, \mathbf{Y})$ with $U(s, \mathbf{Y}) = u_{s0} + \Sigma_{i \in s} u_{si} Y_i$, for a linear ξ-unbiased predictor of $\bar{\mathscr{Y}}$, we set $\mathscr{E}(U - \bar{\mathscr{Y}}_{\tilde{s}}) = 0$ identically in β. Recalling that $\bar{\mathscr{Y}} = f_s \bar{\mathscr{Y}}_s + (1 - f_s) \bar{\mathscr{Y}}_{\tilde{s}}$ this yields,

$$u_{s0} = 0\,, \qquad \sum_{i \in s} u_{si} X_i = \bar{\mathbf{x}}_{\tilde{s}}\,. \tag{10.33}$$

To minimize $\mathscr{E} E_p (e - \bar{\mathscr{Y}})^2$, for a given sampling design, we try to minimize $\mathscr{E}(e - \bar{\mathscr{Y}})^2$ for every sample s with $P(s) > 0$. This is equivalent to minimizing $\mathscr{E}(U - \bar{\mathscr{Y}}_{\tilde{s}})^2$ subject to $\mathscr{E}(U - \bar{\mathscr{Y}}_{\tilde{s}}) = 0$ that is, $\mathscr{E}(U) = \beta \bar{\mathbf{x}}_{\tilde{s}}$. We may rewrite $\mathscr{E}(U - \bar{\mathscr{Y}}_{\tilde{s}})^2$ as

$$\mathscr{E}(U - \beta \bar{\mathbf{x}}_{\tilde{s}} + \beta \bar{\mathbf{x}}_{\tilde{s}} - \bar{\mathscr{Y}}_{\tilde{s}})^2 = \mathscr{E}(U - \beta \bar{\mathbf{x}}_{\tilde{s}})^2 + \mathscr{E}(\bar{\mathscr{Y}}_{\tilde{s}} - \beta \bar{\mathbf{x}}_{\tilde{s}})^2$$
$$- 2\mathscr{E}\{(U - \beta \bar{\mathbf{x}}_{\tilde{s}})(\bar{\mathscr{Y}}_{\tilde{s}} - \beta \bar{\mathbf{x}}_{\tilde{s}})\}\,.$$

Now we observe that $\bar{\mathcal{Y}}_{\bar{s}}$ is a linear function of $\{\mathcal{Y}_i | i \not\in s\}$ and $U(s, \mathbf{Y})$ is again a linear function of $\{\mathcal{Y}_i | i \in s\}$. Under the model (10.29), $\mathcal{C}ov(\mathcal{Y}_i, \mathcal{Y}_j) = 0$, $i \neq j$. Therefore, in the above, covariance term vanishes and, hence,

$$\mathcal{E}(U - \bar{\mathcal{Y}}_{\bar{s}})^2 = \mathcal{V}(U) + \mathcal{V}(\bar{\mathcal{Y}}_{\bar{s}}) . \tag{10.34}$$

The second term above namely, $\mathcal{V}(\bar{\mathcal{Y}}_{\bar{s}}) = \sigma^2 \sum_{i \not\in s} v(X_i)((N - n(s))^{-2})$, is independent of the choice of $U(s, \mathbf{Y})$. Hence we have to minimize $\mathcal{V}(U)$ subject to $\mathcal{E}(U) = \beta \bar{\mathbf{x}}_{\bar{s}}$ by choosing an appropriate homogeneous linear function $U = \sum_{i \in s} u_{si} Y_i$. This is a standard problem in linear regression and using the familiar terminology of best linear unbiased estimator (blue), we state that the blue of $\beta \bar{\mathbf{x}}_{\bar{s}}$ is $\hat{\beta} \bar{\mathbf{x}}_{\bar{s}}$ where

$$\hat{\beta} = \hat{\beta}(s, \mathbf{Y}) = \left\{ \sum_{i \in s} Y_i X_i / v(X_i) \right\} \left\{ \sum_{i \in s} X_i^2 / v(X_i) \right\}^{-1} . \tag{10.35}$$

This yields the optimal ξ-unbiased linear predictor as $e_0(s, \mathbf{Y}) = f_s \bar{\mathbf{y}}_s + (1 - f_s) \hat{\beta} \bar{\mathbf{x}}_{\bar{s}}$. This completes the proof. □

We derive an expression for the minimum value of $E_p \mathcal{E}(e - \bar{\mathcal{Y}})^2$ as follows. Using the fact that $e - \bar{\mathcal{Y}} = (1 - f_s)(U - \bar{\mathcal{Y}}_{\bar{s}})$, we obtain

$$\begin{aligned}
\mathcal{E} E_p(e - \bar{\mathcal{Y}})^2 &= E_p[(1 - f_s)^2 \mathcal{E}(U - \bar{\mathcal{Y}}_{\bar{s}})^2] \\
&= E_p[(1 - f_s)^2 \{\mathcal{V}(U) + \mathcal{V}(\bar{\mathcal{Y}}_{\bar{s}})\}] \\
&\geq E_p\left[(1 - f_s)^2 \left\{ \bar{\mathbf{x}}_{\bar{s}}^2 \mathcal{V}(\hat{\beta}) + \sigma^2 \sum_{i \not\in s} v(X_i)(N - n(s))^{-2} \right\}\right] \\
&= E_p\left[\sigma^2 N^{-2} \left\{ \left(\sum_{i \not\in s} X_i \right)^2 \left(\sum_{i \in s} X_i^2 / v(X_i) \right)^{-1} + \sum_{i \not\in s} v(X_i) \right\}\right] \\
&= \sigma^2 N^{-2} E_p\left[\left(\sum_{i \not\in s} X_i \right)^2 \left(\sum_{i \in s} X_i^2 / v(X_i) \right)^{-1} + \sum_{i \not\in s} v(X_i) \right] .
\end{aligned} \tag{10.36}$$

The expression in (10.36) coincides with $\mathcal{E} E_p(e_0 - \bar{\mathcal{Y}})^2$ since e_0 is the optimal linear ξ-unbiased predictor of $\bar{\mathcal{Y}}$.

The following result due to Brewer (1963) relates to optimality of the ratio estimator.

Corollary 10.8. *Under the linear regression model* {(10.29), (10.30) *with* $g = 1$}, *the best linear* ξ*-unbiased predictor of* $\bar{\mathcal{Y}}$ *under* SRS(N, n) *sampling procedure is given by the ratio estimator* $\hat{\bar{\mathbf{Y}}}_R = (\bar{\mathbf{y}}/\bar{\mathbf{x}})\bar{\mathbf{X}}$.

So far we have not imposed the condition of p-unbiasedness on the predictor of $\bar{\mathcal{Y}}$. To get the optimal strategy for prediction of $\bar{\mathcal{Y}}$, we may or may not insist on p-unbiasedness of the predictor. Accordingly, we end up with the following results.

Theorem 10.9. *Under the linear regression model* (10.29) *an optimal strategy for ξ-unbiased linear prediction of $\bar{\mathcal{Y}}$ based on fixed-size* (n) *designs consists of a degenerate sampling design p_0: $P(s_0) = 1$ where the sample s_0 of size n includes only those n units of the population for which $(\Sigma_{i \notin s_0} X_i)^2(\Sigma_{i \in s_0} X_i^2/v(X_i))^{-1} + (\Sigma_{i \notin s_0} v(X_i))$ is the least. The predictor is accordingly given by $e_0(s_0, \mathbf{Y})$ as in* (10.32).

Proof. We recall that $\mathscr{E}E_p(e - \bar{\mathcal{Y}})^2 \geq \mathscr{E}E_p(e_0 - \bar{\mathcal{Y}})^2$ as shown in Theorem 10.7. By the nature of the sample s_0 the bracketed expression in (10.36) attains its least value when $s = s_0$. Consequently, for the sampling design d_0, $\mathscr{E}E_p(e_0 - \bar{\mathcal{Y}})^2 \geq \mathscr{E}E_{p_0}(e_0 - \bar{\mathcal{Y}})^2$. Hence the result. □

Corollary 10.10. *Under the linear regression model* (10.29) *with*

$$v(x) \textit{ nondecreasing and } \{v(x)/x^2\} \textit{ nonincreasing in } x\,, \qquad (10.37)$$

the best ξ-unbiased linear predictor of the population mean based on fixed-size (n) *sampling designs is given by $e_0(s_0, \mathbf{Y})$ where s_0 corresponds to the n units with largest X-values and $e_0(s_0, \mathbf{Y})$ is of the form $e_0(s, \mathbf{Y})$ in* (10.32) *with s replaced by s_0.*

Proof. When (10.37) holds, the expression $\{(\Sigma_{i \notin s_0} X_i)^2/(\Sigma_{i \in s_0} X_i^2/v(X_i))\} + \Sigma_{i \notin s_0} v(X_i)$ attains its minimum when s_0 consists of units with the largest X-values. Hence the result is proved. □

Remark 10.3. In Theorem 10.9 as also in Corollary 10.10, the optimal strategy suggested has a degenerate sampling design associated with it. The support of the design consists of only one sample (s_0) of size n. This is in sharp contrast to the general definition of a sampling design we started with in Chapter One. According to the latter, a sampling design is required to cover the whole of $\mathcal{U}$ and not just a proper subset of it. It may be argued that the general definition of a sampling design is geared towards achieving p-unbiasedness of estimators. Clearly, neither in Theorem 10.9 nor in Corollary 10.10, we insist on p-unbiasedness. This explains why we may possibly relax the condition that a sampling design should be defined on the whole of $\mathcal{U}$.

For the next result, we introduce the following definition.

Definition 10.3. A fixed-size (n) sampling design is said to be a *generalized* ΠPS (abbreviated as GΠPS(n)) design if for every sample s in the support of the design, $\Sigma_{i \in s} X_i^{1-g/2} = nN\bar{\mathbf{X}}/\Sigma_{i=1}^N X_i^{g/2}$ and, moreover, $\pi_i \propto X_i^{g/2}$ for each i, $1 \leq i \leq N$, where $g \in [0, 2]$.

A GΠPS strategy consists of a GΠPS(n) design and the HTE. T. J. Rao (1971) deduced that $\mathscr{E}E_p(\hat{\bar{\mathscr{Y}}} - \bar{\mathscr{Y}})^2 \propto (\Sigma_1^N X_i^{g/2})^2 - n\,\Sigma_1^N X_i^g$ for such a strategy. Pedgaonkar and Prabhu-Ajgaonkar (1978) made a relative comparison of GΠPS strategies and other related strategies. Padmawar (1981) made some further comparisons of certain sampling strategies again based on the superpopulation model given in (10.29) and (10.30). Below we prove an optimality result due to Ramachandran (1978).

Theorem 10.11. *A GΠPS strategy* (GΠPS(n), HTE) *is optimal in the class of all linear p-unbiased strategies having the same expected sample size n, under the linear regression model given by* (10.29) *and* (10.30).

Proof. Consider a competing strategy (p, e) with $e(s, \mathbf{Y}) = \Sigma_{i\in s} b_{si} Y_i$ where the coefficients satisfy $\Sigma_{s\ni i} b_{si} P(s) = 1/N$, $1 \le i \le N$. The expression $\mathscr{E}E_p(e - \bar{\mathscr{Y}})^2$ simplifies to $\mathscr{E}V_p(e) + \mathscr{V}E_p(e)$. The second term is $\mathscr{V}(\bar{\mathscr{Y}})$ and this is independent of the choice of (p, e). The first term simplifies to

$$\mathscr{E}\left[\sum_s P(s)\left\{\left(\sum_{i\in s} b_{si}^2 Y_i^2\right) + \sum_{\substack{i\neq j\\ i,j\in s}}\sum b_{si} b_{sj} Y_i Y_j\right\} - \bar{\mathscr{Y}}^2\right]. \qquad (10.38)$$

Under the given model, the relevant terms involving b_{si} in the above expression simplify to $\sigma^2[\Sigma_s P(s)(\Sigma_{i\in s} b_{si}^2 X_i^g)] + \beta^2\,\Sigma_s P(s)(\Sigma_{i\in s} b_{si} X_i)^2$. We can now show that both the terms $\Sigma_s P(s)(\Sigma_{i\in s} b_{si}^2 X_i^g)$ and $\Sigma_s P(s)(\Sigma_{i\in s} b_{si} X_i)^2$ are simultaneously minimized by the GΠPS strategy. We leave this verification to the reader (Exercise 10.15). □

Except when $g = 2$, it may *not* be an easy task to come up with a GΠPS design in a given situation. T. J. Rao (1971) has given some hypothetical examples.

We now specialize to the particular model

$$\mathscr{E}(\mathscr{Y}_i) = \beta X_i\,, \qquad \mathscr{V}(\mathscr{Y}_i) = \sigma^2 X_i^2\,, \qquad \mathscr{C}ov(\mathscr{Y}_i, \mathscr{Y}_j) = 0\,, \qquad 1 \le i \ne j \le N\,, \qquad (10.39)$$

and characterize an optimal linear p-unbiased strategy.

Corollary 10.12. *Under the linear regression model* (10.39), *in the class of all linear p-unbiased strategies, having the same expected sample size n, the strategy* (p_0, e_0) *given in* (10.12) *is* $p\xi$*-optimal for prediction of* $\bar{\mathscr{Y}}$.

The proof follows from that of Theorem 10.11 upon observing that for $g = 2$, every GΠPS design is indeed a ΠPS design.

The above result was established by Godambe (1955). In Godambe and Joshi (1965), the result is extended to the class of all p-unbiased strategies, relaxing the condition of linearity. This is done by replacing the condition of zero covariance by independence in (10.39).

In the above, the sampling design p_0 corresponds to a fixed-size (n) ΠPS sampling design. Sometimes, however, the sampling design is chosen in advance and an optimal predictor is sought under a specified superpopulation model.

A study on the robustness of ratio estimator for variation in the model is desirable. The interested reader is referred to Mukhopadhyay (1986). For another related study on the optimality of the mean-of-ratios, we refer to Kott (1986). A useful review is given in Chaudhuri (1988).

10.7. CONCLUDING REMARKS

In this chapter we have introduced some superpopulation models and initiated a study of finite population inference based on such models. In this context, the following result, due to Fuller (1970), may also serve as an interesting observation towards comparison of different predictors based on the same sampling design.

Consider a superpopulation model $\xi(\cdot)$ according to which

$$\mathscr{E}(\mathscr{Y}_i) = \mu\,, \qquad \mathscr{V}(\mathscr{Y}_i) = \sigma^2\,, \qquad \mathscr{C}ov(\mathscr{Y}_i, \mathscr{Y}_j) = 0\,, \qquad 1 \le i \ne j \le N\,.$$

Let d be a sampling design and s a sample chosen and surveyed with the results: $\{(i, Y_i) \mid i \in s\}$. Let $\hat{\mu}_1(s)$ and $\hat{\mu}_2(s)$ be two different estimators of μ based on the survey results. Suppose for every s with $P(s) > 0$, it so happens that $\mathscr{E}(\hat{\mu}_1(s) - \mu)^2 \le \mathscr{E}(\hat{\mu}_2(s) - \mu)^2$. Then the predictor $\hat{\bar{\mathbf{Y}}}_1(s) = f_s\bar{\mathbf{y}}_s + (1 - f_s)\hat{\mu}_1(s)$ is better than the predictor $\hat{\bar{\mathbf{Y}}}_2(s) = f_s\bar{\mathbf{y}}_s + (1 - f_s)\hat{\mu}_2(s)$ for the purpose of prediction of $\bar{\mathbf{Y}}$ in the sense that $\mathscr{E}E_p(\hat{\bar{\mathbf{Y}}}_1(s) - \bar{\mathbf{Y}})^2 \le \mathscr{E}E_p(\hat{\bar{\mathbf{Y}}}_2(s) - \bar{\mathbf{Y}})^2$. Here $f_s = n(s)/N$ and $\bar{\mathbf{y}}_s = \Sigma_{i\in s} Y_i/n(s)$. It turns out that the above comparison is valid irrespective of the nature of the sampling design adopted. (However, the above model would suggest consideration of SRS(N, n) sampling design if we are looking for a fixed-size (n) design). Bellhouse (1987) extended this result to cover a model which yields unequal means and variances but again zero covariances. We do not enter into the technical details.

In the context of a superpopulation model, we have only presented results related to estimation and prediction of a finite population mean. There have been studies on estimation and prediction of a finite population variance as well. We refer to Mukhopadhyay (1978, 1982) and Sengupta (1988) for results in this area.

As against one-stage sampling, superpopulation models have also been considered in the setup of two-stage sampling designs. Bellhouse et al. (1977) considered an exchangeable prior distribution while Padmawar and Mukhopadhyay (1985) considered an RP model to derive various results on optimal strategies in two-stage sampling.

Kalton (1983) and Thomsen and Tesfu (1988) studied the use of superpopulation models in the practice of survey sampling. It is true that most surveys are aimed at providing estimates of simple descriptive parameters (like mean, variance) of the survey population. However, typically, surveys are multipurpose in nature and different parts of multiresponse survey data may need to be analyzed for different purposes. Therefore, even if a model is suitable for one survey variable, it may not be realistic for others. Complete reliance on a particular superpopulation model is not, therefore, advisable in a multiresponse multipurpose survey, but it may be used as a guiding factor for selecting a reasonable sampling strategy.

Godambe and Thompson (1986) have discussed in an interesting manner, the relation between parameters of a superpopulation and the actual survey population. They have explained how the descriptive parameters of a survey population may be viewed in a natural way as the finite population counterparts of the parameters of a superpopulation.

The use of particular superpopulation models help to provide guidance in the choice of sampling designs but it is difficult to determine how far it should dictate the choice of an estimator. Any violation from the true model would make it difficult to validate the survey estimates unless the estimators are robust in an appropriate sense. Iachan (1984) has attempted to integrate some dispersed results in this direction with reference to the sampling strategies based on the use of auxiliary information. Pereira and Rodrigues (1983) have discussed the use of the general theory of linear models in studying robustness of linear predictors of finite population means.

EXERCISES

10.1. In the context of Theorem 10.3, show that $\mathscr{C}ov(U, \bar{\mathscr{Y}}_{\tilde{s}}) = \rho\sigma^2\bar{\mathbf{a}}_{\tilde{s}}^2$.

10.2. Show that under the general transformation model $\mathscr{E}((e_0 - e^*)(e^* - \bar{\mathscr{Y}})) = 0$ for every sample s with $P(s) > 0$ where $e_0(s, \mathbf{Y}) = \bar{\mathbf{b}} + \bar{\mathbf{a}}\bar{\mathbf{z}}_s$ and $e^*(s, \mathbf{Y}) = f_s\bar{\mathbf{y}}_s + (1 - f_s)(\bar{\mathbf{z}}_s\bar{\mathbf{a}}_{\tilde{s}} + \bar{\mathbf{b}}_{\tilde{s}})$ in the usual notation.

10.3. In the context of Theorem 10.5, show that for a p-unbiased predictor e of $\bar{\mathscr{Y}}$,

$$\mathscr{E}E_p(e - \bar{\mathscr{Y}})^2 \geq \mathscr{E}E_{p_0}(e_0 - \bar{\mathscr{Y}})^2$$

where p is any fixed-size (n) design and (p_0, e_0) are as in (10.3) and (10.4), respectively.

10.4. Consider homogeneous linear p-unbiased estimators of $\bar{\mathbf{Y}}$ of the form $e(s, \mathbf{Y}) = \Sigma_{i\in s} b_{si}Y_i$ where the coefficients satisfy the unbiasedness condition: $\Sigma_{s\ni i} b_{si}P(s) = 1/N$, $1 \leq i \leq N$. The estimator $e(s, \mathbf{Y})$ is said to be linear invariant whenever $\Sigma_{i\in s} b_{si} = 1$ for every sample s with positive probability. Show that under an RP model, the optimal estimator is the sample mean when we adopt an SRS(N, n) sampling design.

10.5. Show that in the above result, restriction to linear invariant estimators is *not* necessary.

10.6. Show that the same result can be obtained by minimizing $\mathscr{E}E_{p_0}(e - \bar{Y})^2$ when we restrict to linear invariant estimators which are *not* necessarily p_0-unbiased. Here $p_0 \equiv$ SRS(N, n) sampling design.

10.7. Make a list of all available optimality results concerning the sampling strategy: (SRS(N, n), sample mean).

10.8. In the general transformation model, suppose $b_k \propto a_k$. Show that this leads to optimality of the HTE coupled with a ΠPS sampling scheme.

10.9. Explain how a purposive sampling design may turn out to be optimal for linear prediction of a finite population total.

10.10. Consider the superpopulation model {(10.29), (10.30)} and compare the performances of the HTE and the Rao–Hartley–Cochran Sampling Strategy.

10.11. Compute $\mathscr{E}V_p$(HTE) for a fixed-size (n) sampling design with respect to each of the models: transformation model, exchangeable model, random permutation model and in each case suggest the nature of the sampling design to couple with the HTE to produce the best strategy in a suitable class.

10.12. Let e be any homogeneous linear p-unbiased estimator of the population total with respect to a given sampling design. Specify a superpopulation model $\xi(\cdot)$ for which $E_p V_\xi(e) \geq \Sigma_{i=1}^N \sigma_i^2((1/\pi_i) - 1)$ where $\sigma_i^2 = V_\xi(Y_i)$ and π_i = inclusion probability of unit i, $1 \leq i \leq N$. Examine the case of equality.

10.13. Examine, after postulating a suitable superpopulation model, how the usual ratio estimator, based on a suitable sampling design, turns out to be the best linear unbiased predictor for a finite population total.

10.14. Make a list of all available optimality results involving the HTE.

10.15. With reference to Theorem 10.11, resolve the minimization problems involved in its proof.

10.16. Consider an exchangeable model in two random variables $\mathscr{X}_1$ and $\mathscr{X}_2$ with the parameters μ, σ^2, and ρ as in the usual notation. Let the sampling design adopted be given by: $P((1)) = 0.6$ and $P((1,2)) = 0.4$. Let $h(\cdot)$ be a nonzero nonconstant function of $\mathscr{X}_1$ or $\mathscr{X}_2$ with finite second moment. Define the estimator e^* as follows:

$$e^*((1), \mathscr{X}_1) = \mathscr{X}_1 - 4h(\mathscr{X}_1),\ e^*((1,2), \mathscr{X}) = \frac{\mathscr{X}_1 + \mathscr{X}_2}{2} + 3\{h(\mathscr{X}_1) + h(\mathscr{X}_2)\}\,.$$

Prove the following:

(a) $E_p \mathscr{E}(e^*) = \mu$

(b) $E_p \mathscr{E}\{(e^* - e_0)(e_0 - \mu)\} \neq 0$ when $h(z) = z$ and e_0 is the estimator defined as $e_0((1), \mathscr{X}_1) = \mathscr{X}_1$, $e_0((1,2), \mathscr{X}) = (\mathscr{X}_1 + \mathscr{X}_2)/2$.

(c) The estimator e_0 does not minimize $E_p \mathscr{E}(e_0 - \mu)^2$ uniformly in the parameters of the distribution among all $p\xi$-unbiased estimators of μ.

(d) There is at least one other form of $h(\cdot)$ for which the statement in (b) is true.

10.17. Analyze the data in Exercise 6.13 using a suitable superpopulation model.

REFERENCES

Arnab, R. (1986). Optimal prediction for a finite population total with connected designs and related model-based results. *Metrika*, **33**, 79–84.

Basu, D. (1971). An essay on the logical foundations of survey sampling, part one. In: *Foundations of Statistical Inference* (Godambe, V. P., and Sprott, D. A., Eds.), 203–242. Holt, Rinehart and Winston, Toronto.

Bellhouse, D. R. (1987). Model-based estimation in finite population sampling. *Amer. Statistician*, **41**, 260–262.

Bellhouse, D. R., Thompson, M. E., and Godambe, V. P. (1977). Two-stage sampling with exchangeable prior distributions. *Biometrika*, **64**, 97–103.

Brewer, K. R. W. (1963). Ratio estimation and finite populations: some results deducible from the assumption of an underlying stochastic process. *Austr. J. Statist.*, **5**, 93–105.

Cassel, C. M., Särndal, C. E., and Wretman, J. H. (1976). Some results on generalized difference estimation and generalized regression estimation for finite populations. *Biometrika*, **63**, 615–620.

Chaudhuri, A. (1988). Optimality of sampling strategies. In: *Handbook of Statistics, Vol. 6, Sampling* (Krishnaiah, P. R., and Rao, C. R., Eds.), 47–96. North-Holland, Amsterdam.

Cochran, W. G. (1946). Relative accuracy of systematic and stratified random samples for a certain class of populations. *Ann. Math. Statist.*, **17**, 164–177.

Fraser, D. A. S. (1957). *Nonparametric Methods in Statistics*. Wiley, New York.

Fuller, W. A. (1970). Simple estimators for the mean of skewed populations. Tech. Report, Iowa State University.

Godambe, V. P. (1955). A unified theory of sampling from finite populations. *J. Roy. Statist. Soc.*, **B17** 269–278.

Godambe, V. P., and Joshi, V. M. (1965). Admissibility and Bayes estimation in sampling finite populations I. *Ann. Math. Statist.*, **36**, 1707–1722.

Godambe, V. P., and Thompson, M. E. (1973). Estimation in sampling theory with exchangeable prior distributions. *Ann. Statist.*, **1**, 1212–1221.

Godambe, V.P., and Thompson, M. E. (1986). Parameters of superpopulation and survey population. *Internat. Statist. Rev.*, **54**, 127–138.

Iachan, R. (1984). Sampling strategies, robustness and efficiency: the state of the art. *Internat. Statist. Rev.*, **52**, 209–218.

Kalton, G. (1983). Models in the practice of survey sampling. *Internat. Statist. Rev.*, **51**, 175–188.

Kempthorne, O. (1969). Some remarks on statistical inference in finite sampling. In: *New Developments in Survey Sampling* (Johnson, N. L., and Smith, H. Jr., Eds.), 671–695. Wiley, New York.

Kott, P. S. (1986). When a mean-of-ratios is the best linear unbiased estimator under a model. Amer. Statistician, **40**, 202–204.

Mukerjee, R., and Sengupta, S. (1989). Optimal estimation of finite population total under a general correlated model. *Biometrika*, **76**, 789–794.

Mukhopadhyay, P. (1978). Estimating the variance of a finite population under a superpopulation model. *Metrika*, **25**, 115–122.

Mukhopadhyay, P. (1982). Optimum strategies for estimating the variance of a finite population under a superpopulation model. *Metrika*, **29**, 143–158.

Mukhopadhyay, P. (1986). Estimation under linear regression models. *Metrika*, **33**, 129–134.

Padmawar, V. R. (1981). A note on the comparison of certain sampling strategies. *J. Roy. Statist. Soc.*, **B43**, 321–326.

Padmawar, V. R., and Mukhopadhyay, P. (1985). Estimation under two-stage random permutation models. *Metrika*, **32**, 339–349.

Pedgaonkar, A. M., and Prabhu-Ajgaonkar, S. G. (1978). Comparison of sampling strategies. *Metrika*, **25**, 149–154.

Pereira, C. A. deB., and Rodrigues, J. (1983). Robust linear prediction in finite populations. *Internat. Statist. Rev.*, **51**, 293–300.

Ramachandran, G. (1978). Choice of strategies in survey sampling. Unpublished Ph.D. Thesis, Indian Statistical Institute.

Ramakrishnan, M. K. (1970). Optimum estimators and strategies in survey sampling. Unpublished Ph.D. Thesis, Indian Statistical Institute.

Rao, C. R. (1971). Some aspects of statistical inference in problems of sampling from finite populations. In: *Foundations of Statistical Inference* (Godambe, V. P., and Sprott, D. A., Eds.), 177–202. Holt, Rinehart and Winston, Toronto.

Rao, J. N. K. (1975). On the foundations of survey sampling. In: *A Survey of Statistical Design and Linear Models* (Srivastava, J. N., Ed.), 489–505. North-Holland, Amsterdam.

Rao, J. N. K, and Bellhouse, D. R. (1978). Estimation of finite population mean under generalized random permutation model. *J. Statist. Plann. Infer.*, **2**, 125–141.

Rao, T. J. (1971). Π*Ps* sampling designs and the Horvitz–Thompson estimator. *J. Amer. Statist. Assoc.*, **66**, 872–875.

Rao, T. J. (1984). Some aspects of random permutation models in finite population sampling theory. *Metrika*, **31**, 25–32.

Royall, R. M. (1970). On finite population sampling theory under certain linear regression models. *Biometrika*, **57**, 377–387.

Royall, R. M. (1988). The prediction approach to sampling theory. In: *Handbook of Statistics, Vol. 6, Sampling* (Krishnaiah, P. R., and Rao, C. R., Eds.), 399–413. North-Holland, Amsterdam.

Sengupta, S. (1988). Optimality of a design-unbiased strategy for estimating a finite population variance. *Sankhyā*, **B50**, 149–152.

Thomsen, I., and Tesfu, D. (1988). On the use of models in sampling from finite populations. In: *Handbook of Statistics, Vol. 6, Sampling* (Krishnaiah, P. R., and Rao, C. R., Eds.), 369–397. North-Holland, Amsterdam.

CHAPTER ELEVEN

Randomized Response: A Data-Gathering Tool for Sensitive Characteristics

This chapter is devoted to a study of the methods of sampling and inference when our aim is to gather information on *sensitive* characteristics. It is feared that a direct questionnaire technique for collecting information in such situations may result in untruthful or misleading responses or perhaps even refusal. This calls for special sampling techniques which ensure a high degree of confidentiality to the respondents.

11.1. SURVEYS ON SENSITIVE ISSUES

In some sample survey studies the study variable is of a sensitive nature either personally or legally for the respondent. Examples of such survey variables are numerous: addiction to alcohol or a controlled drug, having a history of abortions, having a clandestine income or a history of income tax evasion, being a gambler or a spouse beater, and the like. Clearly, collecting complete and correct information on such sensitive issues is very difficult if not impossible. For the interviewer it is a delicate affair, if not an invasion of privacy, to bluntly ask a stranger questions about sensitive or highly personal matters. Even if the interviewee does not refuse to answer such sensitive queries, we might wonder whether the replies provided are honest. This means that the collected data will be questionable and very likely incomplete.

In order to avoid excessive refusals or misleading responses to sensitive questions we need new statistical methodologies capable of (i) allowing interviewee's anonymity or at least a satisfactory degree of privacy, thus providing a basis which encourages greater cooperation, and (ii) providing valid estimates of the population parameters under study. Various ingenious statistical methods have been devised by statisticians for dealing with such situations. For reasons to be seen shortly all such techniques are known as randomized response methods (RRM), the first study of its kind being undertaken by Warner (1965) who developed a clever interviewing procedure designed to reduce or eliminate non-sampling bias in sample surveys of human populations.

11.2. RANDOMIZED RESPONSE TECHNIQUES FOR ONE DICHOTOMOUS QUALITATIVE SENSITIVE CHARACTERISTIC

Let Y be a qualitative characteristic with two versions, A and $\bar{A}$. For convenience we let $Y_i = 0$ if the population unit i has the nonsensitive version A, otherwise $Y_i = 1$. With respect to this characteristic the population $\mathcal{U}$ can be divided into two subpopulations $\mathcal{U}_0$ and $\mathcal{U}_1$ with sizes N_0 and N_1, respectively. Thus all units in $\mathcal{U}_1$ have the sensitive version of the characteristic and the goal of the survey is to make a statistical study of the unknown parameter $\phi = N_1/N$, the proportion of the units in the population having the sensitive version of the qualitative characteristic Y.

Suppose we pose the following question to the ith population unit: Do you belong to $\mathcal{U}_1$? Then, if Y_i is the response variable, it is not unreasonable to postulate the following mode of response:

$$P(Y_i = 0 | i \in \mathcal{U}_0) = 1 , \qquad P(Y_i = 1 | i \in \mathcal{U}_0) = 0 \tag{11.1}$$

$$P(Y_i = 1 | i \in \mathcal{U}_1) = q_{i1} , \qquad P(Y_i = 0 | i \in \mathcal{U}_1) = q_{i2} \tag{11.2}$$

$$P(\text{no response from Unit } i | i \in \mathcal{U}_1) = q_{i0} .$$

with

$$q_{i0} + q_{i1} + q_{i2} = 1 .$$

In other words, the unit i will tell the truth $(Y_i = 0)$ if he belongs to $\mathcal{U}_0$. However, if he belongs to $\mathcal{U}_1$, then he might tell the truth $(Y_i = 1)$ or lie $(Y_i = 0)$ or simply refuse to answer (no response) and each unit in $\mathcal{U}_1$ will have his own way of responding to the question. This model allows different likelihoods for telling the truth, telling a lie, or no response for different units in $\mathcal{U}_1$. Unfortunately, this reasonable assumption on the response complicates the structure of the data collected on the sensitive issue and deprives the survey statistician from making any meaningful inference on the unknown parameter.

Uncontrolled responses by the sample units will generate quite a number of unknown parameters in our data. There are no known statistical techniques which can handle such complicated data loaded with many unknown parameters. To overcome this difficulty, several clever techniques have been discovered to collect data on sensitive questions. They all share one common feature. The sample units are allowed to keep their privacy by randomizing their responses, but the freedom of individual randomization is replaced by one (or two) common randomization(s). Since a randomized response is less revealing than a direct answer, it is reasonable to expect greater cooperation for certain surveys on sensitive issues. Some of these methods will be explained below.

It may be mentioned that there are situations where one or more of the sample units are willing to give the true response without caring about randomization. The problem of blending direct and randomized responses in a given random sample has been discussed in Chaudhuri and Mukerjee (1985). We will not examine this possiblity here. An easy (but certainly not efficient) way of getting around this problem is acting as a proxy for the ones who report directly. Namely, we randomize the responses on their behalf.

Method 1: Related Question Procedure

To protect the privacy of the interviewee and to attract his collaboration we do not measure the sensitive characteristic Y, rather we measure a randomized version of it, say Z. To do this each interviewee is presented with two complementary questions:

$$Q_0\text{:}\quad \text{Do you belong to } \mathcal{U}_0?$$

$$Q_1\text{:}\quad \text{Do you belong to } \mathcal{U}_1?$$

Answer to both questions is either "yes" ($Z = 1$) or "no" ($Z = 0$), and if the reply to Q_0 is "yes" it would be "no" to Q_1 and vice versa. In deciding which question to answer each interviewee is provided with an identical randomization device, for example a spinner with its face divided into two mutually exclusive and exhaustive parts labeled $\mathcal{U}_0$ and $\mathcal{U}_1$. Unseen by the interviewer, each interviewee spins his spinner. He is told to answer Q_0 or Q_1 depending on whether or not his spinner stops at $\mathcal{U}_0$ or $\mathcal{U}_1$. Therefore, the outcome of each interview is either "1" or "0" without, of course, the interviewer knowing which question has been answered.

To be able to utilize the collected data it is necessary to assume that the answers are truthful. This means that if the random selection resulted in Q_0, then

$$P(Z_i = 1 | i \in \mathcal{U}_0) = 1\,, \qquad P(Z_i = 0 | i \in \mathcal{U}_1) = 1\,. \tag{11.3}$$

If the random selection resulted in Q_1, then

$$P(Z_i = 1 | i \in \mathcal{U}_1) = 1\,, \qquad P(Z_i = 0 | i \in \mathcal{U}_0) = 1\,. \tag{11.4}$$

The above are assumed to be satisfied for every population unit i interviewed with the corresponding score Z_i.

We will now pass on to the problem of estimation of ϕ based on the randomized response Z_i's. Let q be the probability that the spinner stops at $\mathcal{U}_1$, i.e., Q_1 is selected for answering. The choice for q is under the control of the interviewer and its value is assumed to be known to the data analyst. Under strict randomization, $q \neq 1$ or 0. Further, it is chosen to be different from 0.5. Then the Bernoulli random variable Z_i associated with the population unit i is characterized by the following randomized response distribution to be denoted as $P_R(.)$.

$$\begin{aligned} P_R(Z_i = 1) &= P_R(Q_0 \text{ is selected}).\ P(Z_i = 1 | Q_0 \text{ is selected}) \\ &\quad + P_R(Q_1 \text{ is selected}).\ P(Z_i = 1 | Q_1 \text{ is selected}) \\ &= P_R(Q_0 \text{ is selected}).\ P(i \in \mathcal{U}_0) + P_R(Q_1 \text{ is selected}).\ P(i \in \mathcal{U}_1) \\ &= (1-q)(1-\phi_i) + q\phi_i = (2q-1)\phi_i + (1-q) = \theta_i\,, \\ P_R(Z_i = 0) &= 1 - \theta_i \end{aligned} \tag{11.5}$$

where $\phi_i = 1$ if $i \in \mathcal{U}_1$ and $\phi_i = 0$ otherwise. We can rewrite the above expression for $P_R(Z_i = 1)$ in terms of ϕ_i as

$$\phi_i = [P_R(Z_i = 1) - (1-q)]/(2q-1)\,, \qquad \text{as } q \neq 0.5\,. \tag{11.6}$$

Therefore, if our data allow a response-based unbiased estimation of $\theta_i = P_R(Z_i = 1)$ by $\hat{\theta}_i$, then a response-based unbiased estimator of ϕ_i based on $\hat{\theta}_i$ will be

$$\hat{\phi}_i = [\hat{\theta}_i - (1-q)]/(2q-1)\,. \tag{11.7}$$

Our main objective is, of course, to provide an unbiased estimator of $\phi = \Sigma_1^N \phi_i/N$ by choosing an appropriate sampling design d and carrying out the randomized response experiments for the units included in a sample s.

In the following we will provide an answer to this question in a general set up. Writing $E_R(.)$, $V_R(.)$ and $\mathrm{Cov}_R(.,.)$ for the operators of mean, variance, and covariance with reference to the randomized response distribution (11.5), we have

$$\begin{aligned} &E_R(Z_i) = \theta_i\,, \quad V_R(Z_i) = \theta_i(1-\theta_i)\,, \quad \mathrm{Cov}_R(Z_i, Z_j) = 0\,, \quad i \neq j \\ &E_R(\hat{\phi}_i) = \phi_i\,, \quad V_R(\hat{\phi}_i) = \theta_i(1-\theta_i)/(2q-1)^2\,, \\ &\mathrm{Cov}_R(\hat{\phi}_i, \hat{\phi}_j) = 0\,, \quad i \neq j\,. \end{aligned} \tag{11.8}$$

At this stage we recall that the sample s has been selected according to the sampling design d. Noting that a sampling design provides $\pi_i > 0$ for every unit of the population, it is indeed possible to derive a response-based (design-) unbiased estimator of ϕ. The precise results are stated and proved below.

Theorem 11.1. *Under a fixed-size* (n) *sampling design* d *and under the randomized response model* (11.5), *based on the labelled data set* $\{(i, Z_i) \mid i \in s)\}$, *an unbiased estimator of* ϕ *is given by*

$$\hat{\phi} = \Big[N^{-1} \sum_{i \in s} Z_i/\pi_i - (1-q)\Big]\Big/(2q-1)\,. \tag{11.9}$$

The variance of this estimator is given by

$$V(\hat{\phi}) = \Big[\sum\sum_{i<j} (\pi_i\pi_j - \pi_{ij})\Big(\frac{\theta_i}{\pi_i} - \frac{\theta_j}{\pi_j}\Big)^2 + \sum_i \frac{\theta_i(1-\theta_i)}{\pi_i}\Big]\Big/N^2(2q-1)^2\,. \tag{11.10}$$

Proof. Given the sample s, we first note that according to (11.8), $E_R(Z_i) = \theta_i$ for every $i \in s$. Writing $E_p(.)$ for the operator of design-based expectation, we have

$$\begin{aligned} E_pE_R\Big(\sum_{i \in s} Z_i/\pi_i\Big) &= \sum_s P(s)\Big(E_R \sum_{i\in s} Z_i/\pi_i\Big) \\ &= \sum_s P(s)\Big[\sum_{i \in s} E_R(Z_i/\pi_i)\Big] = \sum_s P(s) \sum_{i \in s} \theta_i/\pi_i \\ &= \sum_i (\theta_i/\pi_i) \sum_{s \ni i} P(s) = \sum_i \theta_i = N\bar{\theta} = N[\phi(2q-1) + (1-q)]\,. \end{aligned}$$

From this the result on unbiasedness of $\hat{\phi}$ in (11.9) follows readily. As regards the variance of the estimate we denote by $V_R(.)$ the design-based variance. Then, by definition,

$$\begin{aligned} V(\hat{\phi}) &= E_pE_R(\hat{\phi} - \phi)^2 \\ &= E_pV_R(\hat{\phi}) + V_pE_R(\hat{\phi})\,. \end{aligned} \tag{11.11}$$

We now note that

$$E_R(\hat{\phi}) = \Big[\sum_{i\in s} (\theta_i/\pi_i)/N - (1-q)\Big]\Big/(2q-1)\,.$$

Therefore,

$$\begin{aligned} V_pE_R(\hat{\phi}) &= \Big[V_p\Big(\sum_{i\in s} \theta_i/\pi_i\Big)\Big]\Big/N^2(2q-1)^2 \\ &= \Big[\sum\sum_{i<j} (\pi_i\pi_i - \pi_{ij})\Big(\frac{\theta_i}{\pi_i} - \frac{\theta_j}{\pi_j}\Big)^2\Big]\Big/N^2(2q-1)^2\,. \end{aligned} \tag{11.12}$$

On the other hand,

$$V_R(\hat{\phi}) = V_R\Big[\Big(\sum_{i\in s}(Z_i/\pi_i)\Big)\Big/N(2q-1)\Big]$$
$$= \Big(\sum_{i\in s}\frac{\theta_i(1-\theta_i)}{\pi_i^2}\Big)\Big/N^2(2q-1)^2 .$$

Hence,

$$E_pV_R(\hat{\phi}) = \Big[\sum_{i=1}^{N}\frac{\theta_i(1-\theta_i)}{\pi_i}\Big]\Big/N^2(2q-1)^2 . \tag{11.13}$$

Combining (11.12) and (11.13) we finally arrive at the expression for the variance of the estimator as given in (11.10). □

Corollary 11.2. *Under* SRS(N, n) *sampling design, based on the labelled randomized data set, an unbiased estimator of* ϕ *is given by*

$$\hat{\phi} = [\bar{Z}(s) - (1-q)]/(2q-1) . \tag{11.14}$$

The variance of this estimator is given by

$$V(\hat{\phi}) = \Big[(N-n)\mathbf{S}_\theta^2 + \sum_{i=1}^{N}\theta_i(1-\theta_i)\Big]\Big/Nn(2q-1)^2 \tag{11.15}$$

where $\mathbf{S}_\theta^2 = \Sigma_{i=1}^{N}(\theta_i - \bar{\theta})^2/(N-1)$, $\bar{\theta} = \Sigma_{i=1}^{N}\theta_i/N$.

Remark 11.1. It is not essential to restrict to a fixed-size (n) design for the purpose of estimating ϕ. Thus, if d is *any* design, then the expression for $V_pE_R(\hat{\phi})$ in (11.12) will be replaced by

$$\Big[\sum_{i=1}^{N}\theta_i^2\Big(\frac{1}{\pi_i}-1\Big) + \sum\sum_{1\le i<j\le N}\theta_i\theta_j\Big(\frac{\pi_{ij}}{\pi_i\pi_j}-1\Big)\Big]\Big/N^2(2q-1)^2 . \tag{11.16}$$

Accordingly, the expression (11.10) for $V(\hat{\phi})$ has to be modified.

Remark 11.2. Suppose the sampling design d corresponds to an SRS with replacement scheme of sampling with parameters N and n. As explained in Section 4.8, this sampling design has the representation $d = \Sigma_\nu p_\nu d_\nu$, where d_ν corresponds to SRS(N, ν) and $p_\nu = \binom{N}{\nu}\Delta^\nu O^n/N^n$, $\nu = 1, 2, \ldots, n$. Therefore, proceeding in the same fashion as with direct response data based on SRS with replacement scheme of sampling we may deduce the following results (see (4.30) and (4.31) in this context).

$$\hat{\phi} = [\bar{Z}(\nu) - (1-q)]/(2q-1) , \qquad \text{if } n(s) = \nu . \tag{11.17}$$

$$V(\hat{\phi}) = \left[\mathbf{S}_{\theta}^2 \left\{ E\left(\frac{1}{\nu}\right) - \frac{1}{N} \right\} + \frac{1}{N} E\left(\frac{1}{\nu}\right) \sum_{i=1}^{N} \theta_i(1-\theta_i) \right] \Big/ (2q-1)^2 . \tag{11.18}$$

All the above results are derived under the crucial assumption that the survey statistician has recorded the labelled randomized data set $\{(i, Z_i) \mid i \in s\}$. However, particularly with SRS with and without replacement schemes, in view of the fact that π_i's and π_{ij}'s are the same for all units or for all pairs of units, the survey statistician might be provided with the unlabelled data set: No. of "yes" responses $= n_1$ and No. of "no" responses $= n_0$. We denote by X_i, $i = 1, 2, \ldots, n$, the randomized response reported by the sampled elements in a sample of size n. Specificially, we will treat the "yes" response as $X_i = 1$ and the "no" response as $X_i = 0$. Warner (1965) observed that under an SRS with replacement scheme of sampling, the unlabelled response variables $X_1, X_2, \ldots, X_n$ are all identically distributed Bernoulli random variables with distribution

$$P_R(X_i = 1) = (1-q)(1-\phi) + q\phi = \theta . \tag{11.19}$$

Therefore, $n_1 = \Sigma_{i=1}^{n} X_i$ has the binomial distribution with parameters n and θ. This yields

$$\hat{\theta} = n_1/n , \tag{11.20}$$

and hence

$$\hat{\phi} = [(n_1/n) - (1-q)]/(2q-1) , \qquad q \neq 0.5 \tag{11.21}$$

with

$$\begin{aligned} V(\hat{\phi}) &= \theta(1-\theta)/n(2q-1)^2 \\ &= \frac{\phi(1-\phi)}{n} + \frac{q(1-q)}{n(2q-1)^2} . \end{aligned} \tag{11.22}$$

Further,

$$\hat{V}(\hat{\phi}) = \hat{\theta}(1-\hat{\theta})/(n-1)(2q-1)^2 \tag{11.23}$$

is an unbiased estimator of $V(\hat{\phi})$. Note that the term $\phi(1-\phi)/n$ in (11.22) corresponds to the variance of the direct nonrandomized estimator of ϕ if truthful responses are obtained. The term $q(1-q)/n(2q-1)^2$ represents the additional contribution to variance introduced by the randomization procedure, which was necessary to obtain truthful responses.

Remark 11.3. At this stage we present an interesting observation. In the direct (i.e., nonrandomized) response case, that is, when $q = 1$ (so that

$\theta_i = \phi_i$, $1 \le i \le N$), it is evident that $V(\hat{\phi})$ in (11.18) and (11.22) would reduce, respectively, to

$$\frac{N\phi(1-\phi)}{N-1}\left\{E\left(\frac{1}{\nu}\right) - \frac{1}{N}\right\} \tag{11.18$'$}$$

$$\frac{\phi(1-\phi)}{n} \tag{11.22$'$}$$

Now it is well known that the expression in (11.18′) is smaller than the expression in (11.22′) for all $n > 2$ with equality when $n = 1$ or 2. In the case of randomized response, however, the expressions in (11.18) and (11.22) do not admit a clear and conclusive comparison in favor of the mean of distinct units. Quite contrary to what is expected by most of us, the mean of all units (including repetitions) may outperform the mean of the distinct units! This is particularly true when $n = 2$ since in that case (11.18) and (11.22) may be seen to reduce, respectively, to

$$\frac{\phi(1-\phi)}{2} + \frac{(N+1)q(1-q)}{2N(2q-1)^2} \tag{11.18$''$}$$

$$\frac{\phi(1-\phi)}{2} + \frac{q(1-q)}{2(2q-1)^2}\,. \tag{11.22$''$}$$

It is now readily observed that the former expression exceeds the latter for any value of q, $0 < q < 1$!

In general, first observe from (11.5) that since ϕ_i is 0 or 1, it is clear that $\theta_i(1-\theta_i) = q(1-q)$ for all $i, 1 \le i \le N$. Moreover, the relation $\theta_i = (2q-1)\phi_i + (1-q)$, $1 \le i \le N$, yields $\mathbf{S}_\theta^2 = (2q-1)^2\mathbf{S}_\phi^2$ while $\mathbf{S}_\phi^2 = (N/(N-1))\phi(1-\phi)$. Therefore, the variance expression in (11.18) simplifies to

$$\begin{aligned} V(\hat{\phi}) &= \frac{N}{N-1}\,\phi(1-\phi)\left\{E\left(\frac{1}{\nu}\right) - \frac{1}{N}\right\} + \frac{q(1-q)E\left(\frac{1}{\nu}\right)}{(2q-1)^2} \\ &= V_1 + V_2\,. \end{aligned}$$

Clearly, V_1 corresponds to the part $\phi(1-\phi)/n$ and V_2 corresponds to the part $q(1-q)/n(2q-1)^2$ in the expression for $V(\hat{\phi})$ in (11.22) corresponding to the usual mean-per-unit estimator. It is now evident that

(i) $V_1 \le \phi(1-\phi)/n$ with equality if and only if $n = 1$ or 2 for any ϕ, $0 < \phi < 1$ and

(ii) $V_2 \ge q(1-q)/n(2q-1)^2$ with equality if and only if $n = 1$ for any q, $0 < q < 1$.

This explains why we do not expect $V_1 + V_2$ to be smaller than the corresponding sum under all circumstances (i.e., for all values of ϕ and q).

Indeed, it can be shown that for any given value of $q, 0 < q < 1$, we can always find at least some value of ϕ, $0 < \phi < 1$, such that the following holds:

$$V_1 + V_2 > \frac{\phi(1-\phi)}{n} + \frac{q(1-q)}{n(2q-1)^2} .$$

Thus the negative aspect of inadmissibility of the usual mean-per-unit estimator in SRS with replacement is absent in the setup of randomized response. One may argue that the randomization device allows the same person to respond differently on different occasions, thereby providing possibly different results (and more information)! Moreover, the analysis of data is simplified when all the observations are used. These seem to be the main reasons for most of the work in this area to be centered around the SRS with replacement scheme of sampling and the unlabelled data set. One may still wonder whether or not people would be somewhat suspicious of having to respond more than once!

Remark 11.4. We mentioned that the choice of the value of q, which represents the chance of facing the question Q_1 on the part of respondents, is at the discretion of the survey statistician. It must be noted, however, that this choice must be made very judiciously, taking the following points into consideration.

(i) As the expression (11.22) shows, the variance of $\hat{\phi}$ is symmetric about $q = 0.5$ and this is minimized by taking q as far from 0.5 as practically possible.

(ii) To minimize the suspicion of the respondents, the value of q should not be close to zero or one, as otherwise the respondents might get a strong feeling of being identified as the ones belonging to the sensitive group in case the response is "yes" for q large or "no" for q small. From the point of view of the respondent, a choice of q close to 0.5 would be most acceptable. However, this produces a very large variance for the estimate of ϕ as can be easily seen from expression (11.22). This suggests that a choice of q in the interval $(0.6, 0.7)$ or $(0.2, 0.3)$ would be most reasonable.

(iii) Referring to the expression for $\hat{\phi}$ in (11.21), it is evident that regardless of the choice of q, $\hat{\phi}$ may assume values outside the interval $[0, 1]$. This raises another difficulty. A study by Bourke and Dalenius (1974) indicates that the chance of $\hat{\phi}$ being outside the interval $[0, 1]$ is substantial when q is close to 0 or 1. For example, with $n = 50$, under Warner's method, the chance of $\hat{\phi}$ being negative is approximately equal to 0.35 when $q = 0.65$.

Method 2: Unrelated Questions Procedure

In the related question procedure described above, the population under study is divided into two mutually exclusive groups $\mathcal{U}_1$ and $\mathcal{U}_0$, one with and the other without the stigmatizing version of the sensitive characteristic; and based on this division of the population, each respondent is asked to respond "yes" or "no" depending on whether or not he belongs to the group indicated by the one of the two complementary questions, Q_1 or Q_0, which is selected for him to answer. This selection is made with known probabilities q and $1-q$, respectively. To increase the respondent's likelihood of cooperation the related question procedure has been revised by replacing Q_0 with a harmless dichotomous question as described formally below. Naturally, the revised form is called the *unrelated question* procedure.

In the unrelated question procedure, the population is divided into two *overlapping* groups $\mathcal{U}_1$ and $\mathcal{U}_2$. The group $\mathcal{U}_1$ contains all units with the stigmatizing version of the sensitive characteristic and the group $\mathcal{U}_2$ contains all units having a version of some other dichotomous characteristic which is completely innocuous and unrelated to the sensitive characteristic under study. Therefore, in this procedure, $\mathcal{U}_1 \cup \mathcal{U}_2 \neq \mathcal{U}$ and for an obvious reason, $\mathcal{U}_2 \subset \mathcal{U}_1$ is *not* allowed. For example, Q_1 and Q_2 can be the following questions:

Q_1: Do you use illegal drugs?
Q_2: Where you born in January?

It seems reasonable to expect more cooperation from the sample units under this procedure in comparison with the related question procedure. Whether or not the unrelated question procedure actually does increase the respondent's likelihood of truthful answering will have to be verified in actual surveys. However, we shall shortly show that under truthful reporting in both procedures, the unrelated question procedure can be made to provide a *more efficient* estimator of the proportion of units, ϕ, associated with the stigmatizing version of the sensitive characteristic under study.

Let N_1 and N_2 be the number of units in $\mathcal{U}_1$ and $\mathcal{U}_2$, respectively. As before, we are interested in estimating $\phi = N_1/N$. We distinguish the two cases: δ known and δ unknown, $\delta = N_2/N$.

Case 1

$\delta = N_2/N$ is known. In this case we select n units via a sampling design of our choice. Each interviewee is provided with the following two questions:

Q_1: Do you belong to $\mathcal{U}_1$?
Q_2: Do you belong to $\mathcal{U}_2$?

Unobserved by the interviewer, each respondent selects randomly Q_1 or Q_2 with known probabilities q and $1-q$, respectively. The interviewer provides the randomization device. Each respondent's report is either "yes" or "no" depending on which question he randomly selected. Suppose n_1 is the number of "yes" responses reported. The theory we developed for estimating ϕ in the related question procedure can be duplicated for this case by replacing $1-\phi$ by δ in (11.19). For example, if we let $X_i = 1$ if the ith sample unit responds "yes" and $X_i = 0$, otherwise, we have the following result.

Theorem 11.3. *Under an SRS with replacement scheme of sampling of size n, an unbiased estimator of ϕ is given by*

$$\hat{\phi} = [(n_1/n) - (1-q)\delta]/q \tag{11.24}$$

where n_1 is the total number of "yes" responses reported. Further,

$$\begin{aligned} V(\hat{\phi}) &= \alpha(1-\alpha)/nq^2 \\ &= \frac{\phi(1-\phi)}{n} + \{(1-q)/nq^2\}K_q(\delta, \phi) \end{aligned} \tag{11.25}$$

where $\alpha = q\phi + (1-q)\delta$ and $K_q(\delta, \phi) = q(1-2\delta)\phi + \delta[1-(1-q)\delta]$. The variance is unbiasedly estimated by

$$\hat{V}(\hat{\phi}) = \hat{\alpha}(1-\hat{\alpha})/(n-1)q^2\,, \qquad \hat{\alpha} = n_1/n\,. \tag{11.26}$$

Proof. Let $X_i = 1$ if the ith sample unit reports "yes" and $X_i = 0$ otherwise. Then $n_1 = \Sigma_{i=1}^{n} X_i$ has the binomial distribution with parameters n and α where

$$\alpha = P_R(X_i = 1) = q\phi + (1-q)\delta\,. \tag{11.27}$$

Therefore, n_1/n is an unbiased estimator of α and consequently the results of (11.24)–(11.26) can be easily deduced. □

It is interesting to compare the performance of the related question procedure with the unrelated question procedure. One aspect of this comparison with respect to the variances has been made by Dowling and Shachtman (1975). They deduced the following result.

Theorem 11.4. *Under SRS with replacement scheme of sampling of size n, denoting by $V_U(\hat{\phi})$ and $V_W(\hat{\phi})$, respectively, the variance of $\hat{\phi}$ under unrelated question procedure and under Warner's model, $V_U(\hat{\phi}) < V_W(\hat{\phi})$ for all $\phi \in [0, 1]$ and $\delta \in [0, 1]$ as long as $q \not< q_0$ or $q \not> q_{00}$ where $q_0 = 0.33933$ is* *the unique solution in* $[0, 0.5]$ *of the equation* $4q(1-q)(1+q^2)=1$ *and* $q_{00}=0.81615$ *is likewise another solution in* $[0.5, 1]$.

Proof. Comparing the expressions in (11.22) and (11.25) for $V_W(\hat{\phi})$ and $V_U(\hat{\phi})$, respectively, we easily deduce that $V_U(\hat{\phi}) < V_W(\hat{\phi})$ if and only if

$$K_q(\delta, \phi) < q^3/(2q-1)^2 \tag{11.28}$$

where $K_q(\delta, \phi)$ simplifies to

$$K_q(\delta, \phi) = \delta(1-\delta) + q\{\delta^2 + \phi(1-2\delta)\} . \tag{11.29}$$

Clearly, (11.28) holds for some q provided

$$\max_{\phi,\delta} K_q(\delta, \phi) < q^3/(2q-1)^2 . \tag{11.30}$$

We now distinguish between two possibilities.

(i) $0 < \delta < 1/2$: It can be seen that the maximum of $K_q(\delta, \phi)$ occurs at $\phi = 1$ and $\delta = \delta_0 = (1-2q)/2(1-q)$. Further, (11.30) reduces to

$$\{1/4(1-q)\} < \{q^3/(1-2q)^2\} . \tag{11.31}$$

(ii) $1/2 \le \delta < 1$: Clearly the maximum of $K_q(\delta, \phi)$ occurs at $\phi = 0$ and $\delta = \delta_1 = 1/2(1-q)$. Further, this time again, (11.30) reduces to (11.31). Therefore, $V_U(\hat{\phi}) < V_W(\hat{\phi})$ provided our choice of q satisfies (11.31) which simplifies to $1 < 4q(1-q)(1+q^2)$. It is not difficult to deduce that this inequality holds for all $q \in (q_0, q_{00})$ where q_0 is the unique solution of $4q(1-q)(1+q^2)=1$ in $[0, 0.5]$ and q_{00} is the unique solution of the same in $[0.5, 1]$. It turns out that $q_0 = 0.33933$ and $q_{00} = 0.81615$. Hence the result is proved. □

Case 2

δ is unknown but an estimator $\hat{\delta}$ is available with known variance estimator $\hat{V}(\hat{\delta})$. In this case we can follow the same data collection procedure as in Case 1. The results concerning an estimator of ϕ and its variance can be obtained quite easily by replacing (11.24)–(11.26) by the following:

$$\hat{\phi} = [(n_1/n) - (1-q)\hat{\delta}]/q , \tag{11.24'}$$

$$V(\hat{\phi}) = \alpha(1-\alpha)/nq^2 + (1-q^2)V(\hat{\delta})/q^2 , \tag{11.25'}$$

$$\hat{V}(\hat{\phi}) = \hat{\alpha}(1-\hat{\alpha})/(n-1)q^2 + (1-q)^2\hat{V}(\hat{\delta})/q^2 . \tag{11.26'}$$

We leave the verification to the reader (Exercise 11.1).

Case 3

δ is unknown and no prior estimate of it is available. We shall present a solution due to Greenberg et al. (1969) for the case of an SRS with replacement scheme of sampling with the total sample size n. As in Case 1, if the same randomization device is followed in collecting data from all n sample units, the expression (11.27) will continue to hold. Having an unbiased estimator of α in the usual manner will not permit us to extract exactly one unbiased estimator of ϕ from the right hand side of (11.27) since we have *one* equation and *two* unknowns. So, the idea is to generate two consistent and independent linear equations of the type (11.27). This can be achieved as follows. We split the total sample size n into two parts, n_1 and n_2. Next we perform the randomized response survey on a group of n_1 sample units selected according to SRS with replacement scheme using probabilities q_1 and $1 - q_1$ of selecting the questions Q_1 and Q_2, respectively. We carry out a similar survey on another group of n_2 sample units selected independently from the entire population again using SRS with replacement scheme of sampling. For this group, however, the randomization device is made to select the questions Q_1 and Q_2 with probabilities q_2 and $1 - q_2$, respectively, with $q_1 \neq q_2$.

We note the above randomization procedure results in two equations of the form (11.27), namely,

$$\alpha_1 = q_1\phi + (1 - q_1)\delta \,, \qquad \alpha_2 = q_2\phi + (1 - q_2)\delta \,. \tag{11.32}$$

We denote by n_{11} and n_{21} the number of "yes" responses from the above two groups of samples. Then, it is evident that n_{i1}/n_i estimates α_i unbiasedly with variance $V(\hat{\alpha}_i) = \alpha_i(1 - \alpha_i)/n_i$ and moreover, $\hat{V}(\hat{\alpha}_i) = \hat{\alpha}_i(1 - \hat{\alpha}_i)/(n - 1)$. It is now enough to use the equations in (11.32) to solve for ϕ as well as δ. The precise results are stated below. The proof is straightforward and is left as an exercise to the reader (Exercise 11.2).

Theorem 11.5. *Based on two independent SRS with replacement samples of sizes n_1 and n_2 and randomized response surveys with respective selection probabilities q_1 and q_2 for selecting the stigmatizing version of the sensitive characteristic, the following serves as an unbiased estimator of ϕ*

$$\hat{\phi} = \{\hat{\alpha}_1(1 - q_2) - \hat{\alpha}_2(1 - q_1)\}/(q_1 - q_2) \tag{11.33}$$

with variance

$$V(\hat{\phi}) = [\{(1 - q_2)^2\alpha_1(1 - \alpha_1)/n_1\} + \{(1 - q_1)^2\alpha_2(1 - \alpha_2)/n_2\}]/(q_1 - q_2)^2 \,. \tag{11.34}$$

Further, an estimator of the variance is given by

$$\hat{V}(\hat{\phi}) = [\{(1 - q_2)^2\hat{\alpha}_1(1 - \hat{\alpha}_1)/(n_1 - 1)\} + \{(1 - q_1)^2\hat{\alpha}_2(1 - \hat{\alpha}_2)/(n_2 - 1)\}]/(q_1 - q_2)^2 \,. \tag{11.35}$$

We note that the efficiency of this method depends on five parameters n_1, n_2, q_1, q_2, and δ. This is evident from the expression (11.34). By a judicious selection of these parameters, the survey statistician may increase the efficiency of the estimator of ϕ. Greenberg et al. (1969) and Moors (1971) have studied some aspects of these problems. It is *not* difficult to verify that for a fixed total number of observations $n = n_1 + n_2$, and for fixed values of q_1, q_2 and δ, $V(\hat{\phi})$ in (11.34) will be minimized if

$$n_1/n_2 = [\alpha_1(1-\alpha_1)(1-q_2)^2/\alpha_2(1-\alpha_2)(1-q_1)^2]^{1/2}\,. \qquad (11.36)$$

Further, for this choice of n_1 and n_2, $V(\hat{\phi})$ reduces to

$$[(1-q_1)\sqrt{\alpha_2(1-\alpha_2)} + (1-q_2)\sqrt{\alpha_1(1-\alpha_1)}]^2/n(q_1-q_2)^2\,. \quad (11.37)$$

With good prior estimates of α_1 and α_2, the expression (11.36) may be used as a guiding factor in the choice of n_1 and n_2. As regards the choice of q_1 and q_2, it is evident from (11.37) that one of them should be chosen as close to zero and the other as close to one as feasible. For example, a choice of q_1 near 0.20 ± 0.10 or 0.80 ± 0.10 and a choice of $q_2 = 1 - q_1$ seems reasonable.

We now discuss the choice of the unrelated question which affects the precision of $\hat{\phi}$ through the parameter δ. From (11.37), it is evident that $V(\hat{\phi})$ decreases as we take α_1 and α_2 away from 0.5. This is achieved by choosing the unrelated question so that the corresponding δ is on the same side of 0.5 as ϕ and moreover, $|\delta - 0.5|$ is a maximum. It is expected, however, that the stigmatizing characteristic will have a low incidence in actual populations. Thus, a choice of the unrelated question with low incidence is appropriate from theoretical considerations. In some cases, however, this may encourage the respondents to indulge in false reporting.

Remark 11.5. Horvitz et al. (1967) have suggested an alternative method of data collection and analysis for Case 3. According to this method each respondent in an SRS with replacement sample of size n is provided with two randomization devices with the parameters q_1 and q_2, $q_1 \neq q_2$ and is asked to respond to both the trials. The inference aspects of this procedure are similar to those presented above and will not be discussed here (Exercise 11.3).

Remark 11.6. Horvitz et al. (1976) have mentioned that in the unrelated questions procedure there is yet another technique available which deals with three separate questions. The randomized response mechanism is so formulated that the probability of a "yes" response becomes a known linear function of ϕ and a *known* quantity. In a way, this technique yields data corresponding to Case 1. See Exercise 11.4 for more details.

Example 11.1. The purpose of this example is to familiarize the reader with the classical Warner model. Suppose in a residential community of 2000 adults we are interested in estimating the proportion of adults who are drug additcts. Admittedly this is a sensitive issue and it seems useless to attempt a direct response survey. It is then decided to sample 50 adults by SRSWR procedure and use the following RR technique (we conveniently replace the experiment of spinning a colored wheel by that of drawing colored chips).

Ten chips of which six are colored red and the rest white, are placed in a box and each selected individual is asked to pick one chip at random and notice its color. If the color is red, then the respondent is requested to truthfully answer the question Q_1: Are you a drug addict? If the color of the selected chip is white, then the respondent is requested to truthfully answer the question Q_0: Are you free from addiction to drug? The interviewer is provided only with the final answer "yes" or "no", without any knowledge as to which question has been answered.

Suppose this survey has been carried out for all sampled individuals (those with repetitions having provided independent repeated responses) and it resulted in 22 "yes" and 28 "no" responses. Then an estimate of ϕ, the proportion of drug addicts in the community, may be obtained using formula (11.21). Clearly, $q = 0.6$, $n_1 = 22$, and $n = 50$. Then we obtain $\hat{\phi} = (0.44 - 0.40)/0.20 = 0.20$. The estimated s.e. of this estimate can be calculated from (11.23) and is seen to be equal to $\sqrt{\{0.44 \times 0.56/49 \times 0.04\}} \doteq 0.35$.

In the above, suppose it is known that 45 different individuals were selected in the sample and each of them responded only once, thereby providing a total of 22 "yes" responses. In that case, the estimate $\hat{\phi}$ can be computed from (11.17) and this yields $\hat{\phi} = (0.49 - 0.40)/0.20 = 0.45$.

Example 11.2. In this example we will take up the same problem as above but this time we will replace the question Q_0 by a harmless question, for example Q_2: Are you fond of watching wrestling? In doing so, we will assume for simplicity that in the community under consideration, only 40% of the adults are fond of watching wrestling. Suppose again that the responses are: 22 "yes" and 28 "no" in a group of 50 individuals. Then an estimate of ϕ is obtainable from (11.24) as $\hat{\phi} = (0.44 - 0.4 \times 0.4)/0.6 \doteq 0.47$. The estimated s.e. of the estimate can be computed from (11.26) and is given by $\sqrt{\{0.44 \times 0.56/49 \times 0.36\}} \doteq 0.12$.

11.3. RANDOMIZED RESPONSE TECHNIQUES FOR ONE POLYCHOTOMOUS QUALITATIVE SENSITIVE CHARACTERISTIC

So far we have discussed some statistical methodologies for collecting and analyzing data on one sensitive dichotomous qualitative characteristic. In practice it is not uncommon to encounter situations where the sensitive characteristic has more than two versions. The techniques developed in the previous section can be generalized without much difficulty to cope with the new situation. Abul-Ela et al. (1967), Greenberg et al. (1969), and Erikson (1973), among others, have developed the needed modifications.

In this section we present some of these techniques. We organize the presentation as follows. First, we discuss the related question procedure when the response is either "yes" or "no". Next, we present results for the unrelated question procedure. This time the respondent's answer could be of one of the two types: "yes" or "no" and multiple choice.

The methodologies to be discussed all have the same common feature, namely, we build up a system of linear equations involving the unknown proportions in the different categories based on the observed responses. In the process we design our response technique in such a way that the system of linear equations is consistent and admits a unique solution to the unknown parameters.

We start by setting the notations to be followed. As before, $\mathcal{U}$ will denote the finite population under study. We assume that the sensitive characteristic has t versions (some of these being stigmatizing in nature) and accordingly the population has a partition into t mutually exclusive and exhaustive subpopultions $\mathcal{U}_1, \mathcal{U}_2, \ldots, \mathcal{U}_t$. We denote by $\phi_1, \phi_2, \ldots, \phi_t$ the true unknown proportions of units in these subpopulations. Clearly, $\Sigma_{i=1}^{t} \phi_i = 1$.

Method 1: Related Question Procedure with "Yes" or "No" Responses

Several methods are available for data collection and analysis. We present one of them. Additional methods are included in the list of exercises at the end. We prepare $t-1$ randomization devices denoted by $D_1, D_2, \ldots, D_{t-1}$. Under any device, the respondent is supposed to respond to one of the questions $Q_1, Q_2, \ldots, Q_t$ where Q_j reads

$$Q_j\text{:}\quad \text{Do you belong to } \mathcal{U}_j\text{?}$$

However, under the ith device, the chance of selection of Q_j is taken to be q_{ij} where q_{ij}'s are chosen so that

(i) $0 \le q_{ij} \le 1\,, \qquad \sum_{j=1}^{t} q_{ij} = 1 \qquad$ for all $i = 1, 2, \ldots, t-1\,;$

(ii) the matrix

$$\mathbf{A} = ((q_{ij} - q_{it}))\,, \qquad 1 \le i,\, j \le t-1 \tag{11.38}$$

is nonsingular.

We select $t-1$ independent samples of sizes $n_1, n_2, \ldots, n_{t-1}$, respectively, each drawn according to an SRS with replacement scheme of sampling. Each member of the first set of n_1 selected respondents is given the randomization device D_1 and requested to report the response of "yes" or "no", depending on whether or not he belongs to the category referred to in the randomly chosen question. Similarly the other randomization devices are used to collect data from the other groups of respondents.

We denote by n_{i1} the number of "yes" responses from the group of n_i sample units using the randomization device D_i, $1 \le i \le t-1$. The following theorem provides desired inferences on the unknown proportions $\phi_1, \phi_2, \ldots, \phi_t$. Clearly it is enough to estimate $\phi_1, \phi_2, \ldots, \phi_{t-1}$ since $\phi_t = 1 - \phi_1 - \phi_2 - \cdots - \phi_{t-1}$. We denote by Φ the column vector $(\phi_1, \phi_2, \ldots, \phi_{t-1})'$ and by ξ the column vector $(\xi_1, \xi_2, \ldots, \xi_{t-1})'$ where $\xi_i = \alpha_i - q_{it}$, $1 \le i \le t-1$. Here $\alpha_i = \Sigma_{j=1}^{t} q_{ij}\phi_j$, $1 \le i \le t-1$.

Theorem 11.6. *Based on the above set of data, an unbiased estimator of Φ is given by*

$$\hat{\Phi} = \mathbf{A}^{-1}\hat{\xi}, \qquad \hat{\xi} = (\hat{\xi}_1, \hat{\xi}_2, \ldots, \hat{\xi}_{t-1})', \quad \hat{\xi}_i = \hat{\alpha}_i - q_{it} = \frac{n_{i1}}{n_i} - q_{it}, \quad 1 \le i \le t-1 \tag{11.39}$$

with

$$\operatorname{Cov}(\hat{\Phi}) = \mathbf{A}^{-1}\operatorname{Diag}(V_1, V_2, \ldots, V_{t-1})\mathbf{A}^{-1}, \quad V_i = \alpha_i(1-\alpha_i)/n_i . \tag{11.40}$$

Further an unbiased estimator of this covariance matrix is given by

$$\widehat{\operatorname{Cov}}(\hat{\Phi}) = \mathbf{A}^{-1}\operatorname{Diag}(\hat{V}_1, \hat{V}_2, \ldots, \hat{V}_{t-1})\mathbf{A}^{-1}, \qquad \hat{V}_i = \hat{\alpha}_i(1-\hat{\alpha}_i)/(n_i - 1). \tag{11.41}$$

Proof. First observe that

$$E(n_{i1}/n_i) = \sum_{j=1}^{t} q_{ij}\phi_j = \alpha_i, \qquad 1 \le i \le t-1. \tag{11.42}$$

Using the fact that $\Sigma_{j=1}^{t} \phi_j = 1$, we can rewrite the system of linear equations in (11.42) into the matrix form: $\mathbf{A}\Phi = \xi$. Here $\mathbf{A}$ is the square matrix of order $t-1$ defined in (11.38) (ii). This system of equations has a unique solution since $\mathbf{A}$ is of full rank. Therefore, to unbiasedly estimate Φ, it is enough to have unbiased estimators of the components of ξ. This is precisely what is provided by the combined data set. Since $\hat{\alpha}_i$ is an unbiased estimator of α_i, $\hat{\xi}_i = \hat{\alpha}_i - q_{it}$ is an unbiased estimator of ξ_i. Therefore, an unbiased estimator of Φ is provided by the solution given in (11.39). The results on the covariance matrix and its estimator are easy to establish and are left to the reader. (Exercise 11.5). □

Remark 11.7. In the above, we explicitly assumed that our randomization devices $D_1, D_2, \ldots, D_{t-1}$ result in non-singularity of the square matrix $\mathbf{A} = ((q_{ij} - q_{it}))$, $1 \le i, j \le t-1$. Indeed, there are plenty of q_{ij}'s satisfying the above requirements. For example, we may choose $q_{it} = 0$, $1 \le i \le t-1$ and choose the other q_{ij}'s so as to make $\mathbf{A}$ a circulant matrix of the form $\mathbf{A} = ((a_1, a_2, \ldots, a_{t-1}))$ whose successive rows are $(a_1, a_2, \ldots, a_{t-1})$, $(a_{t-1}, a_1, \ldots, a_{t-2}), \ldots, (a_2, a_3, \ldots, a_{t-1}, a_1)$. Here, $0 \le a_j \le 1$, $\Sigma_{j=1}^{t} a_j = 1$, with at least two of the a_j's being strictly positive. Simplicity in the analysis of data might be a governing factor underlying such a choice of devices. See Exercises 11.6–11.7 in this context for two additional choices.

Remark 11.8. It must be noted that in the above, a solution of the form $\hat{\Phi} = \mathbf{A}^{-1}\hat{\xi}$ might turn out to be unacceptable in the sense that some $\hat{\phi}_j$ might go outside the interval [0, 1]. This problem requires the careful attention of any survey sampling practitioner dealing with such an approach.

Remark 11.9. To differentiate among various possible choices of the randomization devices $D_1, D_2, \ldots, D_{t-1}$ it is necessary to formulate a criterion for comparing the relative performances of the estimators $\hat{\Phi}$ based on the various choices. Since the precision of the estimators depends on the choice of $\mathbf{A}$ through (11.40), a rational way of choosing $\mathbf{A}$ would have been to minimize the determinant of the covariant matrix of the estimators. This important topic has been mentioned but not satisfactorily pursued in the literature.

Method 2: Unrelated Questions Procedure with "Yes" or "No" Responses

This time we again assume that there are t stigmatizing versions of the sensitive characteristic thereby partitioning the population $\mathcal{U}$ into t subpopulations $\mathcal{U}_1, \mathcal{U}_2, \ldots, \mathcal{U}_t$. However, there is one unrelated dichotomous characteristic with unknown incidence rate δ for one of its versions say U. Below we describe a procedure due to Greenberg et al. (1969) which is based on "yes" or "no" response from the respondents. The procedure is almost the same as in the related question procedure. Using SRS with replacement scheme of sampling, we draw t independent samples of sizes $n_1, n_2, \ldots, n_t$, respectively. Each member of the ith group of n_i sample units will be asked to select one of the set of t alternative questions $Q_1, Q_2, \ldots, Q_t$ with selection probabilities $q_{i1}, q_{i2}, \ldots, q_{it}$, respectively, $1 \le i \le t-1$:

$$Q_i\text{: Do you belong to } \mathcal{U}_i?$$

and

$$Q_t\text{: Do you belong to } U?$$

Denoting, as before, by n_{i1} the number of "yes" responses among the ith sample units, we have in the same way as above,

$$E(n_{i1}/n_i) = \sum_{j=1}^{t-1} q_{ij}\phi_j + q_{it}\delta = \alpha_i^*, \qquad 1 \le i \le t. \qquad (11.43)$$

From these equations we solve for ϕ_j, $1 \le j \le t-1$, and δ by inverting the matrix $\mathbf{A}^* = ((q_{ij}))$, $1 \le i, j \le t$. Clearly,

$$\begin{pmatrix} \hat{\Phi} \\ \delta \end{pmatrix} = (\mathbf{A}^*)^{-1}\hat{\alpha}^*, \qquad \hat{\alpha}_i^* = n_{i1}/n_i, \qquad 1 \le i \le t. \qquad (11.44)$$

Remarks similar to Remarks 11.7–11.9 continue to hold for this case as well. Further details of this case are left for an exercise (Exercise 11.8).

Method 3: Unrelated Questions Procedure with Multiple Responses

Finally we describe a procedure due to Bourke (1974) applicable in situations where the respondents are willing to participate in a multiple choice survey provided an unrelated question with the same number of multiple choices can be successfully blended with the sensitive characteristic. Suppose that there are, as before, t stigmatizing versions of the sensitive characteristic as well as of the unrelated characteristic. Let $\mathscr{U}_i$ and $\mathscr{U}_i^*$ denote, respectively, the subpopulations corresponding to the ith component of the sensitive and the innocuous characteristics, $1 \le i \le t$. It is assumed that $\delta_1, \delta_2, \ldots, \delta_t$ relating to the relative sizes of $\mathscr{U}_1^*, \mathscr{U}_2^*, \ldots, \mathscr{U}_t^*$ are known to the survey statistician. It is desired to estimate the proportions $\phi_1, \phi_2, \ldots, \phi_t$ based on data collected from an SRS sample with replacement of size n. The survey is carried out as follows.

We prepare a randomization device using, for example, a deck of cards. A known proportion p of the cards are of type 1 while the rest are of type 2. Each respondent is supposed to select one card at random. If the selected card is of type 1, then the respondent has to pick up the sensitive issue and is assumed to provide a truthful response to which category he belongs. If the selected card is of type 2 then the respondent has to pick up the unrelated question and provide a similar true response. Thus, for example, a respondent belonging to either of the categories $\mathscr{U}_i$ and $\mathscr{U}_i^*$ would simply respond with the serial number i for data recording.

In the following theorem we summarize the result on data analysis.

Theorem 11.7. *Suppose n_i is the frequency of the category i among the n responses received, $1 \le i \le t$. Then an unbiased estimator of ϕ_i is given by*

$$\hat{\phi}_i = \frac{\hat{\alpha}_i - \delta_i(1-p)}{p}, \qquad \hat{\alpha}_i = \frac{n_i}{n}. \qquad (11.45)$$

Further the variance of the estimator $\hat{\phi}_i$ is given by

$$V(\hat{\phi}_i) = \frac{\alpha_i(1-\alpha_i)}{np^2}, \qquad \alpha_i = \delta_i(1-p) + p\phi_i, \tag{11.46}$$

along with an unbiased variance estimator of the form

$$\hat{V}(\hat{\phi}_i) = \frac{\hat{\alpha}_i(1-\hat{\alpha}_i)}{(n-1)p^2}. \tag{11.47}$$

The proof is straightforward and we leave it to the reader (Exercise 11.9).

Remark 11.10. Once again in using this methodology the survey statistician should be aware of the possibility of $\hat{\phi}_i$ in (11.45) being outside the interval [0, 1].

Remark 11.11. Apart from this simple methodology, Bourke (1974) also developed several other *designs* for data collection. Some of these are presented in the form of exercises at the end. As we have already noticed, the process of collection of data on sensitive issues involves *technical gimmicks* from the perspective of the respondents. It is interesting to note the use of additional mysteries in the form of card games and drawing beads in other methods suggested by Bourke (1974).

Example 11.3. In this example we acquaint the reader with an application of RR technique in estimating the cell probabilities for multiple choice answers to the following highly sensitive question: Are you a spouse beater? We categorize the responses as: $\mathcal{U}_1$, Never; $\mathcal{U}_2$, Rarely; $\mathcal{U}_3$, Occasionally; $\mathcal{U}_4$, Frequently. In the set up of Theorem 11.5, suppose we decide to utilize the following selection probabilities q_{ij}:

$$((q_{ij})) = \begin{bmatrix} 0.4 & 0.3 & 0.2 & 0.1 \\ 0.2 & 0.4 & 0.3 & 0.1 \\ 0.3 & 0.2 & 0.4 & 0.1 \end{bmatrix}$$

Next assume three independent batches of 25 (married) individuals each are selected and the RR technique is applied accordingly. Thus, for example, each individual of the first batch is presented a wheel having 40% area colored red, 30% area blue, 20% area green, and 10% area white. Suppose the person spins the wheel and the marker stops on the green area. Then he or she is requested to truthfully answer the question Q_3: Do you occassionally beat your spouse? The whole experiment is carried out in this fashion. Suppose we end up with the following summary table:

Batch no.	1	2	3
No. of "yes" responses	5	7	8

We are now in a position to apply the result of Theorem 11.5 in order to estimate the cell probabilities ϕ_1 to ϕ_4. In the notation of (11.38), the matrix **A** is given by

$$\mathbf{A} = \begin{bmatrix} 0.3 & 0.2 & 0.1 \\ 0.2 & 0.3 & 0.1 \\ 0.1 & 0.2 & 0.3 \end{bmatrix}$$

Further, $\hat{\xi} = (0.20, 0.28, 0.32)'$. Then formula (11.39) yields $(\hat{\phi}_1, \hat{\phi}_2, \hat{\phi}_3) = (0.01, 0.14, 0.68)$ and, hence, $\hat{\phi}_4 = 0.17$.

In Remark 11.8 it has been rightly pointed out that it is not unlikely to come across situations where $\hat{\phi}$ turns out to be negative! If instead of the data set (5 7 8), we have observed (10 7 8), we would face this undesirable phenomenon.

Example 11.4. We refer to the hypothetical problem studied in Example 11.3. Suppose this time we want to bring into the picture one unrelated (harmless) dichotomous characteristic such as the age of the respondent. We can do so by interviewing from batches of people in the following manner. First, we may choose the matrix of selection probabilities as

$$((q_{ij})) = \begin{bmatrix} 0.4 & 0.2 & 0.2 & 0.2 \\ 0.2 & 0.4 & 0.2 & 0.2 \\ 0.2 & 0.2 & 0.4 & 0.2 \\ 0.2 & 0.2 & 0.2 & 0.4 \end{bmatrix}$$

To each individual of the first batch, for example, we may ask one of the questions:

Q_1: Don't you ever beat your spouse?
Q_2: Do you rarely beat your spouse?
Q_3: Do you occasionally beat your spouse?
Q_4: Are you aged 40 or more?

with the respective selection probabilities 0.4, 0.2, 0.2, and 0.2. As before, let us assume for simplicity that each batch is of size 25 and the responses are summarized as follows:

Batch no.	1	2	3	4
No. of "yes" responses	7	6	8	7

Then applying formula (11.44), we obtain $(\hat{\phi}_1, \hat{\phi}_2, \hat{\phi}_3, \hat{\delta}) = (0.28, 0.08, 0.48, 0.28)$ and, hence, $\hat{\phi}_4 = 0.16$.

In this case as well it is not difficult to come across situations where the estimates may assume negative values!

11.4. RANDOMIZED RESPONSE TECHNIQUES FOR ONE QUANTITATIVE SENSITIVE CHARACTERISTIC

We will denote by Y the sensitive characteristic assumed to be quantitative in nature. As usual, $Y_1, Y_2, \ldots, Y_N$ will denote the values of the survey variable and it is required to estimate the population total $T(\mathbf{Y})$ based on a random sample of size n. We can extend the unrelated question technique to the study of sensitive quantitative characteristics in the following way. We suppose that the population total $T(\mathbf{X})$ of an unrelated innocuous auxiliary variate X is available and we suggest the use of the usual randomized response technique. Thus, for example, every selected sample unit has to respond to either the sensitive survey variable or to the innocuous auxiliary variate, with respective selection probabilities q and $1-q$. We omit the data analysis aspect of this procedure due to its similarity with the procedures discussed before. If $T(\mathbf{X})$ is unknown the randomization procedure can be suitably modified to generate two equations with two unknowns. In this context, we refer to the works of Greenberg et al. (1971), Dalenius and Vitale (1974), Poole (1974), Pollock and Bok (1976), Godambe (1980), and Eichhorn and Hayre (1983).

For the remainder of this section we shall present in considerable detail results relating to the estimation of $T(\mathbf{Y})$ under a unified setup, incorporating the labels of the sample units. Under this setup we shall examine the effect of sampling designs. This kind of study was initiated by Erikson (1973). Later, Bellhouse (1980) extended some of these results to other situations. Godambe (1980) and Adhikary et al. (1984) studied optimality aspects of the derived estimators. Another useful reference is Sengupta and Kundu (1989).

We start by assuming that the survey statistician has a set of M distinct values $K_1, K_2, \ldots, K_M$ available and these are meant to cover the whole set of values under study. In other words, each Y_i is assumed to match with one and only one of the given quantities $K_1, K_2, \ldots, K_M$. We may write $Y_i = \Sigma_{j=1}^{M} I_{ij}K_j$ where $I_{ij} = 1$ if $Y_i = K_j$ and is 0 otherwise. Clearly, $\Sigma_{j=1}^{M} I_{ij} = 1$ for $i = 1, 2, \ldots, N$.

Next suppose a sampling design has been adopted and a sample $s = (i_1, i_2, \ldots, i_n)$ has been drawn. Each respondent will now be asked to choose independently *and* with replacement m values out of the quantities $K_1, K_2, \ldots, K_M$. However, the selection probabilities will depend on the true value associated with the respondent involved. Thus, the respondent numbered i will be asked to use the selection probabilities $q_{i1}, q_{i2}, \ldots, q_{iM}$ where $q_{ij} = q_j + I_{ij}c$, $1 \le j \le M$, $1 \le i \le N$. The quantities $q_1, q_2, \ldots, q_M$ are known proportions satisfying $0 < q_j < 1$ and $\Sigma_{j=1}^{M} q_j < 1$. The quantity c is chosen as $c = 1 - \Sigma_{j=1}^{M} q_j$. For the ith unit, let f_{ij} be the frequency of K_j in the above collection. Then the entire data set based on the sample units may be given a convenient symbolic representation

$$\text{Data:} \quad \{(i; K_1^{f_{i1}}, K_2^{f_{i2}}, \ldots, K_M^{f_{iM}}) | i \in s\}\,.$$

With reference to the above data set, it must be remembered that there are two sources of randomness. The first one is the usual design-based randomness generating the random sample s according to the sampling design used. The second source of randomness is the multinomial experiment which, for every selected unit in the sample, results in the random frequency vector $(f_{i1}, f_{i2}, \ldots, f_{iM})$ where $0 \le f_{ij} \le m$, $\Sigma_{j=1}^{M} f_{ij} = m$; this latter source of randomness is generated by randomized response.

As usual, $E_p(.)$, $V_p(.)$ will stand for the design-based expectation and variance operators, respectively. We will denote by $E_R(.)$, $V_R(.)$, $\mathrm{Cov}_R(.,.)$ the operators of mean, variance, and covariance with reference to the randomized response model. It is assumed that the sampling design is noninformative in the sense explained in Section 10.2. The implication of this is that the operators E_p and E_R can be exchanged conveniently.

Assume for the moment that the true Y_i values can be obtained through direct response. Then an estimator of the form

$$e(s, Y) = b_{s0} + \sum_{i \in s} b_{si} Y_i \tag{11.48}$$

where the coefficients satisfy

$$\sum_{s \ni i} b_{si} P(s) = 1, \qquad 1 \le i \le N; \qquad \sum_{s} b_{so} P(s) = 0\,, \tag{11.49}$$

would serve as a linear design-unbiased estimator of $T(\mathbf{Y})$. The purpose of the randomized response technique described above is to make available an estimator of the true value Y_i through the randomized response variable Z_i attached to unit i. Clearly, our randomization procedure yields m iid observations on Z_i, $i \in s$. Define

$$\bar{\mathbf{Z}}_i = \sum_{l=1}^{m} Z_{il}/m \qquad \text{and} \qquad \mathbf{s}_i^2 = \sum_{l=1}^{m} (Z_{il} - \bar{\mathbf{Z}}_i)^2/(m-1)\,.$$

Then

$$\bar{\mathbf{Z}}_i = \sum_{j=1}^{M} f_{ij} K_j/m\,, \qquad i \in s\,. \tag{11.50}$$

It is easy to verify that

$$E_R(\bar{\mathbf{Z}}_i) = cY_i + \sum_{j=1}^{M} q_j K_j\,,$$

whence

$$\hat{Y}_i = \left(\bar{\mathbf{Z}}_i - \sum_{j=1}^{M} q_j K_j\right)\Big/ c \tag{11.51}$$

serves as an unbiased estimator of Y_i in the sense that $E_R(\hat{Y}_i) = Y_i, i \in s$. We are now in a position to state and prove the main results.

Theorem 11.8. *The estimator* $e^*(s, \mathbf{Y}) = b_{s0} + \Sigma_{i \in s} b_{si} \hat{Y}_i$ *is an unbiased estimator of the population total, where* $\hat{Y}_i$ *is as in* (11.51).

Proof. Recall that the coefficients in the expression for $e(s, \mathbf{Y})$ in (11.48) satisfy the conditions listed in (11.49). Since $\hat{Y}_i$ in (11.51) is an unbiased estimator of Y_i for every unit i in the sample, it is clear that $E_p E_R[e^*(s, Y)] = E_p[e(s, \mathbf{Y})] = T(\mathbf{Y})$. Hence the result. □

We will now compute an expression for the variance of the estimator of $T(\mathbf{Y})$. Using the symbolic formula $V = V_1 E_2 + E_1 V_2$ and also using exchangeability of the operators E_R and E_p, we may write

$$\begin{aligned} V(e^*(s, \mathbf{Y})) &= E_p V_R(e^*(s, \mathbf{Y})) + V_p E_R(e^*(s, \mathbf{Y})) \\ &= \sum_s P(s) \Big[\sum_{i \in s} b_{si}^2 V_R(\hat{Y}_i) \Big] + V_p(e(s, \mathbf{Y})) . \end{aligned}$$

Using (11.51), we then have

$$V(e^*(s, \mathbf{Y})) = V_p(e(s, \mathbf{Y})) + \frac{1}{c^2} \sum_s P(s) \Big\{ \sum_{i \in s} b_{si}^2 V_R(\bar{\mathbf{Z}}_i) \Big\} . \tag{11.52}$$

We now deduce an explicit expression for $V_R(\bar{\mathbf{Z}}_i)$. Recall that the frequencies f_{ij} follow the multinomial distribution so that

$$\begin{aligned} E_R(f_{ij}) &= m q_{ij} \\ V_R(f_{ij}) &= m q_{ij}(1 - q_{ij}) \\ \operatorname{Cov}(f_{ij}, f_{ij'}) &= -m q_{ij} q_{ij'} , \qquad 1 \le j \ne j' \le M . \end{aligned} \tag{11.53}$$

Using these results it can be readily verified that

$$V_R(\bar{\mathbf{Z}}_i) = \Big[\sum_{j=1}^{M} q_{ij} K_j^2 - \Big(\sum_{j=1}^{M} q_{ij} K_j \Big)^2 \Big] \Big/ m . \tag{11.54}$$

Substituting the expression (11.54) in the expression (11.52), we obtain the expression for $V(e^*(s, \mathbf{Y}))$. We want to put this in yet another convenient form. We define

$$\begin{aligned} \mu_i &= E_R(Z_i) = \sum_{j=1}^{M} q_{ij} K_j \\ \sigma_i^2 &= V_R(Z_i) = \sum_{j=1}^{M} q_{ij} K_j^2 - \Big(\sum_{j=1}^{M} q_{ij} K_j \Big)^2 . \end{aligned}$$

In terms of σ_i^2's the final expression for $V(e^*(s, \mathbf{Y}))$ is given by

$$V(e^*(s, \mathbf{Y})) = V_p(e(s, \mathbf{Y})) + \frac{1}{mc^2}\left[\sum_{i=1}^{N} \sigma_i^2\left(\sum_{s \ni i} b_{si}^2 P(s)\right)\right]. \qquad (11.55)$$

We now turn towards estimation of $V(e^*)$. We propose to estimate the two terms separately. As regards the first term, it is seen that the expression for $V_p(e)$ involves terms of the form Y_i^2 and $Y_i Y_j$ for $1 \le i \ne j \le N$. Therefore, we must be able to estimate both the square and the product terms. On the other hand, the second term involves the unknown quantities σ_i^2's. Assuming momentarily that these are known constants, we can estimate $\Sigma_{i=1}^{N} \sigma_i^2(\Sigma_{s \ni i} b_{si}^2 P(s))$ simply by $\Sigma_{i \in s} b_{si}^2 \sigma_i^2$. Thus, here the essential problem is to estimate σ_i^2 from the sample data, for every $i \in s$. This part is quite simple. Recall that we have available m observations on the Z-distribution specific to each of the units in the sample. Clearly, the sample mean $\bar{\mathbf{Z}}(s)$ serves as an unbiased estimator for $E(Z)$ and the sample variance $\mathbf{s}_Z^2$ with divisor $(m-1)$ provides an unbiased estimator for σ_Z^2. Therefore, the second term can be estimated using

$$\frac{1}{mc^2}\left[\sum_{i \in s} b_{si}^2 \mathbf{s}_i^2\right] \qquad (11.56)$$

where

$$\mathbf{s}_i^2 = \left[\sum_{j=1}^{M} f_{ij} K_j^2 - \left(\sum_{j=1}^{M} f_{ij} K_j\right)^2 \Big/ m\right] \Big/ (m-1), \qquad i \in s. \qquad (11.57)$$

We now analyze this result more closely. Observe that the population variance σ_i^2 can be represented as

$$\begin{aligned}
\sigma_i^2 &= \sum_j q_{ij} K_j^2 - \left(\sum_j q_{ij} K_j\right)^2 \\
&= \sum_j q_j K_j^2 + cY_i^2 - \left(\sum_j q_j K_j\right)^2 - c^2 Y_i^2 - 2cY_i\left(\sum_j q_j K_j\right) \\
&= \left[\sum_j q_j K_j^2 - \left(\sum_j q_j K_j\right)^2\right] + c(1-c)Y_i^2 - 2cY_i\left(\sum_j q_j K_j\right)
\end{aligned}$$

where $\Sigma_j q_j K_j^2 - (\Sigma_j q_j K_j)^2$ and $\Sigma_j q_j K_j$ are known constants. Therefore, in effect, $\mathbf{s}_i^2$ estimates $c(1-c)Y_i^2 - 2cY_i \Sigma q_j K_j$ unbiasedly which is the same as saying that we have an unbiased estimator for Y_i^2 in terms of $\bar{\mathbf{Z}}_i$ and $\mathbf{s}_i^2$, since $\bar{\mathbf{Z}}_i$ is used to estimate Y_i. Finally, we have to estimate the product term $Y_i Y_{i'}$. This is straightforward since $\bar{\mathbf{Z}}_i$ and $\bar{\mathbf{Z}}_{i'}$ are, under randomization, independent for any pair (i, i'), $i \ne i' \in s$. We summarize the basic results below.

$$\hat{Y}_i = \left(\bar{\mathbf{Z}}_i - \sum_j K_j q_{ij}\right) \Big/ c \tag{11.58}$$

$$\hat{Y}_i^2 = \left[\mathbf{s}_i^2 - \sum_j q_j K_j^2 + \left(\sum_j q_j K_j\right)^2 + 2\left(\sum_j q_j K_j\right)\left(\bar{\mathbf{Z}}_i - \sum_j q_j K_j\right)\right] \Big/ c(1-c) \tag{11.59}$$

$$\widehat{Y_i Y_{i'}} = \hat{Y}_i \hat{Y}_{i'} = \left(\bar{\mathbf{Z}}_i - \sum_j q_j K_j\right)\left(\bar{\mathbf{Z}}_{i'} - \sum_j q_j K_j\right) \Big/ c^2 \tag{11.60}$$

$$\hat{V}(e^*) = \hat{V}_p(e) + \left[\sum_{i \in s} b_{si}^2 \mathbf{s}_i^2\right] \Big/ mc^2 , \tag{11.61}$$

where $\hat{V}_p(e)$ is to be obtained from the usual design-based unbiased estimator of $V_p(e)$ by replacing Y_i^2 and $Y_i Y_{i'}$ for i, i' in the sample by $\hat{Y}_i^2$ and $\hat{Y}_i \hat{Y}_{i'}$, respectively.

We now briefly mention the extension suggested by Bellhouse (1980). It is argued that in some situations there may be some reporting/recording errors in respect of the values on the sensitive characteristic even under randomization of the response. Thus it is assumed that $E_R(Z_i)$ is a function of a certain value Y_i^* which undergoes a model: $E_R(Y_i^*) = Y_i$, $1 \le i \le N$. Subsequently, he discussed superpopulation models as in Erikson (1973) and Godambe (1980). See also Warner (1989) and Lakshmi and Raghavarao (1991).

So far, the competing estimators were restricted to linear unbiased estimators. Adhikary et al. (1984) considered (design-) unbiased estimators in general terms and extended the well-known optimality results in the usual superpopulation setup situations involving randomized response techniques.

For those interested in further details on this fascinating topic in survey sampling, we recommend the monograph by Chaudhuri and Mukerjee (1987).

EXERCISES

11.1. Verify formulae (11.24′)–(11.26′).

11.2. Prove Theorem 11.5.

11.3. A sensitive character has two versions A and $\bar{A}$ with true unknown proportions ϕ and $1 - \phi$, respectively. Consider a completely harmless character with two versions B and $\bar{B}$ with the corresponding true unknown proportions δ and $1 - \delta$, respectively. To estimate ϕ suppose a statistician utilizes the following RR technique: Draw an SRS with replacement sample of size n and requrest for truthful response to Trial 1 and Trial 2 described below, performed by the respondents.

Trial 1: Opt for Q_1 with probability q_1
Opt for Q_2 with probability $1 - q_1$

Trial 1: Opt for Q_1 with probability q_2
Opt for Q_2 with probability $1 - q_2$

In the above, $0 < q_1, q_2 < 1$, $q_1 \neq q_2$, and Q_1 and Q_2 refer to

Q_1: Do you possess the characteristic A?
Q_2: Do you possess the characteristic B?

Perform a statistical analysis of the resulting data and suggest an unbiased estimator of ϕ, its variance and a variance estimator.

11.4. This exercise is related to Example 11.1. The purpose is to estimate the proportion of adults (in the residential community) who are drug addicts. In the RR technique employed for this purpose suppose M_1 chips are colored red, M_2 are colored white and M_3 are colored blue so that there are altogether $M = M_1 + M_2 + M_3$ chips in the box. If the color of the chip drawn is red, the respondent is requested to truthfully answer the question Q_1. If it is white or blue, then instead of the question Q_0, we set the question Q_0^*: Is the chip selected white?

Find an unbiased estimate of ϕ, its variance and a variance estimate based on data collected by the above method. Work out the numerical example with $M_1 = 6$, $M_2 = 3$, and $M_3 = 1$. Comment on the merit of this technique.

11.5. With reference to Theorem 11.6, establish (11.40) and (11.41).

11.6. This exercise is related to Theorem 11.6. Consider the choice

$$q_{it} = a\,, \qquad 1 \leq i \leq t-1$$

$$q_{ij} = \begin{Bmatrix} c\,, & \text{if } i \neq j \\ b + c\,, & \text{if } i = j \end{Bmatrix}, \qquad 1 \leq i,\ j \leq t-1\,.$$

Deduce an explicit expression for $\hat{\Phi}$ in (11.39) and examine the conditions for its nonnegativity. Show in particular that a necessary condition for $\hat{\Phi}$ to be nonnegative is that $\bar{\hat{\alpha}} \geq a$ where $\bar{\hat{\alpha}} = \Sigma\, \hat{\alpha}_i / (t-1)$. Taking numerical values for a, b, c, and t, examine when this method results in positive and negative $\hat{\Phi}$.

11.7. In the previous exercise, consider the particular choice $c = 0$ and work out an explicit expression for $\hat{\Phi}$. Also examine conditions for its nonnegativity.

11.8. This exercise has reference to Method 2 for estimating the cell probabilities of a t-class qualitative sensitive characteristic. Consider the choice of q_{ij}'s as follows:

(a) $$q_{ij} = \begin{cases} b\,, & \text{if } i \neq j \\ a+b\,, & \text{if } i = j \end{cases}\,, \qquad 0 < b,\ a+b < 1,\ a+bt = 1\,;$$

(b) $$q_{ii} = a\,, \qquad q_{it} = 1-a\,, \qquad 1 \le i \le t-1$$

$$q_{ti} = 1/t\,, \qquad 1 \le i \le t\,.$$

In each case, deduce explicit expressions for $\hat{\Phi}$ and examine the conditions for its nonnegativity.

11.9. Prove Theorem 11.7.

11.10. A sensitive characteristic has three versions with unknown proportions ϕ_1, ϕ_2, and ϕ_3; $\phi_1 + \phi_2 + \phi_3 = 1$. Two independent SRS with replacement samples of sizes n_1 and n_2 are drawn from the population and the survey is conducted as follows. Each respondent in the first sample selects one of the two questions Q_1 and Q_3 with respective probabilities q_1 and $1 - q_1$,

Q_1: Do you belong to $\mathcal{U}_1$?
Q_3: Do you belong to $\mathcal{U}_3$?

where $\mathcal{U}_i$ is the subpopulation whose relative size is ϕ_i. Each respondent in the second sample selects one of the two questions Q_2 and Q_3 with respective probabilities q_2 and $1 - q_2$ where Q_3 is the same as above and Q_2 states:

Q_2: Do you belong to $\mathcal{U}_2$?

In both the cases the responses are simply "yes" or "no".
Let n_i be the total number of "yes" responses in the ith sample, $i = 1, 2$.

(i) Express the probability of a "yes" response in the ith sample in terms of q_i's and ϕ_j's, $i = 1, 2$; $j = 1, 2, 3$.

(ii) Characterize conditions on q_1 and q_2 under which ϕ_j's can be estimated unbiasedly with values remaining in [0, 1].

(iii) Compute the variances of the estimators suggested in (ii) above. Obtain expressions for the corresponding unbiased variance estimators.

(iv) Are the variance estimators in (iii) nonnegative?

11.11. Generalize the sampling scheme and the interviewing procedure of Exercise 11.10 for the case where the sensitive character has t versions.

11.12. Consider the setup of Exercise 11.10 with the following change in the interviewing procedure. Each respondent in the first sample selects with respective probabilities q_1 and $1 - q_1$ one of the following questions and responds "yes" if the respondent belongs to the subpopulation mentioned in the question and "no" otherwise.

$$Q_1\text{: Do you belong to } \mathcal{U}_1\text{?}$$
$$Q_1^*\text{: Do you belong to } \mathcal{U}_2 \cup \mathcal{U}_3\text{?}$$

Similarly, each respondent in the second sample selects with respective probabilities q_2 and $1 - q_2$ one of the following two questions and responds "yes" or "no":

$$Q_2\text{: Do you belong to } \mathcal{U}_2\text{?}$$
$$Q_2^*\text{: Do you belong to } \mathcal{U}_1 \cup \mathcal{U}_3\text{?}$$

(i) Express the probability of a "yes" response in terms of q_i's and ϕ_j's for the ith sample, $i = 1, 2$, $j = 1, 2, 3$.

(ii) Deduce conditions on q_1 and q_2 so that ϕ_1, ϕ_2, ϕ_3 can be estimated unbiasedly with values remaining in [0,1].

(iii) Compute variances of the estimators suggested in (ii) above and also their unbiased estimators.

(iv) Are the variance estimators in (iii) nonnegative?

11.13. Generalize the sampling scheme and the interviewing procedure of Exercise 11.12 for the case where the sensitive character has t versions and the responses are still "yes" or "no".

11.14. The purpose of this exercise is to familiarize the reader with the statistical analysis of data arising out of a simple but quite uncommon RR experiment, suggested by Bourke (1974). Consider a deck of M cards of which M_i are of type i, $i = 1, 2, 3$ so that $M = M_1 + M_2 + M_3$ and a card of Type i has the following appearance:

Type 1	(1) A	(2) $\bar{A}$	(1) $\bar{B}$	(2) B
Type 2	(1) B	(2) $\bar{B}$	(1) A	(2) $\bar{A}$
Type 3	(1) A	(2) $\bar{A}$	(1) $\bar{A}$	(2) A

Suppose a sensitive character has two versions A and $\bar{A}$ and a harmless character has two versions B and $\bar{B}$. Assume further that the two characters are independently distributed in a given population with ϕ and δ as the true unknown proportions of individuals possessing the attributes A and B, respectivley. Every respondent in an SRS with replacement sample of size n is requested to pick up a card at random and examine its contents. Then the respondent is requested to provide a response of the type $(1,1)$, $(1,2)$, $(2,1)$, $(2,2)$ where the first number in the parentheses refers to the response based on his or her true classification with respect to the attribute appearing on the first part of the card and a similar explanation holds for the second number in the response under parentheses. (Thus if a respondent picks up a card of Type 2 and the true state possessed is $(A, \bar{B})$, then the corresponding response is $(2,1)$). Analyze the resulting data and come up with an unbiased estimator of ϕ, treating both the cases of δ known and δ unknown. In each case, compute an expression for the variance of the estimate and provide an unbiased variance estimate.

11.15. The reader will find here yet another fascinating technique for randomizing the responses. This technique is suggested by Chow and Liu (1973). Consider a bottle with a long transparent neck on which some 10 beads can stand one after another. The bottle has 50 beads of which 20 are red and the rest blue. Each respondent in an SRS with replacement sample of size 40 shakes the bottle thoroughly and reports the number of red beads in the neck if he or she possesses the version A of the sensitive character. Otherwise, the number of blue beads in the neck is reported. The responses are summarized below:

Number of beads reported	0	1	2	3	4	5	6	7	8	9	10
Frequency	–	–	1	3	7	14	12	3	–	–	–

Estimate the proportion ϕ of people possessing the version A and compute the value of estimated s.e.

REFERENCES

Abul-Ela, Abdel-Latif A., Greenberg, B. G., and Horvitz, D. G. (1967). A multiproportions *RR* model. *J. Amer. Statist. Assoc.*, **62**, 990–1008.

Adhikary, A. K., Chaudhuri, A., and Vijayan, K. (1984). Optimum sampling strategies for *RR* trial (with discussion). *Internat. Statist. Rev.*, **52**, 115–126.

Bellhouse, D. R. (1980). Linear models for *RR* designs. *J. Amer. Statist. Assoc.*, **75**, 1001–1004.

Bourke, P. D. (1974). Multi-proportions *RR* using the unrelated question technique. Unpublished manuscript. Inst. Statist., Univ. Stockholm.

Bourke, P. D., and Dalenius, T. (1974). *RR* models with lying. Tech. Report 71, Inst. Statist., Univ. Stockholm.

Chaudhuri, A., and Mukerjee, R. (1985). Optionally randomized response techniques. *Cal. Statist. Assoc. Bull.*, **34**, 225–229.

Chaudhuri, A., and Mukerjee, R. (1987). *Randomized Response: Theory and Practice*. Marcel Dekker, New York.

Chow, L. P., and Liu, P. T. (1973). A new *RR* technique: the multiple answer model. Unpublished manuscript. Dept. Popln. Dynamics, Johns Hopkins University.

Dalenius, T., and Vitale, R. A. (1974). A new *RR* design for estimating the mean of a distribution. Tech. Report #78, Brown University.

Dowling, T. A., and Shachtman, R. (1975). On the relative efficiency of *RR* models. *J. Amer. Statist. Assoc.*, **70**, 84–87.

Eichhorn, B. H., and Hayre, L. S. (1983). Scrambled *RR* models for obtaining sensitive quantitative data. *J. Statist. Plann. Infer.*, **7**, 307–316.

Erikson, S. (1973). A new model for *RR*. *Internat. Statist. Rev.*, **41**, 101–113.

Godambe, V. P. (1980). Estimation in *RR* trial. *Internat. Statist. Rev.*, **48**, 29–32.

Greenberg, B. G., Abul-Ela, Abdel-Latif A., Simmons, W. R., and Horvitz, D. G. (1969). The unrelated question *RR* model: theoretical framework. *J. Amer. Statist. Assoc.*, **64**, 520–539.

Greenberg, B. G., Kubler, R. R., Abernathy, J. R., and Horvitz, D. G. (1971). Applications of the *RR* technique in obtaining quantitative data. *J. Amer. Statist. Assoc.*, **66**, 243–250.

Horvitz, D. G., Greenberg, B. G., and Abernathy, J. R. (1976). Randomized response: A data-gathering device for sensitive questions. *Internat. Statist. Rev.*, **44**, 181–196.

Lakshmi, D., and Raghavarao, D. (1991). A test for detecting untruthful answering in randomized response procedure. Tech. Report, Dept. Statistics, Temple Univ., Philadelphia.

Moors, J. J. A. (1971). Optimization of the unrelated question *RR* model. *J. Amer. Statist. Assoc.*, **66**, 627–629.

Pollock, K. H., and Bok, Y. (1976). A comparison of three *RR* models for quantitative data. *J. Amer. Statist. Assoc.*, **71**, 884–886.

Poole, W. K. (1974). Estimation of the distribution function of a continuous type random variable through *RR*. *J. Amer. Statist. Assoc.*, **69**, 1002–1005.

Sengupta, S., and Kundu, D. (1989). Estimation of finite population mean in randomized response surveys. *J. Statist. Plann. Infer.*, **23**, 117–125.

Warner, S. L. (1965). Randomized response: a survey technique for eliminating evasive answer bias. *J. Amer. Statist. Assoc.*, **60**, 63–69.

Warner, S. L. (1989). Quick randomized response. In *Proceedings of 47th Session of Internat. Stat. Inst.*, 17–19.

CHAPTER TWELVE

Special Topics: Small Area Estimation, Nonresponse Problems, and Resampling Techniques

In this concluding chapter we briefly address a few special topics. First we focus our attention on the problem of estimation for small subgroups. Next we discuss the problem of nonresponse. Finally, we mention several resampling techniques. All three topics deserve special mention as most survey practitioners have to deal with one or the other in actual surveys. Sampling techniques are finding more and more applications in a variety of fields. We have listed a few of them at the end.

12.1. SMALL AREA ESTIMATION

At the outset we begin by stating that most sample surveys are multipurpose in nature. The survey practitioners are likely to derive information on various aspects of the population as far as practicable from one and the same survey. One such important aspect is that of *small area estimation or local area estimation*. It is also known in the literature as *subdomain estimation* or *small domain estimation* or as *estimation for small subgroups*. The purpose is to gainfully utilize the results of a survey over the whole population to derive corresponding results on a subpopulation characterized by a meaningful attribute. Thus, for example, a socioeconomic survey may have been conducted in a county and an estimate of the mean income per family for the county as a whole may have been made available. At the same time, it may also be of interest to have estimates of mean income per family for some segments of the population such as those with three or more children, those with two or more earners, those having owned houses/apartments and the like.

U.N. Subcommission (1950) first pointed out the need for estimation of the parameters for domains of study (subpopulations). Yates (1953) discussed the problems of estimation of the domain totals, means and proportions based on an SRS(N, n) sample data. Durbin (1956), Hartley (1959), and Kish (1961) presented analogous results for stratified sampling situations. Subsequently, Kish (1968) also discussed the problem of optimum allocation in stratified sampling while estimating the subdomain means. Scott and Smith (1971) examined the use of prior information about domain sizes in problems of subdomain estimation. The use of inverse probability sampling in this context was examined by Tripathi (1973). Tripathi (1988) also studied the problem of estimation for domains in sampling on two occasions. Bayesian techniques are also being explored.

Our purpose is to acquaint the reader with some preliminary results concerning small area estimation. We will deal with data arising out of an SRS(N, n) sampling procedure with or without stratification and propose estimators for subdomain totals and means. Clearly, in the case of stratified random sample data, if one or more strata correspond entirely to a subdomain of study, then the standard results for SRS are adequate for the estimation of subdomain mean or total. However, if the subdomain cuts across all or some of the strata, which is usually referred to as a *crossclass*, we need separate formulae. In Figure 12.1, we depict the possible situations.

In dealing with the problem of small area estimation, first we consider the case where an SRS(N, n) sampling procedure has been adopted and a sample s has been drawn. Let $\mathcal{U}(A)$ be the subdomain of interest (defined by a meaningful attribute A) with $N(A)$ the number of population units in it. It is quite possible that the subdomain $\mathcal{U}(A)$ exists in the population only in an unidentified manner. In that case the survey practitioner may or may not know the size $N(A)$. Whereas this knowledge is not essential for estimation of the subdomain mean, we will see shortly that an estimator of the subdomain total may be profitably based on this value, if known. The ratio $N(A)/N$ will be denoted as $\phi(A)$ and this represents the relative size of the subdomain of interest. We will write $\phi(\bar{A})$ for $1-\phi(A)$.

As usual, we will denote by Y the study variable with value Y_i on unit i, $1 \le i \le N$. The subdomain mean will be denoted by $\bar{\mathbf{Y}}(A)$ and the subdomain total by $\mathbf{T}(A)$. Further, $\mathbf{S}^2(A)$ will denote the subdomain variance with divisor $N(A)-1$. Let $s(A)$ denote the subsample of units from $\mathcal{U}(A)$. Thus technically $s(A)$ is the collection of elements common to s and $\mathcal{U}(A)$. Let $n(A)$ be the cardinality of $s(A)$. For $n(A) \ge 1$, we denote by $\bar{\mathbf{y}}(A)$ the mean of Y-values for the units in $s(A)$:

$$\bar{\mathbf{y}}(A) = \frac{1}{n(A)} \sum_{i \in s(A)} Y_i \,. \tag{12.1}$$

For $n(A) \ge 2$, we may also define the sample variance of Y-values for the units in $s(A)$ as

```
  1     2     3      4
----   ---   ***   ------
----   ---   ***   ------
----   ---   ***   ------
----   ---   ***   ------
----   ---   ***   ------
```

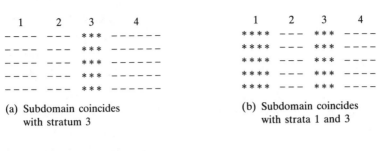

(a) Subdomain coincides with stratum 3

```
  1     2     3     4
****   ---   ***   ----
****   ---   ***   ----
****   ---   ***   ----
****   ---   ***   ----
****   ---   ***   ----
```

(b) Subdomain coincides with strata 1 and 3

```
  1     2      3      4
****   ---   -----   ----
****   ---   -----   ----
****   ***   -----   ----
****   ***   -----   ----
****   ***   -----   ----
```

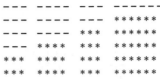

(c) Subdomain coincides with stratum 1 and a part of stratum 2

```
---   ----   ---   ------
---   ----   ---   ******
---   ----   ***   ******
---   ****   ***   ******
***   ****   ***   ******
***   ****   ***   ******
```

(d) Subdomain cuts across all the four strata: crossclass situation

```
 1      2      3      4
---   -----   ---   -----
---   -----   ---   -----
***   -----   ---   -----
***   *****   ---   -----
***   *****   ***   -----
***   *****   ***   -----
```

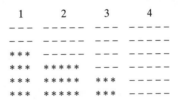

(e) Subdomain cuts across strata 1, 2 and 3: crossclass situation

Figure 12.1

$$\mathbf{s}^2(A) = \frac{1}{n(A) - 1} \sum_{i \in s(A)} (Y_i - \bar{\mathbf{y}}(A))^2 . \tag{12.2}$$

We now present the following results related to subdomain estimation. In the sequel, f denotes the sampling fraction n/N.

Theorem 12.1. *Based on an* SRS(N, n) *sample data,*

(a) *if $N(A)$ is known in advance, the estimator*

$$\hat{\mathbf{T}}(A) = \begin{cases} \bar{\mathbf{y}}(A)N(A) , & \text{if } n(A) \geq 1 \\ 0 , & \text{if } n(A) = 0 \end{cases} \tag{12.3}$$

is asymptotically unbiased for $\mathbf{T}(A)$ *with large sample variance*

$$V(\hat{\mathbf{T}}(A)) \doteq \frac{(1-f)N^2(A)\mathbf{S}^2(A)}{n\phi(A)} ; \tag{12.4}$$

(b) *if $N(A)$ is not known in advance, the estimator*

$$\hat{\hat{\mathbf{T}}}(A) = N\bar{\mathbf{y}}^*, \qquad \bar{\mathbf{y}}^* = \frac{1}{n}\sum_{i\in s} Y_i^*, \qquad Y_i^* = \begin{cases} Y_i, & \text{if } i \in \mathcal{U}(A) \\ 0, & \text{otherwise} \end{cases} \tag{12.5}$$

is (exactly) unbiased for $\mathbf{T}(A)$ *with large sample variance*

$$V(\hat{\hat{\mathbf{T}}}(A)) \doteq \frac{(1-f)}{n\phi(A)}[\mathbf{S}^2(A) + \bar{\mathbf{Y}}^2(A)\phi(\bar{A})]N^2(A). \tag{12.6}$$

Proof. (a) Clearly, for every $n(A) \geq 1$, the conditional expectation $E[\bar{\mathbf{y}}(A) \mid n(A)] = \bar{\mathbf{Y}}(A)$ and, consequently, $E(\hat{\mathbf{T}}(A)) = N(A)\bar{\mathbf{Y}}(A)P[n(A) \geq 1] \doteq N(A)\bar{\mathbf{Y}}(A) = \mathbf{T}(A)$. Further, routine calculations yield,

$$\begin{aligned} V(\hat{\mathbf{T}}(A)) &= (V_1E_2 + E_1V_2)(\hat{\mathbf{T}}(A)) \\ &\doteq N^2(A)E_1\left[\left(\frac{1}{n(A)} - \frac{1}{N(A)}\right)\mathbf{S}^2(A)\right], \qquad n(A) \geq 1 \\ &\doteq N^2(A)\left[\frac{1}{E_1(n(A))} - \frac{1}{N(A)}\right]\mathbf{S}^2(A) \\ &= \left[\frac{1}{n\phi(A)} - \frac{1}{N\phi(A)}\right]N^2(A)\mathbf{S}^2(A) \\ &= \frac{(1-f)N^2(A)\mathbf{S}^2(A)}{n\phi(A)}. \end{aligned}$$

(b) In this case, $E(\bar{\mathbf{y}}^*) = \bar{\mathbf{Y}}^* = \Sigma_{i=1}^N Y_i^*/N = \Sigma_{i\in\mathcal{U}(A)} Y_i/N = \mathbf{T}(A)/N (= \phi(A)\bar{\mathbf{Y}}(A))$ so that $\hat{\hat{\mathbf{T}}}(A) = N\bar{\mathbf{y}}^*$ is (exactly) unbiased for $\mathbf{T}(A)$. Further, it is straightforward to note that

$$\begin{aligned} V(\hat{\hat{\mathbf{T}}}(A)) &= \frac{(1-f)N^2}{n}\left\{\sum_{i=1}^N (Y_i^* - \bar{\mathbf{Y}}^*)^2/(N-1)\right\} \\ &= \frac{(1-f)N^2}{n(N-1)}\left\{\sum_{i\in\mathcal{U}(A)} (Y_i - \phi(A)\bar{\mathbf{Y}}(A))^2 + \sum_{i\notin\mathcal{U}(A)} (0 - \phi(A)\bar{\mathbf{Y}}(A))^2\right\} \\ &= \frac{(1-f)N^2}{n(N-1)}\left\{\sum_{i\in\mathcal{U}(A)} (Y_1 - \bar{Y}(A))^2 + N(A)\bar{\mathbf{Y}}^2(A)\phi^2(\bar{A})\right. \\ &\quad \left. + (N - N(A))\bar{Y}^2(A)\phi^2(A)\right\} \\ &\doteq \frac{(1-f)N^2}{n}\{\phi(A)\mathbf{S}^2(A) + \phi(A)\phi(\bar{A})\bar{\mathbf{Y}}^2(A)\} \\ &= \frac{(1-f)N^2(A)}{n\phi(A)}[\mathbf{S}^2(A) + \phi(\bar{A})\bar{\mathbf{Y}}^2(A)]. \end{aligned}$$

□

Remark 12.1. It is evident that in large samples $\hat{\mathbf{T}}(A)$ is more efficient than $\hat{\hat{\mathbf{T}}}(A)$. To estimate $V(\hat{\mathbf{T}}(A))$, it is enough to use $\mathbf{s}^2(A) = \sum_{i \in s(A)} (Y_i - \bar{\mathbf{y}}(A))^2/(n(A)-1)$ in place of $\mathbf{S}^2(A)$.

It may be noted that an unbiased estimator of $N(A)$ can be obtained from part (b) of Theorem 12.1 as follows. Let X be an auxiliary variable assuming values $X_i = 1$ if $i \in \mathcal{U}(A)$, and=0 if $i \notin \mathcal{U}(A)$. Then $\mathbf{T}(X) = N(A)$ and consequently, based on an SRS(N, n) sample data,

$$\hat{N}(A) = \hat{\mathbf{T}}(X) = N\bar{\mathbf{x}} = N \sum_{i \in s} X_i/n = Nn(A)/n = N\hat{\phi}(A) \qquad (12.7)$$

with

$$V(\hat{N}(A)) \doteq \frac{(1-f)N^2(A)}{n\phi(A)} [\mathbf{S}_X^2(A) + \phi(\bar{A})\bar{\mathbf{X}}^2(A)]\,. \qquad (12.8)$$

However, $\bar{\mathbf{X}}(A) = 1$ and $S_X^2(A) = 0$. Hence,

$$V(\hat{N}(A)) \doteq \frac{(1-f)N^2(A)\phi(\bar{A})}{n\phi(A)} = \frac{N^2(1-f)\phi(A)\phi(\bar{A})}{n}\,, \qquad (12.9)$$

$$V(\hat{\phi}(A)) \doteq \frac{(1-f)\phi(A)\phi(\bar{A})}{n}\,. \qquad (12.10)$$

As regards estimation of the subdomain mean $\bar{\mathbf{Y}}(A)$ of Y-values, we may readily use the estimator in (12.3) to propose

$$\hat{\bar{\mathbf{Y}}}(A) = \begin{cases} \bar{\mathbf{y}}(A)\,, & \text{if } n(A) \geq 1 \\ 0\,, & \text{otherwise} \end{cases} \qquad (12.11)$$

for which

$$E(\hat{\bar{\mathbf{Y}}}(A)) \doteq \bar{\mathbf{Y}}(A)\,, \qquad V(\hat{\bar{\mathbf{Y}}}(A)) \doteq \frac{(1-f)\mathbf{S}^2(A)}{n\phi(A)}\,. \qquad (12.12)$$

On the other hand, even though the estimator in (12.5) is not readily applicable, we may use it in combination with $\hat{N}(A) = N\hat{\phi}(A)$ to propose, for $n(A) \geq 1$,

$$\hat{\hat{\bar{\mathbf{Y}}}}(A) = \hat{\hat{\mathbf{T}}}(A)/\hat{N}(A) \qquad (12.13)$$

as an estimator of $\bar{\mathbf{Y}}(A)$. However, it is observed that $\hat{\hat{\bar{\mathbf{Y}}}}(A)$ simplifies to $\bar{\mathbf{y}}^*/\hat{\phi}(A) = \Sigma_{i \in s(A)} Y_i/n(A) = \bar{\mathbf{y}}(A) = \hat{\bar{\mathbf{Y}}}(A)$. Hence, knowledge of $N(A)$ is not essential for estimation of the subdomain mean. It is interesting to note that in terms of the auxiliary variable X, this latter estimate assumes the form

$$\hat{\hat{\bar{\mathbf{Y}}}}(A) = \begin{cases} \bar{\mathbf{y}}^*/\bar{\mathbf{x}}\,, & \text{for } n(A) > 0 \\ 0\,, & \text{for } n(A) = 0 \end{cases} \qquad (12.14)$$

which can be identified as the ratio estimator discussed in Chapter Six. Accordingly, an alternative derivation for the large sample variance of $\hat{\hat{\mathbf{Y}}}(A)$ is available. To demonstrate this, first observe that

$$E(\bar{\mathbf{x}}) = \bar{\mathbf{X}} = \phi(A)\,, \qquad E(\bar{\mathbf{y}}^*) = \bar{\mathbf{Y}} = \phi(A)\bar{\mathbf{Y}}(A)\,. \tag{12.15}$$

Therefore,

$$\begin{aligned} V(\hat{\hat{\mathbf{Y}}}(A)) &\doteq E\left(\frac{\bar{\mathbf{y}}^*}{\bar{\mathbf{x}}} - \bar{\mathbf{Y}}(A)\right)^2 \\ &\doteq E[\bar{\mathbf{y}}^* - \bar{\mathbf{Y}}(A)\bar{\mathbf{x}}]^2/\phi^2(A) \\ &\doteq (1-f)\left[\sum_{i=1}^{N} (U_i - \bar{\mathbf{U}})^2/(N-1)\right]/n\phi^2(A) \end{aligned} \tag{12.16}$$

where $U_i = Y_i^* - \bar{\mathbf{Y}}(A)X_i$, $1 \le i \le N$ for which $\bar{\mathbf{U}}$ simplifies to zero. Next note that $Y_i = 0 = X_i$ for $i \not\in \mathcal{U}(A)$ so that $U_i = 0$ for $i \not\in \mathcal{U}(A)$. Hence

$$\sum_{i=1}^{N} U_i^2 = \sum_{i\in\mathcal{U}(A)} U_i^2 = \sum_{i\in\mathcal{U}(A)} (Y_i - \bar{\mathbf{Y}}(A))^2 \doteq N(A)\mathbf{S}^2(A)\,. \tag{12.17}$$

Thus, from (12.16), we obtain

$$\begin{aligned} V(\hat{\hat{\mathbf{Y}}}(A)) &\doteq (1-f)\mathbf{S}^2(A)N(A)/n\phi^2(A)(N-1) \\ &\doteq (1-f)\mathbf{S}^2(A)/n\phi(A)\,. \end{aligned}$$

It may be noted that the expression in (12.12) for $V(\hat{\mathbf{Y}}(A))$ is based on the approximation $E((1/n(A)) \doteq 1/E(n(A))$, while in the above derivation we used the approximation $\bar{\mathbf{x}} \doteq \bar{\mathbf{X}} = \phi(A)$. With regard to estimation of $V(\hat{\hat{\mathbf{Y}}}(A))$, it is enough to estimate $\mathbf{S}^2(A)$ by the sample variance $\mathbf{s}^2(A) = \Sigma_{i\in s(A)} (Y_i - \bar{\mathbf{y}}(A))^2/(n(A)-1)$ and to estimate $\phi(A)$ by $\hat{\phi}(A)$, if needed.

We now consider an illustrative example.

Example 12.1. A survey has been conducted in a small county of 500 families. A simple random sample of 50 families has been drawn and among other facts and figures, monthly overall expenditure (averaged over the past five years) has been noted for the sampled families. It is observed that 15 families in the sample reside in owned houses/apartments and the rest reside in rented houses/apartments. Naturally, most of the families in the county have to pay monthly rent while others have to spend money in taking care of their own houses. It is subsequently pointed out that separate estimates of the average monthly overall expenditure for the above two

Table 12.1

Sample	Families Living in Apartments/Houses	
	Rented	Owned
Size	35	15
Mean expenditure	\$2050.00	\$1850.00
Standard deviation	\$14.50	\$15.80

subgroups are needed. To achieve this we may utilize survey data in the following manner. We split the data in two parts (Table 12.1).

Suppose it is known that 40% of the families reside in owned houses/apartments in the county as a whole. Then the estimated average monthly expenditure of such families is \$1850.00 and estimated s.e. of the estimate is given by $\sqrt{\{(1-0.10)/(50 \times 0.40)\}} \times \$15.80 \doteq \$3.35$. An estimate for the other subgroup can be given in a similar manner.

We now turn to the problem of small area estimation based on stratified random sample data. Generally stratification leads to a relatively small number of subdivisions of the population compared to the number of subdomains or small areas of interest to the survey user. While dealing with human population, stratification may be based on age, sex, marital status, nationality, and the like while the subdomains of interest could be various geographical locations. We will consider the problem of small area estimation for *one* subdomain only. As stated earlier, if the subdomain can be identified as being composed of one or more strata (see Figure 12.1(a), (b)), then standard results for stratified random sample data apply. We, therefore, confine ourselves to the crossclass situation below.

Consider a population $\mathcal{U}$ of N labelled units divided into L strata $\mathcal{U}_1, \mathcal{U}_2, \ldots, \mathcal{U}_L$ of sizes $N_1, \ldots, N_L$, respectively. Let a subdomain $\mathcal{U}(A)$ correspond to a meaningful attribute A. As before $N(A)$ will denote the cardinality of $\mathcal{U}(A)$ whose components in different strata will be taken as $\mathcal{U}_1(A), \mathcal{U}_2(A), \ldots, \mathcal{U}_L(A)$ of sizes $N_1(A), N_2(A), \ldots, N_L(A)$, respectively. In general, $N_h(A) \geq 0$, $\Sigma_{h=1}^{L} N_h(A) = N(A)$. A stratified random sample s of size n has the components $s_1, s_2, \ldots, s_L$ of (fixed) sizes $n_1, n_2, \ldots, n_L$ attached to the different strata. We denote by $n_h(A)$ the number of units of the subdomain $\mathcal{U}_h(A)$ in the sample component $s_h(A)$ of s, $1 \leq h \leq L$. The purpose is to estimate the overall subdomain mean $\bar{\mathbf{Y}}(A)$ defined as

$$\bar{\mathbf{Y}}(A) = \sum_{h=1}^{L} \sum_{i \in \mathcal{U}_h(A)} Y_{hi}/N(A) = \frac{\Sigma_{h=1}^{L} N_h(A)\bar{\mathbf{Y}}_h(A)}{N(A)} . \qquad (12.18)$$

We will examine separately the following cases:

Case I. Units in the subdomain are completely identifiable in advance in all the strata,

Case II. Units in the subdomain are not identifiable in advance in any stratum.

The third possibility of identification of units of the subdomain in some of the strata can also be examined in a similar manner.

In Case I, the subdomain sizes $N_h(A)$, $1 \le h \le L$, are known in advance. To estimate $\bar{\mathbf{Y}}(A)$ we now look for an estimate for $\bar{\mathbf{Y}}_h(A)$ in the hth stratum, $1 \le h \le L$. One such estimate is the analogue of $\hat{\bar{\mathbf{Y}}}(A)$ in (12.11) applied to the hth stratum. This suggests that for the subdomain mean based on stratified random sample data in Case I, we may use the estimate

$$\hat{\bar{\mathbf{Y}}}_{\text{str}}(A) = \sum_{h=1}^{L} N_h(A)\hat{\bar{\mathbf{Y}}}_h(A)/N(A) = \sum_{h=1}^{L} N_h(A)\bar{\mathbf{y}}_h(A)/N(A) \tag{12.19}$$

for which, using (12.12), we obtain

$$E(\hat{\bar{\mathbf{Y}}}_{\text{str}}(A)) \doteq \sum_{h=1}^{L} N_h(A)\bar{\mathbf{Y}}_h(A)/N(A) = \bar{\mathbf{Y}}(A)\,, \tag{12.20}$$

$$V(\hat{\bar{\mathbf{Y}}}_{\text{str}}(A)) \doteq \sum_{h=1}^{L} N_h^2(A)(1-f_h)\mathbf{S}_h^2(A)/n_h\phi_h(A)N^2(A)\,, \tag{12.21}$$

where

$$f_h = n_h/N_h,\ \phi_h(A) = N_h(A)/N_h\,, \qquad 1 \le h \le L\,. \tag{12.22}$$

The expression (12.21) can be written in an alternative form as

$$V(\hat{\bar{\mathbf{Y}}}_{\text{str}}(A)) \doteq \sum_{h=1}^{L} W_h^2(A)(1-f_h)\mathbf{S}_h^2(A)/f_h N_h(A) \tag{12.23}$$

where

$$W_h(A) = N_h(A)/N(A)\,, \qquad 1 \le h \le L\,. \tag{12.24}$$

Remark 12.2. If the subdomain of study is known to have cross-classification with only some of the strata, then in (12.19), (12.21), and (12.23), the summation extends over the values of h for which $N_h(A) > 0$. Further, an estimate of the variance can be provided by replacing $\mathbf{S}_h^2(A)$ by its sample analogue $\mathbf{s}_h^2(A)$, assuming $n_h(A) \ge 2$.

Remark 12.3. In a report of the U.S. National Center for Health Statistics (issued by the U.S. Department of Health, Education and Welfare in 1968) an alternative estimator of the population total, namely $\Sigma_{h=1}^{L} N_h(A)\bar{\mathbf{y}}_h$ has been proposed and termed as a *synthetic* estimator.

Clearly, it is a biased estimator. Subsequently, Gonzalez (1973), Gonzalez and Waksberg (1973), and Erickson (1973) have studied such aspects as the bias and the mse of the synthetic estimator, illustrating some situations where such estimators are useful. We omit the details.

We now examine Case II for which the subdomain sizes $N_h(A)$, $1 \leq h \leq L$, are not known in advance. First we observe that $\bar{\mathbf{Y}}(A)$ in (12.18) can be represented as

$$\bar{\mathbf{Y}}(A) = \sum_{h=1}^{L} \mathbf{T}_h(A)/N(A) = \sum_{h=1}^{L} \mathbf{T}_h(A) / \sum_{h=1}^{L} N_h(A) . \qquad (12.25)$$

Next using part (b) of Theorem 12.1, we observe that $\mathbf{T}_h(A)$ can be estimated as

$$\hat{\hat{\mathbf{T}}}_h(A) = N_h \bar{\mathbf{y}}_h^* \qquad (12.26)$$

with

$$E(\hat{\hat{\mathbf{T}}}_h(A)) \doteq \mathbf{T}_h(A) , \qquad V(\hat{\hat{\mathbf{T}}}_h(A)) \doteq \frac{(1-f_h)}{n_h \phi_h(A)} [\mathbf{S}_h^2(A) + \bar{\mathbf{Y}}_h^2(A)\phi_h(\bar{A})]N_h^2(A) \qquad (12.27)$$

where $\phi_h(\bar{A}) = 1 - \phi_h(A)$ and the other notations have their usual significance. Again using (12.7), we obtain

$$\hat{N}_h(A) = N_h n_h(A)/n_h = N_h \hat{\phi}_h(A) , \qquad 1 \leq h \leq L . \qquad (12.28)$$

These suggest that for estimating the subdomain mean based on stratified random sample data in Case II, we may use the estimator

$$\begin{aligned}
\hat{\hat{\mathbf{Y}}}_{\text{str}}(A) &= \sum_{h=1}^{L} \hat{\hat{\mathbf{T}}}_h(A) / \sum_{h=1}^{L} \hat{N}_h(A) = \sum_{h=1}^{L} N_h \bar{\mathbf{y}}_h^* / \sum_{h=1}^{L} N_h \hat{\phi}_h(A) \\
&= \sum_{h=1}^{L} N_h \bar{\mathbf{y}}_h^* / \sum_{h=1}^{L} N_h \bar{\mathbf{x}}_h = \sum_{h=1}^{L} W_h \bar{\mathbf{y}}_h^* / \sum_{h=1}^{L} W_h \bar{\mathbf{x}}_h \\
&= \hat{\bar{\mathbf{Y}}}_{\text{str}}^* / \hat{\bar{\mathbf{X}}}_{\text{str}} . \qquad (12.29)
\end{aligned}$$

In the above $\hat{\bar{\mathbf{Y}}}_{\text{str}}^*$ is the stratified random sample estimate of $\bar{\mathbf{Y}}^*$ and $\hat{\bar{\mathbf{X}}}_{\text{str}}$ is the corresponding estimate for $\bar{\mathbf{X}}$. Recall that X is an auxiliary variable assuming the values 1 and 0 depending on whether the underlying unit is in the subdomain of interest or not. We now proceed to compute a large sample expression for the variance of $\hat{\hat{\mathbf{Y}}}_{\text{str}}(A)$. Towards this end, first note that $X_{hi} = 1$ for $i \in \mathcal{U}_h(A)$, $1 \leq h \leq L$, so that

$$\bar{\mathbf{X}}(A) = 1 \qquad \text{and} \qquad \mathbf{S}_X^2(A) = 0 \qquad (12.30)$$

as also

$$\bar{\mathbf{X}}_h(A) = 1 \qquad \text{and} \qquad \mathbf{S}^2_{X,h}(A) = 0, \bar{\mathbf{x}}_h(A) = 1, \mathbf{s}^2_{X,h}(A) = 0, 1 \le h \le L\,. \tag{12.31}$$

Further, $E(\hat{\bar{\mathbf{X}}}_{\text{str}}) \doteq \phi(A)$. Hence, we obtain

$$\hat{\bar{\bar{\mathbf{Y}}}}_{\text{str}} - \bar{\mathbf{Y}}(A) = \frac{\hat{\bar{\mathbf{Y}}}^*_{\text{str}}}{\hat{\bar{\mathbf{X}}}_{\text{str}}} - \bar{\mathbf{Y}}(A) \doteq \frac{1}{\phi(A)}\,[\hat{\bar{\mathbf{Y}}}^*_{\text{str}} - \hat{\bar{\mathbf{X}}}_{\text{str}}\bar{\mathbf{Y}}(A)]$$

$$= \hat{\bar{\mathbf{U}}}_{\text{str}}/\phi(A) \tag{12.32}$$

where $U_{hi} = Y^*_{hi} - \bar{\mathbf{Y}}(A)X_{hi}$ and $\hat{\bar{\mathbf{U}}}_{\text{str}} = \Sigma\, W_h\hat{\bar{\mathbf{U}}}_h$, $\bar{\mathbf{U}}_h$ being the hth stratum mean of the variable U. This representation yields, using the familiar expression for the variance of the stratified sample mean,

$$V(\hat{\bar{\bar{\mathbf{Y}}}}_{\text{str}}(A)) \doteq \frac{1}{\phi^2(A)}\,V(\hat{\bar{\mathbf{U}}}_{\text{str}})$$

$$= \frac{1}{\phi^2(A)} \sum_{h=1}^{L} \frac{W_h^2(1-f_h)}{n_h}\,\mathbf{S}^2_{U,h}\,. \tag{12.33}$$

Next we simplify the expression for $\mathbf{S}^2_{U,h}$ as follows. We have, using

$$U_{hi} - \bar{\mathbf{U}}_h = (Y^*_{hi} - \bar{\mathbf{Y}}^*_h) - \bar{\mathbf{Y}}(A)(X_{hi} - \bar{\mathbf{X}}_h)\,,$$

$$(N_h - 1)\mathbf{S}^2_{U,h} = \sum_{i\in\mathcal{U}_h} (U_{hi} - \bar{\mathbf{U}}_h)^2$$

$$= \sum_{i\in\mathcal{U}_h} [(Y^*_{hi} - \bar{\mathbf{Y}}^*_h)^2 + \bar{\mathbf{Y}}^2(A)(X_{hi} - \bar{\mathbf{X}}_h)^2 - 2\bar{\mathbf{Y}}(A)(Y^*_{hi} - \bar{\mathbf{Y}}^*_h)(X_{hi} - \bar{\mathbf{X}}_h)]\,. \tag{12.34}$$

We now split $\mathcal{U}_h$ in two parts: $\mathcal{U}_h(A)$ and its complement and notice the obvious facts (a)–(d) stated below.

(a) $$Y^*_{hi} = \begin{cases} Y_{hi}\,, & \text{if } i \in \mathcal{U}_h(A)\,, \\ 0\,, & \text{otherwise}\,. \end{cases}$$

(b) $$\bar{\mathbf{Y}}^*_h = \sum_{i\in\mathcal{U}_h} Y^*_{hi}/N_h = N_h(A)\bar{\mathbf{Y}}_h(A)/N_h\,.$$

(c) $$X_{hi} = \begin{cases} 1\,, & \text{if } i \in \mathcal{U}_h(A)\,, \\ 0\,, & \text{otherwise}\,. \end{cases}$$

(d) $$\bar{\mathbf{X}}_h = \sum_{i\in\mathcal{U}_h} X_{hi}/N_h = N_h(A)/N_h\,.$$

Then

$$\sum_{i\in\mathcal{U}_h} (Y^*_{hi} - \bar{\mathbf{Y}}^*_h)^2 = \sum_{i\in\mathcal{U}_h(A)} \left(Y_{hi} - \frac{N_h(A)\bar{\mathbf{Y}}_h(A)}{N_h}\right)^2 + (N_h - N_h(A))\bar{\mathbf{Y}}^{*2}_h$$

$$= \sum_{i\in\mathcal{U}_h(A)} \left(Y_{hi} - \bar{\mathbf{Y}}_h(A) + \bar{\mathbf{Y}}_h(A) - \frac{N_h(A)\bar{\mathbf{Y}}_h(A)}{N_h}\right)^2 + (N_h - N_h(A))\bar{\mathbf{Y}}^{*2}_h$$

$$= (N_h(A) - 1)\mathbf{S}^2_h(A) + \bar{\mathbf{Y}}^2_h(A)[(N_h - N_h(A))N_h(A)/N_h]$$

$$= (N_h(A) - 1)\mathbf{S}^2_h(A) + \bar{\mathbf{Y}}^2_h(A)N_h(A)\left(1 - \frac{N_h(A)}{N_h}\right). \tag{12.35}$$

Also

$$\sum_{i\in\mathcal{U}_h} (X_{hi} - \bar{\mathbf{X}}_h)^2 = \left(1 - \frac{N_h(A)}{N_h}\right)^2 N_h(A) + (N_h - N_h(A))\left(\frac{N_h(A)}{N_h}\right)^2$$

$$= (N_h - N_h(A))N_h(A)/N_h$$

$$= N_h(A)\left(1 - \frac{N_h(A)}{N_h}\right). \tag{12.36}$$

Finally,

$$\sum_{i\in\mathcal{U}_h} (Y^*_{hi} - \bar{\mathbf{Y}}^*_h)(X_{hi} - \bar{X}_h)$$

$$= \sum_{i\in\mathcal{U}_h(A)} (Y_{hi} - N_h(A)\bar{\mathbf{Y}}_h(A)/N_h)\left(1 - \frac{N_h(A)}{N_h}\right)$$

$$+ (N_h - N_h(A))\frac{N^2_h(A)\bar{\mathbf{Y}}_h(A)}{N^2_h}$$

$$= N_h(A)\bar{\mathbf{Y}}_h(A)\left(1 - \frac{N_h(A)}{N_h}\right)^2 + N_h(A)\bar{\mathbf{Y}}_h(A)\frac{(N_h - N_h(A))N_h(A)}{N^2_h}$$

$$= N_h(A)\bar{\mathbf{Y}}_h(A)(N_h - N_h(A))/N_h. \tag{12.37}$$

Therefore, $\mathbf{S}^2_{U,h}$ simplifies to

$$\mathbf{S}^2_{U,h} \doteq \frac{N_h(A)}{N_h}\mathbf{S}^2_h(A) + \bar{\mathbf{Y}}^2_h(A)\frac{N_h(A)}{N_h}\left(1 - \frac{N_h(A)}{N_h}\right)$$

$$+ \bar{\mathbf{Y}}^2(A)\frac{N_h(A)}{N_h}\left(1 - \frac{N_h(A)}{N_h}\right)$$

$$- 2\bar{\mathbf{Y}}(A)\bar{\mathbf{Y}}_h(A)\frac{N_h(A)}{N_h}\left(1 - \frac{N_h(A)}{N_h}\right)$$

$$= \frac{N_h(A)}{N_h}\mathbf{S}^2_h(A) + \frac{N_h(A)}{N_h}\left(1 - \frac{N_h(A)}{N_h}\right)(\bar{\mathbf{Y}}_h(A) - \bar{\mathbf{Y}}(A))^2. \tag{12.38}$$

Using (12.38) we simplify the expression in (12.33) as follows.

$$V(\hat{\hat{\mathbf{Y}}}_{\text{str}}(A)) \doteq \frac{1}{\phi^2(A)} \sum_{h=1}^{L} W_h^2 \frac{(1-f_h)}{n_h} \left\{ \frac{N_h(A)}{N_h} \mathbf{S}_h^2(A) + \frac{N_h(A)}{N_h}\left(1 - \frac{N_h(A)}{N_h}\right)(\bar{\mathbf{Y}}_h(A) - \bar{\mathbf{Y}}(A))^2 \right\}$$

$$= \frac{1}{\phi^2(A)} \sum_{h=1}^{L} W_h^2 \frac{(1-f_h)}{f_h N_h} \frac{N_h(A)}{N_h} \left\{ \mathbf{S}_h^2(A) + \left(1 - \frac{N_h(A)}{N_h}\right)(\bar{\mathbf{Y}}_h(A) - \bar{\mathbf{Y}}(A))^2 \right\}$$

$$= \frac{1}{\phi^2(A)} \sum_{h=1}^{L} \frac{N_h^2}{N^2} \frac{(1-f_h)}{f_h} \frac{N_h(A)}{N_h^2} \left\{ \mathbf{S}_h^2(A) + \left(1 - \frac{N_h(A)}{N_h}\right)(\bar{\mathbf{Y}}_h(A) - \bar{\mathbf{Y}}(A))^2 \right\}$$

$$= \sum_{h=1}^{L} \frac{(1-f_h)}{f_h N_h(A)} \left(\frac{N_h(A)}{N(A)}\right)^2 \left\{ \mathbf{S}_h^2(A) + \left(1 - \frac{N_h(A)}{N_h}\right)(\bar{\mathbf{Y}}_h(A) - \bar{\mathbf{Y}}(A))^2 \right\}$$

$$= \sum_{h=1}^{L} \frac{(1-f_h)}{f_h N_h(A)} W_h^2(A)\{\mathbf{S}_h^2(A) + (1 - \phi_h(A))(\bar{\mathbf{Y}}_h(A) - \bar{\mathbf{Y}}(A))^2\} \tag{12.39}$$

where, as in (12.24), $W_h(A) = N_h(A)/N(A)$, $1 \le h \le L$. Observe that (12.39) can be put in the following form:

$$V(\hat{\hat{\mathbf{Y}}}_{\text{str}}(A)) \doteq \sum_{h=1}^{L} \frac{(1-f_h)W_h(A)}{f_h N(A)} \{\mathbf{S}_h^2(A) + (1 - \phi_h(A))(\bar{\mathbf{Y}}_h(A) - \bar{\mathbf{Y}}(A))^2\}\,. \tag{12.40}$$

In the particular case when the allocation adopted is proportional which means $n_h = nW_h$, $1 \le h \le L$, the expression (12.40) simplifies to

$$V_{\text{prop}}\left(\hat{\hat{\mathbf{Y}}}_{\text{str}}(A)\right) \doteq \frac{(1-f)}{fN(A)} \sum_{h=1}^{L} W_h(A)\{\mathbf{S}_h^2(A) + (1 - \phi_h(A))(\bar{\mathbf{Y}}_h(A) - \bar{\mathbf{Y}}(A))^2\}\,. \tag{12.41}$$

In Exercise 12.4 the reader is asked to verify that (12.41) can also be expressed in the form

$$V_{\text{prop}}(\hat{\hat{\mathbf{Y}}}_{\text{str}}(A)) \doteq \frac{(1-f)}{fN(A)} \left[\mathbf{S}^2(A) - \sum W_h(A)\phi_h(A)(\bar{\mathbf{Y}}_h(A) - \bar{\mathbf{Y}}(A))^2\right] \tag{12.42}$$

where

$$(N(A) - 1)\mathbf{S}^2(A) = \sum_{h=1}^{L} \sum_{i \in \mathcal{U}_h(A)} (Y_{hi} - \bar{\mathbf{Y}}(A))^2\,. \tag{12.43}$$

An estimate of the variance in the general case is given by

$$\hat{V}(\hat{\hat{\mathbf{Y}}}_{\text{str}}(A)) \doteq \sum_{h=1}^{L} \frac{(1-f_h)w_h^2(A)}{n_h(A)} \left\{ \frac{\sum_{i \in s_h(A)} (Y_{hi} - \hat{\hat{\mathbf{Y}}}_{\text{str}}(A))^2}{n_h(A)} - \frac{n_h(A)}{n_h} (\bar{\mathbf{y}}_h(A) - \hat{\hat{\mathbf{Y}}}_{\text{str}}(A))^2 \right\} \tag{12.44}$$

where $w_h(A) = n_h(A)/n(A)$, $1 \leq h \leq L$. The problem of nonnegative variance estimation poses complications in this case as well. We do not address this issue here.

Example 12.2. The purpose of this example is to familiarize the reader with some of the computations involved. We refer to Example 9.1 relating to annual sales in food stores in a certain county. Suppose in your neighborhood brand A has 20 stores and brand B has 10 while the rest have none. Suppose further that in the stratified random sample, there appeared 3 of brand A and 2 of brand B from your neighborhood with the annual sales figures (in \$1000) as 250, 280, 305 (under brand A) and 380, 440 (under brand B). We can utilize the above data to estimate the average annual sales in food stores in your neighborhood. Recall Remark 12.2 in this context. In the notation of this section,

$$\begin{aligned}
\hat{\mathbf{Y}}_{\text{str}}(A) &= \phi_1(A)\bar{\mathbf{y}}_1(A) + \phi_2(A)\bar{\mathbf{y}}_2(A) \\
&= (2/3)278.3 + (1/3)410.0 \doteq 322.2 .
\end{aligned}$$

$$\begin{aligned}
\hat{V}(\hat{\mathbf{Y}}_{\text{str}}(A)) &= \phi_1^2(A)\left[\frac{1}{n_1(A)} - \frac{1}{N_1(A)}\right] \mathbf{s}_1^2(A) \\
&\quad + \phi_2^2(A)\left[\frac{1}{n_2(A)} - \frac{1}{N_2(A)}\right] \mathbf{s}_2^2(A) \\
&\doteq 175.5
\end{aligned}$$

and estimated $s.e. \doteq 13.25$.

In the above example, suppose the distribution of food stores in your neighborhood is *not* known exactly beforehand and the stratified random sample includes 3 stores of brand A and 2 stores of brand B as stated above. In that case, the estimate of average annual sales in food stores in your neighborhood still remains the same. The s.e. of the estimate can be computed using the formula (12.44).

Remark 12.4. In this section our purpose was to give the reader a taste of the topic of small area estimation. We close this section with some additional references for further study: Durbin (1956), Gonzalez and Hoza

(1978), Holt et al. (1979), Kish (1965), Kish and Frankel (1974), Laake (1979), Nathan (1988), Särndal and Rabäck (1983), Särndal (1984) and Särndal and Hidiroglou (1989).

12.2. NONRESPONSE PROBLEMS

At the early stage of development of the theory of survey sampling attention was focused primarily on methods of sampling and estimation for reduction of sampling errors (standard errors, per se) subject to budgetary and other practical constraints. In actual surveys, however, it was routinely found that some other unavoidable sources of errors also dominate the scenario. Emphasis was then laid on studying the effects of such errors, called *nonsampling* errors. These primarily include *nonresponse* and *measurement* errors or *response* errors. Nonresponse focuses on subjects' *failure to respond* and there are a number of causes attributed to the incidence of nonresponse in actual surveys:

(i) failure to locate the respondent,
(ii) inability on the part of the respondent to provide the required information,
(iii) failure to contact the respondent,
(iv) refusal of the respondent to be interviewed, and the like.

Identifying the causes of nonresponse is not enough. One has to take remedial measures and this calls for additional funds and time. It is generally recommended that at least a second attempt should be made to cover a part of the nonrespondent group. We must emphasize that the problem of nonresponse is very serious and it has received attention of sampling practitioners in almost every field of actual surveys. To highlight the issue, we may refer, for example, to Neyman (1969).

In this section we want to acquaint the reader with some preliminary results in this topic. We formulate the nonresponse problem in terms of estimation of a finite population mean. As usual, we will denote by $\mathcal{U}$ a finite labelled population of N units. This time we will assume that conceptually the population consists of two groups of units: the respondent group $\mathcal{U}_+$ and the nonrespondent group $\mathcal{U}_-$ of sizes N_+ and $N_-(= N - N_+)$, respectively. It is desired to obtain information at the very first attempt but those in the nonrespondent group may need persuasion of various degrees to reveal the information desired. Naturally we are not supposed to know beforehand if a given population unit is in $\mathcal{U}_+$ or in $\mathcal{U}_-$. In other words, the two groups form two strata which are not identifiable before sampling takes place. It is possible that the relative sizes of the two groups $W_+ = N_+/N$ and $W_- = N_-/N$ are known in advance in some cases even though in presenting the results below we will treat them as unknown.

We denote by Y the study variable whose mean $\bar{\mathbf{Y}}$ is to be estimated. Clearly, $\bar{\mathbf{Y}} = W_+\bar{\mathbf{Y}}_+ + W_-\bar{\mathbf{Y}}_-$ where $\bar{\mathbf{Y}}_+$ refers to the mean in the respondent group and $\bar{\mathbf{Y}}_-$ refers to the mean in the nonrespondent group. Suppose an SRS(N, n) sampling procedure has been adopted and this has resulted in a random sample s. Accordingly the units in the sample have been surveyed and it is found that some n_+ units have responded straight away while the others have not. We may say that these n_+ units represent a random sample s_+ of size n_+ from $\mathcal{U}_+$ and the remaining units represent another random sample s_- of size $n_- = n - n_+$ from $\mathcal{U}_-$. We will assume that both W_+ and W_- are away from zero and the sample size n is moderate to make sure (with a high probability) that n_+ and n_- are both sizeable.

It is now proposed that a second attempt should be made to persuade at least a fraction of the nonrespondents to provide the desired information. We will assume that this sort of attempt results in success (of course, at the expense of effort, time, and money) and we can successfully enter the *hard core*, as it is called. Suppose it is decided to sample a fixed proportion γ of the nonrespondents in the sample. We assume for simplicity that $n' = \gamma n_-$ is an integer. (Otherwise, we choose n' as the integer nearest to γn_-). Then we select n' units out of the n_- sample nonrespondents by adopting an SRS(n_-, n') procedure and denote by s' the resulting sample of size n'.

We are now in a position to propose an estimator of $\bar{\mathbf{Y}}$ based on the available data. Observe that n_+/n serves as an unbiased estimator for W_+ and likewise n_-/n serves as an unbiased estimator for W_-. Further, we may use the sample mean $\bar{\mathbf{y}}(s_+)$ and $\bar{\mathbf{y}}(s')$ to estimate $\bar{\mathbf{Y}}_+$ and $\bar{\mathbf{Y}}_-$, respectively. Hence we may state the following result. Below $\mathbf{S}^2$ represents the overall population variance while $\mathbf{S}_-^2$ represents the same in the nonrespondent group.

Theorem 12.2. *In the nonresponse setup based on simple random sampling in both the attempts and with a fixed proportion γ of the nonrespondents sampled, the estimator*

$$\hat{\bar{\mathbf{Y}}} = \{n_+\bar{\mathbf{y}}(s_+) + n_-\bar{\mathbf{y}}(s')\}/n \tag{12.45}$$

is approximately unbiased for $\bar{\mathbf{Y}}$ with large sample variance given by

$$V(\hat{\bar{\mathbf{Y}}}) \doteq \frac{(1-f)\mathbf{S}^2}{n} + \left(\frac{1}{\gamma} - 1\right) W_-\mathbf{S}_-^2/n\,. \tag{12.46}$$

Proof. First observe that conditional on s_-, we readily obtain

$$E(\bar{\mathbf{y}}(s')) = \bar{\mathbf{y}}(s_-)\,, \qquad V(\bar{\mathbf{y}}(s')) \doteq \left(\frac{1}{\gamma} - 1\right) \mathbf{S}_-^2/n_-\,. \tag{12.47}$$

Next observe that conditional on n_+, $n_+ \ll n$,

$$E(\bar{\mathbf{y}}(s_+) = \bar{\mathbf{Y}}_+, V(\bar{\mathbf{y}}(s_+) = \left\{\frac{1}{n_+} - \frac{1}{N_+}\right\} \mathbf{S}_+^2 \tag{12.48}$$

$$E(\bar{\mathbf{y}}(s_-)) = \bar{\mathbf{Y}}_- , \tag{12.49}$$

$$\begin{aligned} V(\bar{\mathbf{y}}(s_-)) &= V_1(E(\bar{\mathbf{y}}(s'))) + E_1(V(\bar{\mathbf{y}}(s'))) \\ &= V_1(\bar{\mathbf{y}}(s_-) + E_1\left(\frac{1}{\gamma} - 1\right)\mathbf{S}^2(s_-)/n_- \\ &= \left\{\frac{1}{n_-} - \frac{1}{N_-}\right\}\mathbf{S}_-^2 + \left(\frac{1}{\gamma} - 1\right)\mathbf{S}_-^2/n_- \\ &= \left\{\frac{1}{\gamma n_-} - \frac{1}{N_-}\right\}\mathbf{S}_-^2 . \end{aligned} \tag{12.50}$$

These results could be derived straight away by using the fact that the sample s' can be regarded as a simple random sample of size γn_- from $\mathcal{U}_-$ (Recall Theorem 4.1). Using (12.48) and (12.49), conditional on n_+, $n_- \ll n$, we obtain

$$\begin{aligned} E(\hat{\bar{\mathbf{Y}}}) &= \{E(n_+)/n\}\bar{\mathbf{Y}}_+ + \{E(n_-)/n\}\bar{\mathbf{Y}}_- \\ &= \mathrm{W}_+\bar{\mathbf{Y}}_+ + W_-\bar{\mathbf{Y}}_- = \bar{\mathbf{Y}} . \end{aligned} \tag{12.51}$$

We now proceed to compute an expression for the large sample variance of $\bar{\mathbf{Y}}$. We will use the formula $V = V_1E_2 + E_1V_2$. Conditional on n_+, $n_- + \ll n$, we already have from (12.51),

$$E_2(\hat{\bar{\mathbf{Y}}}) = \{n_+\bar{\mathbf{Y}}_+ + n_-\bar{\mathbf{Y}}_-\}/n = n_+\{\bar{\mathbf{Y}}_+ - \bar{\mathbf{Y}}_-\}/n + \bar{\mathbf{Y}}_-$$

so that

$$V_1E_2(\hat{\bar{\mathbf{Y}}}) \doteq (1-f)W_+W_-\{\bar{\mathbf{Y}}_+ - \bar{\mathbf{Y}}_-\}^2/n , \qquad f = n/N . \tag{12.52}$$

Again, conditional on n_+, $n_- \ll n$, using relevant results from (12.48), (12.49), and (12.50), we obtain

$$\begin{aligned} n^2V_2(\hat{\bar{\mathbf{Y}}}) &= n_+^2\left[\frac{1}{n_+} - \frac{1}{N_+}\right]\mathbf{S}_+^2 + n_-^2\left[\frac{1}{\gamma n_-} - \frac{1}{N_-}\right]\mathbf{S}_-^2 \\ &= \left[n_+ - \frac{n_+^2}{N_+}\right]\mathbf{S}_+^2 + \left[\frac{n_-}{\gamma} - \frac{n_-^2}{N_-}\right]\mathbf{S}_-^2 . \end{aligned} \tag{12.53}$$

Therefore,

$$\begin{aligned} E_1V_2(\hat{\bar{\mathbf{Y}}}) = &\left[\frac{1}{n}\,E\left(\frac{n_+}{n}\right) - \frac{1}{N_+}\,E\left(\frac{n_+}{n}\right)^2\right]\mathbf{S}_+^2 \\ &+ \left[\frac{1}{n\gamma}\,E\left(\frac{n_-}{n}\right) - \frac{1}{N_-}\,E\left(\frac{n_-}{n}\right)^2\mathbf{S}_-^2\right] \end{aligned}$$

$$= \left[\frac{W_+}{n} - \frac{W_+^2}{N_+} - \frac{(1-f)W_+W_-}{nN_+} \right] \mathbf{S}_+^2$$

$$+ \left[\frac{W_-}{n\gamma} - \frac{W_-^2}{N_-} - \frac{(1-f)W_+W_-}{nN_-} \right] \mathbf{S}_-^2$$

$$= [(1-f)\{W_+\mathbf{S}_+^2 + W_-\mathbf{S}_-^2\}/n] + W_-\mathbf{S}_-^2 \left(\frac{1}{\gamma} - 1\right)/n$$

$$- (1-f)W_+W_-[\{\mathbf{S}_+^2/N_+\} + \{\mathbf{S}_-^2/N_-\}]/n$$

$$\doteq (1-f)\{W_+\mathbf{S}_+^2 + W_-\mathbf{S}_-^2\}/n + W_-\mathbf{S}_-^2 \left(\frac{1}{\gamma} - 1\right)/n . \tag{12.54}$$

Thus combining (12.52) and (12.54), we finally obtain

$$V(\hat{\bar{\mathbf{Y}}}) \doteq (1-f)[W_+\mathbf{S}_+^2 + W_-\mathbf{S}_-^2 + W_+W_-\{\bar{\mathbf{Y}}_+ - \bar{\mathbf{Y}}_-\}^2]/n$$

$$+ W_-\mathbf{S}_-^2 \left(\frac{1}{\gamma} - 1\right)/n . \tag{12.55}$$

We can now represent this expression in an alternative form in tems of the overall population variance $\mathbf{S}^2$. Recall the representation

$$\mathbf{S}^2 \doteq W_+\mathbf{S}_+^2 + W_-\mathbf{S}_-^2 + W_+W_-\{\bar{\mathbf{Y}}_+ - \bar{\mathbf{Y}}_-\}^2 , \tag{12.56}$$

so that we may write

$$V(\hat{\bar{\mathbf{Y}}}) \doteq \{(1-f)\mathbf{S}^2/n\} + \left(\frac{1}{\gamma} - 1\right) W_-\mathbf{S}_-^2/n \tag{12.57}$$

Hence the proof. □

When $\gamma = 1$ so that information is achieved from every member of the nonrespondent group, the estimator reduces to $\bar{\mathbf{y}}(s)$, the simple random sample mean and the variance reduces to $(1-f)\mathbf{S}^2/n$, as it should. The excess term $(\frac{1}{\gamma} - 1)W_-\mathbf{S}_-^2/n$ indicates the effect of subsampling from among the nonrespondents. Obviously if the values of the study variable are more or less the same among units in the nonrespondent group, subsampling with a small sampling fraction γ is justified and the contribution to variance is not significant as $\mathbf{S}_-^2$ will be rather small.

In the above derivation we have used the fact that a fixed proportion of the nonrespondents is to·be sampled. This assumption is certainly questionable. In practice we would be bothered more by nonresponse if the sample contains a good proportion of the nonrespondents. On the other hand, if the sample contains only a few of them, then we would not be worried too much and possibly spend only a little extra effort to recover a small portion of it.

In Exercise 12.8 the reader will find a problem of estimation based on this consideration.

In the nonresponse setup, even though the initial sample size n is chosen in advance, it is clear that the sample size realized in an actual survey is bound to be random. It has two random components namely, n_+ and γn_-. As to the cost of a survey under the present setup, there is a basic cost component C_0. Further we can associate an amount C_1 to the cost of collection of information on an individual contacted at the first attempt and an amount C_2 to the same at the second attempt. Therefore, the total cost required is given by $C_0 + C_1 n_+ + C_2 \gamma n_-$, which is again a random quantity. We may expect to incur an average cost of

$$\bar{C} = C_0 + C_1 n W_+ + C_2 \gamma n W_- \tag{12.58}$$

for an initial sample of size n.

One may now consider the problem of determination of optimal value of the parameter γ while maximizing the precision of the estimator per unit cost. In Exercise 12.9 we pose such a problem for the reader.

Remark 12.5. In case we are interested in estimating a proportion in the framework of nonresponse, the basic results remain the same and can be easily translated in terms of the sample proportions. In this case the added advantage is that the choice of optimum sampling fraction (among the nonrespondents) can be worked out in a reasonable manner and we do not necessarily have to know the population proportions for this situation. See Exercise 12.10 for results in this direction.

We conclude this section by mentioning that the problem of nonresponse is extremely important and it needs the serious attention of researchers. The result presented above is preliminary in nature and it rests on the assumption that the survey sampler can penetrate into the hard core with success in the second attempt. Clearly one must know what happens if this is not the case. We refer the reader to Rubin (1987) for a thorough discussion of available results in the area of *imputation* (which technically means substitution). Another useful well documented reference is "Panel on Incomplete Data", edited by Nisselson et al. (1983, 1984). Also see Cochran (1977) for a discussion on various other similar issues related to nonsampling errors.

12.3. RESAMPLING TECHNIQUES

In many real life survey situations arising in subjects ranging from politics to medicine, the parameters of interst are hardly simple functions like the population mean or the population variance. Quite often the survey data arise out of complex situations and the parameters are complicated nonlinear functions (like the population median or the midrange, for example).

In such cases, though an estimate of the parameter can possibly be suggested based on the data at hand, computing its s.e. becomes almost an impossible task. Even a study of the large sample expression for the variance of the proposed estimate becomes so discouraging that the question of estimating the variance directly using the data does not arise. In an attempt to meet this challenging situation, statisticians have developed some simple procedures like *pseudosampling* (*or half replication*), *bootstrapping and jackknifing*. The central idea is to generate from the given data and the proposed estimate several estimates of the parameter and to use a suitable measure of variation among these values to approximate the s.e. of the proposed estimate. We discuss briefly these methodologies below.

To start with let us assume that the survey population $\mathcal{U}$ consists of N labelled units with Y_i as the value of a survey variable Y on the unit i. Let $\theta(\mathbf{Y})$ be a parameter of interest for which an estimator $e(s, \mathbf{Y})$ has been proposed based on some considerations. The parameter is too complicated a function of $Y_1, Y_2, \ldots, Y_N$ to warrant computation of a manageable analytical expression for the mean squared error (mse) and, hence, of the standard error (s.e.) of the proposed estimate. To overcome this complicated situation, we try to find a way out. We compute, from the given data, several such estimates $e_1(s, \mathbf{Y})$, $e_2(s, \mathbf{Y}), \ldots, e_K(s, \mathbf{Y})$ of the parameter $\theta(\mathbf{Y})$. The estimates are bound to be statistically dependent, however. Nevertheless we approximate the mse of $e(s, \mathbf{Y})$ by the quantity $\Sigma\,(e_i - \bar{\mathbf{e}})^2/K$ where $\bar{\mathbf{e}} = \Sigma\, e_i/K$. This is an intuitively appealing procedure but its success depends on many factors. This represents a common feature of the procedures mentioned above. They differ in the methods of formation of different estimates.

Pseudoreplication Method

According to this method the sample s is split into two half-samples s_1 and $\bar{s}_1$ *of equal size* (taking the sample size $n(s)$ to be even). It is assumed that the estimate $e(s, \mathbf{Y})$ based on a sample of size $n(s)$ has a meaningful version for a sample of size $n(s)/2$. Let us write $n = 2m$. Observe that there are altogether $\binom{2m}{m}$ possible half samples and they occur in $M = \frac{1}{2}\binom{2m}{m}$ pairs of the type $(s_1, \bar{s}_1)$. Denote the half sample pairs by $\{s_i, \bar{s}_i\}$, $1 \le i \le M$. For simplicity in notation, let us write $e(s_i, \mathbf{Y}) = e_i$ and $e(\bar{s}_i, \mathbf{Y}) = \bar{e}_i$, $1 \le i \le M$. Then an estimate of $V(e)$ is computed as

$$\hat{V}_1(e) = \frac{2}{M}\left[\sum_{i=1}^{M}\left(\frac{e_i - \bar{\mathbf{e}}}{n}\right)^2 + \sum_{i=1}^{M}\left(\frac{\bar{e}_i - \bar{\mathbf{e}}}{n}\right)^2\right] \tag{12.59}$$

where $\bar{\mathbf{e}} = \Sigma_{i=1}^{M}(e_i + \bar{e}_i)/2M$. Another formula for computation of an estimate of $V(e)$ is

$$\hat{V}_2(e) = \frac{1}{M}\sum_{i=1}^{M}\left(\frac{e_i - \bar{e}_i}{n}\right)^2 . \tag{12.60}$$

In the particular case when $\theta(\mathbf{Y}) = \bar{\mathbf{Y}}$, suppose the design adopted is SRS(N, n) and we choose $e(s, \mathbf{Y}) = \bar{\mathbf{y}}(s)$ for which $V(e) = (1 - f)\mathbf{S}_Y^2/n$. A version of $e(s, \mathbf{Y})$ for a sample s is the sample mean based on the corresponding half sample. It can now be readily verified that $\hat{V}_1(e) = \hat{V}_2(e) = \mathbf{s}_Y^2/n$ where $(n-1)\mathbf{s}_Y^2 = \Sigma_{i \in s} (Y_i - \bar{\mathbf{y}}(s))^2$. It then follows that either $\hat{V}_1(e)$ or $\hat{V}_2(e)$ provides an unbiased estimate of $V(e)$ when the latter is approximated as $\mathbf{S}_Y^2/n$. In the particular case when $n = 2$, we have $\hat{V}_1(e) = \hat{V}_2(e) = (Y_1 - Y_2)^2/2$.

We now present an extension of this result to the case of stratified simple random sampling when the population is divided into L strata and the sample size is 2 from each stratum. For simplicity of notation, suppose the first two units have been selected from each stratum so that we have a collection of Y-values $\{(Y_{h1}, Y_{h2}), 1 \le h \le L\}$. Since each stratum results in two half samples, we can generate altogether $M = 2^L$ half samples of the type $s(i_1, i_2, \ldots, i_L)$ where each i_h is either 1 or 2. We will write S_M to represent this collection of half samples. Note that an estimate of $\bar{\mathbf{Y}} = \Sigma\, W_h \bar{\mathbf{Y}}_h$ based on the entire stratified sample s is $e(s, \mathbf{Y}) = \hat{\bar{\mathbf{Y}}} = \Sigma\, W_h \bar{\mathbf{y}}_h$ whose version based on a typical half sample $s(i_1, i_2, \ldots, i_L)$ can be taken as $e(s(i_1, i_2, \ldots, i_L), \mathbf{Y}) = \Sigma\, W_h Y_{i_h}$.

Ignoring the finite population correction (fpc), $V(e) = \Sigma\, W_h^2 \mathbf{S}_h^2 / n_h$ and it can be verified that an estimate of this variance is given by

$$\hat{V}_1(e) = \frac{1}{M} \Sigma\, (e(s(i_1, i_2, \ldots, i_L), \mathbf{Y}) - \hat{\bar{\mathbf{Y}}})^2 . \tag{12.61}$$

An equivalent representation for $\hat{V}_1(e)$ is given by

$$\hat{V}_2(e) = \frac{1}{M} \Sigma \left(\frac{e(s(i_1, i_2, \ldots, i_L), \mathbf{Y}) - e(s(3 - i_1, 3 - i_2, \ldots, 3 - i_L), \mathbf{Y})}{2} \right)^2 \tag{12.62}$$

where the summation extends over all the M half samples in S_M in both the expressions above.

While presenting the above results McCarthy (1969) further observed that in the stratified sampling case, we need not compute the estimate based on all possible half samples. A subset of them satisfying the property of *balance* might as well serve the purpose. This has resulted in the concept of *Balanced Repeated Replications* (BRR). We do not discuss this concept here. We refer the interested reader to Kish and Frankel (1974) for a thorough review of results in this area.

Bootstrap Method

The bootstrap method, described as a tool with tremendous applications of computer power to uncertainty in statistics, is believed to draw more information with greater precision from data in complex situations. Almost a decade has been spent refining it since it was introduced by Efron (1979). The article "Theorist Applies Computer Power to Uncertainty In Statistics" prepared by Gina Kolata and reported in the science section of *The New York Times* dated November 8, 1988 contains interesting views of the bootstrap method expressed by renowned statisticians.

Efron (1989) provided a very lucid account of the concept of bootstrapping through an example. The reader, unfamiliar with this concept, will find it illuminating. According to this method, we first attempt to generate many new samples from the given sample. To illustrate the idea, suppose we have a sample s of size n drawn from a population. We then treat this sample s temporarily as a population and start resampling from s applying simple random sampling with replacement and generate samples (as many as we want) of size n. These are known as *bootstrap samples*. Suppose $\theta(\mathbf{Y})$ is the parameter of interest and we have an estimator $e(s, \mathbf{Y})$ available. In a complex survey situation, we are neither able to study any sampling property of the estimator e nor are we able to compute the standard error of it. In a circumstance like this, we may use bootstrapping method. First we generate a large number, say M bootstrap samples of size n and compute the values $e_1, e_2, \ldots, e_M$ of the version of e on these samples. Then we use the empirical distribution generated by $e_1, e_2, \ldots, e_M$ to approximate the sampling distribution of e to be able to compute a measure of variability of the estimator e. This is then used to *attach* a standard error to the estimate e or a confidence interval to $\theta(\mathbf{Y})$.

Bootstrapping is regarded as a nonparametric method mainly for computation of standard errors of estimates and deriving confidence intervals of parameters in complex situations. The goodness of the approximation of the sampling distribution of the estimator e via that generated by the bootstrap samples depends largely on the distribution of the observations in the population, the sampling method used and the form of the estimator e. Many results are available under the set up of iid observations from populations admitting densities. The reader may consult, for example, Efron (1979) for general results and Srivastava and Chan (1989) for some specific results. Unfortunately, these results cannot be utilized in the context of sampling from a finite population where the samples are usually drawn adopting some without replacement sampling scheme. A method of generating *modified* bootstrap samples is described below.

Suppose the sample size n is a factor of N so that $N = kn$. Then we generate an artificial population $\mathcal{U}^*$ of size N in which the values assumed by the artificial variable Y^* on the artificial units are k copies of the values Y_i, $i \in s$, s being the sample of size n actually drawn from the original population $\mathcal{U}$ by following some sampling design. In the second step we sample from this artificial population following the same sampling design as the original one. The sample so generated may be termed as a *modified* bootstrap sample and denoted by s^*. We now repeat this procedure and generate as many modified bootstrap samples as we want. Next we compute the values of the estimator e for these samples and denote them by e_1^*, $e_2^*, \ldots$. Then, as in the bootstrap method, we study the empirical distribution of these estimates and use it to estimate some measures of variability of the distribution of the original estimator e. In Exercise 12.11, the reader will come across a simple problem related to this method.

Some key references in the area of bootstrap and finite population sampling are Beran (1988), Bickel and Freedman (1984), McCarthy and Snowden (1985), Chao and Lo (1985), Gleason (1988), Johns (1988) and J. N. K. Rao and Wu (1988).

Jackknifing Method

This technique is well known to statisticians as a technique for bias reduction. It was suggested by Quenouille (1949, 1956) and rigorously studied, among others, by Tukey (1958), Durbin (1959), and Schucany et al. (1971). Below we describe the procedure in general terms.

To start with we have a sample s of size n and an estimator $e(s, \mathbf{Y})$ for a parameter $\theta(\mathbf{Y})$. It is observed that $e(s, \mathbf{Y})$ is a biased estimator. The purpose is to suggest another estimator $\tilde{e}(s, \mathbf{Y})$ which will possess smaller amount of bias compared to $e(s, \mathbf{Y})$. According to this method we first delete units one by one from the given sample and form *jackknifed* samples $s_{(i)} = s - \{i\}$ for every $i \in s$. Thus we obtain n jackknifed samples each of size $n - 1$. It is assumed that versions of $e(s, \mathbf{Y})$ are well defined when s is replaced by $s_{(i)}$ for every $i \in s$. We denote by $e_{(i)}$ the value of $e(s, \mathbf{Y})$ when $s_{(i)}$ is used in place of s. Next we form the *pseudovalues*

$$\tilde{e}_i = e + (n-1)(e - e_{(i)}) \tag{12.63}$$

and propose

$$\tilde{e} = \sum_{i \in s} \tilde{e}_i / n \tag{12.64}$$

as the revised estimate of the parameter in question. The estimate $\tilde{e}$ is called the jackknife estimate or the jackknife version of e. It can also be expressed as

$$\tilde{e} = ne - (n-1)e_{(.)} \tag{12.65}$$

where

$$e_{(.)} = \sum_{i \in s} e_{(i)} / n \,. \tag{12.66}$$

There is a formula for computation of an estimated mse of e using the pseudovalues. This is given by

$$\text{estimated mse of } e = \sum_{i\in s} (\tilde{e}_i - \tilde{e})^2/n(n-1)\,. \tag{12.67}$$

Under certain assumptions regarding the nature of the bias of the original estimator, it can be shown that the jackknife estimator has reduced bias. This procedure can be repeated in an obvious manner until there is enough reduction in the bias. The reader may recall that we had an occasion to use the jackknife estimate in Chapter Six in the context of ratio method of estimation.

There is substantial literature in the area of jackknife estimation. We refer the reader to Efron (1982) and also to a review article by Miller (1974). Other related references are Majumdar and Sen (1978), Krewski and J. N. K. Rao (1981), Sengupta (1981), Wu (1986) and Sen (1988).

12.4. APPLICATION AREAS

Applications of survey sampling methods are becoming more and more visible in various fields other than the traditional ones. Behavioral sciences, sampling in time and space, study of ecological, environmental, marine and wildlife resources are a few examples of such application areas. An extensive and impressive list of references may be found in the stimulating articles by Binder and Hidiroglou (1988), Boswell et al. (1988), Fienberg and Tanur (1989), Hedayat et al. (1988a, b), Patil et al. (1988), Ramsey et al. (1988), Sukhatme (1988) and Velu and Naidu (1988).

EXERCISES

12.1. Of the estimates $\hat{\mathbf{T}}(A)$ in (12.3) and $\hat{\hat{\mathbf{T}}}(A)$ in (b) of Theorem 12.1, which one, if any, corresponds to the HTE of $\mathbf{T}(A)$? Comment on the result in view of Remark 12.1.

12.2. Show that the HTE for $\mathbf{T}(A)$ based on stratified random sample data does not make use of $N_h(A)$'s, if known. Show further that it is inefficient by comparing its performance with that of an estimator which utilizes known values of $N_h(A)$'s.

12.3. Derive results pertaining to the estimation of a proportion ϕ for an attribute in a subdomain which cuts across the strata in a population.

12.4. Verify formula (12.42) for $V_{\text{prop}}(\hat{\hat{\mathbf{Y}}}_{\text{str}}(A))$ starting from (12.41).

12.5. Compare the estimator $\hat{\mathbf{Y}}_{\text{str}}(A)$ with the estimators based on an SRS($N(A)$, $n\phi(A)$) sample data collected from the subdomain and also with the estimator based on a stratified random sample from the subdomain with parameters $\{N(A), n(A) = N\phi(A), N_h(A), n_h(A), 1 \leq h \leq L\}$ using proportional allocation namely, $n_h(A) = n(A)/N(A)$, $1 \leq h \leq L$.

12.6. Based on SRS(N, n) sample data, suggest an unbiased estimator of the subdomain variance $\mathbf{S}^2(A)$ and hence an unbiased estimator of the variance expression in (12.4).

12.7. Assume the proportion (W_0) of nonrespondents in a population to be known in advance even though such members are not identifiable before sampling took place. Consider the estimator of the population mean $\hat{\mathbf{Y}} = (1 - W_0)\bar{\mathbf{y}}(s_+) + W_0\bar{\mathbf{y}}(s')$. Study its unbiasedness and compute an expression for its large sample variance.

12.8. (a) In offering a choice of γ, it is argued that if a population exhibits nonresponse to a large extent, then one should be prepared to cover a larger subsample of the nonrespondents in the second and subsequent attempts. To formalize this, suppose a survey statistician recommends the random choice $\gamma_r = n_-/n$. Work out the large sample properties of the resulting estimator for this choice of γ.

(b) Work out the same with another choice $\gamma = k\gamma_r$ where $k > 1$ is a fixed number.

12.9. It is desired to suggest the optimal choice of γ in order to maximize the precision of the estimator per unit cost in a nonresponse setup. Consider the representation of $V(\hat{\mathbf{Y}})$ in the form $V(\hat{\mathbf{Y}}) = A_0 + (A_1/n) + (A_2/n\gamma)$ and assume the linear cost function $C = C_0 + C_1 n_+ + C_2 \gamma n_-$.

(a) Find an expression for the optimal value of $\gamma(\gamma_{\text{opt}})$ which maximizes $(V - A_0)(\bar{C} - C_0)$ where $\bar{C}$ is the average cost. Also find the corresponding value for n for a given amount of precision.

(b) Suppose $C_1 = \beta C_2$, $\beta < 1$; $W_- = \theta$, $0 < \theta < 1$; $\mathbf{S}_-^2 = \alpha \mathbf{S}^2$. Show that $\gamma_{\text{opt}} < 1$ whenever $\alpha < 1$. However, γ_{opt} may exceed unity if $\alpha > 1$. What would be your recommendation in such a situation?

12.10. Suppose we are interested in estimating the proportion ϕ of football lovers among high school kids in a certain community. Assume that possible nonresponse (due to vacationing, for example) can be overcome by a second attempt, if needed.

(a) Develop an estimation procedure based on an SRS data for the parameter ϕ, using the results for estimation of $\bar{\mathbf{Y}}$ by $\hat{\bar{\mathbf{Y}}}$ in (12.45).

(b) For the minimization problem formulated in Exercise 12.9, show that the optimum value of γ in estimating ϕ by the estimator in (a), is given by

$$\gamma_{\text{opt}} = \sqrt{\left\{ \frac{C_1 W_+ \phi_-(1-\phi_-)}{C_2[\phi(1-\phi) - W_-\phi_-(1-\phi_-)]} \right\}}$$

where ϕ_- refers to the proportion of football lovers among the nonrespondents.

(c) In case (b), using the fact that $\phi_-(1-\phi_-) < 0.25$, suggest an upper bound to γ_{opt} when ϕ_- is not known in advance but a good prior guess of ϕ is available. Work out the numerical values: $C_2 = 2C_1$, $\phi \doteq 0.40$, $W_+ \doteq 70\%$.

12.11. Let $Y_1^*, Y_2^*, \ldots, Y_n^*$ be a modified bootstrap sample of size n when the original sample s is drawn by an SRS(N, n) sampling procedure from a finite population with mean $\bar{\mathbf{Y}}$ and variance σ^2. Derive expressions for the conditional mean $E(\bar{\mathbf{y}}^* \mid s)$ and the conditional variance $V(\bar{\mathbf{y}}^* \mid s)$ as also the unconditional mean and variance of the modified bootstrap sample mean $\bar{\mathbf{y}}^*$.

12.12. Show that both $\hat{V}_1(e)$ and $\hat{V}_2(e)$ in (12.59) and (12.60), respectively, coincide with $\mathbf{s}_Y^2/n$ when $e(s, \mathbf{Y}) = \bar{\mathbf{y}}(s)$.

12.13. Show that in the half-sample method applied to the stratified sampling situation

$$E(\hat{V}_1(e)) = E(\hat{V}_2(e)) = \sum_{h=1}^{L} (W_h^2 \mathbf{S}_h^2 / n_h)$$

where $\hat{V}_1(e)$ and $\hat{V}_2(e)$ are defined in (12.61) and (12.62), respectively, and e is the stratified sample mean.

REFERENCES

Beran, R. (1988). Bootstrap view of asymptotic refinements. *J. Amer. Statist. Assoc.*, **83**, 687–697.

Bickel. P. J., and Freedman, D. A. (1984). Asymptotic normality and the bootstrap in stratified sampling. *Ann. Statist.*, **12**, 470–482.

Binder, D. A., and Hidiroglou, M. A. (1988). Sampling in time. In: *Handbook of Statistics, Vol. 6, Sampling* (Krishnaiah, P. R., and Rao. C. R., Eds.), 187–211. North-Holland, Amsterdam.

Boswell, M. T., Burnham, K. P., and Patil, G. P. (1988). Role and use of composite sampling and capture-recapture sampling in ecological studies. In: *Handbook of Statistics, Vol. 6, Sampling* (Krishnaiah, P. R., and Rao, C. R., Eds.), 469–488. North-Holland, Amsterdam.

Chao, M. T., and Lo, S. H. (1985). A bootstrap method for finite populations. *Sanakhyā*, **A47**, 399–405.

Cochran, W. G. (1977). *Sampling Techniques*. Third Edition. Wiley. New York.

Durbin, J. (1956). Sampling theory for estimates based on fewer individuals than the number selected. *Bull. Internat. Statist. Inst.*, **36**, 113–119.

Durbin, J. (1959). A note on the application of Quenouille's method of bias reduction to the estimation of ratios. *Biometrika*, **46**, 477–480.

Efron, B. (1979). Bootstrap methods: another look at the jackknife. *Ann. Statist.*, **7**, 1–26.

Efron, B. (1982). *The Jackknife, the Bootstrap and other Resampling Plans*. CMBS-NSF Regional Conference Series in Applied Mathematics. SIAM Publication.

Efron, B. (1989). Bootstrap and other resampling methods. *I.M.S. Bulletin*, **18**, 406–408.

Efron, B. and Gong, G. (1983). A leisurely look at the Bootstrap, Jackknife and cross validation. *Amer. Statistician*, **37**, 36–48.

Erickson, E. P. (1973). Recent developments in estimation for local areas. *Proc. Soc. Statist. Sec., Amer. Statist. Assoc.*, 37–43.

Fienberg, S. E., and Tanur, J. M. (1989). Combining cognitive and statistical approaches to survey design. *Science*, **243**, 1017–1022.

Gleason, J. R. (1988). Algorithms for balanced bootstrap simulation. *Amer. Statistician*, **42**, 263–295.

Gonzalez, M. E. (1973). Use and evaluation of the synthetic estimates. *Proc. Soc. Statist. Sec., Amer. Statist. Assoc.*, 33–36.

Gonzalez, M. E., and Hoza, C. (1978). Small-area estimation with application to unemployment and housing estimates. *J. Amer. Statist. Assoc.*, **73**, 7–15.

Gonzalez, M. E., and Waksberg, J. (1973). Estimation of the error of synthetic estimates. Unpublished Manuscript. U.S. Department of Health, Education and Welfare.

Hartley, H. O. (1959). *Analytic Studies of Survey Data*. Inst. di Statistica, Rome (volume in honour of Corrado Gini).

Hedayat, A. S., Rao, C. R., and Stufken, J. (1988a). Designs in survey sampling avoiding contiguous units. In: *Handbook of Statistics, Vol. 6, Sampling* (Krishnaiah, P. R., and Rao, C. R., Eds.), 575–583. North-Holland, Amsterdam.

Hedayat, A. S., Rao, C. R., and Stufken, J. (1988b). Sampling plans excluding contiguous units. *J. Statist. Plann. Infer.*, **19**, 462–491.

Holt, D., Smith, T. M. F., and Tomberlin, T. J. (1979). A model based approach to estimation for small subgroups of a population. *J. Amer. Statist. Assoc.*, **74**, 405–410.

Johns, M. V. (1988). Importance sampling for bootstrap confidence intervals. *J. Amer. Statist. Assoc.*, **83**, 709–714.

Kish, L. (1961). Efficient allocation of a multiresponse sample. *Econometrika*, **29**, 363–385.

Kish, L. (1968). Design and estimation for subclasses, comparisons and analytical statistics. In: *New Developments in Survey Sampling* (N. L. Johnson, and H. Smith, Jr., Eds.), 416–438. Wiley, New York.

Kish. L. (1965). *Survey Sampling*. Wiley, New York.

Kish, L. and Frankel, M. R. (1974). Inference from complex samples (with discussion). *J. Roy. Statist. Soc.*, **B36**, 1–37.

Krewski, D., and Rao, J. N. K. (1981). Inference from stratified samples: properties of the linearization, Jackknife and balanced replication methods. *Ann. Statist*, **9**, 1010–1019.

Laake, P. (1979). A prediction approach to subdomain estimation in finite populations. *J. Amer. Statist. Assoc.*, **74**, 355–358.

Madow, W. G., and Olkin, I. (Eds.) (1984). *Panel on Incomplete Data, Incomplete Data in Sample Surveys 3: Symposium on Incomplete Data*. Academic Press, New York.

Majumdar, H., and Sen, P. K. (1978). Invariance principles for jackknifing *U*-statistics for finite population sampling and some applications. *Comm. Statist. Theor. Methods*, **A7**, 1007–1025.

McCarthy, P. J. (1969). Pseudo-replication: Half samples. *Rev. Internat. Statist. Inst.*, **37**, 239–264.

McCarthy, P. J., and Snowden, C. B. (1985). The bootstrap and finite population sampling. *Vital and Health Statist.* (Ser. 2, No. 95), Pub. Health Service Publication 85–1369, Washington DC, U.S. Govt. printing Office.

Miller, R. G. (1974). The Jackknife—a review. *Biometrika*, **61**, 1–17.

Nathan, G. (1988). Inference based on data from complex sample designs. In: *Handbook of Statistics, Vol. 6, Sampling* (Krishnaiah, P. R., and Rao, C. R., Eds.), 247–266. North-Holland, Amsterdam.

Neyman, J. (1969). Bias in surveys due to nonresponse. In: *New Developments in Survey Sampling* (Johnson, N. L., and Smith, H., Jr., Eds.), 712–732. Wiley, New York.

Nisselson, H., Madow, W. G., and Olkin, I. (Eds.) (1983). *Panel on Incomplete Data, Incomplete Data in Sample Surveys 1: Report and Case Studies*. Academic Press, New York.

Patil, G.P., Babu, G.J., Hennemuth, R. C., Myers, W. L., Rajarshi, M. B., and Taillie, C. (1988). Data-Based sampling and model-based estimation for environmental resources. In: *Handbook of Statistics, Vol. 6, Sampling* (Krishnaiah, P. R., and Rao, C. R., Eds.), 469–488. North-Holland, Amsterdam.

Quenouille, M. (1949). Approximate tests of correlation in time series. *J. Roy. Statist. Soc.*, **B11**, 18–84.

Quenouille, M. (1956). Notes on bias in estimation. *Biometrika*, **43**, 353–360.

Ramsey, F. L., Gates, C. E., Patil, G. P., and Taillie, C. (1988). On transect sampling to assess wildlife populations and marine resources. In: *Handbook of Statistics, Vol. 6, Sampling* (Krishnaiah, P. R., and Rao, C. R., Eds.), 515–532. North-Holland, Amsterdam.

Rao, J. N. K., and Wu, C. F. J. (1988). Resampling inference with complex Survey data. *J. Amer. Statist. Assoc.*, **83**, 231–241.

Rubin, D. B. (1987). *Multiple Imputation for Nonresponse in Surveys*. Wiley, New York.

Rubin, D. B., Madow, W. G., and Olkin, I. (Eds.) (1983). *Panel on Incomplete Data, Incomplete Data in Sample Surveys 2: Theory and Bibliographies*. Academic Press, New York.

Särndal, C. E., and Räback, G. (1983). Variance reduction and unbiasedness for small domain estimators. *Statist. Rev.* (Essays in honour of T. E. Dalenius), **21**, 33–40.

Särndal, C. E. (1984). Design-consistent versus model-dependent estimation for small domains. *J. Amer. Statist. Assoc.*, **79**, 624–631.

Särndal. C. E., and Hidiroglou, M. A. (1989). Small domain estimation: a conditional analysis. *J. Amer. Statist. Assoc.*, **84**, 266–275.

Scott, A., and Smith, T. M. F. (1971). Bayes estimates for subclasses in stratifed sampling. *J. Amer. Statist. Assoc.*, **66**, 834–836.

Schucany, W., Gray, H., and Owen, O. (1971). On bias reduction in estimation. *J. Amer. Statist. Assoc.*, **66**, 524–533.

Sen, P. K. (1988). Asymptotics in finite population sampling. In: *Handbook of Statistics, Vol. 6, Sampling* (Krishnaiah, P. R., and Rao, C. R., Eds.), 291–331. North-Holland, Amsterdam.

Sengupta, S. (1981). Jackknifing the ratio and the product estimators in double sampling. *Metrika*, **28**, 245–256.

Srivastava, M. S., and Chan. Y. M. (1989). A comparison of bootstrap method and Edgeworth expansion in approximating the distribution of sample variance-one sample and two sample cases. *Comm. Statist. Simulation Computation*, **18**, 339–362.

Sukhatme, P. V. (1988). Observational errors in behavioural traits of man and their implications for genetics. In: *Handbook of Statistics, Vol. 6, Sampling* (Krishnaiah, P. R., and Rao, C. R., Eds.), 555–573. North-Holland, Amsterdam.

Tripathi, T. P. (1973). Estimation of proportion in domains using inverse equal probability sampling. Tech. Report 2/73. Stat-Math. Division, Indian Statistical Institute, Calcutta.

Tripathi, T. P. (1988). Estimation of domains in sampling on two occasions. *Sankhyā*, **B50**, 103–110.

Tukey, J. (1958). Bias and confidence in not quite large samples. *Ann. Math. Statist.*, **29**, 614.

U.N. Statistical Office (1950). The preparation of sample survey reports. Papers in Series C, No. 1.

Velu, R., and Naidu, G. M. (1988). A review of current survey sampling methods in marketing research (Telephone, mall intercept and panel surveys). In: *Handbook of Statistics, Vol. 6, Sampling* (Krishnaiah, P. R., and Rao, C. R., Eds.), 533–554. North-Holland, Amsterdam.

Wu, C. F. J. (1986). Jackknife, bootstrap and other resampling methods in regression analysis. *Ann. Statist.*, **9**, 1261–1295.

Yates, F. (1953). *Sampling Methods for Censuses and Surveys*. Second edition. Charles Griffin, London.

Author Index

Subject Index

Applied Probability and Statistics (Continued)

HOEL and JESSEN • Basic Statistics for Business and Economics, *Third Edition*
HOGG and KLUGMAN • Loss Distributions
HOLLANDER and WOLFE • Nonparametric Statistical Methods
HOSMER and LEMESHOW • Applied Logistic Regression
IMAN and CONOVER • Modern Business Statistics
JACKSON • A User's Guide to Principle Components
JESSEN • Statistical Survey Techniques
JOHN • Statistical Methods in Engineering and Quality Assurance
JOHNSON • Multivariate Statistical Simulation
JOHNSON and KOTZ • Distributions in Statistics
Discrete Distributions
Continuous Univariate Distributions—1
Continuous Univariate Distributions—2
Continuous Multivariate Distributions
JUDGE, GRIFFITHS, HILL, LÜTKEPOHL, and LEE • The Theory and Practice of Econometrics, *Second Edition*
JUDGE, HILL, GRIFFITHS, LÜTKEPOHL, and LEE • Introduction to the Theory and Practice of Econometrics, *Second Edition*
KALBFLEISCH and PRENTICE • The Statistical Analysis of Failure Time Data
KASPRZYK, DUNCAN, KALTON, and SINGH • Panel Surveys
KAUFMAN and ROUSSEEUW • Finding Groups in Data: An Introduction to Cluster Analysis
KEENEY and RAIFFA • Decisions with Multiple Objectives
KISH • Statistical Design for Research
KISH • Survey Sampling
KUH, NEESE, and HOLLINGER • Structural Sensitivity in Econometric Models
LAWLESS • Statistical Models and Methods for Lifetime Data
LEAMER • Specification Searches: Ad Hoc Inference with Nonexperimental Data
LEBART, MORINEAU, and WARWICK • Multivariate Descriptive Statistical Analysis: Correspondence Analysis and Related Techniques for Large Matrices
LEVY and LEMESHOW • Sampling of Populations: Methods and Applications
LINHART and ZUCCHINI • Model Selection
LITTLE and RUBIN • Statistical Analysis with Missing Data
McNEIL • Interactive Data Analysis
MAGNUS and NEUDECKER • Matrix Differential Calculus with Applications in Statistics and Econometrics
MAINDONALD • Statistical Computation
MALLOWS • Design, Data, and Analysis by Some Friends of Cuthbert Daniel
MANN, SCHAFER, and SINGPURWALLA • Methods for Statistical Analysis of Reliability and Life Data
MASON, GUNST, and HESS • Statistical Design and Analysis of Experiments with Applications to Engineering and Science
MILLER • Survival Analysis
MILLER, EFRON, BROWN, and MOSES • Biostatistics Casebook
MONTGOMERY and PECK • Introduction to Linear Regression Analysis, *Second Edition*
NELSON • Accelerated Testing, Statistical Models, Test Plans, and Data Analyses
NELSON • Applied Life Data Analysis
OCHI • Applied Probability and Stochastic Processes in Engineering and Physical Sciences
OSBORNE • Finite Algorithms in Optimization and Data Analysis
OTNES and ENOCHSON • Digital Time Series Analysis
PANKRATZ • Forecasting with Univariate Box-Jenkins Models: Concepts and Cases
PANKRATZ • Forecasting with Dynamic Regression Modes
POLLOCK • The Algebra of Econometrics
RAO and MITRA • Generalized Inverse of Matrices and Its Applications
RÉNYI • A Diary on Information Theory
RIPLEY • Spatial Statistics
RIPLEY • Stochastic Simulation
ROSS • Introduction to Probability and Statistics for Engineers and Scientists
ROUSSEEUW and LEROY • Robust Regression and Outlier Detection
RUBIN • Multiple Imputation for Nonresponse in Surveys

*Now available in a lower priced paperback edition in the Wiley Classics Library.